Chemical Looping Partial Oxidation

Gasification, Reforming, and Chemical Syntheses

This is the first comprehensive guide to the principles and techniques of chemical looping partial oxidation. With authoritative explanations from a pioneer of the chemical looping process, you will:

- Gain a holistic overview of metal oxide reaction engineering, with coverage of ionic diffusion, nanostructure formation, morphological evolution, phase equilibrium, and recyclability properties of metal oxides during redox reactions.
- Learn about the gasification of solid fuels, the reforming of natural gas, and the catalytic conversion of methane to olefins.
- Understand the importance of reactor design and process integration in enabling metal oxide oxygen carriers to produce desired products.
- Discover other applications of catalytic metal oxides, including the production of maleic anhydride and solar energy conversions.

Aspen Plus® simulation software and results accompany the book online. This is an invaluable reference for researchers and industry professionals in the fields of chemical, energy, and environmental engineering, and students studying process design and optimization.

Liang-Shih Fan is Distinguished University Professor and the C. John Easton Professor in Engineering at The Ohio State University. He is a Member of the U.S. National Academy of Engineering, Chinese Academy of Engineering, and Australian Academy of Technological Sciences and Engineering, and an Academician of Academia Sinica.

Cambridge Series in Chemical Engineering

Books in Series

Deen, *Introduction to Chemical Engineering Fluid Mechanics*
Baldea and Daoutidis, *Dynamics and Nonlinear Control of Integrated Process Systems*
Chau, *Process Control: A First Course with* MATLAB
Cussler, *Diffusion: Mass Transfer in Fluid Systems, Third Edition*
Cussler and Moggridge, *Chemical Product Design, Second Edition*
De Pablo and Schieber, *Molecular Engineering Thermodynamics*
Denn, *Chemical Engineering: An Introduction*
Denn, *Polymer Melt Processing: Foundations in Fluid Mechanics and Heat Transfer*
Duncan and Reimer, *Chemical Engineering Design and Analysis: An Introduction*
Fan and Zhu, *Principles of Gas-Solid Flows*
Fox, *Computational Models for Turbulent Reacting Flows*
Franses, *Thermodynamics with Chemical Engineering Applications*
Leal, *Advanced Transport Phenomena: Fluid Mechanics and Convective Transport Processes*
Lim and Shin, *Fed-Batch Cultures: Principles and Applications of Semi-Batch Bioreactors*
Marchisio and Fox, *Computational Models for Polydisperse Particulate and Multiphase Systems*
Mewis and Wagner, *Colloidal Suspension Rheology*
Morbidelli, Gavriilidis, and Varma, *Catalyst Design: Optimal Distribution of Catalyst in Pellets, Reactors, and Membranes*
Nicoud, *Chromatographic Processes*
Noble and Terry, *Principles of Chemical Separations with Environmental Applications*
Orbey and Sandler, *Modeling Vapor-Liquid Equilibria: Cubic Equations of State and their Mixing Rules*
Petyluk, *Distillation Theory and its Applications to Optimal Design of Separation Units*
Rao and Nott, *An Introduction to Granular Flow*
Russell, Robinson, and Wagner, *Mass and Heat Transfer: Analysis of Mass Contactors and Heat Exchangers*
Schobert, *Chemistry of Fossil Fuels and Biofuels*
Shell, *Thermodynamics and Statistical Mechanics*
Sirkar, *Separation of Molecules, Macromolecules and Particles: Principles, Phenomena and Processes*
Slattery, *Advanced Transport Phenomena*
Varma, Morbidelli, and Wu, *Parametric Sensitivity in Chemical Systems*

Chemical Looping Partial Oxidation

Gasification, Reforming, and Chemical Syntheses

LIANG-SHIH FAN

Ohio State University

CAMBRIDGE
UNIVERSITY PRESS

University Printing House, Cambridge CB2 8BS, United Kingdom

One Liberty Plaza, 20th Floor, New York, NY 10006, USA

477 Williamstown Road, Port Melbourne, VIC 3207, Australia

4843/24, 2nd Floor, Ansari Road, Daryaganj, Delhi – 110002, India

79 Anson Road, #06-04/06, Singapore 079906

Cambridge University Press is part of the University of Cambridge.

It furthers the University's mission by disseminating knowledge in the pursuit of education, learning, and research at the highest international levels of excellence.

www.cambridge.org
Information on this title: www.cambridge.org/9781107194397
DOI: 10.1017/9781108157841

First published 2017

Printed in the United Kingdom by TJ International Ltd. Padstow Cornwall

A catalog record for this publication is available from the British Library.

Library of Congress Cataloging-in-Publication Data
Names: Fan, Liang-Shih.
Title: Chemical looping partial oxidation : gasification, reforming, and chemical syntheses / Liang-Shih Fan, Ohio State University.
Description: Cambridge : Cambridge University Press, 2017. | Includes bibliographical references and index.
Identifiers: LCCN 2017014544 | ISBN 9781107194397 (Hardback : alk. paper)
Subjects: LCSH: Synthesis gas. | Biomass energy.
Classification: LCC TP243 .F36 2017 | DDC 662/.88–dc23 LC record available at https://lccn.loc.gov/2017014544

ISBN 978-1-107-19439-7 Hardback

Additional resources for this title are available at www.cambridge.org/fan

In Memory of

My Sister, Liang-Chi Fan Subisak
Whose Passion, Devotion, and Selflessness as a Teacher
Inspired Generations of Students
in the Pursuit of Their Successful Future Endeavors

Dedicated to

Fan club members
For their past and continuing devotion
to the demonstration of the Syngas Chemical Looping Pilot Operation
at the National Carbon Capture Center in Alabama

Contents

Color plate section between pages 196 and 197.

In Cooperation with Professor Fan's Research Group Members

William Wang

Mandar Kathe

Niranjani Deshpande

Ankita Majumder

Elena Chung

Dikai Xu

Siwei Luo

Lang Qin

Zhuo Cheng

Dawei Wang

Sourabh Nadgouda

Qiang Zhou

Alan Wang

Cheng Chung

Tien-Lin Hsieh

Charles Fryer

Preface

This book is written as a sequel to an earlier book, entitled "Chemical Looping Systems for Fossil Energy Conversions," published in 2010 by Wiley/AICHE. For the earlier book, the motivation was to elucidate the rationale for the resurgence of chemical looping technology research and development related to the ease in CO_2 emission control and the enhancement in exergy conversion efficiency for combustion of carbonaceous fuels. The earlier book clearly indicated that the success of chemical looping technology depends strongly on the viability of the metal oxide materials for its redox applications. Knowledge of fundamental properties of these materials such as redox phase behavior, reactivity, recyclability, and metal oxide support is essential for characterizing chemical looping system performance for the conversion of coal, natural gas, petrochemicals, and biomass. Furthermore, it elaborated gasification or reforming processes involving syngas generation from traditional coal gasifiers, and the use of syngas as feedstock for hydrogen production through a steam–iron chemical looping reaction scheme. It also covered traditional methane–steam reforming applications that are coupled with chemical looping heating schemes, followed by water–gas shift reactions for hydrogen generation. Chemical looping process simulations based on Aspen Plus$^{\circledR}$ utilizing reactors such as gasifier, reducer, oxidizer, combustor, and processes such as conventional gasification and chemical looping for electricity and liquid fuel production were presented.

My motivation for writing this book was precipitated by the exciting recent revelation of direct, one-step, chemical looping partial oxidation techniques in gasifiers and reformers using carbonaceous feedstock. These techniques can produce syngas of a high quality, leading to process efficiencies far greater than any traditional gasification or reforming techniques and other chemical looping techniques. The implications of this discovery are significant in that syngas compositions can reach to near the thermodynamic conversion limit with a H_2:CO molar ratio that can readily be used for direct downstream chemicals or liquid fuels synthesis. The uniqueness of these gasification and reforming techniques is that the syngas stream from the chemical looping reducer reactor will contain little CO_2, yielding a process of high carbon utilization efficiency. Even higher carbon utilization efficiencies can be achieved in a chemical looping process scheme for chemicals or liquid fuels production when the CO_2 generated from the process system can be fully recycled to the reducer reactor, yielding a CO_2 neutral chemical looping process system. Furthermore, in a CO_2 negative chemical looping process system, both the recycled CO_2 from the process stream and the fresh CO_2 can be used as feedstock in the operation of the reducer. The CO_2 recycling along with the fresh CO_2 intake operation scheme also allows a reducer modularization strategy to be implemented, in order to optimize the downstream product synthesis functions in the gasification or reforming operation. Opportunities for direct synthesis of chemicals using natural gas, bypassing syngas formation, also exist and are of high academic and industrial interest. The consequence of process intensification with either the indirect or direct route for synthesis of chemicals can appreciably impact the process system economics. Also, recent advances in the science of mixed metal oxide materials have provided valuable insights into their chemical and physical behavior during redox reactions, and have assisted in their effective formulation and synthesis for chemical looping applications. Accounting for the characteristics of the mixed metal oxides as oxygen carrier at the molecular level, such as ion, defect, and electron transport during redox reactions and their effects on morphological transformation and solid phase equilibrium is further motivation for writing this book. The lattice oxygen diffusion discussed in the context of metal oxides serving purely as an oxygen carrier or as an oxygen carrier as well as a catalyst, referred to as "catalytic metal oxide," is also illustrated.

The book contains six chapters with accompanying supplemental material. Chapter 1 presents the underlying theme of the book and provides an update on the world and U.S. energy outlook based on the latest reports. It introduces the essence of redox reactions that are accompanied by the transport of ions and electrons, as well as defects and morphological variations in mixed metal oxides. It also highlights the importance of reactor design and process integration in successfully enabling mixed metal oxide oxygen carriers to produce the desired products. The use of Aspen Plus® simulation software as a valuable tool in process synthesis and techno-economic analysis of various chemical looping process schemes is also described. Chapter 2 provides details on the ionic diffusion, nanostructure, morphology, and phase equilibrium of complex mixed metal oxide based materials, their recyclability and physical properties, the effect of pressure on reaction kinetics, and oxygen carrier enhancement techniques. Chapter 3 discusses reforming processes represented by the oxidative coupling of methane (OCM) that produces chemicals such as olefins. The OCM reaction mechanism, catalytic metal oxide properties, and process concept and simulations using Aspen Plus® are given. Chapter 4 describes the chemical looping gasification and reforming techniques that are characterized by a one-step generation scheme of high purity syngas and desired $H_2:CO$ ratios for immediate downstream chemicals or fuel synthesis, and the simplicity in their process schemes. They can be carried out in a co-feed mode with solid or gaseous feedstock in the presence or absence of H_2O. This chapter also discusses the unique oxygen carrier and feedstock contact mode that allows syngas quality to be easily maintained and novel CO_2 neutral or negative processes that can take recycled CO_2 and fresh CO_2 as feedstock in the reducer operation. Chapter 5 discusses examples of catalytic metal oxides used for other partial oxidation applications that were significant in the past or will be so in the future. These processes include DuPont's butane oxidation to produce maleic anhydride, solar energy conversions, partial oxidation of methane to formaldehyde, and partial oxidation of propylene and alcohols. Although some of these examples are not based directly on the chemical looping concept per se, the oxidation mechanisms of catalytic metal oxides in the presence of molecular oxygen that are discussed elucidate the origin of the oxygen that participates in the selective or partial oxidation reactions, either from a metal oxide lattice or molecular oxygen. Chapter 6 provides general process simulation methodology and techno-economic analysis. The simulations through chemical looping gasification or reforming processes including coal to methanol, natural gas to liquid fuels, and biomass to olefins, in comparison with conventional gasification or reforming processes are also given. Supplemental material that contains various Aspen Plus® process simulation results is available on the book's website.

Like the first book, this book can be used as a research reference and/or used for teaching in courses at undergraduate and graduate levels related to energy and environmental engineering, reaction engineering, and process system engineering. It can also be used as a good technology and process example in teaching a senior design, optimization and strategy course. The book offers students a degree of appreciation of the depth and breadth of knowledge required to understand chemical looping technology for a wide spectrum of process applications. The subject matter for the knowledge required includes thermodynamics, reaction and reactor engineering, metal-oxide reaction engineering, catalysis, process system engineering and simulation, particle science and technology, and density functional theory.

The book was written in collaboration with some of the members of my research team who have worked or are working on chemical looping research and process development in my lab at The Ohio State University. They include one undergraduate student, 11 MS and PhD graduate students, and four post-doctoral research associates. This is an entirely different group of team members from those who participated in writing the earlier book. The current team members are William Wang, Dr. Mandar Kathe, Dr. Niranjani Deshpande (currently at Shell, India), Dr. Ankita Majumder (currently at Dow Chemical), Dr. Elena Chung (currently at EcoCatalytic Technologies), Dikai Xu, Dr. Siwei Luo (currently at Bay Environmental Technology, China), Dr. Dawei Wang, Sourabh Nadgouda, Dr. Qiang Zhou (currently Assistant Professor at Xi'an Jiaotong University, China), Dr. Lang Qin, Dr. Zhuo Cheng, Dr. Yao Wang (currently at ExxonMobil), Cheng Chung, Tien-Lin Hsieh, and Charles Fryer. Completion of this book would not have been possible without their extensive involvement and knowledge in the field. The technical contribution to this book from each one of them is apparent from the citations given in the text. Their devotion to the writing of this book is deeply appreciated. I am, however, solely responsible for the content and the presentation of the book, including the scope, topic selection, logic sequence, format, and style.

My team and I are grateful to Professor Andrew Tong, Fanhe Kong, Deven Baser, Mingyuan Xu, Yu-Yen Chen, Yaswanth Pottimurthy, Yitao Zhang, Mengqing Guo, Abbey Empfield, Elena Blair, and Peter Sandvik, who each reviewed one or more chapters of the book and provided valuable suggestions and comments. We are also grateful to our research collaborators, Dr. John Sofranko of EcoCatalytic Technologies, Professor Steven Chuang of University of Akron, and Professor Jonathan Fan of Stanford University, for their helpful insights into some aspects of chemical looping technology that have been incorporated in the text of this book. We are thankful to the sponsors of our chemical looping gasification and reforming research and development work on which much of the contents of this book are based. They include U.S. Department of Energy, both ARPA-E and NETL, National Science Foundation, Ohio Department of Services Agency, and The Ohio State University. We have also benefitted from useful technical discussion and considerable engineering support for our demonstration activities from our industrial collaborators. They include Babcock and Wilcox Power Generation Group, WorleyParsons Group, American Electric Power, Southern Company, CONSOL Energy Inc, Particulate Solid Research Inc, First Energy Corporation, Clariant Corporation, IWI Incorporated, and Duke Energy Corporation. Our deep thanks are extended to Aining Wang, Yitao Zhang, and Mengqing Guo for their excellent figure drawings.

Liang-Shih Fan
Columbus, Ohio

Nomenclature

$1-\theta$	fraction of anion vacancies
a	total amount of active centers
A	surface area
A_i	attrition index
A_{tot}	attrition rate over 1 hr test period
C_{gT}	overall gas concentration
C_i	concentration of component i
C_{io}	concentration of component i at the grain surface
C_i^*	dimensionless concentration of component i
C_{io}^*	dimensionless concentration of component i at the grain surface
C_o	ion concentration in grain
C_o^*	normalized ion concentration
$C_{o,B}$	unreacted concentration of the diffusing ion in grain of solid B
C_{pi}	specific heat of component i
d	particle diameter
d_m	ion migration distance
$d_{o,B}$	number of the diffusing ion in the chemical formula of solid B
D	diffusivity
D_e	effective diffusivity constant
D_e^*	dimensionless effective diffusivity constant
D_i	diffusivity of component i
D_{int}	interstitial diffusivity
D_0	ionic diffusivity constant
D_o^*	dimensionless ionic diffusivity constant
D_{vac}	vacancy diffusivity
D^*	diffusion pre-exponential factor
D_{int}^*	interstitial diffusion pre-exponential factor
D_{vac}^*	vacancy diffusion pre-exponential factor
E_a	activation energy
$E_{a,int}$	activation energy of interstitial diffusion path
$E_{a,vac}$	activation energy of vacancy diffusion path
E_{ads}	adsorption energy
E_{DFT}	total energy calculated using DFT
$E(i)$	energy of system i

f_A	molar fraction of gas A at grain surface
f_C	molar fraction of gas C at grain surface
F_g	volumetric flow rate of gas
$F_{g,c}$	volumetric flow rate of gas in combustor
F_s	volumetric flow rate of oxygen carrier
G	Gibbs free energy
G_i	Gibbs free energy of component i
h	Planck constant
H_i	enthalpy of component i
$\Delta H_{ad,X}$	enthalpy of adsorption for compound X
J	flux of atoms
k	reaction rate constant
k_B	Boltzmann constant
k_s	reaction rate constant on the surface of the grain
k_s^*	normalized reaction rate constant on the surface of the grain
K	adsorption/desorption equilibrium constant
K_e	equilibrium constant
K_e^*	dimensionless equilibrium constant
L	characteristic dimension of crystallite
m,n	reaction order
$\dot{n}$	molar flow rate of gas
N	atom number in model system
ϵ	voidage of reactor
ϵ_c	voidage of combustor
O_O^X	lattice oxygen
O/C	oxygen to carbon ratio
p	pressure
p^o	standard pressure (1 atm)
pp	partial pressure
Q_{ae}	actual aeration gas flow rate
Q_{ext}	flow rate of external aeration gas
Q_{sp}	flow rate of leaked gas into standpipe
r_{ER}	rate of reaction for Eley–Rideal mechanism
r_i	rate of reaction i or component i
r_{MVK}	rate of reaction for Mars–Van Krevelen mechanism
r_o	radial coordinate of grain
r^*	dimensionless radial coordinate of particle
r_o^*	dimensionless radial coordinate of grain
R	universal gas constant
R^2	coefficient of determination
Re	rate of attrition
R'	particle radius
R_o	grain radius

s	surface adsorption site
S	entropy
S_c	cross-sectional area of combustor
S_i	entropy of component i
S_r	cross-sectional area of reducer
S_{vib}	vibrational entropy
S#	stoichiometric number
t^*	dimensionless time coordinate
T	temperature
T_C	temperature of the combustor
T_m	melting point of bulk solid
T_R	temperature of the reducer
u_{mf}	minimum fluidization velocity
u_t	terminal velocity of oxygen carrier
v_i	normal mode vibrational frequency in i mode/state
V	volume
V_c	volume of combustor
$V_x^{..}$	vacancy of component x
V_r	volume of reducer
$x_{n,i}$	Barin equation parameters where x varies from "a" to "h" for component i where n is a function of temperature
X	reaction/reduction conversion
X_i	vibrational partition function
y	gas concentration
Y_i	mole fraction of component i
α	phase of the component
α_C	ratio between overall gas concentration and unreacted diffusing ion concentration in solid B
α_L	ratio between pellet size and grain size
γ	surface energy
ε	void fraction
θ_f	fraction of surface oxide ions
θ_X	fractional coverage of surface by compound X
θ	ratio between chemical reaction rate constant and the gas diffusivity in pellets
λ	global stoichiometric air:fuel ratio
μ	chemical potential
μ_i^o	standard chemical potential
ξ	dimensionless coordinate
σ	oxidizability of the bulk of crystallite
σ_s	oxidizability of the crystallite surface
τ	characteristic time for pore diffusion
τ_c	oxygen carrier residence time for combustor reactions
τ_t	contact time

τ_g	gas residence time for reducer reactions
τ_s	oxygen carrier residence time for reducer reactions
ψ	ratio between chemical reaction rate constant and the diffusivity of ions in grain
Ψ	air factor
∇_o^*	dimensionless gradient operator relative to grain size
∇^*	dimensionless gradient operator relative to pellet size

Abbreviations

ACFBG	atmospheric circulating fluidized bed gasification
AGR	acid gas removal
ALD	atomic layer deposition
AR	air reactor
ARCO	Atlantic Richfield Company
ASTM	American Society for Testing and Materials
ASU	air separation unit
ATR	autothermal reforming/er
bbl	barrels
B&W	Babcock and Wilcox
BEP	Brønsted–Evans–Polanyi
BET	Brunauer–Emmett–Teller
BFB	bubbling fluidized bed
BFW	boiler feed water
BGL	British Gas Lurgi
BGLS	British Gas Lurgi slagging
BJH	Barrett–Joyner–Halenda
bpd	barrels per day
BTO	biomass to olefins
BTS	biomass to syngas
BTU	British thermal unit
C_2 or C_{2s}	hydrocarbon compounds with two carbons, mainly ethylene and ethane
C_{2+}	higher hydrocarbons starting from C_2
C_3 or C_{3s}	hydrocarbon compounds with three carbons
C_{3+}	higher hydrocarbons starting from C_3
C_4 or C_{4s}	hydrocarbon compounds with four carbons
C_5 or C_{5s}	hydrocarbon compounds with five carbons
C_7 or C_{7s}	hydrocarbon compounds with seven carbons
CAS	Chinese Academy of Sciences
CCF	capital charge factor
CCR	carbonation–calcination reaction
CCUS	carbon capture, utilization, and sequestration
CDCL	coal direct chemical looping
CF	capacity factor

CFB	circulating fluidized bed
CI-NEB	climbing image nudged elastic band
CLC	chemical looping combustion
CLG	chemical looping gasification
CLOU	chemical looping oxygen uncoupling
CLPO	chemical looping partial oxidation
CLR	chemical looping reforming
CMBCR	countercurrent moving bed chromatographic reactor
COE	cost of electricity
CO_x	carbon oxides
CPBMR	conventional packed bed membrane reactor
CPOX	catalytic partial oxidation
CRP	CO_2 reaction parameter
CSP	concentrating solar power
CSTR	continuous stirred tank reactor
CTM	coal to methanol
CTS	coal to syngas
CTS-A	coal to syngas process with a co-feed of natural gas and coal
CVD	chemical vapor deposition
CW	cooling water
DCFB	dual circulating fluidized bed
DFT	density functional theory
DP	deposition–precipitation
EDI	electrodeionization
EDS	energy dispersive spectroscopy
EPR	electron paramagnetic resonance
FBR	fixed bed reactor
FCC	fluid catalytic cracking
FGD	flue gas desulfurization
FR	fuel reactor
F–T	Fischer–Tropsch
GC	gas chromatography
GDC	gadolinium doped ceria
GE	General Electric
GHSV	gas hourly space velocity
GIEC	Guangzhou Institute of Energy Conversion
GTG	gas to gasoline
GTL	gas to liquid(s)
HHV	higher heating value
HRSG	heat recovery steam generator
HTDS	high temperature desulfurization
HTF	heat transfer fluid
HTS	high temperature superconductor

HTW	high temperature winkler
HYGAS	hybrid gas
IACMO	iron–aluminum composite metal oxide
IDGM	ionic diffusion grain model
IGCC	integrated gasification combined cycle
IGT	Institute of Gas Technology
ITCMO	iron–titanium composite metal oxide
KBR	Kellogg, Brown, and Root
LED	light emitting diode
LHHW	Langmuir–Hinshelwood–Hougen–Watson (mechanism)
LHV	lower heating value
LNB	low NO_x burner
LPMeOH	liquid phase methanol
LSF	lanthanum strontium ferrite
MAN	maleic anhydride
MB-A	methanol baseline case with CO_2 capture
MB-B	methanol baseline case with no CO_2 capture
MDEA	monodiethanolamine
MEA	monoethanolamine
MIESR	matrix isolation electron spin resonance
MMgal	one million gallons
MMgal/yr	million gallons per year
MMSCFD	million standard cubic feet per day
MOGD	Mobil's olefin to gasoline and distillate
MPO	methane partial oxidation
MTO	methanol to olefins
Mtoe	million tonnes of oil equivalent
NCCC	National Carbon Capture Center
NDIR	non-dispersive infrared
NEB	nudged elastic band
NGCC	natural gas combined cycle
NREL	National Renewable Energy Lab
NSPS	new source performance standards
O_2-TPD	temperature programmed desorption of oxygen
O&M	operating and maintenance
OC	oxygen carrier
OC_{FIX}	fixed operating cost
OC_{VAR}	variable operating cost
OCM	oxidative coupling of methane
OMS	octahedral molecular sieve
OSU	Ohio State University
PM	particulate matter
POX	partial oxidation

PPBMR	proposed packed bed membrane reactor
PR	production rate
PRB	Powder River Basin
PS	precipitated silica
PSA	pressure swing adsorption
PSRI	Particulate Solid Research Inc.
PV	photovoltaic
REF	reference/baseline case
REMPI	resonance enhanced multiplication ionization
RO	reverse osmosis
RSP	required selling price
SCGP	Shell coal gasification plant/process
SCL	syngas chemical looping
SCMCR	simulated countercurrent moving bed chromatographic reactor
SEM	scanning electron microscopy
SEU	Southeast University
SLPM	standard liters per minute
SMDS	Shell middle distillate synthesis
SMR	steam methane reformer/ing
SMR–CLC	steam methane reforming–chemical looping combustion
SNG	synthetic/supplemental natural gas
STP	standard temperature and pressure
STS	shale gas to syngas
STWS	solar thermal water splitting
SVUV–PIMS	synchrotron vacuum ultraviolet–photoionization mass spectrometer
T&S	transportation and storage
TAD	temperature accelerated dynamics
TAP	temporal analysis of products
TEM	transmission electron microscopy
TES	thermal energy storage
TGA	thermogravimetric analysis/zer
THF	tetrahydrofuran
TOC	total overnight costs
TPC	total plant cost
tpd	metric tons (tonnes) per day
TPR	temperature programmed reduction
TS&M	transportation storage and management
TU Berlin	Technische Universität Berlin
USDOE	United States Department of Energy
USEPA	United States Environmental Protection Agency
VLS	vapor–liquid–solid
VPO	vanadium phosphorus oxide

VUT	Vienna University of Technology
WGS	water–gas shift
WHSV	weight hourly space velocity
WSA	wet-gas sulfuric acid
XRD	X-ray diffraction
YSZ	yttria/um stabilized zirconia

1 Overview

L.-S. Fan

1.1 Introduction

The term "chemical looping" was first used by Ishida et al. (1987) to describe a high exergy efficiency combustion process that engages redox reactions employing metal oxides as a reaction intermediate.[1] In an earlier book of Fan,[2] the term "chemical looping" was used in a broad sense to describe the nature of a reaction scheme in which a given reaction is decomposed into multiple sub-reactions using chemical intermediates that are reacted and regenerated. Further, the sub-reactions are ideally configured in such a manner that the exergy loss of the process as a result of this reaction scheme can be minimized. The term in this broad sense, which has been of interest to industrial practice, can be dated to the late nineteenth century, and is used again in this book. The chemical intermediates considered in the chemical looping reactions are mainly metal oxides. The concept of chemical looping is rooted in the second law of thermodynamics as applied to the reduction of process irreversibility and hence the enhanced exergy efficiency of the process. The application of this exergy concept for combustion systems was reported earlier by Knoche and Richter (1968, 1983).[3,4] In the work of Ishida et al. (1987), they employed metal oxides such as NiO, CuO, Mn_2O_3, and Fe_2O_3 as chemical intermediates for combustion reaction applications.[1] The unique feature of using a metal oxide in a chemical looping combustion scheme is that CO_2 becomes an easily separable combustion product and thus, the parasitic energy requirement in the CO_2 separation is low in this combustion system. Such a CO_2 separation feature has been deemed to be important in recent years, particularly in light of global concerns over the increasing CO_2 concentration in the atmosphere.

Aside from electricity generation from chemical looping combustion, chemicals can also be generated from chemical looping routes that have considerable process advantages. Chemicals such as propylene, butadiene, benzene, and xylene are currently produced mainly from petroleum through catalytic and/or thermal cracking processes of crude oil to naphtha, butane, and propane. When the feedstock is coal or natural gas, the production processes are different. For coal, chemicals are produced via two routes. One route is through coal gasification to generate syngas, then synthesized to chemicals; the other route is through coal pyrolysis to generate volatiles, then refined to chemicals. For natural gas which contains methane and ethane, methane can be reformed to syngas or hydrogen and then converted to chemicals, while ethane can be cracked to ethylene and then converted to other chemicals. The use of shale gas as feedstock is attractive

1

today in the United States due to its abundant supply and low cost.[5,6] In many of the coal and natural gas conversion processes, molecular oxygen and/or steam are required. The use of metal oxides in chemical looping processes for chemical production through selective oxidation can circumvent the step of separating molecular oxygen from air, while creating a highly efficient reaction system that can also minimize steam consumption. The resulting process intensification can lead to significant economic benefits. Other than chemical production, chemical looping gasification and reforming can also result in more efficient electricity, hydrogen, syngas, and/or liquid fuel production than traditional gasification and reforming schemes.[7]

This chapter provides an overview of chemical looping redox reactions and processes that are of direct relevance to energy utilization and CO_2 capture methods. In this connection, chemical looping combustion can achieve a higher efficiency for fuel conversion to electricity, with full or near full CO_2 capture, than conventional combustion processes.[2,8] In gasification or reforming applications, on the other hand, CO_2 generation is intended to be minimized and can reach a neutral or even negative process operating condition.[9] There are two key technical areas that constitute core competency in enabling the chemical looping concept for combustion, gasification, or reforming applications. They are metal oxide reaction engineering and particle science and technology. Metal oxides are used as oxygen carriers in these applications. The synergistic effect of these two subject areas addresses such essential metal oxide issues as crystal structures, ionic transport phenomena, redox mechanism, and reaction kinetics. Understanding these metal oxide issues allows the metal oxide materials to be formulated effectively, synthesized, and used in a sustainable manner for desired chemical looping reaction applications. The synergistic effect of these two subject areas also addresses such essential particulate engineering issues as gas–solid flow stability, gas–solid transport phenomena, non-mechanical solid flow devices, and multiphase reactor design. Understanding these particulate engineering issues allows redox reactors to be optimally configured, and heat integration and process intensification to be properly achieved. One aspect of chemical looping reactions that is of considerable application versatility is concerned with the partial or selective oxidation of feedstock. That is, through chemical looping gasification or reforming routes, the feedstock can be directly converted to syngas, chemicals, or other products. Metal oxide based chemical looping gasification and reforming processes, particularly those developed at Ohio State University (OSU) that involve a wide spectrum of feedstock yielding a variety of products, are described as examples in this chapter to illustrate the novelty and the commercial potential of the chemical looping technology.

As the key interest of this book lies in the subjects of chemical looping partial or selective oxidation and its associated metal oxide reactions and process systems, subsequent chapters in the book discuss the fundamental and applied nature of these subjects.

Specifically, Chapter 2 provides details of the molecular behavior of metal oxide materials with respect to their ionic diffusion mechanisms and the effect on morphology evolution. It illustrates the oxygen carrier deactivation property and effectiveness of countermeasures, from both experimental and theoretical approaches. The chemical and physical properties of complex and mixed metal oxides including their phase equilibrium, cyclic redox kinetics, and attrition behavior are given.

Chapter 3 focuses on the development of the oxidative coupling of methane (OCM) process operating in a chemical looping mode in relation to a pure catalytic co-feed mode. Simulation of the reactor performance for the catalytic co-feed mode based on reported kinetic models is given. The role of catalytic metal oxide is illustrated in the context of chemical looping OCM applications. The advantages and challenges of the OCM process using catalytic metal oxides are presented.

Chapter 4 further discusses oxygen carriers and the configurations of reducer reactor or fuel reactor required for effective performance of chemical looping partial oxidation for syngas production. This chapter underscores the effectiveness of the streamlining of the syngas generation step and the use of CO$_2$ as feedstock in a novel integrated chemical looping process system for the downstream generation of various products.

Chapter 5 presents additional examples on catalytic metal oxides for partial oxidation reactions using lattice oxygen transfer, and their reaction mechanisms in connection with the role of lattice oxygen and gaseous oxygen. These examples include Dupont's two-step process for the production of maleic anhydride through tetrahydrofuran intermediate, molybdenum trioxide for selective oxidation of hydrocarbons, selective oxidation of propylene over bismuth molybdate, and homogeneous gas phase and heterogeneous liquid phase selective oxidation reactions over other group IV catalytic metal oxides. These examples indicate how essential catalytic metal oxides are to each of these reactions. Syngas and/or hydrogen generation through the use of solar or nuclear energy that aid in the thermochemical processes using metal oxides is also described.

Chapter 6 describes the process constraints and methodology used to develop various chemical looping processes through simulations using Aspen Plus® (AspenTech). With the simulation, the process techno-economics of chemical looping is compared to the traditional baseline process to determine the feasibility of the chemical looping process. The results presented in this chapter are provided along with the chemical looping partial oxidation examples given in Chapter 4 on the book's website as supplemental material.

1.2   Energy Scenarios and CO$_2$ Capture Methods

1.2.1   Energy Scenarios

Energy source and energy value are two important factors when discussing energy supply and demand. The primary energy source, which is energy that has not been subjected to transformation or conversion and is obtained directly from natural resources, can be broadly classified as either fossil fuel or non-fossil fuel. Fossil fuels include coal (anthracite, bituminous, sub-bituminous, and lignite), oil (crude oil, gas condensate, and natural gas liquids), and natural gas, while non-fossil fuels include everything else such as nuclear and renewables.

Table 1.1 Energy conversion factors.

Oil equivalent	Fuel	Energy
1 tonne oil equivalent	7.3 barrels oil	3.968×10^7 BTU
1 tonne oil equivalent	1,111 m^3 natural gas	
1 tonne oil equivalent	1.5 tonnes hard coal	
1 tonne oil equivalent	3 tonnes lignite coal	

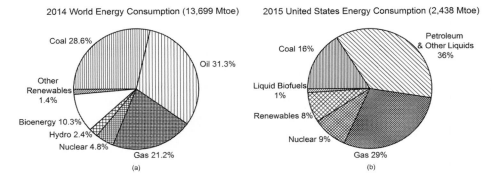

Figure 1.1 Current energy consumption by source in (a) the world in 2014 and (b) United States in 2015.[10,11]

Figure 1.1 shows the breakdown for both world and USA energy consumption. Current world energy consumption from primary energy sources is 13,699 Mtoe (5.4×10^{17} BTU) (Figure 1.1a), with fossil fuels comprising 81%, whereas in the United States, about 2,438 Mtoe (0.97×10^{17} BTU) (Figure 1.1b) is currently being consumed, with fossil fuels comprising 83%;[10,11] 1×10^9 m^3 of natural gas is equivalent to 0.90 Mtoe.[12] Coal is divided into hard coal (anthracite and bituminous) and lignite, where 1.5 tonnes of hard coal is 1 tonne oil equivalent (toe) and 3 tonnes of lignite is 1 tonne oil equivalent.[12] Table 1.1 provides a summary of the conversion factors.

Energy consumption projections into 2040 are given in Figure 1.2. Projected world energy consumption from primary energy sources is estimated to increase by 33% over current consumption to 18,260 Mtoe (7.2×10^{17} BTU) (Figure 1.2a), with fossil fuels comprising about 75%, whereas in the United States only an 11% increase to 2,700 Mtoe (1.07×10^{17} BTU) (Figure 1.2b) is expected, with fossil fuels comprising 78%. The projected forecasts for the world and United States differ greatly in terms of expected growth and energy source distribution, yet it is immediately apparent that the reduction of fossil fuels as the primary energy source is not expected.

1.2.2   CO$_2$ Capture Methods

Carbon capture is a key step of the carbon emission control strategy that includes carbon capture, utilization, and sequestration (CCUS). Current carbon capture technologies

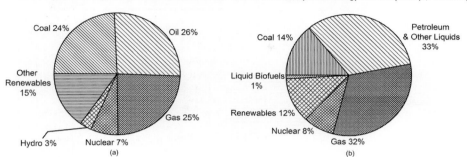

Figure 1.2 Projected energy consumption in 2040 by source in (a) the world and (b) the United States.[10,11]

when integrated into power plants are plagued by both high parasitic energy requirements and poor economics, with the end result being a decrease in thermal-to-electric efficiency, increased fuel consumption, and increased electricity prices with CCUS implementation. The United States Department of Energy (USDOE) has published a roadmap that describes the current available technologies, including amines, Rectisol, Selexol, and cryogenic techniques, and the time evolution, relative to their cost reduction benefits, of a number of techniques that are in development, such as advanced physical or chemical solvents and sorbents, ionic liquids, metal organic frameworks, and membrane systems.[13,14] At the end of the roadmap, there are three techniques considered to be the most desirable. They are biological processes, oxygen transport membranes, and chemical looping. Of the thermochemical approaches, the oxygen transport membrane is considered to be a second generation technology while chemical looping is considered to be third generation. That is, chemical looping represents the eventual goal of the technology to be developed for carbon capture.[13,14] Chemical looping technology is attractive since the CO_2 separation process, as noted earlier, is inherent to the process reaction and operation, such that there is no additional step required for further CO_2 separation. It is important to note that for both oxygen transport membrane and chemical looping technologies, the success of their development depends strongly on the success of the deployment of metal oxide materials.

1.3 Metal Oxide Reaction Engineering

1.3.1 Metal Oxide Applications and Properties

Metal oxides are enabling materials for a variety of applications in energy systems including those exemplified in Figure 1.3.[15–20] Perovskite type metal oxides with mixed electrical and ionic conductivity are used as membranes in air separation to transfer oxygen ions at elevated temperatures while also providing a cost reduction benefit when used in the CCUS process (Figure 1.3a).[15] In solid oxide fuel cells, ionic conductive

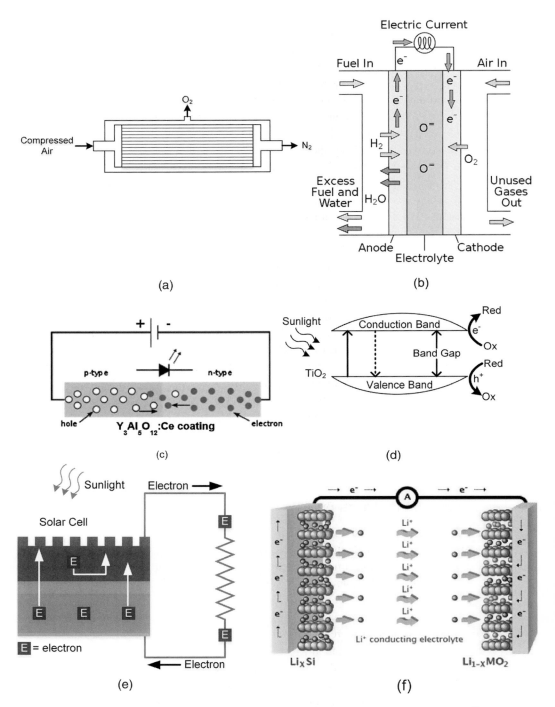

Figure 1.3 Examples of metal oxide applications: (a) oxygen transport membranes;[15] (b) solid oxide fuel cells;[16] (c) phosphor based LEDs;[17] (d) photocatalysts;[18] (e) solar cells;[19] (f) lithium batteries (adapted from Nexeon).[20]

metal oxides such as yttrium-stabilized zirconia (YSZ) and gadolinium-doped ceria (GDC) are used as solid electrolytes to conduct oxygen ions (Figure 1.3b).[16] During fuel oxidation, the ions are transported to the anode side where combustion occurs and electricity is generated. Synthetic crystalline garnets are applied in phosphor based light emitting diodes (LEDs) that convert blue light into white light, which is the most popular method for making high-intensity white LEDs (Figure 1.3c).[17] In addition, semiconducting metal oxide materials can generate electron-hole pairs to directly utilize solar energy; among these, titanium oxide is used as a photocatalyst in water electrolysis to perform reduction–oxidation reactions (Figure 1.3d).[18] Zinc oxide (ZnO), an n-type semiconductor, is widely used in organic solar cells and hybrid solar cells where photon energy is harvested from incoming sunlight to generate electricity (Figure 1.3e).[19] Moreover, in lithium-ion batteries, $Li_{1-x}MO_2$ (where M = Ni, Co, or Mn) functions as a cathode material that stores lithium ions to generate electricity (Figure 1.3f).[20]

The wide range of applications of metal oxides is closely related to their structures. As an insulator, $BaTiO_3$ (Figure 1.4a) possesses a perovskite structure and is used in capacitors of high volumetric efficiency.[21] As a semiconductor or semi-insulator (Figure 1.4b), ZnO has a wide bandgap with high electron mobility.[22] As a conductor, high-temperature superconducting (HTS) material was discovered by J. Georg Bednorz and K. Alex Müller at IBM, for which they were awarded the 1987 Noble Prize.[23] To date, HTS materials have been observed with transition temperatures as high as 138 K and can be cooled down to superconductivity using liquid nitrogen. Superconductive metal oxides generally contain cuprate (CuO_2) planes that are considered to be an electronically active layer. One example is $YBa_2Cu_3O_{7-x}$ (Figure 1.4c), in which yttrium and barium provide the opportunity for partial covalent bonds to form between copper and oxygen, where holes are formed in the conduction band allowing electrons to move freely.

1.3.2   Source and Cost of Metal Oxides

Metal oxides are derived through the oxidation reaction of their constituent base metals. Elemental metals include the transition metals, incorporating the d-block as well as the f-block lanthanide and actinide series, the alkali metals (Group I), and the alkaline earth metals (Group II). While any metal material can show desirable properties, it is essential to consider the cost of the metal if it is to offer any potential for industrial usage. Based on 2015 USD, metal materials range from the very low cost, such as sodium, calcium, and iron (<$1/kg), to more expensive ones such as cobalt, copper, manganese, and nickel ($1/kg–$100/kg), and finally very expensive ones such as silver, gold, and platinum (>$100/kg). Metal oxides used in industrial applications are geared towards abundant and low cost metals.[24–27] Specifically, five transition metals (iron, cobalt, copper, manganese, and nickel) will be examined more closely in the context of chemical looping technology. The selection of metal oxides is based on the modified Ellingham diagram, shown in Figure 1.5,[7] instead of the original Ellingham diagram, which is expressed in terms of the standard Gibbs free energy of formation and its

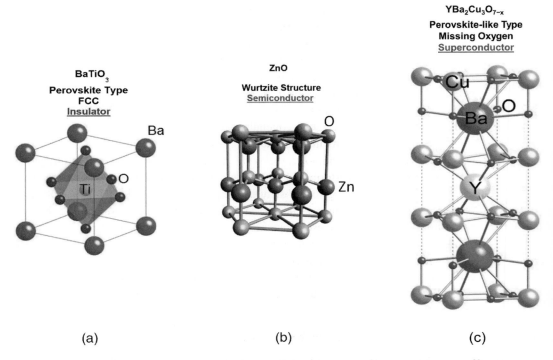

Figure 1.4 Electronic properties of metal oxides: (a) insulator;[21] (b) semiconductor;[22] and (c) superconductor.[23] A black and white version of this figure will appear in some formats. For the color version, please refer to the plate section.

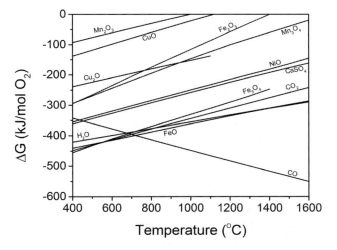

Figure 1.5 Modified Ellingham diagram for metal oxides.

variation with temperature for metal oxides and carbonaceous fuels. In the modified Ellingham diagram, the Gibbs free energies for the reduction–oxidation pair that coexists at equilibrium are represented by a pair of a metal and its adjacent metal oxide or a metal oxide and its adjacent metal oxide at a higher oxidation state. As the redox involves metal oxides at any of their oxidation states, the information provided in the Ellingham diagram of its modified form can be more readily used for a prompt screening of the metal oxides in their selection for the reduction reaction with the intended carbonaceous fuels. Other metallic elements in the periodic table can also be used as dopants or supports in the form of oxides.

In chemical looping technology, the metal oxide material provides lattice oxygen or molecular oxygen to perform the oxidation reaction for the fuel. The success of chemical looping technology is substantially dependent on the characteristics of metal oxides, which require robustness in cyclic reactivity and physical strength. To develop viable metal oxides with these properties, it is necessary to understand them at the molecular level regarding how electrons, ions, and defects are transported in the metal oxide redox process. It is also necessary to understand the effects of dopants, supports, additives, and morphological variations on metal oxide robustness. Here, iron oxide is used as an example to illustrate the metal oxide molecular properties.

1.3.3   Ionic Diffusion Mechanism in Metal Oxide

Ionic diffusion is an important fundamental factor in determining the rate process for metal oxide reactions. Ionic diffusion, which can be considered as the net number of ions to pass through a plane of unit area per unit time, plays a critical role in solid-state reactions. This process is driven by the concentration or chemical gradient, where ions move from a region of high concentration to a region of low concentration. The mathematical description of diffusion was given by Fick (see Equation (1.3.1)) for one spatial dimension:[27]

$$J = -D\frac{\partial C}{\partial x},\tag{1.3.1}$$

where J is the flux of atoms, D is the diffusivity, and $\frac{\partial C}{\partial x}$ represents the concentration gradient.

The intrinsic defects in metal oxides are often the vehicles that allow atoms to perform the tiny jumps that constitute diffusion. For each defect type, there is a specific diffusion mechanism, and in many cases, combinations of these point defects lead to more complex diffusion mechanisms. Thus, to understand the diffusion mechanism, the metal oxide lattice structures need to be clarified. What follows is a description of the diffusion mechanisms by which atoms can be transported through a specific structure.

Sequential oxidation of FeO, wüstite, results in Fe_3O_4, magnetite, and then Fe_2O_3, hematite. The reduction process beginning with Fe_2O_3 is the exact reverse. Figure 1.6 shows the crystal structures of the various oxidation states of iron, and it can be seen that each one of the oxidation states of iron has its own crystal structure, varying from the rock salt type, face-centered cubic of FeO (Figure 1.6a) to the inverse spinel structure of Fe_3O_4 (Figure 1.6b), to the corundum structure of Fe_2O_3 (Figure 1.6c). From Figure 1.6, it becomes clear that during reduction–oxidation (redox) reactions, the ions – the iron cation and oxygen anion – move or migrate, which causes physical variations in the crystal structure through the formation and transport of vacancies. Understanding the migration process and reaction pathway in the reduction–oxidation reaction is thus essential in order to assess its product formation process.

Typically, there are two types of diffusion mechanisms:

(a)   The vacancy mechanism is facilitated by the presence of vacancy point defects. In the process, an atom on a lattice site hops into a neighboring vacancy site.

(b)   The interstitial diffusion mechanism occurs when an ion moves from one interstitial site to a neighboring interstitial site. If the diffusion energy is too high to facilitate migration via a direct interstitial mechanism, an interstitial ion can move onto an occupied lattice site. The atom formally occupying the lattice site is forced to form a new interstitial, which can then continue the process.

At a macro level, ionic transport processes can be directly examined using the reaction between a dense iron pellet and oxygen, whose reaction is shown in Reaction (1.3.2).

$$2\,Fe + \tfrac{3}{2}\,O_2 \rightarrow Fe_2O_3. \tag{1.3.2}$$

When oxidized with oxygen, the iron pellet develops a layer of iron oxide on the outside of the iron pellet sample. This is clearly visualized by a marker experiment, where a platinum marker is placed outside the original structure. The results of the platinum marker experiment, given in Figure 1.7, indicate that iron is diffusing dominantly outward to react with oxygen on the outside surface to form an iron oxide layer.[28] The mechanism of the outward diffusion of Fe ions for the Fe oxidation reaction is similar to that of the outward diffusion of Ca^{2+} and O^{2-} for the CaO–SO_2 reaction based on a similar marker and the isotope experiments.[29,30] The support plays an important role in the ionic diffusion of the metal oxide. To elucidate the effect of a support in conjunction with the iron pellet, identical platinum marker experiments to those in Figure 1.7 were performed using a supported, specifically titanium dioxide supported, iron pellet. Figure 1.8 is an SEM image, and the platinum mapping results indicate that when a supported iron pellet is used, the inward diffusion of oxygen into the iron structure occurs without the formation of an iron oxide layer on the outside surface of the sample.[31]

However, pellet experiments cannot fundamentally explain the ionic diffusion mechanism at the micro level. Density functional theory (DFT) can be used in conjunction with pellet experimental observation and analysis to explore the mechanism of ionic diffusion in supported and unsupported metal oxide. DFT is a quantum mechanical modeling tool that has been widely used in solid-state physics and physical chemistry for atomic, molecular, and other multiple-body system calculations. For example, in transition metal oxides, the migrating cation has to squeeze between two adjacent oxygen anions, a process that requires energy as the atom is activated

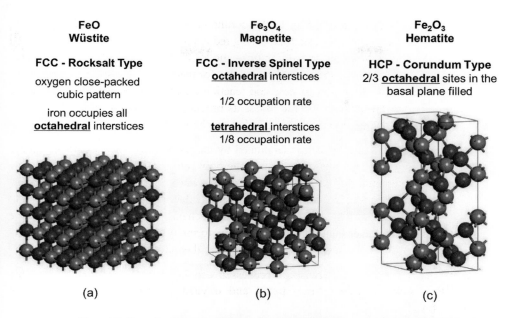

FeO	Fe$_3$O$_4$	Fe$_2$O$_3$
Wüstite	**Magnetite**	**Hematite**

FCC - Rocksalt Type

oxygen close-packed cubic pattern

iron occupies all **octahedral** interstices

FCC - Inverse Spinel Type
octahedral interstices

1/2 occupation rate

tetrahedral interstices
1/8 occupation rate

HCP - Corundum Type
2/3 **octahedral** sites in the basal plane filled

(a)   (b)   (c)

Figure 1.6 Crystal structures of iron oxide: (a) FeO: face-centered cubic; (b) Fe$_3$O$_4$: inverse spinel; (c) Fe$_2$O$_3$: hexagonal close-packed.

into a transition state. This energy is usually greater than the thermal energy, k_BT, and consequently these activated events are infrequent. At a finite temperature, all atoms in a lattice will oscillate around their equilibrium positions. As they do not possess sufficient energy to reach the transition state and hence complete the jump, they simply continue to oscillate within their initial position or initial state (IS). However, as a result of thermal activation, these atoms can have sufficient energy to reach the transition state along a given trajectory, completing the jump to their adjacent equilibrium positions or final state (FS). The nudged elastic band (NEB) method within a DFT framework can calculate the classical rate constant for escape from an initial state, IS, to a final state, FS, which enables the stationary points to be mapped out along the minimum energy path and identifies the transition state with the activation energy for the diffusion process.

Diffusivity is a transport parameter that relates the concentration gradient of a given species to the flux at a given temperature, as illustrated in Equation (1.3.1). As there are two types of diffusion mechanism, vacancy and interstitial, by which matter can be transported through the lattice, the overall diffusivity can be given as the sum of these two contributions, shown in Equation (1.3.3):

$$D = D_{vac} + D_{int}, \qquad (1.3.3)$$

where D represents the overall diffusivity of a species while D_{vac} and D_{int} are the contributions of the vacancy and interstitial diffusion processes, respectively. Because

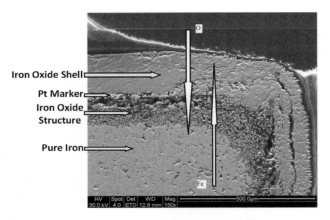

Figure 1.7 Platinum marker study with iron pellet.[28]

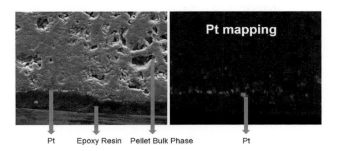

Figure 1.8 SEM of partially oxidized iron oxide with support and platinum mapping.[31]

diffusivity, D, exhibits an Arrhenius temperature dependence, Equation (1.3.3) can be expressed by Equation (1.3.4),

$$D = D^*_{vac} \exp\left(\frac{-E_{a,vac}}{k_B T}\right) + D^*_{int} \exp\left(\frac{-E_{a,int}}{k_B T}\right), \tag{1.3.4}$$

where D^*_{vac} and D^*_{int} are vacancy diffusion pre-exponential factor and interstitial diffusion pre-exponential factor, respectively; $E_{a,vac}$ and $E_{a,int}$ are the activation energy via the vacancy diffusion path and interstitial diffusion path, respectively; k_B is the Boltzmann constant.

According to Vineyard,[32] for a system containing N atoms, the diffusion pre-exponential factor D^* can be expressed by Equation (1.3.5),

$$D^* = \varphi d_m^2, \tag{1.3.5}$$

where d_m is the ion migration distance and φ is the frequency of ion hopping to a neighboring site and can be evaluated using Equation (1.3.6),

$$\varphi = \frac{\prod_{i=1}^{3N} v_i^{IN}}{\prod_{j=1}^{3N-1} v_j^{TS}}, \tag{1.3.6}$$

where v_i^{IN} and v_j^{TS} are the normal mode vibrational frequencies at the initial and transition states, respectively. Both D^* and E_a are constants for a given diffusion path. Thus, the contribution of two diffusion mechanisms can be compared by the migration distance and activation energy, and the dominant diffusion mechanism can be determined.

Bulk $FeTiO_3$ is based on the higher symmetry corundum structure (space group $R^{\overline{3c}}$). The metal atoms all lie on threefold axes parallel to the c axis, forming the sequence of face-shared pairs of octahedral alternating with vacant octahedral sites in the middle. According to the above method, the iron and titanium ion diffusion path can be depicted step by step with energy profile, and then the diffusivity can be calculated. At 900 °C, the diffusivity for titanium in $FeTiO_3$ is 1.35×10^{-10} cm²/s while the diffusivity for iron in $FeTiO_3$ is 2.21×10^{-7} cm²/s, consistent with the two-dimensional Fickian model coupled with experiments that calculated the iron diffusivity to be 1.30×10^{-7} cm²/s. Based on the significant difference in diffusivity, iron is noted to diffuse at a much higher rate compared to titanium. In a practical sense, when titanium dioxide is used as a support for iron oxide, the titanium dioxide can be treated as being stationary relative to iron.[33]

Figure 1.9 shows the role of ionic diffusion in a crystal both with and without oxygen or metal vacancies in the iron oxide formation mechanism using an Fe_2O_3/Fe model system. Specifically, the activation energies for ionic diffusion obtained from DFT calculations are given. It is seen from the calculations that without the oxygen vacancy, the oxidation process will result in the formation of an iron oxide product layer outside the original surface since the diffusion energy barrier for oxygen ion inward diffusion is larger than that for metal ion outward diffusion. However, with the oxygen vacancy, the calculations indicate that the diffusion energy barrier for oxygen ion inward diffusion is lower than that for metal ion outward diffusion and thus, the formation of the iron oxide product layer occurs at the Fe_2O_3–Fe interface. Figure 1.9 serves to provide the molecular interpretation of the macroscopic diffusion phenomena given in Figures 1.7 and 1.8 as the DFT calculations to be shown later indicate that the titanium dioxide support can decrease the oxygen vacancy formation energy in the iron oxide, hence facilitating the oxygen vacancy formation.[33]

The support material is thus important to vacancy formation in active metal oxides and hence the ionic and defect transport properties, which ultimately alters the morphology of the material upon redox reactions. The effect of the type of support material on the behavior of the base metal oxides is, however, complex. For example, when aluminum oxide is used as the support for iron oxide, and it undergoes 50 reduction–oxidation cycles using hydrogen gas as the reducing agent and air as the oxidant, the initially well-mixed iron–aluminum oxide particle segregates into two distinct phases, with iron oxide situated on the shell of the particle and aluminum oxide forming the core. Figure 1.10 visualizes this process by showing a well-mixed iron–aluminum oxide particle at the beginning of the experiment and a core–shell structure with iron on the surface and aluminum on the inside at the end of 50 redox cycles.[31]

However, when titanium dioxide is used in place of aluminum oxide as a support material, the iron oxide–titanium dioxide particle has an identical surface composition

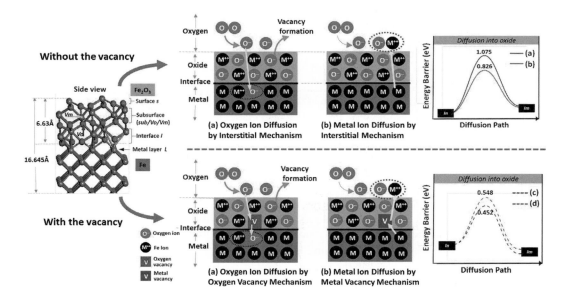

Figure 1.9 Ionic diffusion mechanism with diffusion activation energies obtained from DFT calculations. A black and white version of this figure will appear in some formats. For the color version, please refer to the plate section.

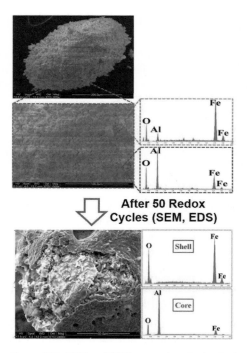

Figure 1.10 SEM and EDS spectral analysis of mixed iron oxide–aluminum oxide at the beginning of the experiment and at the completion of 50 reduction–oxidation cycles.[31] A black and white version of this figure will appear in some formats. For the color version, please refer to the plate section.

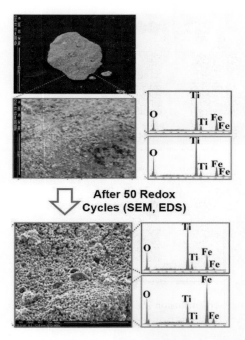

After 50 Redox
Cycles (SEM, EDS)

Figure 1.11 SEM and EDS spectral analysis of mixed iron oxide–titanium dioxide at the beginning of the experiment and at the completion of 50 reduction–oxidation cycles.[31] A black and white version of this figure will appear in some formats. For the color version, please refer to the plate section.

to the interior after 50 reduction–oxidation cycles, with the SEM and EDS spectral analysis shown in Figure 1.11.[31] The results also show that the iron oxide–titanium dioxide particle can maintain its particle structure after multiple reduction–oxidation cycles. These experimental results underscore the intricate interplay of vacancy formation, ionic diffusion, and reaction rates due to the presence of the support on the base metal oxide with respect to particle morphology when applied to a chemical looping reaction system, and will be further illustrated in the following sections.

1.3.4   Ionic Diffusion in Microscale: Nanostructure Formation

The oxidation of iron micro-particles involves consumption of oxygen vacancies, as expressed by Reaction (1.3.7). The reaction will lead to particle volume expansion and a porous center due to vacancy condensation at dislocations:

$$\tfrac{1}{2}O_2 + V_{\ddot{O}} = O_O^X + 2h. \qquad (1.3.7)$$

Volume expansion is the driving force for surface morphological transformation, including the growth of nanowires or nanopores. Closer inspection of the iron micro-particle reveals that convex and concave surfaces are formed during the oxidation reaction since the iron oxide product undergoes volume expansion, and the iron diffuses outward because of the confinement along the grain boundary to form nanowires on the

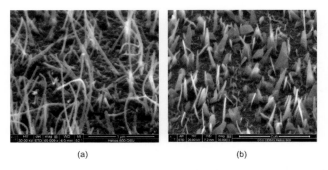

(a)   (b)

Figure 1.12 SEM images of nanowire and nanospike formation from two-dimensional thin film oxidation reactions with iron: (a) at 900 °C; (b) at 700 °C.

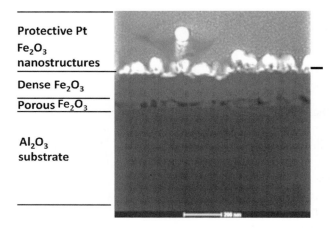

Protective Pt
Fe_2O_3
nanostructures

Dense Fe_2O_3

Porous Fe_2O_3

Al_2O_3
substrate

Figure 1.13 Cavity formation during iron thin film oxidation.

convex surfaces and nanopores in the concave surfaces by the same mechanism, as elaborated in Chapter 2.

Other nanostructures are also present in lower-dimensional systems such as oxidation of iron thin films. When a nanoscale iron thin film is exposed to oxygen, nanowires or nanospikes can form depending on the reaction temperature, where higher temperatures favor the formation of nanowires and lower temperatures favor nanospikes. In addition to the formation of nanowires and nanospikes, cavities below the dense iron oxide layer also form, indicating the outward diffusion of iron. Figure 1.12 shows the formation of the nanowires (Figure 1.12a) and nanospikes (Figure 1.12b), and Figure 1.13 shows the cavity formation from iron oxide thin film oxidation.

The nanowire formation can be explained from its reaction mechanism in conjunction with DFT calculations.[33,34] Using $FeTiO_3$ as an example, oxygen first adsorbs onto the titanium site because titanium has higher adsorption energy than the iron site. It then undergoes an activation step where electron transfer occurs and radical oxygen is formed. This particular step is confirmed from DFT calculations, where there are two possible termination surfaces for the metal oxide, either titanium or iron. Termination on the titanium surface results in a lower activation energy as compared with

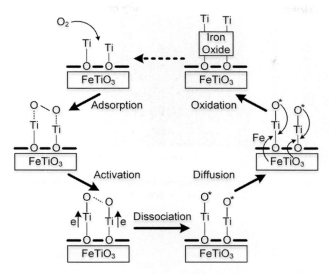

Figure 1.14 Iron oxidation reaction mechanism through adsorption, activation, dissociation, diffusion, and oxidation.[33,34]

termination on the iron surface, indicating that oxidation will preferentially occur on titanium. The dissociation step results in electron transfer and bond breakage to form an oxygen ion terminated surface. Finally, iron diffuses outwards while oxygen diffuses inwards to form iron oxide. Figure 1.14 outlines the iron oxidation reaction mechanism.[33,34] It shows the inter-relationship between the adsorption site and the reaction mechanism with respect to the metal oxide in the context of the oxidation reaction. When an oxygen molecule is adsorbed into surface titanium sites to form a –Ti–O–O–Ti– complex, it dissociates into two oxygen ions with electron transfer from the $FeTiO_3$ surface. The activated oxygen ions then diffuse inward and react with the iron ions that diffuse outward due to their higher diffusivity as compared with titanium ions.

Another important parameter affecting the surface morphology is the reaction temperature. At higher temperatures, the surface tends to be more porous than at lower temperatures. This can be illustrated based on the vacancy formation energy obtained from DFT calculations, shown in Equation (1.3.8), with results given in Figure 1.15[33,34] for temperatures ranging from 400 °C to 900 °C:

$$\Delta G(V_{\ddot{O}}) = E(FeTiO_3, V_{\ddot{O}}) - E(FeTiO_3) + \frac{1}{2}\mu_O(T,p). \quad (1.3.8)$$

It is seen in Figure 1.15 that the vacancy formation energy is lower at higher temperatures, thereby easing porosity formation at higher temperatures. Figure 1.15 also shows vacancy formation energy on the outermost surface layers and the subsurface for $FeTiO_3$ and $\alpha\text{-}Fe_2O_3$. For both metal oxides, the vacancy formation energy on the outermost surface layers is lower than the subsurface over the entire temperature range, so oxygen vacancies have a tendency to form on the outermost surface layers. Also, the vacancy formation energies in the $FeTiO_3$ surface are lower than that in the $\alpha\text{-}Fe_2O_3$ surface at the same temperature, which results in the $FeTiO_3$ surface having more defects than the pure Fe_2O_3 surface, as seen in Figure 1.16.[33]

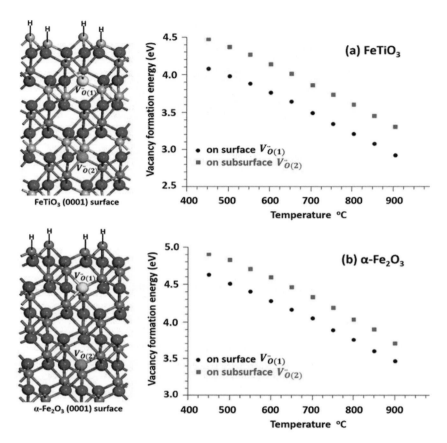

Figure 1.15 Vacancy formation energies on the outermost surface layers and on the subsurface as a function of temperature at pressure 1 atm for (a) FeTiO$_3$ and (b) α-Fe$_2$O$_3$[33,34](Qin et al. *J. Mater. Chem. A*, 2015, 3, 11302–11312 – reproduced by permission of The Royal Society of Chemistry). A black and white version of this figure will appear in some formats. For the color version, please refer to the plate section.

1.3.5   Iron Oxide Reduction Mechanism

Oxygen vacancies can facilitate the partial oxidation of methane by iron oxide, and the reaction pathway with an energy profile can be obtained from the DFT+U method.[34] Figure 1.17 shows the adsorption geometries of CH$_x$ fragments. First, methane adsorbs onto the surface iron site with subsequent dissociation of methane to methyl, methylene, methine, and carbon radical as they adsorb onto the surface of the oxygen vacancy sites, though the adsorption of methylene and methane onto the Fe site adsorption have a similar adsorption energy.

Two dissociated hydrogen ions are then adsorbed onto the same surface iron site to form hydrogen which then desorbs from the iron oxide surface. Simultaneously, the lattice oxygen in the subsurface diffuses to the vacancy site to form an association with

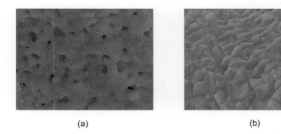

(a)   (b)

Figure 1.16 Pore formation in (a) FeTiO$_3$ and (b) Fe$_2$O$_3$[33] (Qin et al. *J. Mater. Chem. A*, 2015, 3, 11302–11312 – reproduced by permission of The Royal Society of Chemistry).

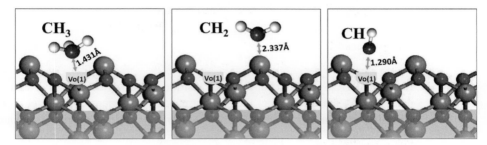

Figure 1.17 CH$_x$ fragments adsorption. Vo(1) denotes oxygen vacancy.[34] (Cheng Z. et al. *Phys. Chem. Chem. Phys.*, 2016, 18, 16423–16435 – reproduced by permission of The Royal Society of Chemistry). A black and white version of this figure will appear in some formats. For the color version, please refer to the plate section.

adsorbed carbon radical and results in the formation of carbon monoxide. This particular pathway reveals the mechanism of partial oxidation of methane using iron oxide.

As a powerful tool, DFT calculations need to be paired with experimental results. As can be seen from the calculation based on classical thermodynamics, the results of the ab initio molecular thermodynamics calculation discussed previously for partial oxidation are now compared with classical thermodynamics calculations, which show the conversion changes as a function of the iron oxide to methane molar ratio, given in Figure 1.18. From Figure 1.18, classical thermodynamics predicts that increasing the iron oxide to methane molar ratio favors complete oxidation instead of partial oxidation. This result is consistent with that predicted from the ab initio thermodynamics by the DFT+U method. It is noted that an increase in the iron oxide to methane molar ratio results in a deficiency of oxygen vacancies that favors complete oxidation. Similarly, a decrease in the iron oxide to methane molar ratio increases the probability of methane activation and dehydrogenation on surface vacancy sites, which favors partial oxidation.

1.3.6 Transition Metal Oxides for Chemical Looping Applications

Transition metal oxides other than iron, including copper, manganese, cobalt, and nickel, have also received widespread attention as chemical looping materials.[2,35–37]

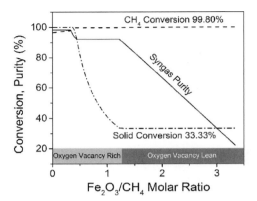

Figure 1.18 Syngas yield as a function of iron oxide to methane molar ratio based on classical thermodynamics.

In particular, structural changes in transition metal oxide particles at a nanoscale during redox reactions affect their performance on a macroscopic level. Upon oxidation, dense Co micro-particles convert to Co_3O_4 with the morphology changed to a porous structure. This is ascribed to processes associated with ion transport, diffusion, and volume expansion. However, upon five reduction and oxidation cycles, the quantity of CoO increases dramatically, indicating that the recyclability of Co_3O_4 deteriorates due to sintering effects. CuO has excellent recyclability as the chemical looping oxygen uncoupling (CLOU) material and is desired to be reduced to Cu_2O during reduction and oxidized to CuO for combustion applications. Atomistic thermodynamics methods and DFT calculations indicate that the activity of Cu is weaker than that of Mn, Co, and Ni. Mn micro-particles are oxidized to a mixed phase of MnO and Mn_3O_4, which have different volume expansion rates and thus causes severe delamination during oxidation. Nevertheless, this delamination can be self-healed during reduction. As CLOU material, manganese oxides are used for combustion and gasification applications. Nickel micro-particles can promptly react with air to form NiO crystals on the surface. These crystals have very few oxygen defects due to a high oxygen diffusion barrier, which leads to a dense NiO surface. Consequently, Ni micro-particles in chemical looping combustion (CLC) applications are mostly limited to surface activities.

For O_2 adsorption on the (100) plane of transition metal systems, the surface metal atom closest to the oxygen molecule is labeled as 1 (M1), the metal atom closest to 1 (M1) as 2 (M2), the metal atom closest to 2 (M2) as 3 (M3), and the metal atom closest to 3 (M3) as 4 (M4). The two oxygen atoms are labeled as O1 and O2. The adsorption energies of the four 3d metal systems follow a periodic trend, i.e. lower chemisorption energy as one proceeds from left to right in the 3d series: Mn (1.54 eV) > Co (1.44 eV) > Ni (1.26 eV) > Cu (0.85 eV), which indicates that the early transition metals tend to have stronger interaction with O_2 than the late transition metals. For the four transition metal (100) surfaces, two transition states are identified in Figure 1.19. The first transition state, TS1, is obtained in the dissociation of the O1–O2 bond of the oxygen

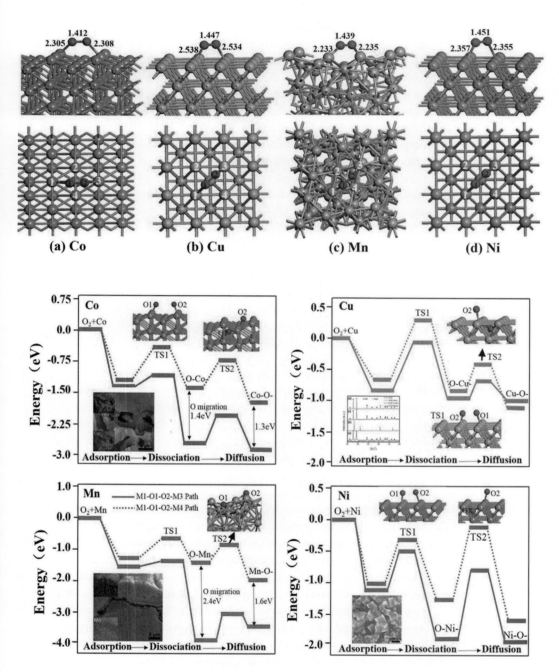

Figure 1.19 Surface structures and energy profiles for the oxidation of transition metals Co, Cu, Mn, and Ni (top: the most stable adsorption structures of O_2 on Co, Cu, Mn, and Ni. The distances of the O–O bond and the O–metal bond are indicated (Å). Bottom: calculated reaction coordinates of O_2 dissociation and diffusion on Co, Cu, Mn, and Ni (100) surfaces. Inset: SEM images of oxidized Co, Mn, and Ni and XRD spectra of CuO during redox reactions). A black and white version of this figure will appear in some formats. For the color version, please refer to the plate section.

molecule, and the second transition state, TS2, is found when the dissociated O ion diffuses into the subsurface of metals. The reaction barriers for the TS1 of Co, Cu, Mn, and Ni are 0.5, 0.85, 0.28, and 0.73 eV, respectively. Therefore, Co, Mn, and Ni are more active for O_2 dissociation and thus more favorable to produce O ion for redox reaction in a CLC system. The oxygen ion diffusion barrier in Cu is lower than the O_2 dissociation barrier, which results in the surface lattice oxygen atoms of Cu oxide being easy to diffuse and release. The Brønsted–Evans–Polanyi (BEP) relationship for reaction energies and total reaction barriers reveals that reactions of early transition metals are more exergonic and have lower oxygen dissociation barriers than those of late transition metals. These findings together with the morphological changes in a transition metal system form a fundamental basis for understanding transformations occurring during the redox process that are of value to metal oxide material selection and design for chemical looping applications.

1.4   Chemical Looping Classification

According to the carbonaceous feedstock reaction conditions in the reducer, chemical looping reactions can be classified into two categories: (1) complete/full oxidation; (2) partial/selective oxidation. Category 1 represents the typical chemical looping combustion applications, while category 2 represents the chemical looping gasification or reforming applications. As the chemical looping reactions can also be applied to systems when the fossil feedstock is absent, the discussion of this book also includes category 3, denoted as non-carbonaceous fuels and represented by solar or nuclear systems or by the use of non-carbonaceous feedstock. Figure 1.20 shows these three categories with reactions for categories 1 and 2 represented by Fe_2O_3 as the metal oxide and CH_4 as the carbonaceous feedstock. These three categories are further elaborated below.

Category 1 Complete/Full Oxidation – Combustion

This chemical looping application using carbonaceous fuels, mainly fossil fuels, has been the subject of recent extensive studies. As noted in Sections 1.1 and 1.2.2, the interest is driven in part by its high exergetic efficiency, where irreversible energy losses are minimal.[2–4,7,38,39] The complete oxidation of fuel is used for heat and power generation and provides maximum cost reduction benefits when used in carbon-constrained scenarios. In the reducer or fuel reactor, metal oxide particles typically provide lattice oxygen to the fossil fuel in order to produce the combustion products of the reduced metal oxide, CO_2, and H_2O, with CO_2 easily separable from H_2O. In the combustor or air reactor, the reduced metal oxide is re-oxidized with air to regenerate the metal oxide particle and high-quality heat.

Category 2 Partial/Selective Oxidation – Gasification and Reforming

This chemical looping application using carbonaceous fuels can produce synthesis gas (syngas), which is primarily a mixture of CO and H_2.[7,40–43] The syngas can then be used as an intermediate for the synthesis of chemicals such as methanol

Category 1: Complete/Full Oxidation – Combustion

Reducer: $4Fe_2O_3 + CH_4 \rightarrow 8FeO + CO_2 + 2H_2O$
       (Oxidized)   (Reduced)

Combustor: $4FeO + Air (O_2) \rightarrow 2Fe_2O_3$

Category 2: Partial/Selective Oxidation – Gasification and Reforming

Reducer: $Fe_2O_3 + CH_4 \rightarrow 2FeO + CO + 2H_2$
       (Oxidized)   (Reduced)

Combustor: $4FeO + Air (O_2) \rightarrow 2Fe_2O_3$

Reducer: $6Mn_2O_3 + 2CH_4 \rightarrow 4Mn_3O_4 + C_2H_4 + 2H_2O$
       (Oxidized)   (Reduced)

Combustor: $4Mn_3O_4 + Air (O_2) \rightarrow 6Mn_2O_3$

Category 3: Non-Carbonaceous Fuels – Solar, Nuclear Systems
          and Non-Carbonaceous Feedstock

Reducer: $2ZnO \rightarrow 2Zn + O_2$
       (Oxidized)   (Reduced)

Oxidizer: $Zn + H_2O \rightarrow ZnO + H_2$

Reducer: $Fe_3O_4 + 4NH_3 \rightarrow 3Fe + 2N_2 + 2H_2 + 4H_2O$
       (Oxidized)   (Reduced)

Oxidizer: $3Fe + 4H_2O \rightarrow Fe_3O_4 + 4H_2$

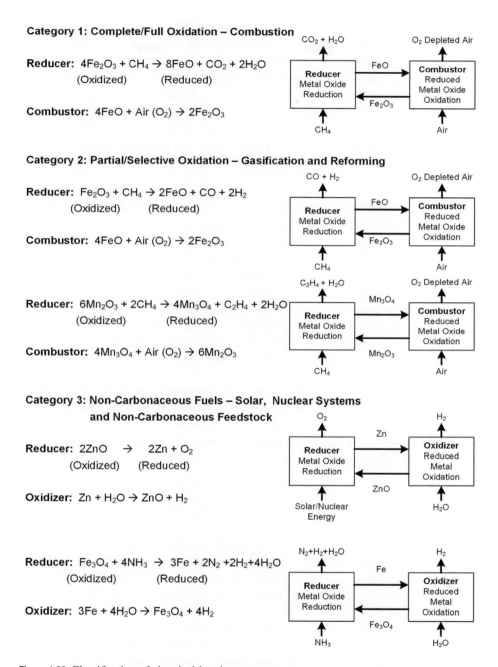

Figure 1.20 Classification of chemical looping systems.

and liquid fuels. In the reducer reactor, the fuel is partially oxidized to syngas by lattice oxygen transfer from metal oxide particles, which are reduced. The combustor reactor then regenerates the reduced metal oxide using air. Category

2 chemical looping can also produce chemicals in one step, which relies on a multifunctional metal oxide that possesses catalytic and oxygen transfer properties to selectively convert hydrocarbon feedstock to chemicals. The oxidative coupling of methane to ethylene and selective oxidation of butane to maleic anhydride are two examples. In the reducer reactor, a catalytic metal oxide reacts with a hydrocarbon feedstock to selectively produce chemicals and reduced catalytic metal oxide. In the combustor reactor, the reduced catalytic metal oxide is regenerated by oxidation with air.[44,45] By directly producing chemicals in the reducer, the generation of syngas as an intermediate is not necessary. Category 2 is the main subject of this book.

Category 3 Non-Carbonaceous Fuel – Solar and Nuclear Systems and Non-Carbonaceous Feedstock

Non-carbonaceous fuel chemical looping relies on sources like solar energy and nuclear energy as the primary energy source for the reducer reactor. When nuclear or solar energy is applied to the reducer reactor, the metal oxide can be directly reduced to produce oxygen and reduced metal oxide. This energy source can also be used to supply endothermic heat for the carbonaceous feedstock reaction in the reducer. In the combustor, oxidants such as H_2O or CO_2 are used to produce H_2 or CO, respectively, with a high solar energy conversion efficiency, while regenerating the metal oxide.[46–49] Non-carbonaceous fuels such as ammonia, hydrogen sulfide, and urea can also be used as feedstock for its conversion to hydrogen through the chemical looping redox route.

1.5 Chemical Looping Partial Oxidation Historical and Recent Developments

The chemical looping concept for technological applications has been known since the late nineteenth century. In 1897, Franz Bergmann filed a German patent where manganese (IV) oxide reacts in the presence of a hydrocarbon fuel (coal, coke, liquid hydrocarbons) and calcium oxide (CaO) to produce calcium carbide (CaC_2). In the calcium carbide furnace, manganese (IV) oxide is reduced to manganese (II, III) oxide, which provides the oxygen for the combustion of hydrocarbon fuel. In a separate reactor, the manganese (II,III) oxide is regenerated to manganese (IV) oxide using air.[50]

In 1903, Howard Lane invented an apparatus, known as the Lane hydrogen producer or Lane process, for the production of hydrogen to be used for military balloons and zeppelins. In the Lane process, shown in Figure 1.21,[51–53] hydrogen was produced in a fixed bed. First, syngas derived from coal gasification reduced iron oxide ores and was followed by the introduction of steam, where the steam-iron reaction produced hydrogen. By 1913, plants throughout Europe and the United States were constructed with a potential annual production capacity of 24 million m^3 of hydrogen.[51] However, the incomplete conversion of reducing gas and low recyclability of the iron ores prevented

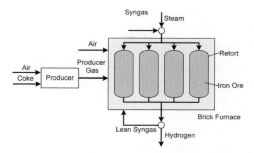

Figure 1.21 Lane hydrogen producer apparatus.[51–53]

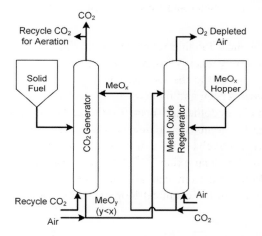

Figure 1.22 Lewis and Gilliland CO_2 production process.[54]

the Lane process from operating efficiently. With the availability of oil and natural gas in the 1940s, the steam-iron process for hydrogen production became less competitive and was eventually phased out.

In the 1950s, Warren Lewis and Edwin Gilliland patented a chemical looping process using iron and copper oxides as the metal oxide to generate CO_2 for use in the beverage industry.[54] Figure 1.22 shows the general process. A countercurrent gas–solid contacting pattern was proposed with the iron oxide circulating between Fe_2O_3 and Fe_3O_4 in either fluidized beds or moving beds. Experimental data using batch reactors and copper oxide were reported, but no experimental data in a continuous process that substantiated the desired gas–solid contact pattern was published. As mentioned previously, copper oxide is one of the metal oxide oxygen carriers for use as a chemical looping oxygen uncoupling (CLOU) material.

In the 1960s and 1970s, the Consolidation Coal Company, which later became the Conoco Coal Development Company, and now CONSOL Energy, developed the CO_2 acceptor process to produce substitute, or synthetic/supplemental, natural gas (SNG).[55] The process, shown in Figure 1.23, begins with coal gasification. In the gasifier, the reaction between coal and steam produces hydrogen, and the addition of

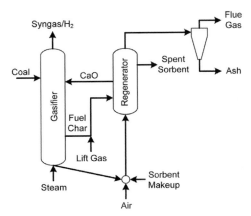

Figure 1.23 CONSOL CO_2 acceptor process for SNG production from coal.[55]

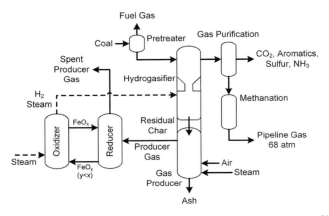

Figure 1.24 IGT HYGAS process for SNG production from coal.[56]

CaO enhances hydrogen generation by reactively removing CO_2 to form $CaCO_3$. The CO_2 acceptor process was operated at the pilot-scale but was not commercially realized, in part due to economic reasons such as the low market price of natural gas and the high capital cost requirements for the process. Furthermore, the processing of lignite coals introduced operational difficulties due to agglomeration caused by the low temperature eutectic formation between fly ash and calcium compounds in the gasifier and the regenerator. It was also realized that the CaS–$CaSO_4$ redox cycle observed in the CO_2 acceptor process was different from the typical redox cycles based on metal/metal oxide.

In the 1970s, the Institute of Gas Technology (IGT) developed the HYGAS process, shown in Figure 1.24, to convert coal to SNG by the methanation reaction. The required hydrogen for the methanation reaction came from the steam–iron reaction, where iron was obtained from iron oxide reduction by syngas produced from coal gasification.[56] Synthetic iron oxide particles consisting of 4% silica, 10% magnesia, and 86% hematite ore were used as the iron oxide looping material. Both the strength and reactivity of the

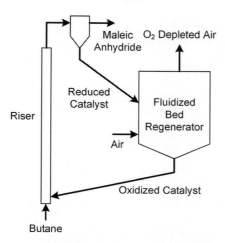

Figure 1.25 DuPont process for maleic anhydride production from butane using VPO.[45]

synthetic iron oxide were an improvement over the iron ore used in the Lane process. The reducer and the oxidizer were both two-stage countercurrent fluidized bed reactors to enhance the fuel gas and iron oxide conversions, as well as heat and mass transfer. Consequently, the efficiency of the steam–iron reaction in the HYGAS process was higher than that of the Lane process. Like the CO_2 acceptor process, the HYGAS process was demonstrated at pilot-scale but was not commercialized. Drawbacks of the HYGAS process included low metal oxide conversions, low redox rates, and incomplete syngas conversions.

In the 1980s, Atlantic Richfield Company (ARCO) developed the gas to gasoline (GTG) process.[57,58] Using oxidative coupling of methane, methane was converted to an ethylene-rich intermediate, which was then converted to gasoline using Mobil's olefin to gasoline and distillate (MOGD) process. The GTG process used two circulating fluidized bed reactors with a catalytic metal oxide active for the oxidative coupling of methane (OCM) reaction reacting with methane in one reactor, and reduced catalytic metal oxide regeneration in the other.[57,58] This process was operated at the pilot-scale and eventually discontinued towards the end of 1980s due to oil prices being comparatively lower than natural gas prices.

In the 1990s, DuPont developed a process, shown in Figure 1.25, to produce maleic anhydride through the selective oxidation of butane using a vanadium phosphorus oxide (VPO) catalytic metal oxide. Referred to as the DuPont process, a lean phase riser was used for butane conversion to maleic anhydride, where a multifunctional VPO metal oxide provided catalytic activity as well as lattice oxygen. In a fluidized bed regenerator, the reduced VPO catalytic metal oxide reacted with air for re-oxidation. This process was scaled to commercial demonstration but failed as the VPO catalytic metal oxide was not reactive enough to provide the oxygen and particle integrity was compromised.[45,59]

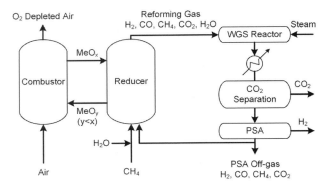

Figure 1.26 Chemical looping partial oxidation of methane for syngas production.

Also in the 1990s, the partial oxidation of methane to syngas using CeO_2 was attempted in a fixed bed reactor system. Using CeO_2 in a chemical looping process to convert methane to syngas is thermodynamically favorable, but experimental results showed extensive carbon deposition, which negatively affected the CeO_2 reactivity, and reduced the reaction kinetics between methane and CeO_2, resulting in a low methane conversion (4% per pass).[60]

In the 2000s a chemical looping process was developed to provide the heat necessary to operate conventional steam methane reforming (SMR) reactions using metal oxide oxygen carriers.[61–63] Heat is supplied to the SMR tubular reactor that is placed inside either the reducer or combustor reactor. An advantage of this approach is the elimination of an amine based carbon capture step since the chemical looping reducer can completely convert the off-gas stream from the pressure swing adsorption (PSA) unit to CO_2. This process requires diverting natural gas in an equal or greater quantity than the conventional approach to fulfill the endothermic heat requirements for the SMR reactions. Furthermore, the transfer of heat from the SMR reactors inside either the reducer or the combustor has drawbacks that are related to the operational complexity and cost, and its long-term viability in high temperature operation due to tube erosion. The overall benefits involved in the process intensification as given in this chemical looping partial oxidation system are minimal.

Figure 1.26 shows a simplified process system that is an improved chemical looping scheme for syngas production. It is seen that the partial oxidation of methane generating the syngas is carried out directly in a reducer. This chemical looping reducer system takes a combination of the PSA off-gas stream and the entire methane leading to a higher syngas product yield and the elimination of methane heating to provide the endothermic heat of reaction from the tubular reformers in SMR. It also leads to a higher H_2 production efficiency on a molar basis. The quality of the syngas generated, however, is strongly dependent on the reducer reactor contact mode. The fluidized bed is commonly used as a reactor for the reducer operation.

Beginning in the early 1980s, several chemical looping based processes utilizing solar energy for H_2 production through water splitting reactions have been proposed.[48] In the reducer reactor, solar energy input releases oxygen from a metal oxide. In the oxidizer reactor, the reduced metal oxide produces hydrogen via the water splitting reaction.[64] The metal oxide particles initially developed had operating temperatures for oxygen release using solar energy that were higher than their melting points. The water-splitting reaction, however, is only feasible at lower temperatures. An efficient reactor design involved re-crystallization of the melted metal oxides for the water splitting reactions, which hindered large-scale demonstrations due to heat transfer and flow difficulties. In principle, metal oxides as a medium to convert solar energy to H_2 holds promise for significant efficiency improvements over conventional electrolysis based approaches.[48,49] Recent developments focus on developing metal oxides that release oxygen at temperatures below their melting point, like $CoFe_2O_4$ doped with ZrO_2, Al_2O_3, $Co_{0.8}Fe_{2.1}O_4$, and $FeAl_2O_4$. These processes currently possess a low steam to H_2 conversion efficiency, but the development of a metal oxide that can also improve steam yields is ongoing.[65]

1.6   Ohio State Chemical Looping Technology Platform

A salient feature of all OSU chemical looping systems is their simplicity, involving only a single loop, compared to almost all other chemical looping systems, for continuous solids circulation from the top of the reducer reactor through the combustor to the riser and return to the reducer. This simplicity is possible for these chemical looping systems through using a moving bed as the reducer. Such system simplicity, however, is difficult to achieve if a fluidized bed is used as the reducer. Through the use of the moving bed, a wide range of process configurations are available for generating a multitude of products in a very efficient manner. For example, as will be illustrated in the process examples given by simply changing the direction of the flow of the feedstock relative to the metal oxide in the reducer from countercurrent to cocurrent, the condition of the reactions can change readily from full oxidation (Category 1) to gasification or reforming (Category 2). The system contains no mechanical devices, with controllable solids circulation accomplished by using non-mechanical valves. Also, due to its simplistic design, the system is easily scalable. Thus, the OSU chemical looping processes form a technology platform that can convert many types of carbonaceous feedstock, including coal, biomass, natural gas, and syngas, into products such as electricity, syngas, hydrogen, and chemicals, as shown in Figure 1.27.[7] The term "technology platform" refers to a set of technologies developed for varying applications but which share a common underlying basic concept. The OSU technologies have been developed over more than 20 years through an evolutionary process from metal oxide material syntheses, through reactivity tests, fixed bed, bench scale, and sub-pilot scale unit tests, and pilot plant demonstration, as shown in Figure 1.28.[7] The technologies developed include both calcium looping technology and chemical looping redox technology. Calcium looping technology is a subject that has been extensively discussed in an earlier book by Fan,[2] and is thus not elaborated on further in this book.

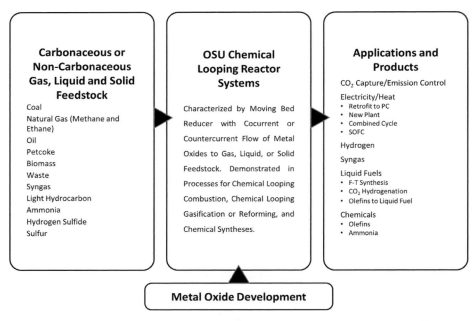

Figure 1.27 Feedstock and products in the OSU chemical looping technology platform.[7]

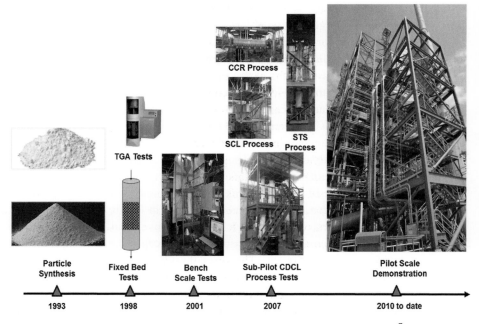

Figure 1.28 Evolution of development of the OSU chemical looping technologies.[7] A black and white version of this figure will appear in some formats. For the color version, please refer to the plate section.

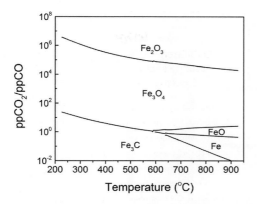

Figure 1.29 Thermodynamic phase diagram of the Fe–CO–CO$_2$ system at 1 atm pressure, generated using FactSage 7.0 software.

1.6.1 Chemical Looping Full Oxidation Technology – Category 1

The chemical looping full oxidation process converts hydrocarbon feedstock to CO$_2$ and H$_2$O using an iron oxide based composite metal oxide in a countercurrent moving bed reducer. The thermodynamic motivation for using a countercurrent reactor system is illustrated by the Fe–CO–CO$_2$ phase diagram, shown in Figure 1.29, that provides the oxidation state of iron in equilibrium with CO–CO$_2$. This phase diagram has been generated using FToxid and FactPS databases in FactSage 7.0 software[66] at a constant pressure of 1 atm. In a fluidized bed reducer, the oxygen carrier conversion is constrained by thermodynamic limitations since the gas inside the fluidized bed and in the outlet of the fluidized bed is rich in CO$_2$, which restricts the extent of reduction of Fe$_2$O$_3$ in the reactor. As an example, in order to achieve 99.9% CO conversion from a fluidized bed reducer, the Fe$_2$O$_3$ cannot be reduced to an oxidation state lower than Fe$_3$O$_4$, which corresponds to a maximum solid conversion of 11.1%.

In contrast, a moving bed reducer using CO as the reducing gas in a countercurrent gas–solid contacting pattern can reduce Fe$_2$O$_3$ to an oxidation state lower than Fe$_3$O$_4$. Under this operating condition, greater than 99.99% of CO can be converted and the solids can be reduced to an FeO/Fe mixture.[2,7] The same principle can be applied to processing other gaseous feedstock, including H$_2$, CH$_4$, and coal/biomass volatiles. It is noted that, as seen in Figure 1.29, there is Fe$_3$C formation that occurs at low temperatures ($<$~600 °C) and partial pressure ratios (ppCO$_2$/pCO) $<$1. The formation of Fe$_3$C needs to be avoided to prevent deactivation of the metal oxide as well as loss of carbon capture efficiency. For a given fuel input, the Fe$_2$O$_3$ circulation rate in a countercurrent moving bed reducer is less than 30% of a fluidized bed reducer.[7] Another important advantage of the countercurrent moving bed processing of solid fuels is the ability to easily separate ash from metal oxide particles due to the particle size difference between the two.[2]

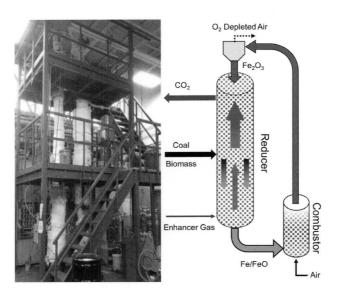

Figure 1.30 Sub-pilot scale demonstration unit for the CDCL process (left); flow schematic of the CDCL process for power production (right).[67] A black and white version of this figure will appear in some formats. For the color version, please refer to the plate section.

Coal Direct Chemical Looping (CDCL) Process

The coal direct chemical looping (CDCL) process for electricity production is shown in Figure 1.30. In the CDCL process, coal is injected into the middle of the moving bed reducer. The coal volatiles flow upwards countercurrently with iron oxide particles entering from the top of the reducer and are combusted with iron oxide particles (category 1 reaction) while the coal char flows downwards cocurrently with the iron oxide particles and is gasified with CO_2 and/or steam as an enhancing gas (category 2 reaction). The outlet gas stream from the reducer is mainly CO_2 and H_2O.

The CDCL process has been demonstrated at the bench scale and 25 kW$_{th}$ sub-pilot scale, with over 600 h of operation at the sub-pilot scale.[67,68] The fully integrated 25 kW$_{th}$ sub-pilot CDCL unit was successfully operated for 200 continuous hours with near complete conversion of coal to CO_2 and steam, as shown in Figure 1.31 and Figure 1.32, smooth solids circulation, gas-sealing using non-mechanical valves, and predicted pressure drops. A 250 kW$_{th}$ pilot plant is under construction by Babcock & Wilcox (B&W) Power Generation Group.

The experimental results from the long-term sub-pilot scale demonstration were the input for a techno-economic analysis of the CDCL process for power generation, shown in Figure 1.33. The integrated CDCL process is compared to a baseline case for a power generation plant with a net output of 550 MW$_e$ and 90% carbon capture using an amine based solvent. Table 1.2 summarizes the results of the process simulation and techno-economic analysis. The process simulation results show that for the same new power output, the net plant efficiency is 35.2% for CDCL as compared to 28.5% for the baseline plant. This higher efficiency translates to a 20% reduction in coal input. The energy penalty as compared to a baseline case without CO_2 capture is 10% for the CDCL plant,

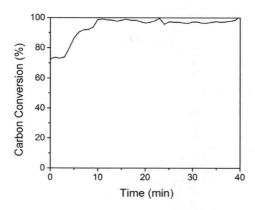

Figure 1.31 Carbon conversion profile for CDCL operation.

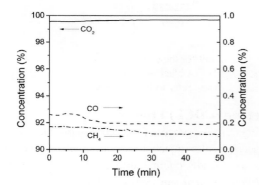

Figure 1.32 Reducer gas outlet composition for long-term CDCL operation.

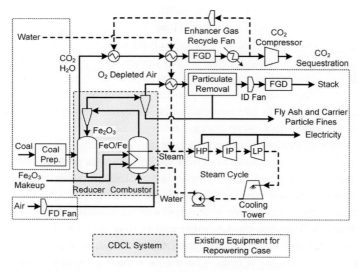

Figure 1.33 Integrated process flow diagram of the CDCL process for power generation.

Table 1.2 Comparative summary of capital costs and cost of electricity for the CDCL plant with 90% carbon capture.

Parameter	Baseline amine plant	CDCL plant
% Carbon capture	90%	97%
Energy penalty (% relative to baseline no-capture case)	27.6	10.6
Net plant efficiency (HHV, %)	28.5	35.2
First year capital cost ($/MWh) (2011 $)	59.6	44.2
First year cost of electricity ($/MWh) (2011 $)	**100.9**	**78.4**

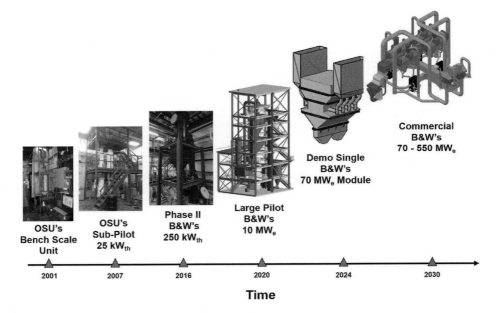

Figure 1.34 Commercialization pathway for CDCL technology by OSU and B&W Power Generation Group. A black and white version of this figure will appear in some formats. For the color version, please refer to the plate section.

which is 17% lower than the baseline case with CO_2 capture using the amine technique. The higher efficiency and lower energy penalty leads to a significant reduction in the capital cost requirements for the CDCL plant. The first year capital cost ($/MWh (2011 $)) for the CDCL plant is 25% lower than the corresponding value for an amine plant, and the first year cost of electricity for a CDCL plant is estimated to be $78.40/ MWh, which is 22% lower than the baseline value of $100/MWh (2011 $).

The United States Department of Energy considers carbon capture technologies that can achieve 90% carbon capture without increasing the cost of electricity by more than 35% to be a transformational technology. The CDCL cost of electricity successfully meets this target and is considered to be a third generation, transformational technology.[8,14,69–72] The commercialization pathway for CDCL technology by OSU

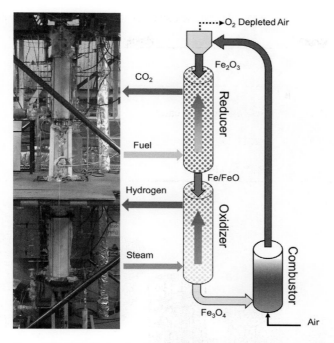

Figure 1.35 Sub-pilot scale demonstration unit for the SCL process for H_2 production (left); flow schematic of the SCL process unit for H_2 production (right).[73] A black and white version of this figure will appear in some formats. For the color version, please refer to the plate section.

and B&W Power Generation Group is shown in Figure 1.34. It is seen that the full commercial scale plant is projected to be in operation in around 2030.

Syngas Chemical Looping (SCL) Process

The syngas chemical looping (SCL) process was developed to convert gaseous fuels like syngas and natural gas in a three-reactor, iron based chemical looping scheme to produce H_2. In the SCL reducer reactor, a metal oxide reacts with the gaseous fuel in a counter-current contact pattern for complete fuel oxidation; given as a category 1 combustion operation in Section 1.4. The countercurrent moving bed reducer reactor yields a reduction of Fe_2O_3 to a mixture of Fe/FeO, which can then proceed with the water splitting function in the SCL process to produce hydrogen, while Fe/FeO are oxidized to Fe_3O_4.[2,7,73] The use of a single stage fluidized bed restricts the oxygen carrying capacity from Fe_2O_3 to Fe_3O_4 or Fe_3O_4 to FeO/Fe, leading to either no feasible H_2 production or unfavorably large solids circulation rates, respectively. The SCL process has been successfully tested at the bench scale and sub-pilot scale, shown in Figure 1.35, with a pilot-scale, pressurized system under demonstration at the National Carbon Capture Center (NCCC) in Wilsonville, Alabama.[73–75] Figure 1.36 is a photo of the 250 kW_{th} to 3 MW_{th} pilot plant. The demonstration results using syngas as feedstock from KBR's transport gasifier are good and under ambient operation, the results are consistent with those obtained in the sub-pilot scale unit.

Figure 1.37 shows a typical gas composition exiting the reducer demonstrating complete syngas conversion to CO_2 that can result in 100% carbon capture for a coal

Figure 1.36 SCL pilot-scale demonstration unit at the National Carbon Capture Center, Wilsonville, AL[7] (reprinted from Fan, L.-S. et al., 2015, *AIChE J.,* 61: 2–22 with permission from John Wiley and Sons). A black and white version of this figure will appear in some formats. For the color version, please refer to the plate section.

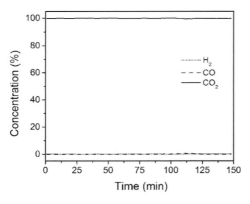

Figure 1.37 Reducer outlet gas composition obtained during sub-pilot scale demonstration of the SCL process.

based H_2 production process.[73–76] Figure 1.38 shows a typical gas composition from the oxidizer, which indicates >99.9% purity H_2 production. A three-day continuous operation of the sub-pilot unit demonstration run was completed showing steady, continuous, and complete syngas conversion in the reducer reactor and high purity H_2 production from the oxidizer reactor.

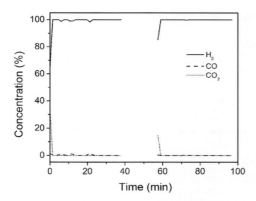

Figure 1.38 Oxidizer outlet gas composition obtained during sub-pilot scale demonstration of the SCL process.

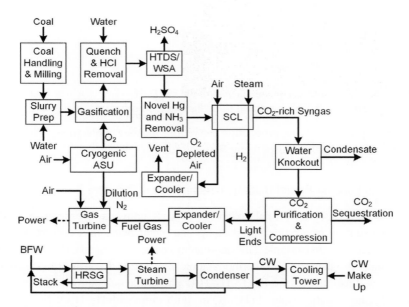

Figure 1.39 Process flow diagram of the SCL process for H_2 production integrated into an IGCC process.

Sub-pilot scale and pilot-scale SCL experimental results were used to develop an integrated process for H_2 production from the SCL system for an integrated gasification combined cycle (IGCC) application. Figure 1.39 shows the block diagram of the SCL process on which a comprehensive techno-economic analysis was performed. The comparison of capital cost investment and cost of electricity is shown in Table 1.3. Based on 2011 USD, the baseline IGCC process has a total plant capital cost investment of \$3,324/kW with a first year cost of electricity of \$143.1/MWh. The total plant capital cost investment and the overall cost of electricity are both lower for the SCL–IGCC system as compared to the baseline IGCC system, by 12% and \$10/MWh, respectively.

Table 1.3 Economic comparison of the SCL plant for H_2 production with $>90\%$ carbon capture in an IGCC configuration.

Parameter (2011 $)	Baseline IGCC	SCL IGCC
Process efficiency (%)	32.4	32.7
Total plant cost ($/kW)	3,324	2,934
Cost of electricity ($/MWh)	143.1	133.8

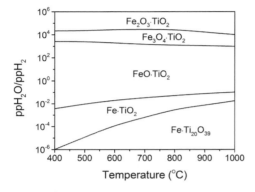

Figure 1.40 Thermodynamic phase diagram of the Fe–Ti–H_2–H_2O system at 1 atm pressure for $0.5 \leq$ Ti/(Fe + Ti) ≤ 0.6667, generated using FactSage 7.0 software.

The SCL process with H_2 production can reduce the cost of electricity with CO_2 capture by 7%–9% over the corresponding baseline case.

1.6.2   Chemical Looping Partial Oxidation Technology – Category 2

The chemical looping partial oxidation process converts hydrocarbon feedstock to syngas using an iron oxide based composite metal oxide in a cocurrent moving bed reducer. Here, as an example, an iron–titanium composite metal oxide (ITCMO) is used for syngas production. The thermodynamic motivation for using a cocurrent reactor system is illustrated by the Fe–Ti–H_2–H_2O phase diagram, shown in Figure 1.40, and provides the oxidation state of ITCMO in equilibrium with H_2–H_2O. The phase diagram is generated using FToxid and FactPS databases in FactSage 7.0 software[66] for a pressure of 1 atm. The composition of ITCMO plotted in this phase diagram is for a molar composition ranging from $0.5 \leq$ Ti/(Fe + Ti) ≤ 0.6667.

The first design aspect consideration when developing a chemical looping reactor system for partial oxidation is to recognize the thermodynamic constraint for the formation of product gases that minimizes the $H_2O:H_2$ ratio at the reactor outlet. For the Fe–Ti–H_2–H_2O system, as given in Figure 1.40, the optimum reactor system can be designed such that the reactor outlet product gases are in a condition of equilibrium between Fe·TiO_2/FeO·TiO_2. If the reactor outlet product gases are in equilibrium with Fe_2TiO_5 and/or Fe_3O_4·TiO_2 instead, a maximum equilibrium $H_2O:H_2$ ratio would be formed and the resulting reactor design would therefore be undesirable for chemical looping partial oxidation applications. The second design aspect consideration is to

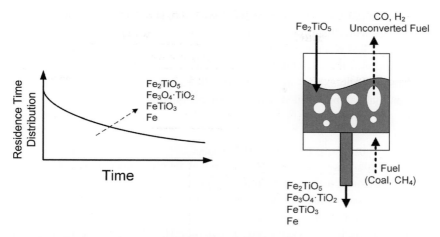

Figure 1.41 Residence time in a fluidized bed reducer system for a pulse of Fe_2TiO_5 oxygen carrier particles.

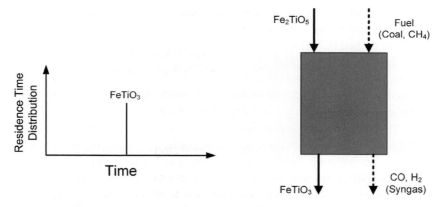

Figure 1.42 Residence time in a cocurrent moving bed reducer system for a pulse of Fe_2TiO_5 oxygen carrier particles.

determine a gas–solid contact mode for the reactors that can achieve the first design aspect. Two possible gas–solid contact modes for the reactor are considered: (1) a cocurrent moving bed reactor; and (2) a fluidized bed reactor. Both reactor systems can be operated by adjusting the stoichiometric ratio of metal oxide to fuel to a reaction condition in favor of partial oxidation. With the stoichiometric ratio for the feedstock input being identical for the fluidized bed and the moving bed, a fluidized bed system will be operated with the gaseous products being contacted with ITCMO in multiple oxidation states, as opposed to the desired oxidation state of $Fe\cdot TiO_2/FeO\cdot TiO_2$. Specifically, a fluidized bed reducer reactor will exhibit mixed oxidation states, as shown in Figure 1.41, due to the mixed residence time of the ITCMO. In contrast, a cocurrent moving bed reactor will exhibit a single oxidation state, corresponding to a single residence time of the ITCMO, as shown in Figure 1.42. That is, a cocurrent moving bed reactor can initially contact the fuel with the highest oxidation state of

ITCMO (Fe_2TiO_5) for fast kinetic conversion while the final stage of the products is in contact with a mixture of $Fe \cdot TiO_2/FeO \cdot TiO_2$ to ensure a low $H_2O:H_2$ ratio. Thus, the moving bed reactor system can synchronize its operation to a desired thermodynamic condition while the fluidized bed cannot. Further, the channeling of the reactants through bubble flows in a fluidized bed system renders an incomplete reactant conversion that can be avoided with a cocurrent moving bed.

Shale Gas-to-Syngas (STS) Process

The shale gas-to-syngas (STS) chemical looping process was developed to convert gaseous fuels to syngas with a high efficiency and flexible composition, so that it is suitable for downstream processing.[40] The STS process uses a cocurrent moving bed reducer reactor for producing syngas and a fluidized bed combustor reactor for regenerating the reduced oxygen carrier material. The gaseous fuel to syngas conversion is obtained in a single step by the use of lattice oxygen from ITCMO. The STS process is designed such that the endothermic heat of reaction for syngas production is balanced by the exothermic reaction of ITCMO regeneration in the combustor reactor.[40] This autothermal operating mode is optimized by considering multiple parameters like temperature swings on the ITCMO particles, support material type and weight fraction added to the ITCMO mixture, and stoichiometric ratio of active component of ITCMO to gaseous fuel. The $H_2:CO$ ratio can be flexibly adjusted by controlling the steam concentration, composition of gaseous fuels, ITCMO circulation rate, and the reaction residence time, in addition to the parameters adjusted to satisfy the autothermal heat balance condition.

The STS process was demonstrated on a sub-pilot scale moving bed reactor, shown in Figure 1.43, using methane as a model gaseous fuel. Comprehensive studies of the STS range of operating conditions including $Fe_2O_3:CH_4$ molar ratios from 0.5 to 1.4, temperatures from 900 to 1050 °C, and steam:CH_4 molar ratios from 0 to 0.4 were conducted. Process parameters were adjusted to ensure a syngas product directly amenable for liquid fuel production through the Fischer–Tropsch (F–T) process. A typical gas composition from the STS sub-pilot scale demonstrations is shown in Figure 1.44.[40] Bench-scale and sub-pilot tests could achieve >99.9% CH_4 conversion, $H_2:CO$ ratio of 1.97, and a syngas purity of 91.3%. With steam injection, the $H_2:CO$ ratio was further improved to 2.3, demonstrating the flexibility of the STS process for downstream process integration.

Experimental results are used to integrate the STS process into several commercial liquid fuels/chemical production plants to validate the economic market feasibility of the STS process. Figure 1.45 shows several STS integration options for liquid fuels production using the Fischer–Tropsch synthesis, methanol production, hydrogen production, and power production. A specific process application for integrating the STS process into a 50,000 bbl/day plant and comparing the process performance to a baseline plant using an autothermal reformer (ATR) for syngas production is presented below.[77] The total plant cost for the baseline 50,000 bbl/day GTL plant using ATR is estimated to be $86,000/bbl (2011 $), with the ATR and air separation unit representing

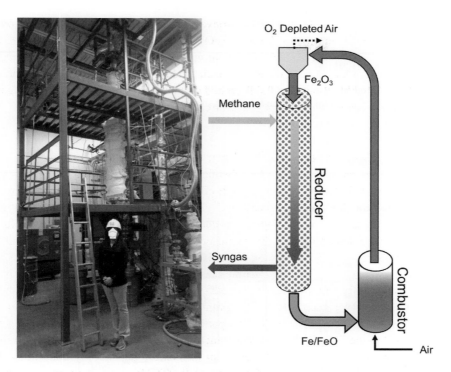

Figure 1.43 Sub-pilot scale demonstration unit for the STS process (left); flow schematic of the STS process unit (right).[40] A black and white version of this figure will appear in some formats. For the color version, please refer to the plate section.

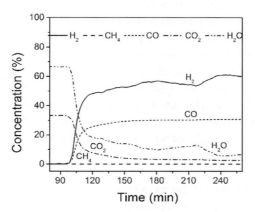

Figure 1.44 Results from the 25 kW$_{th}$ STS sub-pilot scale moving bed demonstration unit.[40]

over 38% of the total plant cost.[77] An economic analysis is performed on the STS process based on Figure 1.45. For the same liquid fuel production rate, the higher syngas yield from the STS process leads to a reduction in the overall natural gas flow by 10% and the overall steam requirement by 50%, leading to significantly higher carbon

Table 1.4 Economic comparison of the STS process to the conventional ATR process when integrated in a 50,000 bpd GTL plant.

Component (2011 $)	ATR[77]	STS
Syngas production unit capital cost ($×1000)	1,030,000	283,000
CO_2 removal, compression, and drying system capital cost ($×1000)	306,000	181,000
Net plant power (kW_e)	40,800	83,000
Total plant cost ($×1000)	2,750,000	1,880,000
Total as-spent cost ($/bbl/day)	86,200	65,100

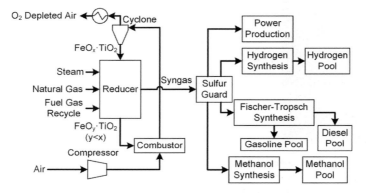

Figure 1.45 STS process and integration options for liquid fuels production using Fischer–Tropsch synthesis, methanol production, hydrogen production, and power production.

utilization efficiencies. Table 1.4 provides an economic comparison for the STS process as compared to the baseline ATR process.

Results from techno-economic analysis show that the total capital cost for the STS process is $65,000/bbl (2011 $) due to a 60% reduction in capital investment for the syngas production unit as compared to ATR. At a natural gas market price of $3/MMBTU, the STS process for liquid fuel production is economically competitive even when Western Texas intermediate crude oil prices are as low as $45/bbl. The STS process is a disruptive technology that, if commercialized, can transform the liquid fuel industry.

Coal-to-Syngas (CTS) Process

The coal-to-syngas (CTS) chemical looping process was developed for high efficiency conversion of solid fuels to syngas.[78] The CTS process uses a cocurrent moving bed reducer for producing syngas and a fluidized bed combustor reactor for regenerating the reduced oxygen carrier material. The CTS process intensifies conventional gasification systems by eliminating the coal drying system, coal gasifier, air separation unit, and the water–gas shift reactor.[79] Oxygen transfer from the ITCMO oxygen carrier to the fuel is controlled to improve the carbon utilization efficiency for partial oxidation process schemes. The conceptual advantages of the CTS process

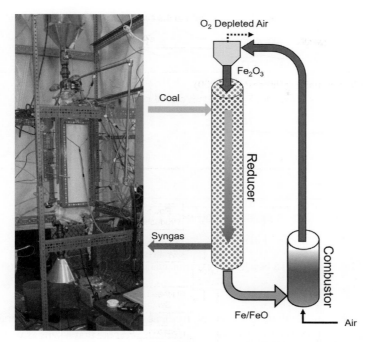

Figure 1.46 Bench-scale demonstration unit for the reducer (left); flow schematic of the CTS process unit (right).[78] A black and white version of this figure will appear in some formats. For the color version, please refer to the plate section.

system can be illustrated further by considering its integration into a methanol production system. The conventional approach gasifies the fuel to obtain a H_2:CO ratio that is close to the intrinsic value of the ratio in the coal used. A methanol processing system requires a H_2:CO ratio close to 2 and a stoichiometric number $(H_2 - CO_2)/(CO + CO_2)$ with a value also close to 2. The H_2:CO ratio is increased to 2 by using a water–gas shift reactor at the expense of carbon utilization. The CTS process can adjust the syngas composition in situ using a precise ratio of ITCMO to coal and steam to coal ratio. This leads to a significant increase in carbon utilization efficiency in addition to the intensification benefits described above. The CTS reducer has been demonstrated successfully at the bench scale, using a variety of coals, including bituminous and sub-bituminous. The bench-scale reducer unit and the CTS flow schematic is shown in Figure 1.46.

Typical gas compositions from CTS bench-scale experiments are shown in Figure 1.47. Comprehensive investigation of CTS for a range of operating conditions including Fe_2O_3:CH_4 molar ratios from 0.5 to 2.5, temperatures from 850 to 1000 °C, steam:C molar ratios from 0 to 2.0, and Fe_2O_3:C molar ratio from 0.3 to 1.4 have been conducted. Here, C corresponds to the carbon content from the coal and methane feedstock. Figure 1.47 shows the results from the CTS reducer for a co-injection of natural gas with sub-bituminous coal without steam under the operating conditions of Fe_2O_3:CH_4 molar ratio of 2.4, Fe_2O_3:C molar ratio of 0.45, and temperature 1000 °C.

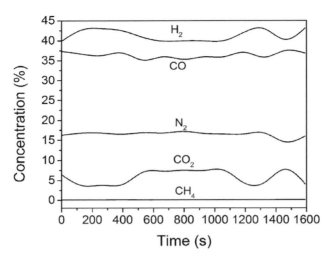

Figure 1.47 Results of the CTS bench-scale experimental unit using a sub-bituminous type coal and natural gas.

A syngas purity greater than 90% (dry, N_2 free), near complete carbon conversion, and a H_2:CO ratio of 1.1 are obtained. Other experiments indicate that a syngas with different compositions can be produced by adjusting feedstock ratios and operating conditions, reflecting the flexibility of the CTS in producing different qualities of syngas for different applications. Syngas with the composition given in Figure 1.47 can be used directly in an iron based Fischer–Tropsch synthesis or an IGCC power generation system.

The overall process flow for integrating the CTS process into several commercial liquid fuel/chemical production plants is shown in Figure 1.48. The experimental results are used to integrate the CTS process with a methanol production plant and the performance is compared to a baseline plant using a conventional coal gasification system.[80] Table 1.5 summarizes the economic comparison of the CTS technology to a conventional gasification technology when integrated into a methanol production plant.

In the baseline case, the total plant capital cost for a 10,000 tpd methanol production system is $4,775 million (2011 $). An economic analysis was performed on the CTS process for methanol production based on Figure 1.48, and it showed that the total plant capital cost for the CTS case was $3,497 million (2011 $). A 28% reduction in capital cost using the CTS process as compared to the baseline coal gasification case is obtained. For the gasification equipment alone, a capital cost reduction of more than 50% can be obtained by replacing it with the CTS process. The thermodynamics of the CTS process allows for coal consumption to be reduced by 14% and a methanol production price 21% lower than the baseline case. When carbon capture costs are considered, the methanol required selling price using the CTS process with 90% carbon capture is still 14% lower than the baseline case without carbon capture,

Table 1.5 Comparative summary of capital costs and cost of methanol production for the CTS case and the conventional baseline case.

Case	Conventional baseline	CTS
Total plant costs (2011 MM$)	4,775	3,497
Total as spent capital (2011 MM$)	6,852	5,003
Capital costs ($/gal, 2011$)	1.23	0.89
Required selling price ($/gal, 2011$)	**1.78**	**1.41**

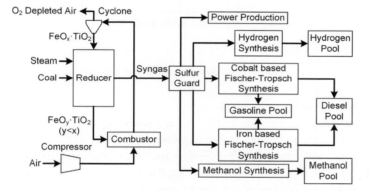

Figure 1.48 CTS process and integration options for liquid fuels production using cobalt based and iron based Fischer–Tropsch synthesis, methanol production, hydrogen production, and power production.

showing that the CTS process can provide a significant economic incentive for production of chemicals from coal.

Biomass-to-Syngas (BTS) Process

The biomass-to-syngas (BTS) chemical looping process was developed for high efficiency conversion of biomass to syngas. The BTS process uses a cocurrent moving bed reducer for producing syngas and a fluidized bed combustor for regenerating the reduced oxygen carrier material. The thermochemical conversion of biomass requires effective handling of tars produced during pyrolysis. The ITCMO material is multifunctional in nature, and it was initially tested for cracking tars and volatiles from a woody biomass. The set-up for investigating the tar conversion properties of the ITCMO material is shown in Figure 1.49. Figure 1.50 compares the product gas chromatogram of biomass pyrolysis product downstream of the fixed bed in the presence and absence of ITCMO material. In the absence of ITCMO material, a wide distribution of higher hydrocarbons, including aromatics like benzene, anthracene, and phenanthrene, is

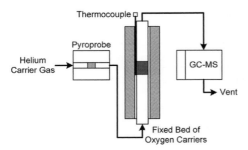

Figure 1.49 Experimental set-up to study the tar-cracking ability of multifunctional ITCMO materials.

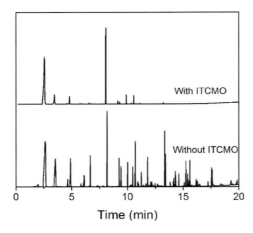

Figure 1.50 Comparison of product gas chromatogram for biomass pyrolysis product in the presence and absence of ITCMO material.

produced. The ITCMO material converts all these to a mixture of CO and CO_2, as seen in Figure 1.50.

The BTS moving bed reducer has been tested successfully at the bench scale for a variety of biomass. Figure 1.51 shows the gas composition from the BTS bench-scale moving bed reducer. The reducer was operated at 1000 °C with a molar ratio of Fe_2O_3: C of 2.87 and a molar ratio of steam:C of 1.16 that yields a syngas product with a H_2:CO ratio of 2.23, as seen in Figure 1.51. Here, C corresponds to the carbon content from the biomass feedstock. Several reaction conditions have been investigated and in these tests, syngas compositions with a H_2:CO ratio ranging from 1.7 to 2.2 and a syngas purity greater than 70% were obtained. The results matched well with the theoretical thermodynamic simulation predicted from Aspen Plus® process modeling software.

The BTS process is integrated into a methanol production system for techno-economic analysis, as shown in Figure 1.52 with results presented in Table 1.6. The BTS process is designed to match the 95 MMgal/yr methanol production rate of the

Table 1.6 Comparative summary of capital costs and cost of methanol production for the BTS case and the conventional baseline case.

Parameter	Indirect gasification	BTS
Total-as-spent cost (2011 million $)	296	224
O&M costs ($/gal, 2011 $)	0.74	0.69
Methanol required selling price ($/gal, 2011 $)	1.77	1.62

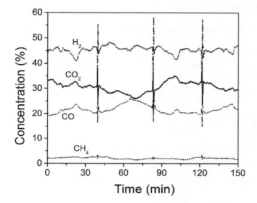

Figure 1.51 Gas composition out of the BTS system reducer reactor.

baseline indirect gasification case using a traditional gasifier. Analysis of the process simulation results indicates that for obtaining the identical quantity of syngas from a BTS system, the biomass requirement decreases by 11% and the steam usage decreases by 22% from the baseline case.

The total plant capital costs for a 95 MMgal/yr methanol production system is $296 million (2011 $) for the baseline study and $224 million (2011 $) for the BTS process. The BTS process reduces operating and maintenance costs by 6.8% and the methanol selling price decreases by 8.5%. The BTS process is an advanced gasification technology that is more efficient and flexible for H_2-rich syngas production compared to traditional gasification technology.

Chemical Looping Technology for Chemicals Production

Direct chemical production using oxidative coupling of methane (OCM) is a promising route that is of widespread interest in the methane to chemicals industry. A chemical looping based schematic for the OCM reaction is shown in Figure 1.53.[81,82] Chemical looping OCM produces a broad range of hydrocarbon products. The separation of ethylene would be more energy intensive than either traditional ethane or even naphtha steam cracking and hence it is being considered as a promising option for liquid fuels production. The operating temperature of ~850 °C is typical of chemical looping OCM reactions. Regeneration of the reduced OCM

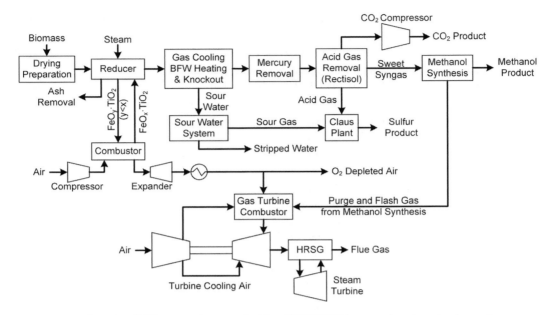

Figure 1.52 BTS process integrated with a 95 MMGAL/year methanol production unit.

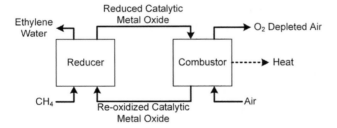

Figure 1.53 Chemical looping OCM reaction process scheme.[81,82]

catalytic metal oxide is exothermic, so the heat of reaction can be used for electricity generation. Currently, OCM is more expensive than traditional ethylene production routes. Economic analyses have suggested that C_2 yields greater than 30% are required to be economically competitive. With improvements in catalytic metal oxide design, reactor design, and process integration, chemical looping OCM has the potential to be economically competitive with current steam cracking processes for ethylene production.

1.7   Concluding Remarks

This chapter provides an overview of the development of chemical looping technology, specifically, in partial/selective oxidation as applied to gasification and reforming for high valued product generation. It also sets the stage for discussion on the other material presented in this book. The potential for chemical looping to transform a carbonaceous feedstock as well as to interrupt non-fossil systems is high, yet challenges remain that are directly associated with metal oxide reaction engineering. The development of viable solid metal oxide chemical looping particles for applications in fuel combustion, carbon capture, syngas generation, direct chemical production, or thermochemical water splitting requires extensive knowledge of metal oxide reaction engineering at the molecular scale, such as electron, ion, and defect transport mechanisms in crystals, morphological variations induced by this transport, thermodynamic properties of metal oxides, support, dopant, and binder effects, and kinetics and reaction pathways in redox reactions. Of equal importance are composite formulation, synthesis methods, and associated physical strength which allow the metal oxide particles to be enabled for their sustained redox functions in the process system while maintaining physical integrity. Thus, metal oxide redox technology encompasses all facets of particle science and technology from particle synthesis to its controlled flow behavior. It also considers the optimal chemical looping reactor design principles and their close connection to the metal oxide phase equilibrium relationship with the desired gaseous product yield. The system engineering that includes process simulation is an important, integral part of successful process development as it significantly affects product purity, energy conversion efficiency, and its economic outcome. This chapter describes the historical accomplishments with chemical looping partial oxidation that have shed light on recent advances in this technology development. The Ohio State chemical looping strategy forms a platform technology for various applications demonstrated at sub-pilot or pilot scales, as in the coal direct chemical looping (CDCL) process, syngas chemical looping (SCL) process, shale gas-to-syngas (STS) process, coal-to-syngas (CTS) process, biomass-to-syngas (BTS) process, and the hydrogen and the chemical production processes. This technology platform is established on the basis of the unique moving bed reducer design. The significant difference in performance of the moving bed reducer design in contrast to that of the widely used fluidized bed reducer design illustrates the critical role that the reactor plays in the versatility of these technology applications.

References

1. Ishida, M., D. Zheng, and T. Akehata, "Evaluation of a Chemical-Looping-Combustion Power-Generation System by Graphic Exergy Analysis," *Energy*, 12(2), 147–154 (1987).
2. Fan, L.-S., *Chemical Looping Systems for Fossil Energy Conversions*, John Wiley & Sons, Hoboken, NJ (2010).
3. Knoche, K. F. and H. Richter, "Improvement of Reversibility of Combustion Processes," *Brennstoff-Wärme-Kraft*, 20, 205 (1968).
4. Richter, H. J. and K. F. Knoche, "Reversibility of Combustion Processes," in Gaggioli, R. A., ed., "Efficiency and Costing," ACS Symposium Series 235, Washington, D.C., 71–86 (1983).

5. Kerr, R. A., "Natural Gas from Shale Bursts onto the Scene," *Science*, 328(5986), 1624–1626 (2010).

6. Hughes, J. D., "Energy: A Reality Check on the Shale Revolution," *Nature*, 494(7437), 307–308 (2013).

7. Fan, L.-S., L. Zeng and S. Luo, "Chemical-Looping Technology Platform," *AIChE Journal*, 61(1), 2–22 (2015).

8. Connell, D. P., L. Zeng, L.-S. Fan, D. A. Lewandowski, and R. M. Statnick, "Process Simulation of Iron-Based Chemical Looping Schemes with CO_2 Capture for Hydrogen and Electricity Production from Coal," Presented at 29th Annual International Pittsburgh Coal Conference, Pittsburgh, PA, October 15–18 (2012).

9. Fan, L.-S., A. Empfield, M. Kathe, and E. Blair, "Chemical Looping Syngas Production from Carbonaceous Fuels," U.S. Patent PCT/US2017/027241 (2017).

10. International Energy Agency, *Key World Energy Statistics*, International Energy Agency, Paris, France (2016).

11. United States Energy Information Administration, *Annual Energy Outlook 2016 with Projections to 2040*, Report No. DOE/EIA-0383(2016), Washington, D.C. (2016).

12. British Petroleum, *BP Statistical Review of World Energy June 2015*, London, United Kingdom (2015).

13. Figueroa, J. D., T. Fout, S. Plasynski, H. McIlvried, R. D. Srivastava, "Advances in CO_2 Capture Technology-the U.S. Department of Energy's Carbon Sequestration Program," *International Journal of Greenhouse Gas Control*, 2, 9–20 (2008).

14. Miller, D. C., J. T. Litynski, L. A. Brickett, and B. D. Morreale, "Toward Transformational Carbon Capture Systems," *AIChE Journal*, 62, 1–10 (2016).

15. Airpack Netherlands, www.airpack.nl/products/nitrogen-generators (Accessed on 18 September 2016).

16. Wikipedia, https://en.wikipedia.org/wiki/Solid_oxide_fuel_cell (Accessed on 18 September 2016).

17. Wikipedia, https://en.wikipedia.org/wiki/Light-emitting_diode (Accessed on 18 September 2016).

18. The Prashant Kamat Laboratory, www3.nd.edu/~kamatlab/research_photocatalysis.html (Accessed on 18 September 2016).

19. Gautam, N., Exploring Solar Electricity as Renewable Energy Source, *EE Times-India*, September 10, 2012.

20. Nexeon, www.nexeon.co.uk/about-li-ion-batteries (Accessed on 18 September 2016).

21. Randall, C. A., R. E. Newnham, and L. E. Cross, *History of the First Ferroelectric Oxide, BaTiO_3*, Electronics Division of the American Ceramic Society (2008).

22. Morkoç, H. and Ü. Özgür, "General Properties of ZnO," Chapter 1 in *Zinc Oxide: Fundamentals, Materials and Device Technology*, Wiley-VCH, Weinheim, Germany (2009).

23. Nobel Prize, www.nobelprize.org/nobel_prizes/physics/laureates/1987/ (Accessed on 18th September 2016).

24. Mineral Fund Advisory, www.mineralprices.com/ (Accessed on 18 September 2016).

25. Investmentmine Mining Markets & Investment, www.infomine.com/investment/ (Accessed on 18 September 2016).

26. United States Geological Survey, www.minerals.usgs.gov/minerals/pubs/commodity/ (Accessed on 18 September 2016).

27. Hammond, C., ed., *CRC Handbook of Chemistry and Physics*, 81st edition, CRC Press, Boca Raton, FL (2000).

28. Li, F., S. Luo, Z. Sun, X. Bao, and L.-S. Fan, "Role of Metal Oxide Support in Redox Reactions of Iron Oxide for Chemical Looping Applications: Experiments and Density Functional Theory Calculations," *Energy & Environmental Science*, 4(9), 3661–3667 (2011).

29. Hsia, C., G. R. St. Pierre, K. Raghunathan, and L.-S. Fan, "Diffusion through $CaSO_4$ Formed During the Reaction of CaO with SO_2 and O_2," *AIChE Journal*, 39, 698–700 (1993).

30. Hsia, C., G. R. St. Pierre, and L.-S. Fan, "Isotope Study in $CaSO_4$ Formed During Sorbent Flue-Gas Reaction," *AIChE Journal*, 41, 2337–2340 (1995).

31. Sun, Z., Q. Zhou and L.-S. Fan, "Reactive Solid Surface Morphology Variation via Ionic Diffusion," *Langmuir*, 28(32), 11827–11833 (2012).

32. Vineyard, G. H., "Frequency Factors and Isotope Effects in Solid State Rate Processes," *Journal of Physics and Chemistry of Solids*, 3, 121–127 (1957).

33. Qin, L., Z. Cheng, J. A. Fan, et al., "Nanostructure Formation Mechanism and Ion Diffusion in Iron–Titanium Composite Materials with Chemical Looping Redox Reactions," *Journal of Material Chemistry A*, 3, 11302–11312 (2015).

34. Cheng, Z., L. Qin, M. Guo, J. A. Fan, D. Xu, and L.-S. Fan, "Methane Adsorption and Dissociation on Iron Oxide Oxygen Carriers: The Role of Oxygen Vacancies," *Physical Chemistry Chemical Physics*, 18, 16423–16435 (2016).

35. Gayán, P., I. Adánez-Rubio, A. Abad, et al., "Development of Cu-based Oxygen Carriers for Chemical-Looping with Oxygen Uncoupling (CLOU) Process," *Fuel*, 96, 226–238 (2012).

36. Lloyd, G. J. and J. W. Martin, "The Diffusivity of Oxygen in Nickel Determined by Internal Oxidation of Dilute Ni-Be Alloys," *Metal Science Journal*, 6(1), 7–11 (1972).

37. Fang, H. Z., S. L. Shang, Y. Wang, Z. K. Liu, D. Alfonso, D. E. Alman, Y. K. Shin, C. Y. Zou, A. C. T. van Duin, Y. K. Lei, and G. F. Wang, "First-Principles Studies on Vacancy-Modified Interstitial Diffusion Mechanism of Oxygen in Nickel, Associated with Large-Scale Atomic Simulation Techniques," *Journal of Applied Physics*, 115(4), 043501 (2014).

38. Bhavsar, S., M. Najera, R. Solunke, and G. Veser, "Chemical Looping: To Combustion and Beyond," *Catalysis Today*, 228, 96–105 (2014).

39. Ohlemuller, P., J. Busch, M. Reitz, J. Strohle, and B. Epple, "Chemical Looping Combustion of Hard Coal: Auto-thermal Operation of a 1 MW_{th} Pilot Plant," *Journal of Energy Resources Technology*, 138, 042203 (2016).

40. Luo, S., L. Zeng, D. Xu, M. Kathe, E. Chung, N. Deshpande, L. Qin, A. Majumder, T.-L. Hsieh, A. Tong, Z. Sun, and L.-S. Fan, "Shale Gas-to-Syngas Chemical Looping Process for Stable Shale Gas Conversion to High Purity Syngas with H_2:CO Ratio of 2:1," *Energy and Environmental Science*, 7(12), 4104–4117 (2014).

41. Rydén, M., A. Lyngfelt and T. Mattisson, "Chemical-Looping Combustion and Chemical-Looping Reforming in a Circulating Fluidized-bed Reactor using Ni-based Oxygen Carriers," *Energy & Fuels*, 22, 2585–2597 (2008).

42. Nalbandian, L., A. Evdou, and V. Zaspalis, "$La_{1-x}Sr_xM_yFe_{1-y}O_{3-\delta}$ Perovskites as Oxygen-Carrier Materials for Chemical-Looping Reforming," *International Journal of Hydrogen Energy*, 36, 6657–6670 (2011).

43. Dai, X. P., Q. Wu, R. J. Li, C. C. Yu, and Z. P. Hao, "Hydrogen Production from a Combination of the Water–Gas Shift and Redox Cycle Process of Methane Partial Oxidation via Lattice Oxygen over $LaFeO_3$ Perovskite Catalyst," *Journal of Physical Chemistry B*, 110, 25856–25862 (2006).

44. Keller, G. E. and M. M. Bhasin, "Synthesis of Ethylene via Oxidative Coupling of Methane. 1. Determination of Active Catalysts," *Journal of Catalysis*, 73, 9–19 (1982).

45. Contractor, R. M., "Dupont's CFB Technology for Maleic Anhydride," *Chemical Engineering Science*, 54, 5627–5632 (1999).

46. Steinfeld, A., P. Kuhn, and J. Karni, "High-Temperature Solar Thermochemistry: Production of Iron and Synthesis Gas by Fe_3O_4-Reduction with Methane," *Energy*, 18, 239–249 (1993).

47. Fletcher, E. A., "Solarthermal Processing: A Review," *Journal of Solar Energy Engineering*, 123, 63–74 (2001).

48. Steinfeld, A., "Solar Thermochemical Production of Hydrogen - A Review," *Solar Energy*, 78, 603–615 (2005).

49. Muhich, C. L., B. W. Evanko, K. C. Weston, et al., "Efficient Generation of H_2 by Splitting Water with an Isothermal Redox Cycle," *Science*, 341, 540–542 (2013).

50. Bergmann, F. J., "Process for the Production of Calcium Carbide in Blast Furnaces," German Patent 29,384 (1897).

51. Hurst, S., "Production of Hydrogen by the Steam-Iron Method," *Journal of the American Oil Chemists' Society*, 16, 29–35 (1939).

52. Gasior, S. J., A. J. Forney, J. H. Field, D. Bienstock, and H. E. Benson, *Production of Synthesis Gas and Hydrogen by the Steam-Iron Process: Pilot Plant Study of Fluidized and Free-Falling Beds*, United States Department of the Interior, BM-RI-5911, Washington, D.C. (1961).

53. Teed, P. L., *The Chemistry and Manufacture of Hydrogen*, Edward Arnold, London, United Kingdom (1919).

54. Lewis, W. K. and E. R. Gilliland, "Production of Pure Carbon Dioxide," *U.S. Patent* 2,665, 971 (1954).

55. Dobbyn, R. C., H. M. Ondik, W. A. Willard, W. S. Brower, I. J. Feinberg, T. A. Hahn, G. E. Hicho, M. E. Read, C. R. Robbins, and J. H. Smith, *Evaluation of the Performance of Materials and Components Used in the CO2 Acceptor Process Gasification Pilot Plant*, United States Department of Energy, DOE-ET-10253-T1, Washington, D.C. (1978).

56. Institute of Gas Technology, *Development of the Steam-Iron Process for Hydrogen Production*, United States Department of Energy, EF-77-C-01–2435, Washington, D.C. (1979).

57. Jones, C. A., J. J. Leonard, and J. A. Sofranko, "Fuels for the Future: Remote Gas Conversion," *Energy & Fuels*, 1, 12–16 (1987).

58. Jones, C. A., J. J. Leonard, and J. A. Sofranko, "The Oxidative Conversion of Methane to Higher Hydrocarbons over Alkali-Promoted Mn/SiO_2," *Journal of Catalysis*, 103, 311–319 (1987).

59. Dudukovic, M. P., "Frontiers in Reactor Engineering," *Science*, 325, 698–701 (2009).

60. Otsuka, K., Y. Wang, E. Sunada, and I. Yamanaka, "Direct Partial Oxidation of Methane to Synthesis Gas by Cerium Oxide," *Journal of Catalysis*, 175, 152–160 (1998).

61. Rydén, M. and A. Lyngfelt, "Using Steam Reforming to Produce Hydrogen with Carbon Dioxide Capture by Chemical-Looping Combustion," *International Journal of Hydrogen Energy*, 31(10), 1271–1283 (2006).

62. Adanez, J., A. Abad, F. García-Labiano, P. Gayan, and L. F. de Diego, "Progress in Chemical-Looping Combustion and Reforming Technologies," *Progress in Energy and Combustion Science*, 38, 215–282 (2012).

63. Pans, M. A., A. Abad, L. F. de Diego, et al., "Optimization of H_2 Production with CO_2 Capture by Steam Reforming of Methane Integrated with a Chemical-Looping Combustion System," *International Journal of Hydrogen Energy*, 38, 11878–11892 (2013).

64. Nakamura, T., "Hydrogen Production from Water Utilizing Solar Heat at High Temperatures," *Solar Energy*, 19, 467–475 (1977).

65. Muhich, C. L., B. D. Ehrhart, V. A. White, et al., "Predicting the Solar Thermochemical Water Splitting Ability and Reaction Mechanism of Metal Oxides: A Case Study of the Hercynite Family of Water Splitting Cycles," *Energy & Environmental Science*, 8, 3687–3699 (2015).

66. Bale, C., E. Bélisle, P. Chartrand, et al., "FactSage Thermochemical Software and Databases - Recent Developments," *Calphad*, 33, 295–311 (2009).

67. Bayham, S. C., H. R. Kim, D. Wang, et al., "Iron-Based Coal Direct Chemical Looping Combustion Process: 200-h Continuous Operation of a 25-kW$_{th}$ Subpilot Unit," *Energy & Fuels*, 27(3), 1347–1356 (2013).

68. Zeng, L., F. He, F. Li, and L.-S. Fan, "Coal-Direct Chemical Looping Gasification for Hydrogen Production: Reactor Modeling and Process Simulation," *Energy & Fuels*, 26(6), 3680–3690 (2012).

69. United States Department of Energy/National Energy Technology Laboratory, *Cost and Performance Baseline for Fossil Energy Plants Volume 1: Bituminous Coal and Natural Gas to Electricity*, United States Department of Energy/NETL, DOE/NETL-2010/1397, Pittsburgh, PA (2010).

70. United States Department of Energy/National Energy Technology Laboratory, *Quality Guidelines for Energy System Studies: Capital Cost Scaling Methodology*, USDOE/NETL, DOE/NETL-341/013113, Pittsburgh, PA (2013).

71. United States Department of Energy/National Energy Technology Laboratory, *Quality Guidelines for Energy System Studies: Process Modeling Design Parameters*, USDOE/ NETL, DOE/NETL-341/051314, Pittsburgh, PA (2014).

72. United States Department of Energy/National Energy Technology Laboratory, *Quality Guidelines for Energy System Studies: CO_2 Impurity Design Parameters*, USDOE/NETL, DOE/NETL-341/011212, Pittsburgh, PA (2012).

73. Li, F., L. Zeng, L. G. Velazquez-Vargas, Z. Yoscovits, and L.-S. Fan, "Syngas Chemical Looping Gasification Process: Bench-Scale Studies and Reactor Simulations," *AIChE Journal*, 56, 2186–2199 (2010).

74. Li, F., H. R. Kim, D. Sridhar, et al., "Syngas Chemical Looping Gasification Process: Oxygen Carrier Particle Selection and Performance," *Energy & Fuels*, 23, 4182–4189 (2009).

75. Tong, A., D. Sridhar, Z. Sun, H. R. Kim, L. Zeng, F. Wang, D. Wang, M. V. Kathe, S. Luo, Y. Sun, and L.-S. Fan, "Continuous High Purity Hydrogen Generation from a Syngas Chemical Looping 25 kW$_{th}$ Sub-Pilot Unit with 100% Carbon Capture," *Fuel*, 103, 495–505 (2013).

76. Tong, A., L. Zeng, M. V. Kathe, D. Sridhar, and L.-S. Fan, "Application of the Moving-Bed Chemical Looping Process for High Methane Conversion," *Energy & Fuels*, 27, 4119–4128 (2013).

77. Goellner, J. F., V. Shah, M. J. Turner, et al., *Analysis of Natural Gas-to Liquid Transportation Fuels via Fischer-Tropsch*, United States Department of Energy/NETL, DOE/NETL-2013/1597, Pittsburgh, PA (2013).

78. Fan, L.-S., S. Luo, and L. Zeng, "Methods for Fuel Conversion," U.S. Patent Application PCT/US2014/014877 (2013).

79. Kathe, M., D. Xu, T.-L. Hsieh, J. Simpson, R. Statnick, A. Tong, and L.-S. Fan, *Chemical Looping Gasification for Hydrogen Enhanced Syngas Production with in-situ CO_2 Capture*, United States Department of Energy, OSTI: 1185194 (2015).

80. Goellner, J. F., N. J. Kuehn, V. Shah, C. W. White, and M. C. Woods, *Baseline Analysis of Crude Methanol Production from Coal and Natural Gas*, United States Department of Energy/NETL, DOE/NETL-341/013114, Pittsburgh, PA (2014) .

81. Chung, E. Y., W. K. Wang, H. Alkhatib, et al., "Process Development of Manganese-Based Oxygen Carriers for Oxidative Coupling of Methane in a Pressurized Chemical Looping System," Presented at 2015 AIChE Spring Meeting and 11th Global Congress on Progress Safety, Austin, TX, April 26–30 (2015).

82. Chung, E. Y., W. K. Wang, H. Alkhatib, et al., "Examination of Oxidative Coupling of Methane by Traditional Catalysis and Chemical Looping with Manganese-Based Oxides," Presented at 2015 AIChE Fall Meeting, Salt Lake City, UT, November 8–13 (2015).

2 Metal Oxide Oxygen Carriers

N. Deshpande, A. Majumder, S. Luo, Z. Cheng, L. Qin, Q. Zhou, D. Wang, C. Chung, S. Nadgouda, C. Fryer, and L.-S. Fan

2.1 Introduction

Partial oxidation of carbonaceous fuels in the chemical looping process uses metal oxide based oxygen carriers for the transfer of oxygen from air to the fuel. As discussed in Chapter 1, in a chemical looping partial oxidation (CLPO) process, oxygen carriers provide oxygen to partially oxidize the fuel in the fuel reactor or the reducer, thereby undergoing reduction themselves. The reduced oxygen carriers are then regenerated by air in the air reactor or the combustor and circulated back to the reducer. The oxidation reactions occurring in the combustor are highly exothermic and the heat generated can be used for power generation. The CLPO process thus generates separate streams of syngas or chemicals from the reducer and an O_2-depleted flue gas from the combustor, which obviates many downstream processing steps for product generation.[1]

Oxygen carrier materials play a key role in determining the product quality and process efficiency for a CLPO process. During reduction, the oxygen carrier donates the required amount of lattice oxygen for fuel conversion and product synthesis. In the oxidation step, the depleted oxygen carriers are replenished with oxygen from the air while heat is released, allowing for autothermal operation. Extensive research has been conducted into the design and development of optimum oxygen carrier materials. Oxygen carriers for successful chemical looping operation need to possess certain properties, like high oxygen carrying capacity, high fuel conversion, good redox reactivity, fast kinetics, good recyclability, long-term stability, high attrition resistance, good heat carrying capacity, high melting point, resistance to toxicity, and low production cost. In addition, the applications for partial oxidation require certain distinct oxygen carrier properties in order to accurately control product selectivity. Various materials have been studied for CLPO applications.[1-3] During the early stages of development, single component materials were used as oxygen carriers. More recently, the development of oxygen carriers has been directed towards using multiple metal based composite materials as oxygen carriers for improved performance. While selecting oxygen carrier materials, an understanding of the underlying mechanisms helps in engineering particles with the desired qualities. The structural evolution and integrity of these particles during reaction has a direct implication on particle strength and reactivity. These, in turn, are governed by intrinsic ionic transfer mechanisms.

Some of the major challenges in the development of oxygen carriers include designing oxygen carriers with both good reactivity and good recyclability. It has been widely reported that most single metal oxide based oxygen carriers suffer steady deterioration in their reactivity and recyclability over multiple redox cycles.[3] Such deactivation is usually a result of decayed morphological properties like decrease in active surface area and pore volume. Because the redox reactions involved in chemical looping processes are mostly conducted at temperatures above 700 °C, this performance decay and morphological deterioration has been attributed to the sintering effect, which fuses pore structure and agglomerates smaller particles (grains) into bigger ones.[4–9] The susceptibility to sintering is related to the melting point of the bulk solid (T_m, K) and characterized by the Hüttig temperature (at which point atoms at defects become mobile, defined empirically as $0.3T_m$) and the Tamman temperature (at which point atoms from the bulk exhibit mobility, defined empirically as $0.5T_m$). The melting points of most metals and metal oxides used in chemical looping applications vary over a wide range, from 500 to 1800 K (227 °C to 1527 °C). Hence, severe sintering is inevitable at the high operating temperatures for almost all oxygen carrier particles, e.g. $Cu/Cu_2O/CuO$, $Fe/FeO/Fe_2O_3/Fe_3O_4$. Another major challenge is developing oxygen carriers with high attrition resistance that can sustain their physical strength over multiple redox cycles in chemical looping systems. It has been shown that cyclic pore opening and closing per se, which features in chemical looping reactions, is an independent deactivating factor and also results in a loss in physical strength of the oxygen carriers.[10] In most published works on oxygen carrier particle synthesis, focus has been placed on the improvement of particle morphological properties, i.e. surface area, pore volume, and pore size distribution. The motivation behind this strategy is to overcome the sintering of the oxygen carrier particle at elevated temperatures.[3,4–9,11] It is expected that particles with modified morphology would be able to sustain the extreme conditions that are detrimental to their longevity. Numerous research efforts have been made towards optimizing the surface area and pore structure of oxygen carrier particles through techniques such as solution combustion, precipitation, addition of physical support, anti-sintering doping, and particle breaking. Some studies suggest adopting naturally occurring materials with high intrinsic ionic conductive capability or doping exotic atoms into the crystal of primary oxygen carrier materials to enhance their ionic conductivity.[6,12] In spite of efforts towards morphological enhancement, the deterioration of surface area and pore volume of oxygen carrier particles is unavoidable; this occurs not only through sintering but also through cyclic redox reactions. It should be noted that cyclic chemical looping reactions are fundamentally different from conventional catalytic reactions, where the improved morphology of solid catalysts is sustainable. Therefore, owing to the different nature of reaction conditions to which oxygen carriers and solid catalysts are subjected, the approach towards their development is different. After decades of imitating the morphology-driven strategy of catalysts and non-cyclic solid reactants syntheses, recent research and development of oxygen carrier particles is being directed towards enhancing the solid-phase ionic conductivity, which is believed to make the oxygen carriers relatively sustainable and cost efficient.

This chapter discusses the thermodynamic principles in metal oxide selection, and the ionic transport phenomena during metal oxide redox processes. It also illustrates the underlying mechanism of the redox reactions, oxygen carrier particle deactivation and the effectiveness of countermeasures, and phase diagrams of mixed metal oxides, along with relevant experimental and theoretical work in connection with solid morphological transformation and ionic diffusion behavior. Different methods for studying particle attrition and optimization strategies for enhancing the physical and chemical properties of oxygen carriers are also discussed.

2.2   Thermodynamic Principles and Reactivity

Metal oxide selection for various chemical looping applications can be conducted using the modified Ellingham diagram (see Section 1.3.2) based on the Gibbs free energy of reactions. This diagram categorizes metal oxides according to their applications as oxygen carriers for combustion or partial oxidation. With the base metal oxides identified, their physical and chemical properties can then be further characterized. For commercial applicability, the oxygen carrier should possess several properties, including redox reactivity, long-term stability, physical strength, toxicity, and appropriate production cost.[13] Many oxygen carrier materials have been studied for CLPO applications.[1,2,13] During the early development of these applications, single metal oxides or sulfates were considered to be the active components in oxygen carrier materials. Recent research has been guided towards the use of multiple metal based composite materials for improved process performance.[13]

During the reduction of the oxygen carrier for partial oxidation processes, the oxygen carrier must provide adequate oxidation capacity to effectively convert carbonaceous fuels such as CH_4 to CO and H_2, with minimal formation of complete combustion products, CO_2 and H_2O. The incomplete conversion of carbonaceous fuels will result in additional post-treatment steps, decreasing the overall process efficiency and increasing the operating costs. The reduction extent of the oxygen carrier determines its oxygen carrying capacity and affects the overall solids circulation rate, which significantly impacts the chemical looping process efficiency. When the reduced oxygen carrier is re-oxidized, the oxygen carrier needs to be regenerated with minimal excess air usage to reduce the associated air compression cost. The redox behavior of the oxygen carrier depends largely on the thermochemical properties of its primary metal oxide.[13] It is also important for the redox behavior of the oxygen carrier to be sustained over multiple cycles without deterioration of its physical and chemical properties.

2.2.1   Modified Ellingham Diagram

The Ellingham diagram is a popular tool in metallurgical processing for determining the relative reduction potentials of metal oxides in metal production, at different temperatures.[14] Figure 2.1 provides more detailed information on the modified Ellingham diagram for different metal oxides as oxygen carriers than that given in Figure 1.5 for various chemical looping applications based on their oxidation capabilities.[13,15]

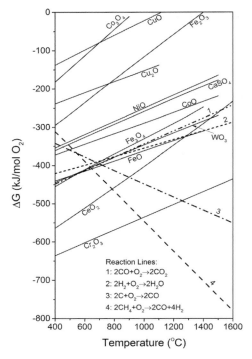

Figure 2.1 Modified Ellingham diagram to determine metal oxide performance as oxygen carriers.

The diagram has been divided into different sections based on the following four key reactions:

$$\text{Reaction line 1}: 2CO + O_2 \leftrightarrow 2CO_2 \tag{2.2.1}$$

$$\text{Reaction line 2}: 2H_2 + O_2 \leftrightarrow 2H_2O \tag{2.2.2}$$

$$\text{Reaction line 3}: 2C + O_2 \leftrightarrow 2CO \tag{2.2.3}$$

$$\text{Reaction line 4}: 2CH_4 + O_2 \leftrightarrow 2CO + 4H_2 \tag{2.2.4}$$

In Figure 2.2, the different sections are classified according to the four reaction lines. Based on these sections, the metal oxides are identified as potential oxygen carriers for different chemical looping processes.

NiO, CoO, CuO, Fe_2O_3, and Fe_3O_4 belong in the combustion section (section A) as they all lie above reaction lines 1 and 2. These metal oxides have good oxidizing properties and can be used as oxygen carriers for chemical looping combustion, gasification, or partial oxidation processes. For the small section between reaction lines 1 and 2, section E, the metal oxides can be used for CLPO, although with significant amounts of H_2O in the syngas product. Metal oxides in this transition region include SnO_2. The third section lies between reaction lines 2 and 3 (section B) and is the one for syngas

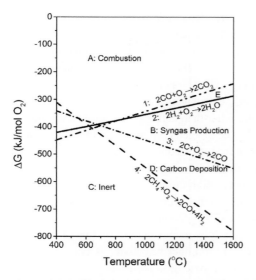

Figure 2.2 Modified Ellingham diagram with sections of chemical looping operation.

production. Metal oxides lying in this region have average oxidation properties and can be used for partial oxidation but not for complete oxidation processes. CeO_2 lies in this section. Metal oxides below reaction line 3 (sections C and D) lack the potential to be used as oxygen carriers and are generally considered to be inert. These include Cr_2O_3 and SiO_2. While metal oxides below reaction line 3 cannot be used as oxygen carriers by themselves owing to their low oxidation capabilities, they can, however, be used as support materials along with active oxygen carrier materials. For example, when TiO_2, which has very low oxidation ability, is combined with FeO, it forms an $FeTiO_3$ complex which is observed to have higher reactivity compared to FeO alone.[16] Thus, the addition of a support material in this case improves the oxygen carrier performance. On the other hand, addition of Al_2O_3 with low oxidation capability to CuO leads to the formation of $CuAl_2O_4$, which is highly undesirable as it results in the loss of the chemical looping oxygen uncoupling (CLOU) property of CuO.[17] Thus, the addition of materials with low oxidation capability as supports does not always enhance the oxygen carrier performance. It becomes essential to understand how different phases behave in the presence of each other during the development of high performing oxygen carriers. Methane (CH_4) is thermodynamically unstable at temperatures higher than 750 °C and spontaneously decomposes to form C and H_2 in the absence of an oxygen source. Hence, reaction line 4 is not considered very important for dividing Figure 2.2 into the different sections.

Maintaining syngas selectivity, quality, and productivity in a mixed reactor such as a fluidized bed in CLPO processes is a tricky affair and one can use oxygen carriers from section B to achieve the desired product quality and yields. Reactor operation requires controlled transfer of oxygen from the oxygen carriers for partial oxidation of the fuel to syngas. For metal oxides in sections A and E, i.e. above reaction line 2, it is difficult to control the syngas composition because they have high oxidation capabilities. That is, the syngas composition is heavily dependent on the oxygen carrier conversion, governed by the oxygen carrier to fuel stoichiometric ratio. The syngas yield decreases with a high oxygen carrier to fuel stoichiometric ratio, promoting the oxidation of syngas to CO_2 and H_2O. In the case of NiO as oxygen carrier in the circulating fluidized bed chemical looping reforming (CLR) operation for a methane-to-syngas CLPO process, the process is designed such that the combustor generates a 70% NiO and 30% Ni mixture after oxidation, rather than a 100% NiO stream, in order to control the syngas yield from the reducer.[18] The oxygen carrier circulation rate in this CLR process is controlled such that the stream from the reducer contains 80% Ni and 20% NiO. The oxygen carrier circulation rate for this process is thus high owing to a low oxygen carrying capacity. In addition to the high circulation rate, this process requires excess steam injection in the reducer in order to suppress carbon deposition, since metallic Ni also serves as a catalyst for the CH_4 decomposition reaction.

For metal oxides in section B, i.e. lying between reaction lines 2 and 3, the predominant product is typically syngas. The product selectivity is better controlled for these metal oxides owing to their moderate oxidation ability and thermodynamic restriction. In a moving bed reactor with oxygen carriers from this section, there is better control over the extent of oxygen carrier conversion and hence syngas composition.[14] Controlling the oxygen carrier conversion to a certain extent, it is even possible to inhibit carbon formation in the operation of this reactor without introducing any external steam to it. For metal oxides that have multiple oxidation states, such as iron oxides, the syngas yield line shows an integrated pattern of individual syngas lines of Fe_2O_3, Fe_3O_4 (section A), and FeO (section B).[13] For CeO_2, syngas yield increases linearly, reaching a maximum and remaining constant as the CeO_2:fuel stoichiometric ratio increases.

It should be noted that the modified Ellingham diagram represents only a thermodynamic analysis of metal oxides to be used as potential oxygen carrier materials. In addition to thermodynamics, many other factors come into play when selecting oxygen carrier materials. Along with the oxygen carrier performance, reactor design and gas–solid flow dynamics are factors that are crucial in determining the success of a CLPO process. With optimization of all of the factors involved, chemical looping systems can then be effectively used to generate a variety of products.

2.2.2   Single Metal Oxides

CeO_2

Ceria has a fluorite structure, which is considered to be favorable for rapid diffusion of oxygen.[13] Otsuka et al.[19] investigated the partial oxidation of methane to syngas with cerium oxide in the 1990s. Their work defined the chemical looping scheme using lattice oxygen from ceria for partial oxidation of methane and its subsequent regeneration. The reduction of oxygen carriers occurs at temperatures greater than 700 °C and about 500 °C for oxidation. Thermodynamically, CH_4 oxidation with CeO_2 had a very high selectivity toward CO and H_2, with full CH_4 conversion.[13] As seen in Figure 2.3, CeO_2 and Ce_2O_3 can be in equilibrium with a high concentration of syngas (>97% CO selectivity). CO selectivity for this case, as well as other single metal oxides discussed

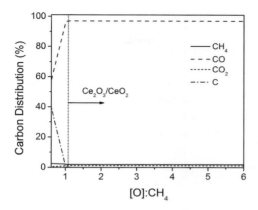

Figure 2.3 Equilibrium carbon balance for Ce_2O_3/CeO_2 oxygen carriers as a function of oxygen-to-methane ratio for syngas generation at 900 °C.

later in this section, refers to the ratio of the moles of CO to that of the total moles of carbon in the product. The molar distribution of carbon species at equilibrium is investigated by varying the molar ratios for reducible oxygen in the active compound to CH_4.[20] The ratio of reducible oxygen provided by the metal oxides to CH_4 ([O]:CH_4) significantly affects the syngas composition and the possibility of carbon formation.[20] High purity syngas is produced at an [O]:CH_4 ratio greater than 1, which thermodynamically inhibits carbon deposition on the Ce-based oxygen carrier. Thus, the thermodynamic property of cerium oxides favors syngas production. However, cerium oxide is thermodynamically constrained from converting more than approximately 98% of CH_4 to syngas. Additionally, the slow reaction kinetics together with the high oxygen carrier material cost makes a cerium oxide based CLPO process commercially impractical.[19–21] Otsuka et al.[19] attempted to determine the rate-limiting step of the partial oxidation reaction, i.e. reaction (2.2.1). They believed that H_2 formation and desorption were slower steps than the CH_4 splitting reaction or lattice oxygen diffusion;

$$CeO_2 + xCH_4 \rightarrow CeO_{2-x} + xCO + 2xH_2. \qquad (2.2.5)$$

A significant amount of work has been conducted to enhance the ionic diffusivity of CeO_2-based materials. Various types of dopants have been incorporated into CeO_2 to increase its reactivity.[13] For example, the addition of Sn ions has the ability to increase reactivity. Gupta et al.[22] used density functional theory (DFT) calculations to study the effect of doping Sn ions on CeO_2 and found that Ce exhibits 4 + 4 coordination, whereas Sn exhibits 4 + 2 + 2 coordination. Because the metal–oxygen bond became weaker owing to the substitution by Sn ions, the oxygen ion diffusivity was enhanced significantly. Furthermore, other types of metal oxides have also been discovered by researchers as promising dopants to increase ionic conductivity of ceria; however, they have not been tested for chemical looping applications.[13,23] Use of γ-Al_2O_3-supported CeO_2 with Zr and Rh promoters could also increase the reaction rate as the addition of Zr and Rh increases the oxygen mobility in CeO_2.[13,24] Introduction of ZrO_2 to the

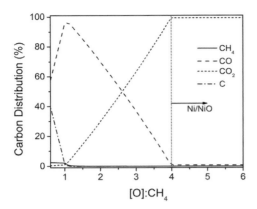

Figure 2.4 Equilibrium carbon balance for Ni/NiO oxygen carriers as a function of oxygen-to-methane ratio for syngas generation at 900 °C.

Pt–CeO$_2$ system was found to increase the reaction rate. It was suggested that Ce$_x$Zr$_{1-x}$O$_2$ ($x > 0.5$) solution had a very high oxygen storage capacity owing to its cubic fluorite structure, which enhanced the reactivity. The Ce–Zr–O solid solution was also a catalyst for the methane reforming reaction.[13,25] In some extreme cases, a higher percentage of ZrO$_2$, PrO$_2$, or TbO$_2$ was incorporated into the CeO$_2$ oxygen carriers. Oxides with a fluorite-type structure were assumed to be critical for superior lattice oxygen reactivity and recyclability.[13,26] Sadykov et al.[27] studied the effect of Sm and Bi dopants for the partial oxidation of methane with CeO$_2$ and found that the Ce–Sm–O solid solutions exhibited much higher reactivity and selectivity. However, Ce–Bi–O solid solutions exhibited lower selectivity toward syngas.[13,26,27] Although considerable research has been conducted along these lines with CeO$_2$ as the base material, the effect of carbon deposition on the metal oxide oxygen carrier remains an unresolved issue.[13,28–31]

NiO

NiO is one of the most extensively tested oxygen carriers for both CLC and CLPO applications.[13] The structure of NiO is similar to that of FeO, which is shown in Section 1.3.3. Because of its selectivity to CO$_2$ and H$_2$O, the reforming reaction, i.e. reaction (2.2.6), has a narrow range for the NiO-to-fuel ratio to generate the syngas with desirable concentrations and negligible carbon deposition,[13]

$$\text{NiO} + \text{CH}_4 \rightarrow \text{Ni} + \text{CO} + 2\text{H}_2. \tag{2.2.6}$$

The thermodynamic evidence of this sensitivity can be seen in Figure 2.4, which shows the high carbon deposition with [O]:CH$_4$ ratio less than one, and a rapid decrease in CO selectivity for [O]:CH$_4$ ratio higher than one.

Since pure metal oxides lose their reactivity within a few cycles, most of the work with NiO as an oxygen carrier for partial oxidation has focused on selecting and synthesizing favorable support materials.[13] Figure 2.5 shows the effect of repeated cycling on the reactivity of pure NiO under redox conditions. Using the freeze

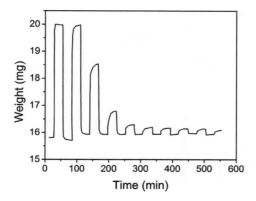

Figure 2.5 Effect of ten cycles of reduction and oxidation on the regeneration property of pure NiO observed in a thermogravimetric analyzer (TGA) under H_2 and O_2 redox conditions.[5]

granulation method, Johansson et al.[32] compared $MgAl_2O_4$ and $NiAl_2O_4$ as supports. They observed that $MgAl_2O_4$ achieved a much higher methane conversion and had a lower tendency for carbon formation. Other oxygen carrier synthesis methods have also been tested. The deposition–precipitation methods were found to produce oxygen carriers more resistant to carbon deposition than those prepared by dry impregnation.[13,33] Spray drying could be a promising alternative, as this method produces oxygen carriers with similar performance to that of oxygen carriers synthesized using freeze granulation.[13,34] Zafar et al.[35] further tested NiO on SiO_2 and $MgAl_2O_4$ supports in a thermogravimetric analyzer. They observed that SiO_2 reacted with primary metal oxides and formed stable unreactive silicates with increasing numbers of redox cycles. The formation of silicates resulted in a poor recyclability with a high reactivity only in the first few cycles. Lowering the reaction temperature alleviated the degradation but could not fully eliminate this characteristic.[13] They concluded that $MgAl_2O_4$-supported NiO performed best for both full oxidation and partial oxidation. Additionally, different phases of Al_2O_3 support have also been tested. NiO with α-Al_2O_3 support demonstrated the highest reactivity; NiO with γ-Al_2O_3 had the lowest reactivity for the partial oxidation of methane. The low reactivity of Ni-γ-Al_2O_3 was attributed to the formation of $NiAl_2O_4$.[36]

Additive materials have also been explored for increasing the performance of Ni-based oxygen carriers. For example, a small amount of $Ca(OH)_2$ doping was found to increase the particle strength and, hence, its attrition resistance.[13] MgO could also work as a promoter to increase fuel conversion in the early stages of the reaction.[13,34] Further discussed in Section 4.4, partial oxidation using NiO-based oxygen carriers have been demonstrated in fluidized bed reactors.[18,37–39] However, because of cost and other reasons, NiO is not likely to be the material used for commercial chemical looping partial oxidation.[13]

Fe_2O_3

Similar to NiO, Fe_2O_3 is another metal oxide oxygen carrier that has been widely tested for chemical looping applications.[13] Unlike NiO, iron oxide has multiple oxidation

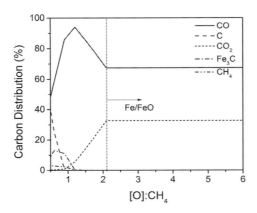

Figure 2.6 Equilibrium carbon balance for Fe_2O_3 oxygen carriers as a function of oxygen-to-methane ratio for syngas generation at 900 °C.

states that provide it with unique advantages for fuel partial oxidation when coupled with a favorable reactor configuration. The thermodynamic equilibrium carbon balance of this system as a function of the [O]:CH_4 ratio is shown in Figure 2.6.

Nakayama et al.[40] found that Fe_2O_3/Y_2O_3 with Rh_2O_3 as a promoter produces syngas of a high purity at 54% methane conversion. Steinfeld et al.[41] proposed the coproduction of iron and syngas via partial oxidation of methane with Fe_3O_4 in a high temperature reactor heated using solar thermal energy. This scheme could be altered to produce syngas and hydrogen if iron was regenerated with steam.[13] Luo et al.[20] demonstrated that an Fe_2O_3-based oxygen carrier can generate syngas at a concentration higher than 90%, balanced by CO_2 and steam, with full fuel conversion. In their unique moving bed reactor configuration, the reactor is designed to operate with minimal carbon deposition and without the use of steam. The feedstock can be methane, biomass, coal, or any other type of carbonaceous fuel, and the H_2:CO ratio may vary from 1:1 to 3:1 depending on the feedstock and operating conditions. Compared with NiO, iron oxides require a higher operating temperature for achieving complete methane conversion (>99%).[13,20]

Iron oxide reactions with fuels are a complex process owing to their stepwise reduction from Fe_2O_3 to Fe_3O_4, to FeO, and to Fe. Fan et al.[13,42–48] reported the characteristics of multiple oxidation states of iron and developed a countercurrent moving bed reactor system for their application in chemical looping systems. Each oxidation state corresponds to a different crystal structure that includes rhombohedral (α-Fe_2O_3), inverse spinel (Fe_3O_4), rock salt cubic (FeO), and body-centered cubic (Fe) structures, as shown in Section 1.3.3. Fe_2O_3, Fe_3O_4, and FeO possess similar close-packed oxygen structures, which render the redox transitions between each phase faster than that between the lower oxidation phases of FeO and Fe.[13,49,50] On the other hand, even though there exists a minor variation from the hexagonal close-packed structure to the cubic close-packed structure, Fe_2O_3 reduction to Fe_3O_4 could lead to an irreversible structure change, such as volume expansion, as the reaction proceeds.[13,51] Coupled with the sintering effects that occur at high temperatures, pure iron oxides can quickly lose reactivity and oxygen-carrying capacity after the first several redox cycles.

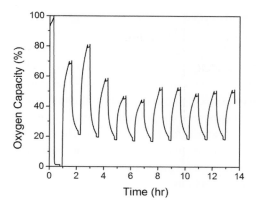

Figure 2.7 Cyclic reduction–oxidation reaction studies in TGA for pure Fe_2O_3 powders at 900 °C using H_2- and O_2- rich gases.

The loss in reactivity of pure Fe_2O_3 over multiple redox cycles is seen in Figure 2.7. When a suitable support material is added, the reactivity and recyclability of iron based oxygen carriers can be drastically improved.[13] For example, Chen et al.[52] studied iron oxide materials mechanically mixed with Al_2O_3 or TiO_2 as support. They showed that both supports provide stable reactivity and mechanical strength over three to ten redox cycles. They concluded that oxygen carriers with 60% Fe_2O_3 and 40% Al_2O_3 presented the best performance. The post-experimental results indicated that there was no formation of inactive $FeAl_2O_4$.

Phase separation is another cause for the degradation of oxygen carrier materials, as the iron phase may segregate and sinter.[13] To prevent this behavior, Liu and Zachariah[13,53] doped potassium to Al_2O_3-supported iron oxides and showed improved reactivity, stability, CO_2 product selectivity, and carbon deposition resistance. The K ions could assist in binding Fe and Al, thereby reducing the potential for phase separation. Li et al.[13,54] investigated TiO_2-supported iron oxide oxygen carriers using an inert marker experiment in combination with DFT calculations. The results indicate that pure iron oxide oxidation is dominated by outward diffusion of Fe cation, whereas the TiO_2-supported iron oxide is dominated by inward diffusion of oxygen anion. These results are due to the different structures of Fe_2O_3 and TiO_2, which create a considerable number of oxygen vacancies, enhancing oxygen ion diffusivity when combined. This phenomenon explains why the reactivity of TiO_2-supported Fe_2O_3 can be maintained over multiple redox cycles while the pore volume and surface area significantly decrease during this cyclic process.[13]

Galinsky et al.[13,55,56] have reported the use of mixed conductive supports like perovskites, specifically lanthanum strontium ferrite (LSF) and $Ca_{0.8}Sr_{0.2}Ti_{0.8}Ni_{0.2}O_3$, to increase the reactivity of iron oxide oxygen carriers. The LSF-supported Fe_2O_3 had a reactivity 5–70 times greater than the reactivity of TiO_2-supported Fe_2O_3.[13,55] The reactivity of Fe_2O_3 with $Ca_{0.8}Sr_{0.2}Ti_{0.8}Ni_{0.2}O_3$ as the support was compared with CeO_2 and $MgAl_2O_4$ supports.[56] The $Ca_{0.8}Sr_{0.2}Ti_{0.8}Ni_{0.2}O_3$ supported Fe_2O_3 was more active and stable than the other two supports. The Fe_2O_3 particles with CeO_2 as support deactivated by 75% within ten redox cycles, and $MgAl_2O_4$-supported iron oxide lost its structural integrity due to filamentous carbon formation. The reason for the high reactivity for both perovskites, LSF and $Ca_{0.8}Sr_{0.2}Ti_{0.8}Ni_{0.2}O_3$, has been attributed to their ability to facilitate both O^{2-} and electron transport to and from iron oxide during reduction and oxidation reactions. The LSF support is however found to migrate towards the exterior of the iron oxide oxygen carriers over multiple redox cycles. Based on this morphology, LSF was thought to act as a mixed conductive membrane on the iron oxide particle, which acts as a source for lattice oxygen. The LSF-supported iron oxide carriers were also found to be resistant to sintering and coke deposition over 50 redox cycles. The use of such perovskite materials in chemical looping applications is discussed further in the following sections.

Other Metal Oxides

In addition to the three most widely studied metal oxides, CeO_2, NiO, and Fe_2O_3, several other metal oxides have been proposed as oxygen carriers for chemical looping partial oxidation, including WO_3 and ZnO.[13] Kodama et al.[13,57] found that WO_3 could oxidize methane to CO and H_2 with high selectivity without carbon deposition or without the formation of WC phase at 1000 °C. When ZrO_2 was used as the support material for WO_3, the reactivity could be improved owing to reduced grain size and improved dispersion of WO_3. They further tested WO_3/ZrO_2 in a solar furnace at temperatures of 900–1000 °C, and the results showed that CH_4 conversion was higher than 80% and CO selectivity was higher than 76%.[13,58] CH_4 oxidation by ZnO also generated syngas with a 2:1 H_2-to-CO ratio, and it was found that the ZnO conversion could be 90% at 1327 °C in a 5 kW prototype solar furnace. The partial oxidation reaction was found to be very fast. However, Zn existed in a gaseous state at high temperatures, and thus a fast quenching step was required.[13,59] Similar studies have been carried out on MgO, SiO_2, and TiO_2; however, none of these has so far been used in any practical applications.[13,57,60,61] Complex metal oxides made up of different single metal oxides have been attempted to try to achieve a more favorable performance than those demonstrated with single metal oxides. These studies are discussed in the following sections.

2.2.3 Complex and Mixed Metal Oxide Based Materials

The surface area and pore volume of pure solid particles tend to decrease during cyclic reactions, which leads to solid reactivity deterioration in most cases. The "pure solid particle" here is defined as untreated solid particles without dopant or support. In application of the chemical looping scheme for partial oxidation of CH_4, along with high reactivity and syngas selectivity, one of the parameters of utmost importance for the success of a commercial scale partial oxidation scheme is the long-term stability of the oxygen carrier particles. The study of complex or supported metal oxide particles is, thus, of considerable interest. The following sections describe the main complex and supported metal oxide materials used for CH_4 partial oxidation applications. Here, CH_4 conversion is defined as the ratio of total moles of product to moles of CH_4 input, whereas selectivity towards CO is the ratio of moles of CO to total moles of carbon oxides.

Perovskite

Perovskite materials have been widely studied for many different applications and have also been considered as promising oxygen carrier materials for CH_4 partial oxidation to syngas.[13] The application of these materials in oxygen coupling of the CH_4 reaction to produce ethylene is described in Chapter 3. Wei et al.[13,62] investigated the performance of $La_{1-x}Sr_xMO_3$ (M = Mn, Ni; x = 0–0.4) and $La_{1-x}Sr_xMnO_{3-\alpha}F_\beta$ (x 0–0.3, $\beta/(3 - \alpha)$ = 0.1), synthesized by the auto-combustion method, as oxygen carrier materials. They obtained 75% CO selectivity at 16% CH_4 conversion with a H_2:CO molar ratio of 2.5:1 at 800 °C. Furthermore, they investigated the effect of partial substitution of La by Sr, which resulted in an increase in reactivity and a decrease in selectivity. Dai et al.[13,63] synthesized $AFeO_3$ (A = La, Nd, Eu) via the sol–gel method and found that there were two stages of reactions for CH_4 partial oxidation. In the first stage, there was a significant amount of CO_2 formation, and in the second stage, the reactions had a high selectivity toward syngas. Among these three perovskite materials, $LaFeO_3$ showed the highest selectivity. The conversion of CH_4 and selectivity toward syngas increased with temperature. At 900 °C, the CH_4 conversion could be as high as 65% with more than 90% selectivity toward syngas with $LaFeO_3$ as an oxygen carrier.[13,63] These findings encouraged further studies of $LaFeO_3$ as the base structure for CH_4 conversion. Various methods for $LaFeO_3$ synthesis have been investigated, mainly by using different complex agents, including citric acid and glycine.[13,64] Mihai et al.[13,65] tested multiple synthesis methods and concluded that $LaFeO_3$ prepared by a DL-tartaric acid aided method possessed a large surface area and good reactivity. With the addition of a small amount of Sr, $La_{0.9}Sr_{0.1}FeO_3$ produced an even higher reactivity. They reasoned that this improvement was due to an increased number of oxygen vacancies induced by heteroatoms. Furthermore, many other types of additives have been incorporated to increase the reactivity. For example, Nalbandian et al.[13,66] found that $La_{0.7}Sr_{0.3}Cr_{0.05-}Fe_{0.95}O_3$ mixed with 5% NiO gave excellent performance for CH_4 partial oxidation.

$M_xFe_{3-x}O_4$ (M = Ni, Co, Zn, Cu; 0<x<1.5)

Fe_3O_4 has attractive reaction properties but also has a low selectivity toward CO production.[13] To improve the performance of Fe_3O_4 as an oxygen carrier, a series of ferrites have been synthesized from Fe_3O_4 and various types of metal oxides. For CH_4 partial oxidation, Kodama et al.[13,67] investigated oxygen carriers made up of three ferrites: $Ni_{0.39}Fe_{2.61}O_4$, $Co_{0.39}Fe_{2.61}O_4$, and $Zn_{0.39}Fe_{2.61}O_4$. Under the same experimental conditions, $Co_{0.39}Fe_{2.61}O_4$ and $Zn_{0.39}Fe_{2.61}O_4$ showed poorer performance compared to Fe_3O_4, but $Ni_{0.39}Fe_{2.61}O_4$ demonstrated a significant improvement in performance with a CO yield of 22.0% and CO selectivity of 72.2% at 827 °C. The CO yield was defined as the ratio of moles of CO to that of methane in the feed to the reactor. The sintering properties of $Ni_{0.39}Fe_{2.61}O_4$ were also studied with ZrO_2 as support for a 2:1 weight ratio. However, low methane conversion and high carbon deposition were observed. Because Ni is also widely used as a catalyst for SMR,

Sturzenegger et al.[13,68] studied Ni-ferrite as both an oxygen carrier and a SMR catalyst. At the early stage of reduction, Ni-ferrite performed as an oxygen carrier, and at the late stage, Ni in the reduced phase acted as a SMR catalyst. Excess steam was added to promote the water–gas shift reaction and ensuring there was minimal CO in the gaseous product. Cha et al.[13,69] studied $Cu_xFe_{3-x}O_4/ZrO_2$ as oxygen carriers, with x varying between 0 and 1.5 for syngas generation. Adding Cu to Fe_3O_4 decreased CH_4 conversion and selectivity. However, carbon deposition was eliminated when $0.3 < x < 0.7$. Their results indicated that Cu was effective in enhancing the gasification reaction rate of deposited carbon and was thus beneficial to preventing coke formation or carbon deposition at the cost of decreasing CH_4 conversion and CO selectivity.[13,69] In addition, by substituting a fraction of the Zr with Ce, the reaction rate for syngas generation could be enhanced. Of the compositions studied, $Cu_{0.7}Fe_{2.3}O_4/Ce-ZrO_2$ (Ce/Zr) had the best performance, with no deactivation observed within ten cycles.[13,70]

The materials discussed above represent the majority of oxygen carrier formulations studied.[13] Another composition that has been investigated is $NiO-Cr_2O_3-MgO$. This composition showed promising behavior for CH_4 partial oxidation for several redox cycles without significant carbon deposition. The relatively low reaction temperature, i.e. 700 °C, was also an advantage. Further, adding Cr_2O_3 assisted in preventing the formation of $NiMgO_2$ and also reduced reaction temperatures.[13,71] To summarize, oxygen carrier materials to be employed in industrial applications are likely to be composite metal oxides because they have shown favorable reaction behavior and can overcome some of the drawbacks of single metal oxides. However, significant research is still needed to develop a robust oxygen carrier that can provide a high fuel conversion and concurrently a high selectivity toward syngas.[13]

2.2.4   Recyclability and Strength

Physical integrity of oxygen carriers throughout multiple redox cycles is a critical requirement in particle design criteria. Chemical looping systems utilizing particles with a high attrition rate need a high particle make-up rate, resulting in a high operating cost. Mass and ionic diffusion, change of lattice structure, heat stress, and impurities also induce external and internal physical stresses that affect the physical integrity of oxygen carriers. To determine whether the extent of make-up of the oxygen carrier particles is feasible or not in process operation, the cost of the particles used is the key factor. For commercial operation of a chemical looping system, the lifetime of the oxygen carriers used in the reactor can vary between days and over a year before they are replaced, corresponding to hundreds to tens of thousands of operational redox cycles in the system. Designing oxygen carriers that can sustain a long period of active redox property in a reactor system remains one of the major challenges for the technology. Given the vast and exotic variety of metal oxide formulations and synthesis methods, recyclability and strength of the particle are two simple yet powerful criteria that can be used to screen the chemical looping particles before reactor application. As an example, the recyclability of commercially-ready iron-titanium complex metal oxide particles that were developed at OSU was tested for 3000 redox reactions in a TGA, and the result is

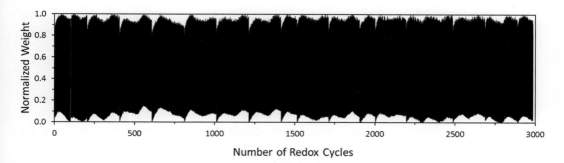

Figure 2.8 Recyclability and reactivity of the OSU-developed ITCMO particles over 3000 redox cycles in a TGA at 1000 °C with H_2/CO and air.

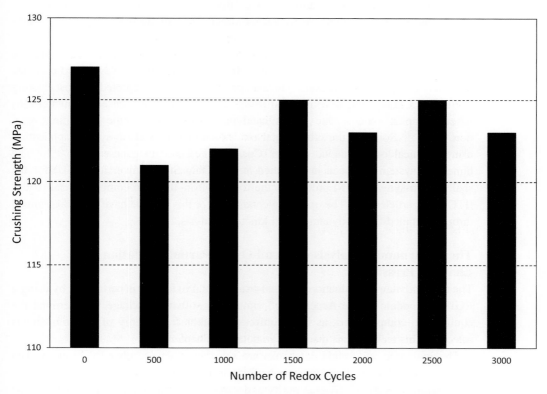

Figure 2.9 Crushing strength of OSU-developed ITCMO particles over 3000 redox cycles with recyclability and reactivity results given in Figure 2.8.

shown in Figure 2.8. The particle crushing strength using compression test for the same particles in the course of 3000 redox reactions is given in Figure 2.9. The redox cycles were conducted at 1000 °C with a mixture of H_2 and CO as reducing gas and air as oxidizing gas. The averaged, steady conversion of the OSU oxygen carriers was 33%, indicating the reduction of Fe_2O_3 to FeO. Over the 3000 cycles, the particle crushing strength and diameter remained constant at 120 MPa and 1.5 mm, respectively. The steady reactivity, recyclability, and high physical strength fulfill the desired screening results for an oxygen carrier since 3000 cycles represents approximately 8 months of operation in a commercial OSU chemical looping moving bed reducer system, and this OSU developed particle is suitable for further testing in flow and reactor systems. It is important to note that the initial strength of an oxygen carrier should not be misinterpreted as a criterion in oxygen carrier design. Only when the physical strength is sustained over long-term redox cycles should the oxygen carrier be considered for further testing in reactors.

2.2.5   Effect of Elevated Pressures

Experimental kinetic investigations on the effect of pressure on this system are relatively sparse.[72] Conventional approaches to chemical looping reforming also identify the need to operate syngas production at elevated pressure even though it is less favorable in thermodynamics.[73] Nickel is the most popular material to be studied for catalytic partial oxidation of CH_4 or for SMR, and development of Ni-based catalysts for reduction of coke deposition at elevated pressures has been reported.[74] Most existing studies on CH_4-to-syngas conversion using metal oxides are focused on nickel (Ni) based complex oxides, due to its catalytic capabilities for the reforming reaction.[36,72,73,75] Some of the other metal oxides studied for CH_4-to-syngas application using chemical looping include copper (Cu), iron (Fe), and manganese (Mn).[72,76,77] The bimetallic system has been investigated at The Ohio State University (OSU) for the shale gas-to-syngas (STS) process in the form of iron–titanium complex metal oxides (ITCMO) particles.[20,72] The operating conditions for this particle have been determined through detailed thermodynamics and kinetics analysis.

Thermodynamic Analysis of Metal Oxide Partial Oxidation at Elevated Pressures

The thermodynamic evaluation of metal oxide partial oxidation is performed by using a RGIBBS module in the Aspen Plus® simulation software package to determine the equilibrium compositions, as it minimizes the Gibbs free energy of the components selected. This technique is discussed in detail in Chapter 6.

The inorganic and solids databanks are selected to ensure physical property data accuracy. The Aspen simulations are performed for pressures of 1, 3, 5, and 10 bar at a temperature of 900 °C. Syngas purity and carbon deposition per mole of CH_4 at the equilibrium state are investigated by varying the molar ratios for reducible oxygen in the active compound to CH_4 ([O]:CH_4). The redox pairs investigated are NiO–Ni, CeO_2–Ce, Fe_2O_3–FeO/Fe and Fe_2TiO_5–Fe/Fe_2TiO_4/$FeTiO_3$. The syngas purity is defined as

$$\text{Syngas purity} = \frac{\text{moles of CO + moles of } H_2}{\text{moles of all gaseous species}} \times 100\%$$

As observed from Figure 2.10a, b, c, and d, the ratio of reducible oxygen provided by the metal oxides to CH_4 significantly affects the syngas composition and the possibility

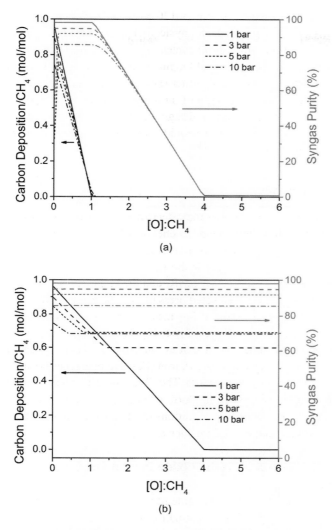

Figure 2.10 Effects of pressure on the syngas purity and carbon deposition for different metal oxide systems at pressures of 1 bar, 3 bar, 5 bar, and 10 bar: (a) NiO, (b) CeO_2, (c) Fe_2O_3, and (d) ITCMO system.

of carbon formation. For all of the metal oxides, the syngas purity is found to decrease with increasing pressure, except for the ITCMO system where the syngas purity increases with pressure at higher $[O]:CH_4$ ratios. Also, for a given pressure the syngas purity is found to decrease with an increase in the $[O]:CH_4$ ratio which is due to higher concentrations of CO_2 and H_2O products, except in CeO_2 system as will be explained later. For a NiO system (Figure 2.10a) the syngas purity is highly sensitive to the $[O]:CH_4$ ratios in the range from 1 to 4, whereas for a Fe_2O_3 system (Figure 2.10c) this range is from 1 to 2.1 after which the syngas purity remains constant at a value of 64%. In the ITCMO system (Figure 2.10d), the syngas purity

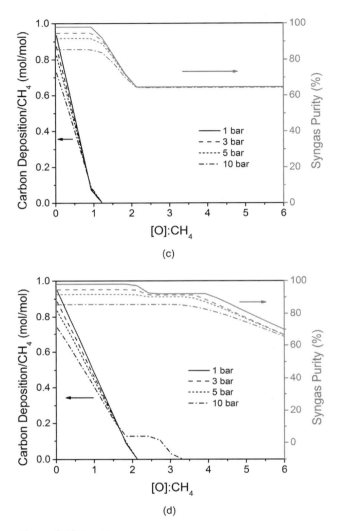

Figure 2.10 (*cont.*)

drops slightly at [O]:CH$_4$ ratios >2, plateaus, and then drops again at [O]:CH$_4$ ratios >3.5 for all operating pressures. The syngas purity is observed to be independent of the [O]:CH$_4$ ratio in the range 0 to 6 at all pressures analyzed for the CeO$_2$ system (Figure 2.10b), which is in contrast to all other metal oxide systems considered here. Thus, CeO$_2$ and ITCMO systems are found to exhibit a high syngas purity over a wider range of [O]:CH$_4$ ratios, which allows for greater operational flexibility in chemical looping systems.

The carbon deposition per mole of CH$_4$ is higher at lower pressures for [O]:CH$_4$ ratios <0.8 for the Fe$_2$O$_3$ and NiO systems and it becomes negligible for ratios greater than 1. For the ITCMO system, the $[O]:CH_4$ ratio at which the carbon deposition becomes negligible increases significantly for 10 bar pressure, but is relatively unaffected at lower pressures. In contrast, the tendency towards carbon deposition begins to reduce at lower values of the $[O]:CH_4$ ratio with increasing pressures for the CeO_2 system. However, at pressures of 3, 5, and 10 bar, C deposition is observed even with increasing values of the $[O]:CH_4$ ratio, as seen from Figure 2.10b. Overall, carbon deposition tends to occur at lower values of $[O]:CH_4$ ratios indicating a higher tendency of reduced metal oxides to catalyze coking. It is in this region of coking that the highest syngas purity possible for a metal oxide system is obtained. In spite of the high syngas purity, the system operation is not recommended to be conducted in this region, as coking would have adverse effects on both the physical and chemical properties of the metal oxides. Therefore, the preferred region of operation would be at $[O]:CH_4$ ratios where a high syngas purity is achieved with little or no carbon deposition. The desired region of operation for syngas production in chemical looping systems is analyzed in detail in Chapter 4.

Sensitivity Analysis

The different single metal oxides can be studied for the effect of temperature on partial oxidation. The important parameters against which the performance of these metal oxides can be measured are syngas purity and overall CH_4 conversion.

Specifically, in the temperature range 900 to 1100 °C as applicable to a partial oxidation system operation, a sensitivity analysis can be performed to elucidate the effect on the above parameters of these metal oxides. An equimolar ratio of active metal oxide to CH_4 was simulated. Pressures of 1 and 10 atm were used.

The Ce–CH_4 system shows excellent syngas purity as well as high CH_4 conversion (Figure 2.11a), making Ce-based systems thermodynamically most suitable for partial oxidation applications. However, the slow kinetics of the Ce-oxides coupled with their high cost impedes further development of Ce-based oxygen carriers for commercial applications. The Ni-based system exhibits lower syngas purity as compared to the Ce system, although the overall CH_4 conversion is high (Figure 2.11b). As compared to Ce and Ni systems, the Fe-based systems show superior CH_4 conversion at the conditions employed here.

The syngas purity as well as CH_4 overall conversion is negatively affected by an increase in pressure because of the volume expansion reactions of partial and complete oxidation of CH_4. The CH_4 conversion increases with temperatures at two pressures (1 atm and 10 atm) considered here. For the Fe and ITCMO system (Figure 2.11c and d), the syngas purity decreases with an increase in temperature, but the total CH_4 conversion increases. That is, at higher temperatures, complete oxidation of CH_4 to CO_2 and H_2O is more favorable than at lower temperatures. The syngas purity obtained in the pure Fe system (Figure 2.11c) is much lower than that of the ITCMO system (Figure 2.11d), indicating a higher propensity for complete oxidation of CH_4 in the presence of Fe-oxide alone.

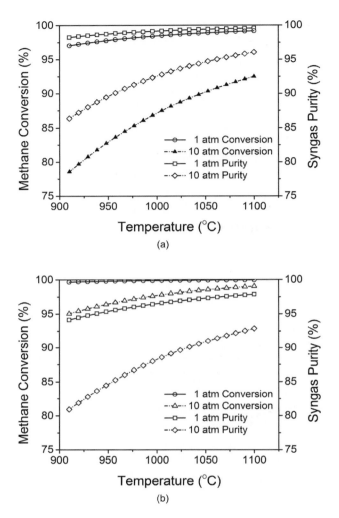

Figure 2.11 Effect of temperature on the CH_4 conversion and syngas purity for different metal oxide systems at pressures of 1 atm and 10 atm: (a) CeO_2, (b) NiO, (c) Fe_2O_3, and (d) ITCMO system.

Effect of Pressure on Reaction Rates

The rates of reactions involved in partial oxidation chemical looping systems are expected to increase at elevated pressures. When the chemical looping partial oxidation system is operated under elevated pressures, the partial pressures of reacting gas also increase proportionally and hence their contribution towards the reaction rate increases. The improved reaction rates due to a pressure increase have been experimentally validated using ITCMO particles.[72] For the Fe–Ti oxide based system, isothermal and isobaric tests carried out at various pressures indicate an increase in reaction rates for various reacting gases, temperatures, and concentrations of the active species.

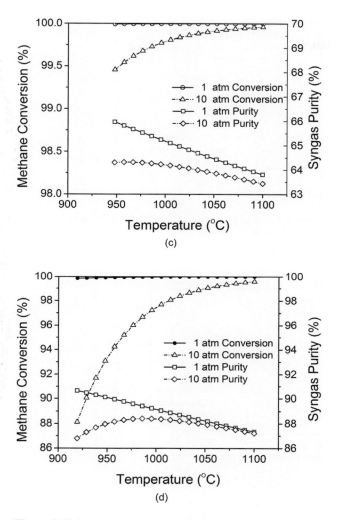

Figure 2.11 (*cont.*)

Constant partial pressure of reducing gas. When the reducing gas partial pressures (pp) are maintained and the overall system pressure is increased, the mole fraction of the reducing gas in the gas environment decreases. Thus, at a constant partial pressure of the reducing gas, an increase in system pressure results in a decrease in the reaction rate of the ITCMO solid particles; noting that the gas diffusivity decreases with an increase in pressure. This result can be seen from Figure 2.12 and Figure 2.13, which show the reaction/reduction conversion (X) curves for ITCMO particles at constant partial pressure values of 1, 1.5, and 3 atm of H_2 as the reducing gas, where the system pressure is varied from 1 to 10 atm. X at a particular time of reaction is defined as the ratio between weight change at that time and the maximum weight change possible for ITCMO particles during reduction. When the partial pressure of the reducing gas is increased,

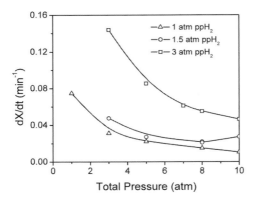

Figure 2.12 The effect of total system pressure on the rates of reduction of ITCMO particles at X = 0.75 for three different ppH$_2$ with T = 900 °C.[72]

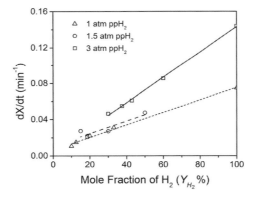

Figure 2.13 The effect of mole fraction of reducing gas (Y_{H_2}%) on the rates of reduction of ITCMO particles at X = 0.75 for three different ppH$_2$ with T = 900 °C.[72]

it has a favorable impact on reaction rate. This can be seen in Figure 2.14, with H$_2$ as the reducing gas, at a fixed system pressure of 5 atm and increasing reducing gas mole fraction.

Constant mole fraction of reducing gas. In order to effectively realize the implications of elevated pressure operation of the partial oxidation chemical looping scheme for a scaled-up system, the reaction rates are analyzed by comparing them at a constant mole fraction of the reducing or active gas, at various pressures. While scaling up with respect to pressure in the case of chemical looping partial oxidation on a large scale scenario, the mole fraction of the active gas, rather than the partial pressure, would be required to be kept constant. Thus, if the reaction rates are compared as a function of the system pressure at a constant mole fraction of reducing gas, the reaction rates are found to proportionately increase with increase in pressure. This can be seen in Figure 2.15,

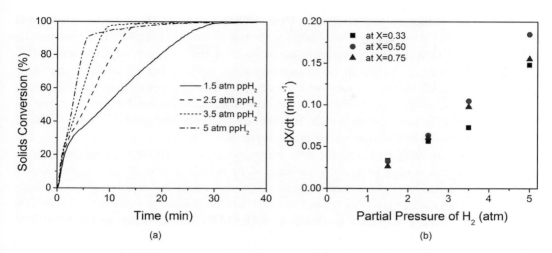

Figure 2.14 The effect of partial pressure of reducing gas ppH_2 on (a) conversion curves obtained and (b) rates of reduction of ITCMO particles at constant system pressure of 5 atm, $T = 900\ °C$.[72]

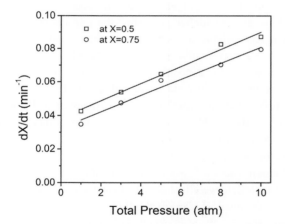

Figure 2.15 The effect of system pressure on rates of reduction of ITCMO particles at X = 0.5 and X = 0.75, and constant mole fraction of reducing gas $Y_{H_2} = 50\%$, $T = 900\ °C$.[72]

with H_2 as the reducing gas, held at a mole fraction of 0.5 between pressures of 1 and 10 atm. The rate is computed at two different values of reduction conversion, 0.5 and 0.75.

Reduction in CH_4. The reduction of oxygen carrier particles with CH_4 as the reducing agent results in the formation of elemental carbon (C). The C deposition is experimentally observed when the oxygen carrier reaches a certain degree of reduction conversion, and is always after an elemental Fe phase has been formed.[72] This experimental finding is consistent with thermodynamics analysis. Specifically, evidence of carbon deposition can be found after the reduction conversion of the iron oxide increases beyond 33%. Thus, in order to study the reaction kinetics of the reduction of oxygen carrier materials in the presence of CH_4, the reaction is arrested at or before the initiation of C deposition.[72] Thermodynamic analysis shows that an increase in system pressure from 1 to 10 atm results in a decrease in overall CH_4 conversion, H_2 and CO formation, along with an increase in the formation of CO_2 and H_2O.[75] However, an increase in the system pressure is found to have a favorable impact on the equilibrium H_2:CO ratio, which is desired to be ~2 for downstream processing such as Fischer–Tropsch synthesis. Carbon deposition can be managed by careful manipulation of the gas:solid ratios in the moving bed reducer reactor system.[72] A more detailed discussion of this aspect is given in Chapter 4.

Considering Fe_2O_3 as the only active metal in the ITCMO, its reduction to Fe proceeds through progressive variation of the oxidation states. For a complete loss of oxygen as the 100% reduction conversion, three distinct stages can be observed. Stage I relates to Fe_2O_3 to Fe_3O_4 transformation, which translates to 11% conversion (or X = 0.11). Stage II relates to Fe_3O_4 to FeO transformation, which translates to 33% conversion (X = 0.33). At conversions higher than 33% the stage III is commenced, which corresponds to the formation of elemental Fe.[74] Unlike reduction in H_2, in the case of CH_4 reduction, these three stages have three distinct reaction rates, as seen in Figure 2.16. Further, the different stages react differently to an increase in pressure in terms of the rate of reaction. The rate of each reaction stage is studied at various pressures between 1 and 10 atm. At higher pressures, the rate disparity between the three stages is less pronounced, giving a faster overall conversion obtained without three distinct rate stages. It was also observed that the reaction halted at lower conversions, owing to the higher amount of C deposition. The evaluation of three separate rate values is warranted for the three distinct stages of reduction. These rate values are plotted in Figure 2.17.[72] It is noted that reaction rates for stages I and III go through a maximum in the range of 1 to 10 atm. However, the rate of stage II increases exponentially with pressure. Since stage II is the slowest reaction stage, it is the overall rate determining step and therefore any change in the rate of stage II overwhelmingly affects the overall rate of the reduction reaction.[72] For example, it can be seen from Figure 2.16 that at 10 atm, 33% reduction is achieved in almost 1/7th of the time taken at 1 atm; and similarly, 60% reduction is achieved in 1/3rd of the time.[72]

An increase in pressure, specifically at constant mole fraction of the reacting gas, results in an increase in concentration of the active species in the gas phase. This increased concentration of gaseous species inevitably results in an increased reaction rate, which has also been reported earlier (for H_2). Specifically, the role of pressure in the reaction rate is observed to be particularly pronounced in the case of reduction of ITCMO particles with CH_4.[72] The effect of pressure on reduction kinetics for the chemical looping combustion (CLC) scenario has previously been investigated by García-Labiano et al.[72,78] They studied the reduction kinetics on Fe, Cu, and Ni based oxygen carriers at pressures up to 30 atm. However, it was noted that the actual reaction rate was influenced to a certain degree by several factors, an important factor being "gas dispersion" which occurs particularly during the initial stage of introduction of the reacting gas to the sample cell. The use of a constant molar flow rate across all pressures resulted in a progressively increasing gas dispersion effect in the reaction cell at increased pressures. To minimize the progressive gas dispersion effect due to an

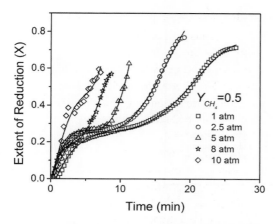

Figure 2.16 Reduction conversion curves of ITCMO particles obtained using CH_4 from the thermogravimetric analysis between 1 and 10 atm at constant mole fraction of reducing gas $Y_{CH_4} = 50\%$, $T = 950$ °C.[72]

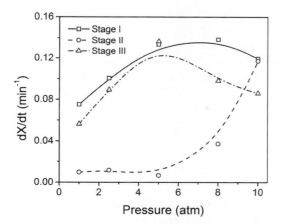

Figure 2.17 The effect of system pressure on reaction rate of ITCMO particles for the three-step reduction with CH_4 as the reducing gas. $Y_{CH_4} = 50\%$, $T = 950$ °C, $p = 1$ to 10 atm.[72]

increase in pressure in the present study, the ITCMO particles were reduced at constant space velocity across all pressures.[72] The use of a constant space velocity leads to an overall increase in reaction rate of the reduction reaction of the ITCMO particle with increasing pressure, as seen in Figure 2.17.[72]

The effect of gas dispersion can be explained using Figure 2.18. This figure shows the evolution of reduction conversion of ITCMO particles between 1 and 10 atm with constant space velocity and constant molar flow rate. The solid line indicates the reduction conversion curve at space velocity the same as that at 1 atm. The dashed line denotes the reduction conversion curve at molar flow rate the same as that at 1 atm. If the space velocity is held constant, the advantage of operating the system at elevated pressure can be easily verified (comparison of the solid line and dotted line). On the

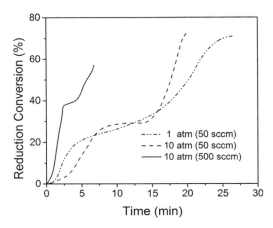

Figure 2.18 Effect of gas dispersion in the reduction kinetics of ITCMO particles in the presence of CH_4.

other hand, if the molar flow rate is held constant, the kinetic advantage is not apparent due to the gas dispersion or end effect. At a constant molar flow rate, a fixed reactor volume equilibrates over a longer duration of time at higher pressure. Therefore, at the beginning the sample "sees" a much lower concentration of the active species than the set point. This results in a trade-off between the positive effect of higher pressure, and the negative effect of lower concentration of the active species. In such experiments, the difference in reaction rate is found to diminish significantly with increasing pressure, confirming that the negligible effect of increased pressure on reduction rate observed by García-Labiano et al.[72,78] is attributed to the "gas dispersion" effect with pressure.

Mechanistic Differences Due to Elevated Pressure

Possible mechanistic differences in the reactions conducted at ambient pressure versus high pressure have been suggested to explain the increased reduction rates for oxygen carrier particles with increasing pressure.[72] When the samples treated at higher pressure were compared to those treated at ambient pressure, certain differences were observed in the morphology as well as formation of complex bimetallic oxide species within the particles. Typically, Ti phase is used as an inert oxide support material, designed to enhance mechanical and morphological properties. As such, they are not expected to participate in the reaction. However, ITCMO particles reduced under H_2 at 10 atm indicated a higher degree of reduction in not only the Fe but also the Ti oxide phase upon XRD analysis. On the other hand, Fe and Ti complex species of Fe_2TiO_5 and $FeTiO_3$ are formed to a lesser extent at high pressure oxidation. X-ray diffraction spectra showing the pressure effect are seen in Figure 2.19.

It has also been reported that reactions carried out at elevated pressures result in morphologically superior particles.[72] Higher pressure results in a higher surface

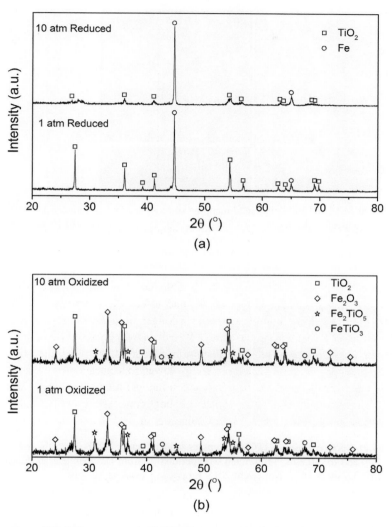

Figure 2.19 XRD analysis of (a) reduced and (b) oxidized samples at 1 and 10 atm, 900 °C. Reducing environment is under H_2 with $Y_{H_2} = 0.5$, and oxidizing environment is air.[72]

porosity compared to ambient pressure, as well as smaller grain size. For example, a typical non-uniform micro-particle reduced at ambient pressure is shown in Figure 2.20a. The particle shows non-homogeneous distribution of Fe and Ti phases, with a denser part of Ti-rich oxides of particle size 40 μm and a porous part containing comparable amount of Fe and Ti oxides with an average grain size of 1–2 μm. In comparison, the oxygen carrier particles reduced at high pressure exhibit a more porous internal structure upon reduction, with an average grain size of 500 nm, which can be seen clearly in the cross-section (Figure 2.20b). Figure 2.21 shows the difference between samples treated under 1 atm and 10 atm. After reduction, the

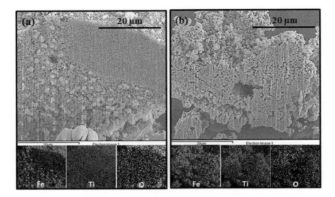

Figure 2.20 SEM and EDS elemental mapping of cross-sections of reduced ITCMO particles. Samples reduced under H_2, $Y_{H_2} = 0.5$ and T = 900 °C. (a) 1 atm and (b) 10 atm.[72] Reprinted with permission from Deshpande et al. *Energy & Fuels*, **2015**, 29(3), 1469–1478. Copyright 2015 American Chemical Society. A black and white version of this figure will appear in some formats. For the color version, please refer to the plate section.

Figure 2.21 Surface grains in ITCMO samples reduced under H_2, $Y_{H_2} = 50\%$ and T = 900 °C. (a) 1 atm and (b) 10 atm.[72] Reprinted with permission from Deshpande et al. *Energy & Fuels*, **2015**, 29(3), 1469–1478. Copyright 2015 American Chemical Society.

particles processed at 1 atm have a grain size of 1 μm and higher pressures lead to smaller grain size of ~500 nm. Consequently, a reaction pressure at 10 atm can largely promote the increase of overall surface area. These samples were further tested for surface area and pore volume measurements using a BET analyzer. The BJH pore size distribution method was used to find the values of total surface area and pore volume. From the analysis, it was discovered that an increase in pressure from 1 to 10 atm resulted in increases in both surface area and pore volume values, for reduced as well as oxidized samples. For the reduced samples, the increase in pressure resulted in a surface area change from 7.036 m^2/g to 7.227 m^2/g, whereas the pore volume values increased from 0.014 cm^3/g to 0.022 cm^3/g. The same trend was observed for oxidized samples, where the change was more pronounced: surface area increased from 4.726 m^2/g to 15.507 m^2/g and the increase in total pore volume was from 0.025 to 0.117 cm^3/g.

2.3   Morphology and Ionic Diffusion

The morphology of oxygen carriers plays a significant role in chemical looping technology because it determines the reaction surface area and structure stability. During redox reactions, the morphology evolves with ionic diffusion in the solid phase. Therefore, an understanding of morphology and its origin is crucial in the development of oxygen carriers. The intent of this section is to provide insights into morphological evolution with ionic diffusion in single and binary metal oxide micro-particles during reduction and oxidation processes.

From the macroscopic view, ionic diffusion can be categorized into three modes: outward diffusion mode, inward diffusion mode, and mixed diffusion mode.[16,79] The inert marker experiment is a popular method for determining the dominating ionic transfer mode. The inert marker experiment was first invented and reported by Ernest Kirkendall during his study of the solid-state diffusion mechanism in brass.[80] Because of its simplicity and effectiveness, it has since become widely used to determine the dominating ionic transfer mode in many gas–solid and solid-state reactions.[81,82] As shown in Figure 2.22, this method involves attaching a piece of thin inert material, e.g. Pt, Au, or α-Al_2O_3, on the surface of a dense solid reactant to mark the initial gas–solid interface. The inert marker experiment is most effective for a volume-expansion or volume-constant reaction, that is, most oxidation reactions. A volume-shrinking reaction would create pores inside the reactant and allow gases to bypass the marker. However, once the ionic diffusion mode during the volume-expansion or volume-constant process is identified, the mechanism during the volume-shrinking process (e.g. reduction reaction) can be extrapolated. When the reaction proceeds through an

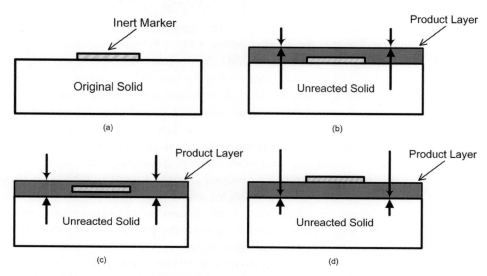

Figure 2.22 Illustration of three ionic diffusion modes in the oxidation reaction and the principle of the inert marker experiment: (a) original solid reactant; (b) outward diffusion mode; (c) mixed diffusion mode; (d) inward diffusion mode.[81]

Table 2.1 Molar volume information of typical oxygen carrier particles at different oxidation states.*

Metal oxide	Molar volume (cm^3/mol)	Metal oxide	Molar volume (cm^3/mol)
Fe	7.1	FeO	12.5
Fe_3O_4	44.7	Fe_2O_3	30.5
Fe_2TiO_5	54.4	$FeTiO_3$	31.7
$FeTi_2O_5$	55.3	Fe_2TiO_4	46.0
Co	6.6	CoO	11.6
Co_3O_4	39.4	Co_2O_3	32.0
Ni	6.6	NiO	10.9
Cu	7.1	Cu_2O	23.7
CuO	12.6		
Mn	7.6	MnO	13.0
Mn_3O_4	47.08	Mn_2O_3	35.11
MnO_2	17.30		

* The molar volume reported in the literature varies from study to study, based on different experimental and theoretical methods.

outward diffusion mode, the solid product would grow outward through the gas–solid interface and the inert marker would be sandwiched between the product layer and the solid reactant. When the inward diffusion of oxygen ions dominates the ionic diffusion step then the product layer forms a new interface with the unreacted solid and the final location of the marker is still at the top of the solid reactant. When the inward and outward diffusion rates are comparable, the product layer would grow both at the gas–solid interface and at the solid–solid interface. As a result, the marker would be buried in the solid product layer. Therefore, based on the final location of the inert marker, the dominating ionic diffusion mode can be identified.

Inert marker experiments have been performed on many oxygen carrier materials. Li et al.[54] studied the effect of adding TiO_2 to Fe during oxidation, and found that the dominating ionic transfer mechanism changes from outward diffusion of Fe ions (for pure Fe) to inward diffusion of oxygen anion (for Fe + TiO_2). This change was attributed to the enhanced oxygen anion diffusivity through the extrinsic oxygen vacancies created by TiO_2. Naturally occurring $FeTiO_3$ also shows a similar inward diffusion pattern for oxygen anions. Day and Frisch[83] used chromium chromate as inert marker to study the Cu oxidation process and found that the reaction proceeds via the outward diffusion of Cu. In the study of Co oxidation, α-Al_2O_3 was used as inert marker. It was found that Co cations and electrons diffuse through the oxide layer for the oxidation of pure Co. While with Mn dopant, the product layer grew inward, leaving the marker at the gas–solid interface.[84] It is generally found that the dominating diffusing ionic species during oxidation of pure metals are cations. Once the ionic diffusion mode of a volume-expansion or volume-constant reaction is determined, morphological changes induced by the oxidation reaction can be deduced. The molar volumes for typical oxygen carrier particles are listed in Table 2.1. While in this section a few phenomenological descriptions related to molar volume changes are being discussed, morphological transformations due to ionic diffusion will be discussed in Sections 2.3.1–2.3.3.

Most oxidation reactions incur volume expansion of the solid product layer and may close the internal pores during solid conversion. Hence, the pore size distribution can significantly affect the overall surface area exposed to the gas phase. Molar volume increase can also induce internal physical stress in the solid product layer. Rather than "abruptly" increasing at certain locations, such stress theoretically follows a gradual increase from the solid product/solid reactant interface to the gas/solid product interface.[85] If the solid product cannot hold this additional stress, such as in the hydration reaction of CaO, the solid product layer cracks and disintegrates into smaller pieces.[86]

Morphology of solids is also affected by sintering along with molar volume changes and solid-phase ionic diffusion.[82] The oxidation reaction involving outward ionic diffusion of solid species often smoothens concave and/or convex structures present on the surface.[80] Specifically, a concave surface is filled via a large outward diffusion of the surface towards the valley of the concave. On the other hand, the convex surface is smoothed out via a lower outward diffusion flux in the vertical direction. Both factors can affect the available surface area during oxidation reactions. Similar phenomena are observed during reduction reactions. Hence, when considering the morphological variation in redox reactions in oxygen carriers, these factors should all be considered.

2.3.1   Core–Shell Structure Formation

As discussed in Section 1.3.3, the cyclic gas–solid reactions in certain physically supported oxygen carrier particles can result in the formation of a core–shell structure. It is observed that if the redox reaction of oxygen carrier material with an inert support proceeds through the outward diffusion of cations, the active oxygen carrier material diffuses from the particle interior to the particle surface and forms a shell covering the inert support core.[87] Figure 2.23 illustrates a homogeneously mixed Fe_2O_3–Al_2O_3 particle which, after numerous reduction and oxidation cycles, would become a

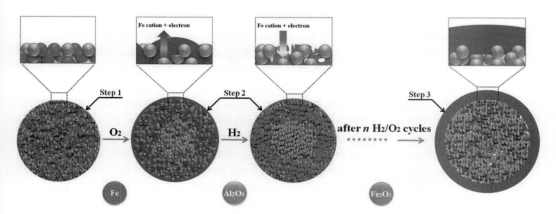

Figure 2.23 Formation of a core–shell structured composite of Fe_2O_3–Al_2O_3 micro-particles via cyclic redox reactions.[87] Reprinted with permission from Sun et al. *Langmuir*, **2013**, 29(40), 12520–12529. Copyright 2013 American Chemical Society. A black and white version of this figure will appear in some formats. For the color version, please refer to the plate section.

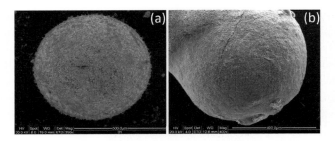

Figure 2.24 SEM analysis on solid surface of alumina supported iron oxide (40 wt%), (a) before and (b) after 100 redox cycles.

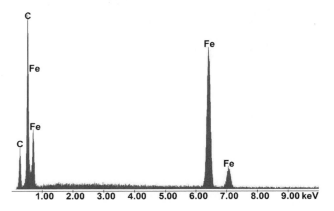

Figure 2.25 EDS analysis on an alumina supported iron oxide (40 wt%) carrier surface after 100 redox cycles.

Fe_2O_3(shell)–Al_2O_3(core) particle. On the other hand, if the redox reaction proceeds through inward diffusion of oxygen anions rather than cations, the active oxygen carrier material would locally expand and shrink without long-distance migration. As a result, the homogeneous mixture form of the particle is maintained. Formation of such a core–shell structure can potentially cause undesired inter-particle aggregation, especially when the active oxygen carrier material has a low sintering temperature, e.g. Cu.[9] On the other hand, this phenomenon can potentially be used to form desired core–shell structures for other chemical and physical applications.

Figures 2.24 and 2.25 show SEM and EDS analysis on a 40 wt% iron oxide particle supported on Al_2O_3 after 100 redox cycles using hydrogen as reducing gas, respectively. Ionic diffusion and sintering effect have smoothed out the particle surface. EDS on the particle surface shows an iron oxide-rich shell.

2.3.2   Morphology Evolution and Nanostructure Formation

The nanoscale morphology of metal oxides often determines their performance and behavior at the macroscopic scale. Many morphological changes at the nanoscale are a

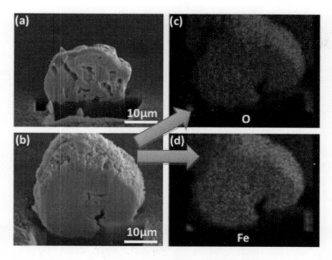

Figure 2.26 Fe micro-particle. (a) SEM image of cross-section of fresh micro-particle; (b) SEM image of cross-section of Fe_2O_3 micro-particle after Fe oxidation at 700 °C; (c) EDS mapping of O from (b); (d) EDS mapping of Fe from (b).[88] Qin et al. *J. Mater. Chem. A*, **2014**, 2(41), 17511–17520. Reproduced by permission of The Royal Society of Chemistry. A black and white version of this figure will appear in some formats. For the color version, please refer to the plate section.

result of solid-state ionic diffusion, which varies from system to system. Single metal oxide systems demonstrate morphological transformations very different from those observed in bimetallic systems. A thorough understanding of the metal oxide system behavior under redox conditions can yield better strategies when selecting oxygen carrier materials. This section discusses some of the nanoscale morphological transformations observed in single, bimetallic, and perovskite-supported systems during chemical looping redox reactions.

Fe System

In the case of Fe micro-particles, oxidation leads to volume expansion and change in pore distribution. Figure 2.26a,b are cross-sectional SEM images of the particle before and after oxidation.[88] The particle changes from having a randomly distributed porosity to a more porous center after oxidation. The change in porosity is accompanied by the formation of two types of nanostructure. Figure 2.27 shows that the surface of the oxidized Fe particle is composed of multiple grains containing the two nanostructures: (a) nanowires and (b) nanopores. The nanowires have relatively uniform diameter of 100 nm ± 30 nm, while the nanopores have diameters in the range 140–300 nm. Both kinds of nanostructures are found to form at the center of the grains.

The formation of nanowires, after thermal oxidation of Fe, has been also reported by Fu et al.[89] and Wen et al.[90] Nanowire formation is also observed in the case of a few other metal oxides, like CuO and ZnO, upon thermal oxidation.[91,92] The oxidized Fe particles on further reduction lose the nanostructure. Figure 2.28 shows the cross-

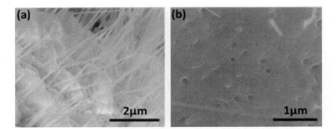

Figure 2.27 SEM image of top surface of Fe_2O_3 micro-particle with (a) nanowires, (b) nanopores.[88] Qin et al. *J. Mater. Chem. A*, **2014**, 2(41), 17511–17520. Reproduced by permission of The Royal Society of Chemistry.

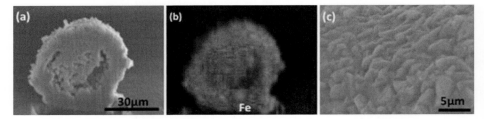

Figure 2.28 Fe micro-particle after one oxidation–reduction cycle. (a) SEM image of cross-section; (b) EDS mapping of Fe; (c) SEM image of surface.[88] Qin et al. *J. Mater. Chem. A*, **2014**, 2 (41), 17511–17520. Reproduced by permission of The Royal Society of Chemistry. A black and white version of this figure will appear in some formats. For the color version, please refer to the plate section.

section of a Fe particle having undergone one oxidation–reduction cycle. The reduced particle retains the porous center from its earlier oxidized form, but loses the nanowires and nanopores. These nanostructure formations are only observed during the first oxidation step. The Fe particle does not show nanostructure formations during any subsequent redox cycles.

For active metal particles, sintering is generally inevitable especially at elevated temperatures and longer heating times. The Fe micro-particle surface after multiple redox cycles shows some sintering effects, like dense surface and decreased surface areas. XRD scans of the particles after multiple cycles show the presence of Fe_3O_4 along with Fe. Fe_3O_4 is most likely present in the form of irreversible aggregates that ultimately lead to particle deactivation in applications that require pure Fe for redox reactions.

Nickel

When pure Ni micro-particles are subjected to oxidation, the reaction is found to occur only on the surface. For instance, in Figure 2.29, the Ni micro-particles have been subjected to five oxidation–reduction cycles at 700 °C in the presence of O_2/

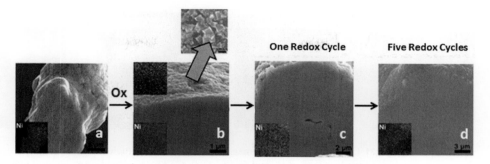

Figure 2.29 Cross-sectional SEM image of a Ni micro-particle subjected to redox cycles at 700 °C in the presence of H_2 and O_2. (a) Fresh Ni, (b) after oxidation, (c) after oxidation and reduction, single cycle, (d) after five redox cycles. A black and white version of this figure will appear in some formats. For the color version, please refer to the plate section.

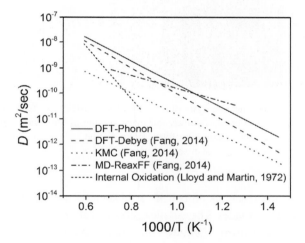

Figure 2.30 Oxygen diffusivity in Ni as a function of temperature; a compilation of literature values.[93,94]

H_2. The EDS mapping of the micro-particle after oxidation (shown as insets in the pictures) indicates the presence of oxygen only on the surface of the micro-particle, whereas the interior of the particle indicates only Ni. This can be attributed to the highly cubic ordered crystals seen in the inset of Figure 2.29b. The lack of surface defects seems to be contributing to the poor diffusion of oxygen ions. Therefore, it may be concluded that chemical looping applications involving pure Ni particles should employ small to moderate particle sizes, as large particles will likely exhibit lower overall conversion due to the lower oxygen diffusion to the interior of the particle. The poor diffusivity of oxygen in Ni reported in various studies in the literature is given in Figure 2.30.[93,94] At ~1000 K (727 °C), the diffusivity value is as low as 10^{-10} m^2/s.

One Redox Cycle   **Five Redox Cycles**

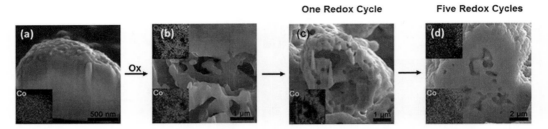

Figure 2.31 Cross-sectional SEM image of the Co micro-particle subjected to redox cycles at 700 °C in the presence of H_2 and O_2: (a) fresh Co, (b) after oxidation, (c) after oxidation and reduction, single cycle, (d) after five redox cycles. A black and white version of this figure will appear in some formats. For the color version, please refer to the plate section.

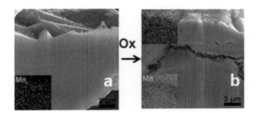

Figure 2.32 Cross-sectional SEM image analysis of the Mn micro-particle: (a) fresh Mn, (b) after oxidation at 700 °C in the presence of O_2. A black and white version of this figure will appear in some formats. For the color version, please refer to the plate section.

Cobalt

The reaction rates for redox reactions on cobalt oxide samples are appreciably slower than the other transition metal oxides. The samples generated after reaction show a more porous internal structure, as shown in Figure 2.31c and d. Nevertheless, the thermodynamics and kinetics of cobalt oxide redox reactions inhibit the material from being considered as a viable candidate for partial oxidation applications.

Manganese

Like Co, Mn reactivity is also very low. In addition to the slower reaction rates, the Mn micro-particle also shows the presence of multiple oxidation states upon oxidation, which may result in physical layer separation, or de-lamination, as can be seen in Figure 2.32b in the oxidized sample. This may result in mechanical strength deterioration of the material.

Apart from observing the morphological changes, the difference in reactivity of the metals with reducing gas can also be observed using DFT calculations. The following Figure 2.33 shows the bond energy for successive C-H bond cleavage during CH_4 partial oxidation on various transition metal oxide materials.[95] Based on this figure, energy barriers for the different dissociation steps and the heat of reaction, whether endothermic or exothermic, can be estimated for the transition

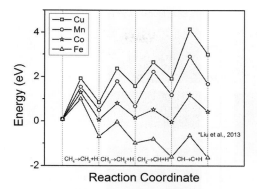

Figure 2.33 CH_4 partial oxidation, bond energy for successive C–H bond cleavage for transition metal oxides, Mn, Co, Cu, and Fe.[95]

metals. Also, it can be observed that the rate determining step, which is the step having the highest energy barrier, varies with the transition metal. For example, CH_4 dissociation is the rate determining step for Fe whereas CH dissociation step is rate determining for Co and Cu. The details of using DFT calculations to gain insight into the reaction mechanism of partial oxidation of CH_4 are given in Section 2.4.

FeNi Bimetallic System

Bimetallic systems behave differently under redox conditions depending on the intrinsic properties of the components involved, as will be seen in the following examples. A binary FeNi micro-particle goes from having a non-porous cross-section to having a porous center upon oxidation, along with some volume expansion, similar to the Fe system. Figure 2.34 shows the SEM images of a binary FeNi micro-particle with a highly crystalline $Fe_{0.64}Ni_{0.36}$ phase, before and after oxidation. Formation of the porous center during oxidation is accompanied by the migration of Fe towards the surface and concentration of Ni in the center, forming an iron oxide-rich shell and a nickel oxide-rich core, as can be seen from the EDS images in Figure 2.34. The iron oxide-rich particle surface consists of nanopores and nanowires, similar to the Fe system, illustrated in Figure 2.35.

Similar to the Fe system, the nanostructures disappear for this bimetallic system upon reduction and the Fe and Ni are uniformly redistributed throughout the particle cross-section, as illustrated in the SEM and EDS images in Figure 2.36. The images in fact show that the morphology after one oxidation–reduction cycle resembles that of the fresh particle, except with a rougher and more porous surface. For the FeNi binary system too, the nanostructures only appear during the first oxidation step and no more during any subsequent redox cycles. Like the Fe micro-particles, the FeNi particles also show sintering effects after multiple redox cycles, which is a commonly observed phenomenon in many active complex metal oxide systems leading to decreased surface area and reduced reactivity.

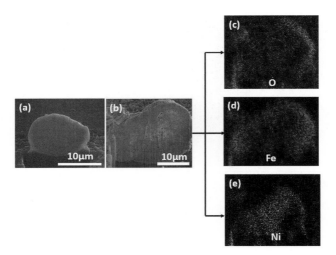

Figure 2.34 FeNi micro-particle. SEM images of the cross-section: (a) fresh micro-particle; (b) oxidized particle after oxidation at 700 °C for 0.5 h; EDS mapping of oxidized micro-particles: (c) O; (d) Fe; and (e) Ni.[88] Qin et al. *J. Mater. Chem. A*, **2014**, 2(41), 17511–17520. Reproduced by permission of The Royal Society of Chemistry. A black and white version of this figure will appear in some formats. For the color version, please refer to the plate section.

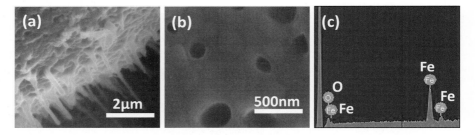

Figure 2.35 FeNi micro-particle after oxidation: (a) SEM image of nanowires; (b) SEM image of nanopores; (c) EDS spectrum of surface.[88] Qin et al. *J. Mater. Chem. A*, **2014**, 2(41), 17511–17520. Reproduced by permission of The Royal Society of Chemistry.

CuNi Bimetallic System

A CuNi binary micro-particle, similar to the Fe and FeNi systems, develops porosity on oxidation.

Figure 2.37 shows SEM images of the fresh CuNi micro-particle with a dense cross-section and the oxidized particle with a porous center. The CuNi system, however, does not form the core–shell structure on oxidation, unlike the FeNi binary system. EDS images of the oxidized CuNi particle show all the elements well dispersed throughout the cross-section. Also, unlike the Fe and FeNi systems, the CuNi binary micro-particle

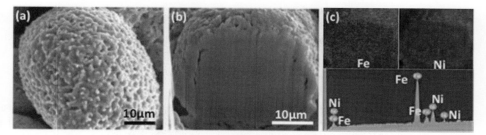

Figure 2.36 FeNi micro-particle after one oxidation–reduction cycle: (a) SEM image of surface; (b) SEM image of cross-section; (c) EDS mapping of cross-section and EDS spectrum of surface.[88] Qin et al. *J. Mater. Chem. A*, **2014**, 2(41), 17511–17520. Reproduced by permission of The Royal Society of Chemistry. A black and white version of this figure will appear in some formats. For the color version, please refer to the plate section.

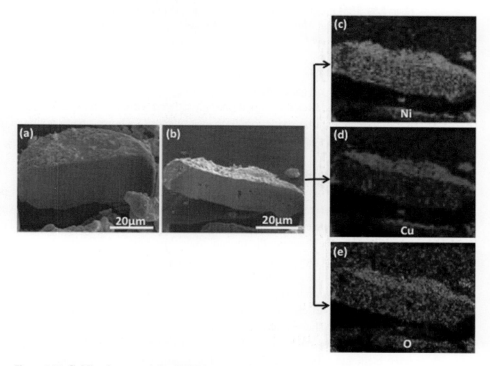

Figure 2.37 CuNi micro-particle. SEM images of the cross-section: (a) fresh micro-particle; (b) oxidized micro-particle after oxidation at 700 °C for 0.5 hour; oxidized micro-particle EDS mapping: of (c) Ni; (d) Cu; and (e) O.[88] Qin et al. *J. Mater. Chem. A*, **2014**, 2(41), 17511–17520. Reproduced by permission of The Royal Society of Chemistry. A black and white version of this figure will appear in some formats. For the color version, please refer to the plate section.

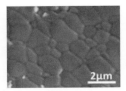

Figure 2.38 SEM image of the top surface of an oxidized CuNi micro-particle.[88] Qin et al. *J. Mater. Chem. A*, **2014**, 2(41), 17511–17520. Reproduced by permission of The Royal Society of Chemistry.

Figure 2.39 SEM image of the cross-section of a CuNi micro-particle after one oxidation–reduction cycle.[88] Qin et al. *J. Mater. Chem. A*, **2014**, 2(41), 17511–17520. Reproduced by permission of The Royal Society of Chemistry.

shows no signs of nanostructure formations upon oxidation. Figure 2.38 does show clear grain boundaries developed on the oxidized particle with grain sizes of 0.3–2 μm. Reduction again, after oxidation, in this case leads to a porous particle like in the case of the Fe and FeNi systems, as can be seen in Figure 2.39.

FeTi Bimetallic System

The FeTi binary system behaves differently compared to all the systems previously discussed. In the cases so far, all of the metals/metal oxides involved were active, whereas in the case of FeTi, Fe is the only active component while the Ti acts as a support material enhancing the morphological properties of Fe and its oxides. Shown in Figure 2.40 are the SEM and EDS images of the cross-section of a binary FeTi micro-particle, showing a non-porous cross-section and the Fe-rich and Ti-rich regions. Oxidation of the micro-particles leads to uniform distribution of all the elements throughout the particle and the formation of nanobelts on the particle surface, as shown in Figure 2.41. These nanobelts, with widths ranging between 150 and 200 nm and width-to-thickness ratios of 5–10, protrude from the surface of the particle and contain all three elements – Fe, Ti, and O.[89–91] The triangular tips of these nanobelts, seen in the SEM image, suggest an anisotropic growth in the initial stages.[96] Reduction of the oxidized micro-particle causes the nanobelts to partially retract with increased porosity throughout the particle, while still maintaining the uniform elemental distribution (Figure 2.42).

The nanostructures disappear completely over multiple redox cycles. Figure 2.43 shows SEM and EDS images of the particle after five redox cycles, still showing well-dispersed elements, but with increased porosity and completely retracted nanobelts. But unlike the previously discussed active metal systems, where the particle

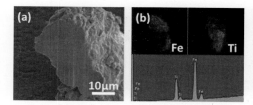

Figure 2.40 FeTi micro-particle: (a) SEM image of cross-section; (b) EDS mapping of Fe and Ti, and EDS spectrum of the surface.[97] Qin et al. *J. Mater. Chem. A*, **2015**, 3(21), 11302–11312. Reproduced by permission of The Royal Society of Chemistry. A black and white version of this figure will appear in some formats. For the color version, please refer to the plate section.

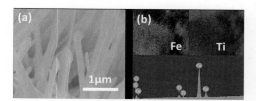

Figure 2.41 Oxidized FeTi micro-particles at 700 °C: (a) SEM image of the surface at higher magnification; (b) EDS mapping of the cross-section and EDS spectrum of the surface.[97] Qin et al. *J. Mater. Chem. A*, **2015**, 3(21), 11302–11312. Reproduced by permission of The Royal Society of Chemistry. A black and white version of this figure will appear in some formats. For the color version, please refer to the plate section.

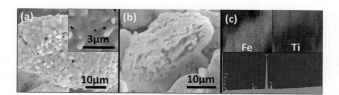

Figure 2.42 FeTi micro-particle after one oxidation–reduction cycle at 700 °C: (a) SEM image of surface; inset: higher magnification SEM image; (b) SEM image of cross-section; and (c) EDS mapping and spectrum of (b).[97] Qin et al. *J. Mater. Chem. A*, **2015**, 3(21), 11302–11312. Reproduced by permission of The Royal Society of Chemistry. A black and white version of this figure will appear in some formats. For the color version, please refer to the plate section.

surface suffers from sintering effects over multiple redox cycles, the FeTi system gains porosity over multiple redox cycles.[88] The particle has higher porosity after the five redox cycles as compared to the particle after a single redox cycle. This implies that addition of Ti to Fe gives the system more stability in terms of reactivity and recyclability. The excellent stability of this redox system has been confirmed previously through cyclic reactivity studies,[4] showing no drop in particle reactivity over 1000 redox cycles, as indicated earlier in Section 2.2.4. It has been previously demonstrated that a severe sintering effect hinders active metals or alloys from being used effectively in high temperature applications, including chemical looping processes.[97] The addition of

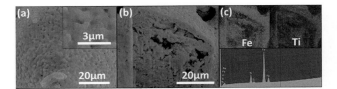

Figure 2.43 FeTi micro-particle after five oxidation–reduction cycles at 700 °C: (a) SEM image of surface; inset: higher magnification SEM image; (b) SEM image of cross-section; and (c) EDS mapping and EDS spectrum of (b).[97] Qin et al. *J. Mater. Chem. A*, **2015**, 3(21), 11302–11312. Reproduced by permission of The Royal Society of Chemistry. A black and white version of this figure will appear in some formats. For the color version, please refer to the plate section.

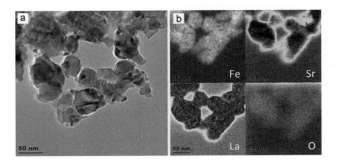

Figure 2.44 (a) TEM image of Fe_2O_3–$La_xSr_{1-x}FeO_3$ core–shell particles and (b) EFTEM mappings of individual elements such as Fe, Sr, La, and O.[98] Reproduced with permission from Shafiefarhood et al. *Chem Cat Chem*, **2014**, 6(3), 790–799. Copyright ©2014 WILEY-VCH Verlag GmbH & Co. KGaA, Weinheim.

certain inert oxides can alleviate the sintering effects by creating micropores on the micro-particle surface. All of the morphological transformations occurring in the systems discussed in this section can be explained with the help of solid phase ionic diffusion phenomena within the particles during the redox reactions. Each of the transformations is explained in detail in Section 2.4.

Perovskite-supported Fe_2O_3

A core–shell structure for Fe_2O_3– $La_xSr_{1-x}FeO_3$ (LSF) system has been investigated for syngas production from CH_4.[98–100] This structure was engineered specifically to improve the reactivity and selectivity of the Fe_2O_3 based oxygen carriers in chemical looping reforming applications. Shafiefarhood et al.[98] synthesized this core–shell redox catalyst using a sol–gel method where the LSF shell was of the order of 10 nm and the Fe_2O_3 core diameter was between 20 and 60 nm. A transmission electron microscopy (TEM) image, shown in Figure 2.44, of the prepared particles showed La ion and Sr ion to be present on the surface, whereas Fe ion and O ion were distributed evenly in the bulk.

The core–shell structure was found to be stable and active for over 100 redox cycles with CH_4 as the reducing gas. The SEM images of the oxidized core–shell particles at

Figure 2.45 SEM images of the oxidized Fe_2O_3–$La_xSr_{1-x}FeO_3$ core–shell particles after (a) sintering, (b) five redox cycles, (c) 50 redox cycles, and (d) 100 redox cycles.[98] Reproduced with permission from Shafiefarhood et al. *Chem Cat Chem*, **2014**, 6(3), 790–799. Copyright ©2014 WILEY-VCH Verlag GmbH & Co. KGaA, Weinheim.

various stages are shown in Figure 2.45. Neal et al.[100] reported that this structure could also be obtained starting from a composite of iron oxide and lanthanum strontium ferrite (LSF) material, prepared by a solid-state reaction method, by subjecting it to continuous redox cycles. There was an increase in the concentration of La and Sr cations on the catalyst surface over the course of 50 redox cycles with methane and air on the Fe–LSF composite. This phenomenon of cation enrichment was found to be similar to that observed for strontium doped perovskites. The mechanism by which the LSF shell helps in modifying the properties of iron oxide materials is further explained in Section 2.7.2.

2.3.3 Impact of Ionic Diffusion on Morphology

The morphological transformations in the different metal systems discussed in the previous section are a consequence of the ionic diffusion behavior of the metal/metal oxide systems in the presence or absence of other species. In the Fe system, during oxidation, reaction on the surface creates a concentration gradient in the particle, which causes Fe ions to continuously diffuse from the core of the particle to the surface. This phenomenon can be attributed to the Kirkendall effect, and it is this solid-state diffusion that leads to the formation of a porous center in the oxidized Fe particle.[101,102] The oxygen atoms are ionized during the diffusion process, which occurs via vacancy consumption, as shown below:

$$\tfrac{1}{2}O_2 + V_{\ddot{O}} = O_O^X + 2h. \tag{2.3.1}$$

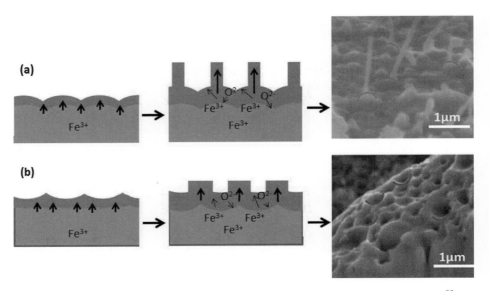

Figure 2.46 Growth mechanisms of (a) iron oxide nanowires and (b) iron oxide nanopores.[88] Qin et al. *J. Mater. Chem. A*, **2014**, 2(41), 17511–17520. Reproduced by permission of The Royal Society of Chemistry. A black and white version of this figure will appear in some formats. For the color version, please refer to the plate section.

The oxygen vacancies are thus consumed during Fe oxidation. The porous center formation is a result of the different diffusion rates of Fe and O in the particle. The outward Fe diffusivity is higher than the inward oxygen diffusivity in Fe_2O_3 at 700 °C, which results in net Fe transport from the center of the micro-particle to the particle edge.[88] This outward Fe transport is further enhanced because the high volume expansion of Fe to Fe_2O_3 creates physical space for the Fe ions to diffuse. As a result, the net directional flow of Fe is balanced by an opposite flow of vacancies, which can condense into pores at dislocations. This phenomenon was first observed in the movement of the interface between a diffusion couple of copper and zinc in brass as a result of the different diffusion rates of these two species at high temperature.[88] There is also the formation of nanowires and nanopores which is driven by the stress caused due to the volume expansion during oxidation. The kinds of nanostructures formed depend on the grain/crystallite surface curvature, which is explained schematically in Figure 2.46.

For grains/crystallites with positive surface curvature (shown in Figure 2.46a), there is a continuous outward diffusion of Fe atoms, which are transformed into Fe^{3+} ions in the course of diffusion. The Fe^{3+} ions are a source for nanowire growth, which continues as long as the compressive stress in the Fe_2O_3 layer due to volume expansion exists. Owing to the mechanical stress that arises between neighboring grains because of volume expansion, nanowire growth occurs perpendicular to the grain surface, as shown in

Figure 2.46a. In the case of grains with negative surface curvature (shown in Figure 2.46b), the oxide grows at the grain boundary because of the mechanical stress perpendicular to the surface. Since ionic diffusion (of both Fe^{3+} and O^{2-}) occurs only in a direction perpendicular to the surface, the material can only expand in the vertical direction.[88] This pushes the grain boundaries upwards, leading to formation of the nanopores.

When oxidation is followed by reduction, there is an outward diffusion of oxygen ions, leaving behind oxygen vacancies $V_{\ddot{O}}$ in the particle, as shown in the following equation:

$$O_O^X = \frac{1}{2}O_2 + V_{\ddot{O}} + 2e. \qquad (2.3.2)$$

The vacancies aggregate to form micropores throughout the particle and subsequently cause all the nanowires to disappear. Nanostructure growth in Fe micro-particles only occurs during the first oxidation because it is mainly a result of a stress-driven mass transport mechanism. During the first oxidation step, grain surfaces experience a large mechanical stress because of volume expansion, which leads to the formation of nanowires and/or nanopores, based on surface morphology. However, after the first oxidation–reduction cycle, the many voids or micropores formed in the particle promote internal ionic diffusion and volume expansion and hence decrease the surface stresses during subsequent reaction cycles. Lower surface stress is the reason why no nanostructure formations are observed during any subsequent redox cycles. There are two other types of growth mechanisms commonly described in the literature in addition to that mentioned above. One is a vapour–liquid–solid (VLS) process, where the metal liquefies and then vaporizes at a temperature higher than its melting point and becomes active for oxidation. The second growth mechanism is a self-catalytic growth mechanism, where the metal acts both as a reactant and a catalyst simultaneously. Similar to the VLS process, this too requires temperatures higher than the melting point of the metal. However, considering the operating temperatures commonly employed for chemical looping partial oxidation schemes, neither of the two above mentioned mechanisms seems relevant.

Morphological transformations in the FeNi system are explained schematically in Figure 2.47. In this binary active metal/metal oxide system, a complex interplay of oxidation rates, ionic diffusion kinetics, and mechanical expansion leads to the formation of the core–shell micro-particle structure through a set sequence. Ni and Fe have significantly different oxidation rates, with the Ni oxidation rate being significantly faster. In the first step, when the FeNi binary micro-particle system is subjected to oxidation, oxygen diffuses inwards through the particle surface consuming the oxygen vacancies, as shown in equation (2.3.1). Oxidation of Ni to NiO leads to volume expansion in the crystal lattice around the Ni atoms, although the expansion is much smaller compared to the volume expansion during Fe_2O_3 formation. Because of the significantly faster oxidation rate, Ni throughout the particle is oxidized to NiO before Fe undergoes any appreciable oxidation. It should be noted that NiO is a p-type semiconductor due to non-stoichiometry of NiO. Hence oxidation to NiO creates excess

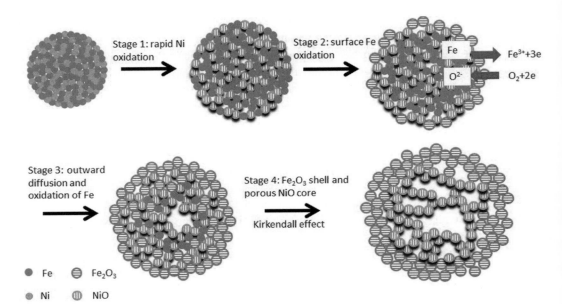

Stage 1: rapid Ni oxidation

Stage 2: surface Fe oxidation

Fe → $Fe^{3+}+3e$

O^{2-} → O_2+2e

Stage 3: outward diffusion and oxidation of Fe

Stage 4: Fe_2O_3 shell and porous NiO core

Kirkendall effect

- Fe
- $\ominus$ Fe_2O_3
- Ni
- NiO

Figure 2.47 Core–shell structure formations in oxidized FeNi.[88] Qin et al. *J. Mater. Chem. A*, **2014**, 2(41), 17511–17520. Reproduced by permission of The Royal Society of Chemistry. A black and white version of this figure will appear in some formats. For the color version, please refer to the plate section.

Ni vacancies $V_{Ni}^{''}$. On the other hand, formation of Fe^{3+} ions occurs through aliovalent substitution which fills the $V_{Ni}^{''}$ and creates negatively charged electrons.

$$e + h = null. \tag{2.3.3}$$

As the electrons consume the holes created during Ni oxidation, as per equation (2.3.3), the reaction shifts to the right side, as per equation (2.3.1), and the Ni oxidation rate is thus further enhanced. Similar to the case of the single Fe system, the Fe atoms move from the particle core to the surface, which has a large concentration of oxygen ions. As the Fe ions are consumed on the surface, there is continuous diffusion of Fe atoms from the core, due to the Fe concentration gradient, and the atoms are ionized to Fe^{3+} during the course of diffusion. As discussed before, outward diffusion of the Fe ions as well as volume expansion occurring during Fe oxidation creates physical space within the micro-particle, which further promotes ionic diffusion. The Fe oxidation proceeds in a way similar to the single Fe system. The third stage involves growing of the Fe_2O_3 nanostructures on the surface and merging them into a continuous layer. It should be noted that Fe has multiple oxidation states and the O^{2-} concentration gradient within the particle leads to the formation of FeO and Fe_3O_4 layers underneath the outermost Fe_2O_3 shell. The fourth stage involves the formation of these underlayers and their conversion into Fe_2O_3 with sufficient oxidation time. It is this sequential series of steps that finally gives the oxidized FeNi particle, with a Fe_2O_3-rich shell and a porous

NiO-rich core. The porous core is a result of the vacancies left behind with the outward diffusion of Fe atoms. The density of the nanowires fluctuates based on the grain size.

In the CuNi binary system, both metals have very similar oxidation rates as well as similar volume expansion during oxidation, which is much lower than that caused during Fe oxidation. As a result, Cu and Ni oxidize almost simultaneously, with very little mechanical stress created during oxidation. Additionally, Cu^{2+} is an isovalent substitution and does not create any electrons or holes while filling the V_{Ni}''. As a result, the oxidation rate of Ni remains unaffected by Cu^{2+} formation, unlike in the FeNi system. The similar oxidation rates of the two metals thus do not promote core–shell structure formation in the CuNi micro-particle. Also, as a result of lower mechanical stress created during oxidation, there is no driving force for nanostructure formation. A second reason for the absence of nanostructures in the oxidized CuNi particle is the lack of any large positive or negative curvatures within individual grains; these play a major role in nanostructure growth for stress-driven mass transport. Micropores in the reduced particles can be explained in a manner similar to that for the Fe system.

In the FeNi and CuNi systems, the presence of the two metals influences the particle morphology during high temperature redox reactions. However, binary transition metal oxides are not formed. Unlike these two systems, oxidation of a FeTi micro-particle leads to the formation of a binary metal oxide complex $FeTiO_3$. Nanobelt formation on the oxidized particle surface and the porous particle center can be explained by the mechanical stress-driven mass transport, as in the other systems mentioned so far. The nanobelt structure suggests an anisotropic growth during the initial stage.[96] The TiO_2 present in the rutile phase in the oxidized particle creates more effective pathways for ionic diffusion than the anatase phase.[103] TiO_2 participates in a solid-state reaction with iron oxide to form the ilmenite phase $FeTiO_3$. Although TiO_2 has been extensively studied, literature on the morphology of $FeTiO_3$ surfaces is limited.[104] The nanostructures disappear over multiple oxidation–reduction cycles, as observed in the other systems. However, the FeTi particle develops a more porous surface over multiple cycles, unlike the other systems which undergo sintering effects.[88] It is difficult to determine the ionic diffusion mechanism in this complex mixed metal oxide system, by experimentation alone. The ionic diffusion mechanism in this system and its influence on particle morphology and surface reaction is better explained with the help of a modeling study, discussed in the following section.

2.4   Ionic Diffusion Behavior Modeling

Numerous chemical reactions and morphological transformations on metal oxide particles take place through the diffusion of ions, as discussed in Section 2.3. Diffusion takes place because of the presence of defects in solids, e.g. vacancies and interstitial ion point defects, which are responsible for lattice diffusion. Diffusion also takes place along surface defects, which include grain boundaries and dislocations. The aim of this section is to provide a fundamental understanding of ionic diffusion behavior at the molecular level using quantum DFT, and at a macroscopic level using the ionic diffusion grain model.

2.4.1   Density Functional Theory (DFT) Calculations

DFT is a quantum mechanical modeling tool, which is widely used in solid physics and physical chemistry for atomic, molecular, and other multiple-body system calculations. DFT is a well-established theory and hence it will not be elaborated on here. In studies of solid-phase ionic transfer mechanisms, DFT is mainly used to calculate diffusion energy barriers of ions in a crystal lattice. The basic methodology is to first create vacancies or possible interstitial sites for the ions of interest to diffuse through.[105] With all the possible diffusion pathways identified, total system energies are calculated, which correspond to the different locations of the ion on its diffusion pathway. The diffusion energy barrier is usually defined as the energy difference of the highest energy state on the pathway and the initial state. Generally, the higher the diffusion energy barrier, the more difficult will be the diffusion of an ion. However, the overall ionic diffusion process is also determined by the availability of interstitial sites and/or vacancies, as discussed previously. Hence, this quantitative diffusion energy barrier can only be used to estimate roughly the dominating ionic diffusion mechanism, when the availability of interstitial sites and/or vacancies is not quantifiable. Application of DFT to the study of reaction mechanisms of some Ca-based and Fe-based chemical looping oxygen carrier particles only began in the late 2000s, due to a lack of recognition of the importance of solid-phase ionic transfer.[81,90,105]

As discussed in previous sections, the initial product layer is formed by surface reaction on the solid-grain surface. Following that, the ionic concentration gradient throughout the particle acts as a driving force for diffusion of the reactant species both into and outwards from the particle core. Because different ions (or ion groups) have different diffusion rates, that are determined by the temperature and the defect concentration, including oxygen vacancies and cation interstitial defects, a favorable diffusion process needs to have a high diffusivity for each individual ion. The overall ionic diffusion mechanism is determined by the diffusion rate of the slowest diffusing ion. For example, in the reaction of CaO and COS, the Ca^{2+} cation has the highest diffusion energy barrier (Figure 2.48), which makes the cocurrent outward diffusion of Ca^{2+} cations and O^{2-} anions a less favorable diffusion process. Therefore, the dominating ionic transfer mechanism in the CaO sulfidation reaction is the countercurrent inward diffusion of S^{2-} anion and outward diffusion of O^{2-} anion.[81] The important factor determining the ion diffusion pathway is the diffusion energy barrier. The diffusion energy barrier can be influenced by several factors, e.g., ion size, crystal structure, and valence state. Ideally, a smaller ion should have a lower diffusion barrier than an ion of larger size, when both have the same charge and crystal structure, such as the O^{2-} anion and S^{2-} anion in Figure 2.48. However, the advantage in size may be limited by structural constraints under certain conditions. For example, in the crystal structure of CaS, diffusion of the Ca^{2+} cation is more difficult than that of the S^{2-} anion, even though the Ca^{2+} cation is much smaller than the S^{2-} anion. In the crystal structure of

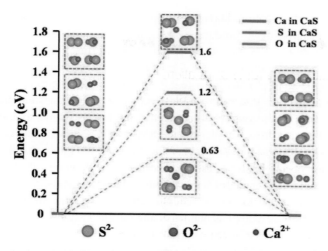

Figure 2.48 The DFT calculated energy barriers of three key diffusing ions in the reaction of COS with CaO.[81] Reproduced with permission from Sun et al. *AIChE J.*, **2012**, 58(8), 2617–2620. Copyright©2011 American Institute of Chemical Engineers (AIChE). A black and white version of this figure will appear in some formats. For the color version, please refer to the plate section.

CaS (rock salt structure), all the S^{2-} anions form close-packed surfaces while Ca^{2+} cations are "loosely" located in the octahedral interstices. In order to diffuse from one site to another, Ca^{2+} cations have to penetrate through the close-packed S^{2-} anion layers. In contrast, the "loosely packed" Ca^{2+} cation layers are much less resistant to S^{2-} anion diffusion.

Based on the diffusion energy barrier obtained for each individual ion, the dominant diffusion pathway can be determined. It is known that ions diffuse in crystal structures through two kinds of defects: (1) vacancies and (2) interstitial point defects. In the case of diffusion through vacancies, ions diffuse through their corresponding vacancies, which are the imperfections in the crystal structure. A good example for this mechanism is that of iron (II) oxide –wüstite. In wüstite, the actual ratio of Fe^{2+} cations and O^{2-} anions is slightly lower than 1:1, which is due to the existence of additional Fe^{3+} cations that create Fe^{2+} vacancies. For undoped metal oxides, the concentration of vacancies is mainly dependent on the solid's intrinsic properties and oxygen partial pressure.[79] But for metal oxides doped with exotic ions, extrinsic vacancies can be created. In the case of interstitial diffusion, ions diffuse through interstitial sites in the crystal structure, which are the intrinsic "free spaces" of the crystal structure. Whether ionic diffusion occurs through a vacancy diffusion mechanism, interstitial diffusion mechanism, or a mix of both, is determined by factors such as the specific structural type, size ratio of cation to anion, and reaction conditions. Generally, if two ions have similar ionic diffusion barriers, their overall diffusion rates are determined by the availability of their diffusion pathway. In the redox cycle of Fe, Ni, Co, Cu, and Mn-based oxygen carrier particles, ionic diffusion involves either the cocurrent diffusion of metallic cations and electrons or the countercurrent diffusion of oxygen anion and electrons. During oxidation of pure Fe, Fe^{x+} cations and electrons diffuse outward from the Fe solid-grain interior through the iron oxide product layer to the solid-grain surface. While in the reduction of pure Fe_2O_3, the formed Fe product layer diffuses from the solid-grain surface to the solid-grain interior in the form of Fe^{x+} and electrons. In some cases, during reduction the solid product layer (e.g. Cu) may not be able to cover the entire solid reactant surface, given that there is a lower molar volume of the solid product than of the solid reactant. This incomplete surface covering leads to direct exposure of the solid reactant sub-layer to the gas phase, which results in a direct chemical reaction. Thus, during reduction, solid conversion via ionic diffusion and direct chemical reaction on uncovered areas occur simultaneously. Experimental work along with DFT calculations help the study of these mechanisms in a comprehensive manner in different metal/metal oxide systems. In the following sections, DFT calculations are used to gain a better understanding of the redox mechanism, using the Fe–Ti system as an example, as well as the mechanism for partial oxidation of methane in iron oxide.

Fe–Ti System Redox Mechanism

In Fe–Ti system, the porous center of the micro-particles is developed due to the outward diffusion of Fe and Ti atoms during oxidation. Ti is oxidized to TiO_2 and participates in a solid-state reaction with iron oxide to form ilmenite phase $FeTiO_3$. After oxidation the iron titanium phases are converted to $FeTiO_3$.[97] Limited studies on $FeTiO_3$ surfaces have revealed that $FeTiO_3$ has a hexagonal close-packed oxygen lattice and its metal atoms occupy two-thirds of the available octahedral sites.[97,105] Additionally, clean $FeTiO_3$ (0001) surfaces are stable and can relax in an unreconstructed termination.[97,104] Along the (0001) direction, $FeTiO_3$ can be regarded as a stack of layers comprised of Fe, Ti, and O. The bulk is based on the higher symmetry corundum structure (space group $R\bar{3}c$). The metal atoms all lie on threefold axes parallel to the c axis, forming a sequence of face-shared pairs of octahedra alternating with vacant octahedral sites in the middle.[97] In the following DFT model, the $FeTiO_3$ bulk is cleaved along the (0001) surface, to represent the surface slab. A 15 Å thick vacuum layer is used to separate the surfaces and their images. The use of periodic models avoids the introduction of edge effects and allows for a more accurate description of surface relaxation. The layers can be described as a recurring sequence of –O–Fe(1)–Fe(2)–O–Ti(1)–Ti(2)–, as shown in Figure 2.49. A semi-infinite slab is used to represent the $FeTiO_3$ (0001) surface. Depending on the cleavage position, six terminations are possible:

(a) O–Fe(1)–Fe(2)–; (b) Fe(1)–Fe(2)–O–; (c) Fe(2)–O–Ti(1)–; (d) O–Ti(1)–Ti(2)–; (e) Ti(1)–Ti(2)–O–; (f) Ti(2)–O–Fe(1)–.[97]

Because the oxidation and reduction of Fe–Ti particles occur in a high temperature process, DFT calculations are combined with statistical mechanics and thermodynamics. DFT total energies and Gibbs free energy $G(T, p)$ calculations are combined to compare the stability of the different surface termination scenarios under experimental conditions. The $FeTiO_3$ (0001) surface energy γ as a function of temperature (T) and pressure (p) can be written as

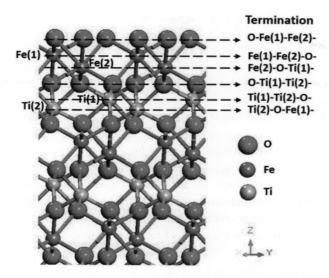

Termination
- O-Fe(1)-Fe(2)-
- Fe(1)-Fe(2)-O-
- Fe(2)-O-Ti(1)-
- O-Ti(1)-Ti(2)-
- Ti(1)-Ti(2)-O-
- Ti(2)-O-Fe(1)-

Figure 2.49 $FeTiO_3$ (0001) surfaces and possible surface termination.[97] Qin et al. *J. Mater. Chem. A*, **2015**, 3(21), 11302–11312. Reproduced by permission of The Royal Society of Chemistry. A black and white version of this figure will appear in some formats. For the color version, please refer to the plate section.

$$\gamma(T,p) = \frac{1}{A}[G(T,p) - N_{Fe}\mu_{Fe} - N_{Ti}\mu_{Ti} - N_O\mu_O], \qquad (2.4.1)$$

where $\mu_{Fe}, \mu_{Ti},$ *and* μ_O are the chemical potentials of Fe, Ti, and O, respectively, A is surface area, and G is given by

$$G(T,p) = E_{DFT} - TS + pV. \qquad (2.4.2)$$

For a solid slab system, the vibrational entropy S_{vib} is the main contribution. It can be obtained directly from the vibrational partition function (X_i), using DFT-estimated frequencies v_i :

$$S_{vib} = \sum_{n=1}^{N}\left[\frac{hv_i}{T}\frac{X_i}{1-X_i} - k_B\ln(1-X_i)\right], \qquad (2.4.3)$$

where $X_i = \exp\left(-{hv_i}/{k_BT}\right)$.
In this slab,
$\mu_{FeTiO_3} = \mu_{Fe} + \mu_{Ti} + 3\mu_O$ and $N_{Fe} = N_{Ti}$.
Hence, the surface energy equation can be simplified to

$$\gamma(T,p) = \frac{1}{A}\left[G(T,p) - \left(N_{Fe}\mu_{FeTiO_3} - 3N_{Fe}\mu_O + N_O\mu_O\right)\right]. \qquad (2.4.4)$$

However, if there are vacancies in the slab, N_{Fe} will be different from N_{Ti}. In that case, Fe_2O_3 can be used as the chemical potential reference for Fe:

$$\mu_{Fe_2O_3} = 2\mu_{Fe} + 3\mu_O. \qquad (2.4.5)$$

The chemical potential of Ti can then be calculated from the μ_{FeTiO_3} expression:

$$\mu_{Ti} = \mu_{FeTiO_3} - 3\mu_O - \left(\frac{\mu_{Fe_2O_3} - 3\mu_O}{2}\right), \qquad (2.4.6)$$

which gives a surface energy based on equation (2.4.1):

$$\gamma(T,p) = \frac{1}{A}\left[G(T,p) - N_{Ti}\mu_{FeTiO_3} - \frac{1}{2}(N_{Fe} - N_{Ti})\mu_{Fe_2O_3} + \frac{3}{2}(N_{Fe} + N_{Ti})\mu_O - N_O\mu_O\right].$$
$$(2.4.7)$$

The chemical potential can be correlated to the actual pressure and temperature conditions based on thermodynamic equilibrium between the surface and the surrounding gas phase (assuming it is an ideal gas):

$$\mu_O(T,p) = \frac{1}{2}\left[E_{O_2,DFT} + \mu_{O_2}^\circ(T,p^\circ) + k_B T \ln\left(\frac{ppO_2}{p^\circ}\right)\right]. \qquad (2.4.8)$$

From the surface energy calculation results, the Ti(1)–Ti(2)–O– termination has the lowest surface energy, 0.211 eV/Å^2, thus it is the most stable FeTiO$_3$ particle surface structure in the oxidation step. The second most stable termination is Fe(1)–Fe(2)–O–, with a surface energy of 0.257 eV/Å^2.[97] The Ti(1)–Ti(2)–O– and Fe(1)–Fe(2)–O– terminations are both nonpolar surfaces with no dipole moment in the repeat unit perpendicular to the FeTiO$_3$ (0001) surface. Therefore, in chemical looping oxidation conditions, the Ti(1)–Ti(2)–O– structure is proposed as the dominant surface termination.

Initial relaxations of the Ti(1)–Ti(2)–O– surface from its cleaved bulk geometry lead to a metastable geometry in which Ti(1) and Ti(2) atoms remain above the first oxygen layer. Substituting one of the surface titanium atoms with one iron atom leads to a less stable geometry with a higher surface free energy.[97]

Understanding the adsorption of the O$_2$ molecule onto the FeTiO$_3$ matrix provides insight into its diffusion pathway. For example, the O–O bond lengths of molecular oxygen are correlated strongly with the binding of O$_2$ to the surface due to the amount of electronic charge transferred to the molecule. The strongest energy of adsorption ($E_{ads} = -0.513$ eV) is observed for O$_2$ adsorbing as a "bridge" between two surface titanium atoms. Upon adsorption, the relaxed O–Ti(1) bonds are found to be slightly elongated to 2.112 Å while the O–Ti(2) bonds are slightly elongated to 1.875 Å. The calculated O–O bond lengths after adsorption are around 1.43 Å, which is larger than the value of the gas phase O–O bond of 1.23 Å.[97]

The oxygen dissociation profile is shown in Figure 2.50. The Ti surface is more favorable for oxygen dissociation than the Fe surface. Therefore, the top Ti layer facilitates oxygen dissociation in order to produce oxygen ion ready for diffusion towards the surface. The stress effect due to thermal oxidation is the driving force for nanobelt growth (discussed in Section 2.3.3). The Ti and Fe ion diffusion has been

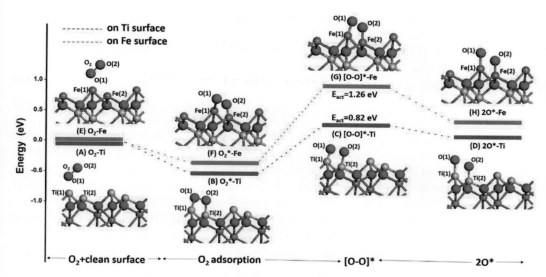

Figure 2.50 O_2 molecule adsorption for Ti–Ti–O and Fe–Fe–O surface terminations.[97] Qin et al. *J. Mater. Chem. A*, **2015**, 3(21), 11302–11312. Reproduced by permission of The Royal Society of Chemistry. A black and white version of this figure will appear in some formats. For the color version, please refer to the plate section.

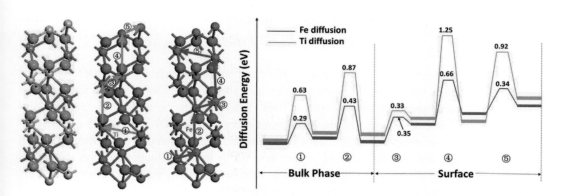

Figure 2.51 Ti and Fe diffusion paths and the associated energies.[97] Qin et al. *J. Mater. Chem. A*, **2015**, 3(21), 11302–11312. Reproduced by permission of The Royal Society of Chemistry. A black and white version of this figure will appear in some formats. For the color version, please refer to the plate section.

investigated further in order to gain a better understanding of nanobelt formation. Diffusion barriers are calculated using CI–NEB. The Ti diffusion and Fe diffusion paths and associated energies are shown in Figure 2.51. From the energy barrier diagram it is observed that the highest barrier for Ti ion diffusion is step 4 with 1.25

eV, while the highest barrier for Fe ion diffusion is 0.66 eV. In fact, except for step 3, the diffusion barrier in all of the other diffusion steps for Fe ions is found to be lower than for Ti ions. This indicates that the outward diffusion of Fe ions is more favorable than Ti ions.[97]

According to the Arrhenius equation, the diffusivity D is

$$D = D_0 \exp\left(E_{a,diffusion}/RT\right), \tag{2.4.9}$$

where D_0 is the ionic diffusivity constant, R is the gas constant, and $E_{a,\text{diffusion}}$ is the highest activation energy for the diffusion. Using equation (2.4.9), the diffusivities at 1173 K of Ti ion and Fe ion are calculated to be 1.35×10^{-10} m²/s and 2.21×10^{-7} m²/s, respectively. Therefore, Fe ion diffusivity is three orders of magnitude larger than Ti ion diffusivity, which serves to explain the formation of just Fe_2O_3 nanobelts.

During the reduction step for the Fe–Ti system being discussed, lattice oxygen is released to the surface where it reacts with the reducing gas which leads to oxygen vacancy formation. The formation energy of the vacancies is determined as follows:

$$E\left(V_{\ddot{O}}\right) = E\left(FeTiO_3, V_{\ddot{O}}\right) - E(FeTiO_3) + \frac{1}{2}E(O_2), \tag{2.4.10}$$

where $E(FeTiO_3)$ is the energy of $FeTiO_3$, $E(FeTiO_3, V_{\ddot{O}})$ is the energy of $FeTiO_3$ with a vacancy, and $E(O_2)$ is the energy of molecular oxygen.

By replacing the energy of the oxygen molecule in equation (2.4.10) with the oxygen chemical potential, the vacancy formation energies at temperatures relevant to chemical looping can be calculated by

$$\Delta G\left(V_{\ddot{O}}\right) = E\left(FeTiO_3, V_{\ddot{O}}\right) - E(FeTiO_3) + \frac{1}{2}\mu_O(T,p). \tag{2.4.11}$$

These vacancies are created at the surface and on the subsurface. When the temperature increases, the vacancy formation energy increases. Vacancy formation on the surface is thermodynamically more favorable than vacancy formation on the subsurface. Compared with Fe_2O_3, the vacancy formation energy on $FeTiO_3$ is about 0.5 eV lower.[97] This comparison indicates that the formation and aggregation of oxygen vacancies ($V_{\ddot{O}}$) through outward diffusion of oxygen ions (O^{2-}) during reduction to form the porous surface is easier on the $FeTiO_3$ particle than on the Fe_2O_3 particle. Consequently, the ionic diffusivity is better in the $FeTiO_3$ particle.

The redox process is illustrated in Figure 2.52. To better understand the evolution of the morphology, consider an infinitesimal element of the micro-particle, as indicated by the dotted frame. The surface of this element can be approximated as being flat. The driving force of the protrusion is volume expansion during oxidation. The diffusion of Fe ions is faster than the Ti ions. Consequently, the nanobelts formed during oxidation are mainly Fe_2O_3, with Ti remaining in the micro-particles. Reduction causes oxygen vacancy generation in the $FeTiO_3$ phase which is substantially easier than in Fe_2O_3. The vacancy-induced pore formation explains the superior recyclability of Fe–Ti micro-particles over pure Fe micro-particles.[97]

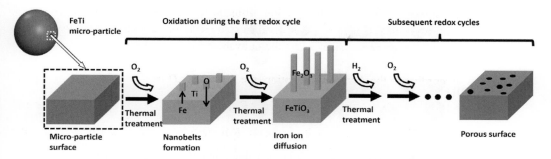

Figure 2.52 Illustration of the redox process and the morphological changes in the Fe−Ti system over multiple cycles.[97] Qin et al. *J. Mater. Chem. A*, **2015**, 3(21), 11302–11312. Reproduced by permission of The Royal Society of Chemistry.

With the diffusion mechanisms responsible for cation diffusion in metal oxide particles discussed for the Fe–Ti system, in moving forward, the principal mechanisms responsible for oxygen diffusion are examined by determining the migration energy and other temperature independent variables, using a similar approach to that for cation diffusion.

Because of the availability of a convenient radioactive tracer isotope of oxygen (^{18}O), there have been some experimental investigations into the diffusion of oxygen in metal oxides. Ando and Oishi[106] conducted radioactive tracer experiments and determined an activation energy of 439 kJ/mol (4.55 eV) for oxygen self-diffusion in a single $MgO \cdot nAl_2O_3$ crystal. Furthermore, they observed only a very slight increase in the activation energy, 443 kJ/mol (4.59 eV), for oxygen transport in a single $MgAl_2O_4$ crystal with Al_2O_3 support, thus implying that the concentration of the mediating oxygen defects is unaffected by incorporation of foreign cations like Al^{3+}; in this case, Oishi and Ando[107] repeated their diffusion measurement experiment on a polycrystalline sample as opposed to a single crystal and they obtained an activation energy of 384 kJ/mol (3.99 eV), which demonstrated the importance of processes such as grain boundary diffusion. By conducting depth profiling experiments on a metal oxide single crystal, Reddy and Cooper[108] obtained oxygen diffusion barriers of ~370 kJ/mol for MgO and ~405 kJ/mol for α-Fe_2O_3. In addition to the activation energy, they also reported a value for the pre-exponential term, D_0, in equation (2.4.9) of about 2.2×10^{-7} m^2/s. Clearly, there is good agreement between the experimental data on the oxygen transport in metal oxide particles. However, due to experimental technique limitations, the degree of inversion present in the crystals used is not reported in any of these studies, thus making a detailed oxygen diffusion process unclear. Uberuaga et al.[109] used the atomistic simulation TAD technique to determine the migration energies for the intrinsic defect processes capable of facilitating oxygen diffusion in $MgAl_2O_4$, $MgGa_2O_4$, and $MgIn_2O_4$. Their work showed that the migration energy for oxygen interstitial migration is lower than for vacancy mediated migration. Erhart and Albe[110] performed ab initio

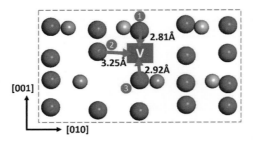

Figure 2.53 Cross-section through FeTiO$_3$ bulk unit cell in the (100) plane, where red spheres: O anions, purple spheres: Fe cations, gray spheres: Fe cations. A black and white version of this figure will appear in some formats. For the color version, please refer to the plate section.

calculations on diffusion of oxygen in zinc oxide and found a migration energy of 0.89 eV for migration of the charge neutral vacancy defects. These values are not directly comparable with the experimental observations made above, as they only represent the migration energy and do not include the energy to form the defects. However, they are useful in terms of exploring the mechanism responsible for oxygen self-diffusion in metal oxide particles.

To further understand oxygen diffusion leading to morphological transformations in metal/metal oxide systems, the Fe–Ti system has been considered for a more detailed study. The energy barriers involved in oxygen ion diffusion are analyzed for the vacancy and the interstitial mechanism. As described previously, FeTiO$_3$ bulk is based on the higher symmetry corundum structure (space group $R\bar{3}c$). The metal atoms all lie on threefold axes parallel to the c axis, forming the sequence of face-shared pairs of octahedra alternating with vacant octahedral sites in the middle.[106] Each oxygen site in FeTiO$_3$ is surrounded by 12 nearest neighboring oxygen sites, therefore, should an oxygen ion be removed to create an oxygen vacancy, $V_{\ddot{O}}$, it is possible for any of these 12 oxygen anions to hop onto the vacant site. However, after removal of the oxygen ion, these 12 anions do not all reside at equivalent distances from the existing vacant site due to relaxation of the cations away from the tetrahedral sites and contraction on the octahedral lattice sites. The DFT optimized model shows that four nearest neighboring oxygen atoms reside at a distance of 2.81 Å from the vacancy site, a further four at a distance of 3.25 Å, and the remainder at 2.92 Å. Figure 2.53 presents a cross-section through the FeTiO$_3$ bulk unit cell in the (100) plane and shows an example of three different diffusion paths for vacancy mechanisms as a result of oxygen positional parameter displacement. Here the red spheres represent oxygen anions while the purple spheres and the gray spheres represent iron cations and titanium cations, respectively. Thus, there are three distinct groups of nearest neighboring oxygen sites, based on their separation from any given vacant oxygen site. The blue arrows therefore represent examples of each of three different Vo migration mechanisms. The distance the oxygen anion has to migrate is 2.81 Å (path 1), 3.25 Å (path 2), or 2.92 Å (path 3) for diffusion into the nearest vacancy site.

The change in the energy of the system, ΔE_{system}, as a function of the reaction coordinate for the vacancy mechanisms is shown in Figure 2.54. Figure 2.54(1) shows

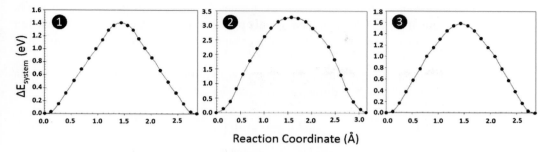

Figure 2.54 Change in the energy of the FeTiO$_3$ bulk system, ΔE_{system}, as a function of the reaction coordinate for the oxygen vacancy migration paths 1, 2, and 3, shown in Figure 2.53.

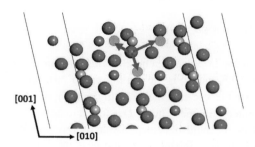

Figure 2.55 Schematic illustration of oxygen interstitial diffusion in FeTiO$_3$ bulk, where red spheres: O anions, purple spheres: Fe cations, gray spheres: Fe cations, dotted red spheres: Fe–Ti interstitial site. A black and white version of this figure will appear in some formats. For the color version, please refer to the plate section.

the migration energy for an oxygen anion to hop 2.81 Å into the vacancy site by path 1. The activation energy for this path is 1.37 eV. By making repeated identical jumps this mechanism is capable of facilitating oxygen diffusion through the lattice. The migration energies as a function of the reaction coordinates for paths 2 and 3 are shown in Figure 2.54(2) and Figure 2.54(3), respectively. The activation energies are 3.36 eV for path 2 and 1.52 eV for path 3. Consequently, the overall diffusion energy for an oxygen ion migrating via a vacancy mechanism is predicted to be 1.37 eV.

If an oxygen atom on an interstitial site moves to one of the neighboring interstitial sites, the diffusion occurs by an interstitial mechanism. This process is shown schematically in Figure 2.55. Such a movement or jump of the interstitial oxygen atom involves a considerable distortion of the lattice. For interstitial migration to occur, one of the oxygen interstitials from the Fe–O$_i$–Ti defect prefers to diffuse to the next Fe–Ti interstitial site. The migration distances all fall within 2.09–3.03 Å. However, the mechanism assumes movement of the oxygen interstitial defect in a one-step diffusion. A migration energy of 0.21 eV was obtained from the most favorable interstitial diffusion path. It is difficult to determine the diffusion barrier if considering random oxygen interstitial defects moving through one-step diffusion. If assuming two neighboring oxygen interstitial defects diffusing in one step, the predicted migration energy is 0.47 eV greater in energy than the one O_i diffusion. Therefore, it is expected that there is only one O_i to diffuse in each step within a unit cell, which can represent the principal mechanism for oxygen interstitial diffusion in metal oxide.

In order to determine which of the two oxygen diffusion mechanisms is dominant in the $FeTiO_3$ bulk, oxygen diffusivity, D_{O_2}, can be calculated from the equation

$$D_{O_2} = D^* \exp\left(\frac{-E_a}{k_B T}\right), \tag{2.4.12}$$

where D^* is the diffusion pre-exponential factor expressed by equation (1.3.5) in Section 1.3.3, k_B is the Boltzmann constant, and E_a is the energy barrier for the diffusion. The oxygen diffusivity, D_{O_2}, is also given by equation:

$$D_{O_2} = \varphi d_m^2 \exp\left(\frac{-E_a}{k_B T}\right), \tag{2.4.13}$$

where φ is the rate coefficient of ion hopping to a neighboring site and can be evaluated by equation (1.3.6) in Section 1.3.3 and d_m is the ion migration distance.[106,111]

At 1000 K, for the most favorable vacancy mechanism (path 1), the diffusivity D is 2.58×10^{-8} cm^2/s, and for the interstitial mechanism D is found to be 8.07×10^{-6} cm^2/s, which suggests that the interstitial mechanism is responsible for oxygen self-diffusion in the $FeTiO_3$ bulk. Similar diffusion mechanisms are found in Fe_2O_3 bulk. Based on the oxygen diffusivity values for the vacancy and interstitial mechanisms, it can be concluded that the O_i defect is the more mobile point defect compared to oxygen vacancy in iron based metal oxide. Using this approach, the mechanism of gas–solid reactions with lattice oxygen diffusion can be further explored. Partial oxidation of methane, as an important reaction in chemical looping for the production of high valued chemicals and liquid fuels, is studied and described below.

Methane Partial Oxidation Mechanism

The partial oxidation of CH_4 on a metal oxide follows a different mechanism from catalytic partial oxidation. To understand the mechanism of partial oxidation of CH_4 on metal oxide surfaces, iron oxide (Fe_2O_3) can be looked at as a model system.[112] The DFT simulation methods used have already been explained in Section 2.4.1. On the surface layer of Fe_2O_3, five different sites for adsorption can be identified; see Figure 2.56. The five sites are as follows:

(a)   Fe atop (on top of the Fe site on the surface layer)
(b)   Fe–O bridge (between Fe and O on the surface layer)
(c)   O atop (on top of the O site on the surface layer)
(d)   O vacancy (on an oxygen vacancy site on the surface layer)
(e)   O atop second layer (on top of the O from the subsurface layer)

Of these five identified sites, the Fe atop site has the highest affinity for CH_4 adsorption. In other words, the adsorption energy released by CH_4 on this site is the highest.

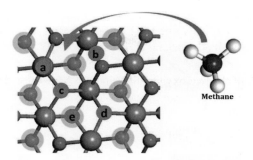

Figure 2.56 Methane adsorption sites on the surface layer of Fe_2O_3 system (top view). A black and white version of this figure will appear in some formats. For the color version, please refer to the plate section.

Therefore, Fe on the surface is the most favorable site for adsorption, and therefore, the preferred site for CH_4 adsorption in this gas–solid system.

If oxygen vacancies exist adjacent to the Fe site of adsorption, the CH_4 adsorption on such sites is weaker. In other words, for Fe sites with adjacent vacancies, the adsorption energy released upon CH_4 adsorption is lower, due to lower available electron density. Thus, the probability of CH_4 adsorption on a vacancy-adjacent Fe site is lower than that on a Fe site with a perfect crystal structure.

Therefore on a Fe_2O_3 system, CH_4 adsorbs on the Fe site, preferably one without any adjacent O vacancy. Based on DFT calculations, the following mechanism is proposed for CH_4 oxidation on the Fe_2O_3 system. In the absence of adjacent vacancies, CH_4 is expected to undergo complete oxidation to CO_2 and H_2O by utilizing adjoining O atoms from the surface layer. This pathway for CH_4 oxidation may be termed "path 1." The vacancies formed as a result are subsequently filled by lattice oxygen diffusion to the surface. Although this pathway has been studied in detail, it is not of relevance to this discussion. Instead, the pathway for partial oxidation of CH_4 to syngas is of interest here. The pathway, termed as "path 2," has been represented in Figure 2.57.

Even though the CH_4 molecule has the strongest adsorption affinity towards the Fe site on the surface, the methyl (CH_3) group, formed after dehydrogenation of CH_4, shows preferential adsorption towards the oxygen vacancy site, while the methylene (CH_2) and methine (CH) groups show preferential adsorption towards both the Fe site and the oxygen vacancy site.[113] In the event of CH_4-rich reaction conditions, the CH_4 molecules have a higher probability of adsorbing onto vacancy-adjacent Fe sites. Upon the first dehydrogenation, the CH_3 fragment formed is transferred to the neighboring oxygen vacancy site, whereas H remains on the original Fe adsorption site. Thereafter, the C–H bonds are successively cleaved at this vacancy site, and the protons are adsorbed onto the neighboring Fe sites. Once an Fe site has two H atoms, a covalent bond is formed between the two and the resulting H_2 molecule is released from the site in gaseous form. After the last C–H bond is cleaved, the lattice oxygen diffuses from the subsurface to the surface vacancy site, to form a C–O complex. The C–O complex transfers to the neighboring Fe site, from which it dissociates as a gaseous CO molecule. This pathway results in partial oxidation of methane. Figure 2.58 shows the energy

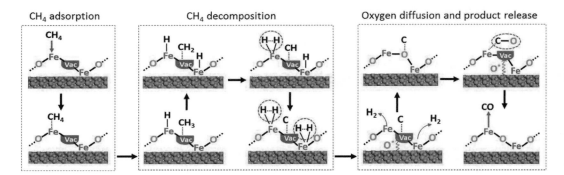

Figure 2.57 Proposed mechanism of CH_4 partial oxidation to syngas on Fe_2O_3 system based on DFT calculations. A black and white version of this figure will appear in some formats. For the color version, please refer to the plate section.

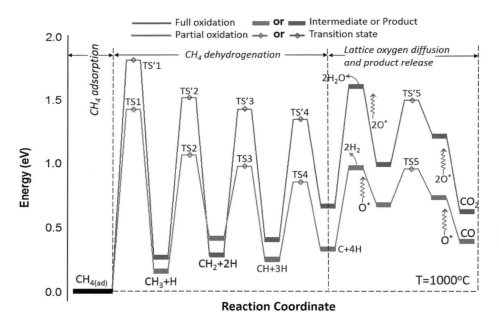

Figure 2.58 CH_4 oxidation energy profile through full oxidation (pathway 1) and partial oxidation (pathway 2).[112] A black and white version of this figure will appear in some formats. For the color version, please refer to the plate section.

levels for the two different pathways – paths 1 and 2, leading to complete and partial oxidation of CH_4, respectively.

In CH_4-lean conditions, path 1 is the preferred pathway wherein CH_4 molecules always preferentially adsorb onto the Fe sites with no neighboring vacancy, or in other words, in the perfect crystal structure region. This results in complete oxidation of the CH_4 molecule to CO_2 and H_2O. However, in CH_4-rich conditions, path 2 is the dominating pathway because of lower reaction barriers resulting in partial oxidation to CO and H_2. These reaction mechanisms support the resulting gas compositions predicted by classical thermodynamics (Figure 2.6). Further, experimental data[39] obtained at varying values of the CH_4 to Fe_2O_3 ratio also serves as additional proof of the proposed mechanisms.

2.4.2   Ionic Diffusion Grain Model

The mechanism of ionic diffusion in oxygen carriers at the molecular level has been explored in Section 2.4.1, using DFT. In this section, the mechanism of ionic diffusion at the macroscopic level is studied further using the grain model for a more comprehensive understanding of ionic diffusion behavior and related reaction kinetics.

Many models for gas–solid reactions have been developed for practical use over the past several decades. The majority of models have been based on the assumption of a nonporous solid reactant, among which a typical representative is the shrinking core model.[113,114] This type of model leads to sharp interfaces between the reacted zone and unreacted zone in the solids. However, in reality, some of the solids are porous, and there is a gradual change in the gas profile throughout the pellets. The structure of the grain inside the pellet has been investigated over several decades. For a detailed microstructure, one can refer to the images captured using SEM by Turkdogan and Vinters.[115] The concept of grain structure has been taken into account by some previous models.[116–122] The ionic diffusion mechanism inside grains has been studied recently and found to be influential on gas–solid reaction kinetics.[82,123] Vargas[123] has developed an oxygen carrier reaction model, in which the ionic diffusion mechanism was applied to entire pellets, not individual grains. The continuum assumption is used for pellets, which ignores the grain structure inside pellets.

An improved grain model was proposed by Zeng[124] to represent the reaction between iron oxide pellets and reducing gases that considered ionic diffusion and gas–solid reaction kinetics. In this model, iron oxide pellets are assumed to be composed of spherical micro-grains of small, constant radius. Structural changes in individual pellets during reduction are negligible. Diffusion of the oxygen ions is taken into account within the crystal lattice. Chemical reaction is considered to take place only at the outermost surface of each grain. Three resistances are considered in this model, which determine the overall reaction progress, including pore diffusion of gases in pellets, oxygen ion diffusion in grains, and chemical reaction on grain surfaces. The model developed, based on the idea depicted above, will be called the "ionic diffusion grain model or IDGM" hereafter.

In the following section, this IDGM is elaborated through its application on a specific reaction of iron oxide reduction with hydrogen and/or carbon monoxide, which is of special interest in chemical looping applications.

Assumptions for the ionic diffusion grain model are as follows:

1.   The reduction is an isothermal process.
2.   The pellet is composed of spherical micro-grains with a small, constant radius.
3.   Iron atoms form the grain lattice, which allows oxygen ions to diffuse out.

4.   There is no structural change of the pellet during the reduction.
5.   A reversible chemical reaction takes place at the outermost surface of each grain and the solid product formed is assumed to have negligible oxygen ion concentration.
6.   The concentrations of bulk flow do not change due to diffusion through the pellet.
7.   The resistance of the reaction due to the gas film around the pellet is negligible.

Governing Equations and Analysis on Limiting Cases of the IDGM

Generally, consider a solid–gas reaction, represented by

$$y_a A_{(g)} + B_{(s)} \leftrightarrow y_c C_{(g)} + D_{(s)}.$$

A material balance for gaseous reactant A gives

$$\varepsilon \frac{\partial C_A^*}{\partial t^*} = \nabla^* \cdot \left(D_{eA}^* \nabla^* C_A^* \right) + \frac{y_a}{\alpha_C} \frac{\partial C_B^*}{\partial t^*}. \tag{2.4.14}$$

Similarly, for gaseous product C, it yields

$$\varepsilon \frac{\partial C_C^*}{\partial t^*} = \nabla^* \cdot \left(D_{eC}^* \nabla^* C_C^* \right) - \frac{y_c}{\alpha_C} \frac{\partial C_B^*}{\partial t^*}, \tag{2.4.15}$$

where ∇^* is a dimensionless gradient operator relative to the particle radius, R'.
   The boundary conditions are

$$\nabla^* C_A^* = \nabla^* C_C^* = 0 \ (r^* = 0) \ \text{ and } C_A^* = C_{A0}^*, C_C^* = C_{C0}^* (r^* = 1).$$

The initial conditions for the gases are

$$C_A^* = 0, \ \ C_C^* = 0 \ (0 \leq r^* < 1, t^* = 0).$$

The initial condition for the solid is

$$C_B^* = C_{B0}^* \ (t^* = 0).$$

Assuming the solid particle is composed of small but highly dense grains, each of which reacts according to the ionic diffusion model proposed by Vargas[123], it yields

$$\frac{\partial C_o^*}{\partial t^*} = \alpha_L^2 \nabla_o^* \cdot \left(D_o^* \nabla_o^* C_o^* \right), \tag{2.4.16}$$

where ∇_o^* is a dimensionless gradient operator relative to the grain radius, R_o. The corresponding boundary conditions for each grain are given as

$$\nabla_o^* C_C^* = \nabla_o^* C_A^* = 0 \ (r_o{}^* = 0)$$

$$-D_o^* \nabla_o^* C_o^* = \left(\frac{\alpha_C}{\alpha_L} \right) k_s^* C_o^{*\beta} \left(f_A^{y_a} - \frac{f_C^{y_c}}{K_e^* C_o^{*\beta}} \right) (r_o^* = 1).$$

The initial condition for each grain is

$$C_o^* = 1 \ (t^* = 0).$$

All the quantities with asterisks are dimensionless. Gas concentrations are undimensionalized by C_{gT}, the overall gas concentration. Solid concentrations and ion concentrations are normalized by the unreacted concentration of the diffusing ion in solid B, which can be denoted by $C_{o,B}$. For the iron oxide reduction, B denotes Fe_2O_3 and thus $C_{o,B}$ becomes C_{o,Fe_2O_3}. The effective pore diffusivities and ionic diffusivities are normalized by (R'^2/τ), where τ is the characteristic time of pore diffusion. Some important dimensionless variables are defined as follows:

$$\alpha_L = R'/R_o$$
$$\alpha_C = C_{gT}/C_{o,B}$$

$$t^* = t/\tau$$

$$r^* = r/R'$$

$$r_0^* = r_o/R_o$$

$$D_{eA}^* = D_{eA}/(R'^2/\tau)$$

$$D_{eC}^* = D_{eC}/(R'^2/\tau)$$

$$D_o^* = D_o/(R'^2/\tau)$$

$$k_s^* = \frac{k_s \tau C_{o,B}^\beta C_{gT}^{y_a-1}}{R'}$$

$$C_o^* = C_o/C_{o,B}$$

$$K_e^* = K_e C_{gT}^{y_a-y_c} C_{o,B}^\beta.$$

For the case where the ionic diffusion in the grain is fast relative to chemical reaction on the grain surface, the ionic distribution within the grain is nearly uniform. Under this condition, the reaction rate for a grain is modeled using the unreacted-core shrinking model. Thus the change in oxygen ion concentration for a single grain yields

$$-\frac{\partial C_o^*}{\partial t^*} = 3\alpha_L \alpha_C k_s^* C_o^{*\beta}\left(f_A^{y_a} - \frac{f_C^{y_c}}{K_e^* C_o^{*\beta}}\right).$$

The rate of concentration change for the entire particle can therefore be expressed by equation (2.4.17),

$$-\frac{\partial C_B^*}{\partial t^*} = 3(1-\varepsilon)(\alpha_C \alpha_L)\frac{(k_s^* C_o^{*\beta})}{d_{o,B}}\left(f_A^{y_a} - \frac{f_C^{y_c}}{K_e^* C_o^{*\beta}}\right). \tag{2.4.17}$$

Consecutively, equation (2.4.14) and equation (2.4.15) can be modified to the following equations:

$$\varepsilon\left(\frac{\partial C_A^*}{\partial t^*}\right) = \nabla^* C(D_{eA}^* \nabla^* C_A^*) - 3(1-\varepsilon)y_a \alpha_L \frac{k_s^* C_o^{*\beta}}{d_{o,B}}\left(f_A^{y_a} - \frac{f_C^{y_c}}{K_e^* C_o^{*\beta}}\right) \tag{2.4.18}$$

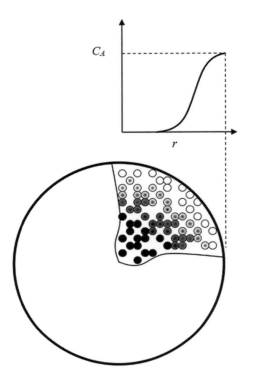

Figure 2.59 Representation of the ionic diffusion grain model for the reaction of a porous solid pellet with a gas, where C_A is the concentration of gaseous reactant, and r is the radius of the pellet.

$$\varepsilon\left(\frac{\partial C_C^*}{\partial t^*}\right) = \nabla^* C\left(D_{eC}^* \nabla^* C_C^*\right) + 3(1-\varepsilon)y_c\alpha_L \frac{k_s^* C_o^{*\beta}}{d_{o,B}}\left(f_A^{y_a} - \frac{f_C^{y_c}}{K_e^* C_o^{*\beta}}\right) \qquad (2.4.19)$$

An important dimensionless number can be defined as

$$\theta = \alpha_L k_s^* / D_{eA}^* = \alpha_L k_s R' C_{o,B}^\beta C_{gT}^{y_a-1} / D_{eA}.$$

This parameter, the ratio between the chemical reaction rate constant and the gas diffusivity in pellets, determines the property of the overall reaction front in the pellet. Generally, if $\theta \gg 1$, the reaction is controlled by pore diffusion. If $\theta \ll 1$, the reaction is controlled by the chemical reaction. If $\theta \approx 1$, the reaction is controlled by a combination of pore diffusion and chemical reaction.

For the reaction on the outermost surface of each grain, the following parameter is defined: $\psi = k_s^* a_C / (a_L D_o^*) = k_s R' C_{gT}^{y_a-1} C_{o,B}^\beta a_C / (a_L D_o)$. This parameter is the ratio between the chemical reaction rate constant and the diffusivity of the ions. Generally, if $\psi \gg 1$, the reaction is controlled by ion diffusion. If $\psi \ll 1$, the reaction is controlled by the chemical reaction. If $\psi \approx 1$, the reaction is mixed, controlled by both the ion diffusion and the chemical reaction.

Considering the effects of both the parameters θ and ψ, four limiting cases can be defined:

(1)   When $\theta \gg 1, \psi \gg 1$: the reaction is controlled by pore diffusion in pellets and ionic diffusion in grains. The typical reaction state under this condition can be illustrated by Figure 2.60a. In this case, a sharp interface exists between the reacted and the unreacted zones. Also, the grains in the reacted zones are not fully reacted since ionic diffusion is too slow to supply enough ions on the outermost

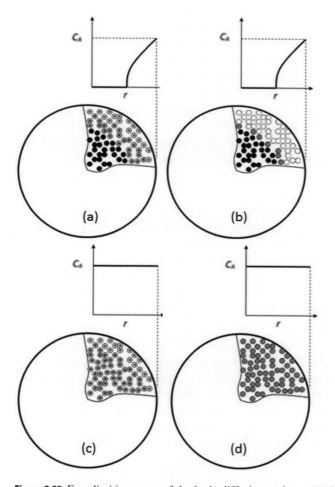

Figure 2.60 Four limiting cases of the ionic diffusion grain model for reaction of a porous solid with a gas. (a) $\theta \gg 1, \psi \gg 1$, the reaction is controlled by both pore diffusion in pellets and ion diffusion in grains; (b) $\theta \gg 1, \psi \ll 1$, the reaction is controlled by both pore diffusion in pellets and the chemical reaction on grain surfaces; (c) $\theta \ll 1, \psi \gg 1$, the reaction is controlled by both the chemical reaction on grain surfaces and ion diffusion in grains; (d) $\theta \ll 1, \psi \ll 1$, the reaction is controlled only by the chemical reaction on grain surfaces.

surfaces of the grains. The core of each grain is still not reacted. To obtain the solution of this case, for ion concentration in grains, equation (2.4.16), and for gases, equations (2.4.14) and (2.4.15) should be solved simultaneously using numerical methods since there is no ready analytical solution for them.

(2)   When $\theta \gg 1, \psi \ll 1$: the reaction is controlled by pore diffusion in pellets and the chemical reaction on grain surfaces. The typical reaction state under this condition can be illustrated by Figure 2.60b. There is a sharp interface between the reacted and unreacted zones. In the reacted zone, the grains are fully converted from the surface to the innermost core. This is because the ion diffusion is so fast that it not only supplies the surface reaction but can also maintain a uniform concentration of ions throughout the grain. The scenario of this case is very similar to that represented by an unreacted shrinking core model.

(3)   When $\theta \ll 1, \psi \gg 1$: the reaction is controlled by chemical reaction on grain surfaces and ionic diffusion in grains. The typical reaction state under this condition can be illustrated by Figure 2.60c. $\theta \ll 1$ indicates that the pore diffusion is a relative fast process. This makes the concentration of the reducing gases uniform throughout the pellet. Therefore, there will be no visible reaction front inside the pellet. In this case, the size of the pellet does not affect the reaction process at all. The only things that really matter are the size of the grain and the ionic diffusivity. In other words, the concept that each pellet is composed of a great number of grains breaks down here. This means every grain can be treated as an individual continuum pellet. The solution for this case can be obtained by solving equation (2.4.16) alone. An analytical solution can also be obtained under some particular condition, e.g., if the reaction is first order with respect to the solid reactant concentration, i.e. $\beta = 1$. For details of methods for analytical solution, one can refer to the work done by Vargas.[123]

(4)   When $\theta \ll 1, \psi \ll 1$: the reaction is controlled only by a chemical reaction on the grain surfaces. The typical reaction state under this condition can be illustrated by Figure 2.60d. This case is very similar to case (3). The only difference is that ionic diffusion is also a fast process. The concentration of the diffusing ions in grains will be kept uniform throughout every grain. This will make solving of the procedure for the gas–solid reaction even simpler and easier. Each grain can be treated as a continuum pellet. The chemical reaction rate will be the only parameter that matters since the concentration of the diffusing ions is always uniform. For details on solving methods, one can refer to the work done by Szekely et al.[120], in which the problem of this case is classified as the "volumetric reaction of a highly porous solid."

The experimental data obtained by Turkdogan and Vinters[115] are used to test the IDGM. From Figure 2.61, it can be seen that there is good agreement between the numerical results and the experimental results. The case with the largest diameter gives the largest deviations. This may be due to the fact that large pellets have a more complex structure which cannot be mimicked by spherical grains. The pore diffusivities of hydrogen found for the pellets of different sizes are listed in Table 2.2. The pore diffusivities are comparable with the measurements performed by Turkdogan and

Table 2.2 The diffusivities of hydrogen calculated for pellets in experiments performed by Turkdogan and Vinters.[115]

	d = 0.8 mm	d = 1.8 mm	d = 3.6 mm	d = 8.0 mm
D_{eA}	1.1×10^{-4}	4.3×10^{-5}	5.8×10^{-5}	7.1×10^{-5}

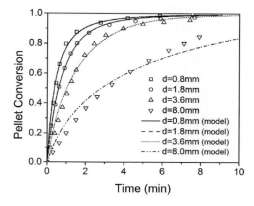

Figure 2.61 Comparison between results from the IDGM and experimental data from the work of Ishida and Wen.[122]

Vinters.[115] The diffusivity in the pellet with d = 0.8 mm does not follow the trend established by the diffusivities in other pellets. This may be because the assumption that the grain sizes are the same in pellets with different sizes does not hold for very small pellets in these experiments. It is intuitive that pellets with different sizes may have different micro-structures. However, the assumption that the grain sizes are the same in pellets with different sizes can be very useful for prediction of the conversion profile without detailed knowledge of the micro-structures.

The IDGM is also used to fit the experimental results obtained at OSU. It can be seen that the model results are in good agreement with the experimental results (Figure 2.62). The pore diffusivities of hydrogen found for pellets of different sizes are listed in Table 2.3. Clearly, the results indicate that the pore diffusivity of hydrogen increases as the pellet size increases. This trend qualitatively matches the measurements performed by Turkdogan and Vinters.[115] Also it is seen that the pore diffusivities of the pellets presented in Table 2.3 are much smaller than those from the work done by Uberuaga et al.[109] The smaller pellets developed at OSU are found to be stronger and have proven their excellent recyclability during long-term testing.

To sum up, in this IDGM model, iron oxide pellets are assumed to be composed of spherical micro-grains with a small constant radius. Structural changes of individual pellets during the reduction are negligible. The diffusion of oxygen ions is taken into account within the grain lattice defined by iron atoms. Chemical reaction is considered to take place only at the outermost surface of each grain. Three resistances to the overall

Table 2.3 The diffusivities of hydrogen in pellets used in experiments for Figure 2.62.

	0 mm < d < 0.5 mm	1.2 mm < d < 1.6 mm	2.0 mm < d < 2.8 mm	2.8 mm < d < 3.4 mm
D_{eA}	3.5×10^{-7}	5.1×10^{-7}	1.1×10^{-6}	1.4×10^{-6}

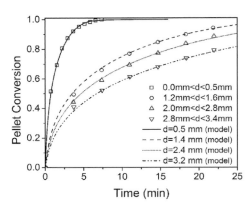

Figure 2.62 Comparison between results from the IDGM and experimental data produced at OSU.

progress of reaction are considered in this model: pore diffusion of gases in pellets, oxygen ion diffusion in grains, and chemical reaction on grain surfaces. Four limiting cases are discussed thoroughly when one or both resistances are dominant. When the ratio between chemical reaction rate constant and diffusivity of the diffusing ions is very small, the governing equations for the present model can be significantly simplified. Numerical solutions can be used to validate the proposed model by means of fitting available experimental data. This model is found to be satisfactory for the interpretation of two individual sets of experimental results. It should be mentioned that these two validation cases both fall into the limiting case (2), where the reaction is controlled by both pore diffusion in pellets and chemical reaction on grain surfaces. From a macroscopic point of view, ionic diffusion is a slow process. However, with respect to the reaction inside a grain, the relative importance of ionic diffusion in the rate process is severely reduced by the size of the grain, which is usually more than three orders of magnitude smaller than a pellet. The dominance of ionic diffusion will appear in reactions in large dense bulk bodies, or in small pellets but with large grains.[82] It is also conjectured that ionic diffusion will become a controlling factor in the oxidation process of metal pellets since grains become denser and larger during reactions. This requires further experimental investigation as well as model refinements.

The effect of ionic diffusion on oxygen carrier morphology and reaction dynamics is examined experimentally and theoretically in Sections 2.3 and 2.4, respectively. In Section 2.5, thermodynamic assessment of oxygen carriers for syngas production applications is focused upon. This assessment is important to the determination of the selectivity of an oxygen carrier towards syngas production, which is a crucial aspect of

CLPO processes. Using methane as feedstock, a high selectivity towards syngas reduces the CH_4 consumption and also minimizes the processing steps for downstream product yield.

2.5   Phase Diagrams for Oxygen Carriers

Phase diagrams have been traditionally applied in the manufacturing and processing industries of alloys and ceramics, where control of the material composition is essential. Generally, a plot of temperature (T) versus composition is used to describe the phase diagram, while the effect of total pressures is not usually considered as a variable as they do not appreciably affect the phases of the liquid and the solid. Phase diagrams provide composition and phase information on the material that is in equilibrium under the given conditions of interest to the operating systems of the process. For example, a temperature versus composition phase diagram for a Fe–C system is used in the steel industry to discern the required C content to exhibit desirable physical and chemical properties in the final steel product. Phase diagrams for a Na–Fe–O–H–C system are used to understand the effect of CO_2 and H_2O on the Na–Fe oxides on the corrosion mechanism in fast breeder nuclear reactors.[125] Phase diagrams are also used to design the metal oxide oxygen carrier material for applications in chemical looping combustion (CLC) of gaseous fuels,[4] CLOU reactions of solid fuels,[126–129] and thermochemical water splitting for hydrogen generation.[130]

The phase diagrams used for chemical looping applications are generally expressed in terms of temperature (T) versus partial pressure of oxygen (ppO_2), T versus composition of oxygen carrier material, and partial pressure ratios of H_2O:H_2 (ppH_2O/ppH_2) or CO_2:CO ($ppCO_2/ppCO$) versus T. In the following, these three types of phase diagram are discussed in an attempt to evaluate oxygen carrier materials for chemical looping partial oxidation (CLPO) applications. All the phase diagrams are presented for a given total pressure of 1 atm, unless otherwise noted. The behavior of oxygen uncoupling metal oxides whose oxidation states are largely affected by T and ppO_2 is the focus of the discussion.

Two major factors are germane to the success of an oxygen carrier in CLPO processes. First, oxygen carriers should be readily reduced to yield a desired level of fuel conversion. Second, the reduced phase should be close to an equilibrium condition with a high concentration of H_2 and CO. These two factors ensure a high syngas purity at the reducer outlet. The first factor relates to the T versus ppO_2 phase diagram for single metal oxygen carriers and T versus composition at a constant ppO_2 phase diagram for bimetallic oxygen carriers. The second factor relates to the ppH_2O/ppH_2 versus T phase diagram. The phase diagrams for these factors are generated in FactSage 7.0 (FactSage) software using FToxide and FactPS databases. Specifically, the phase diagram module is used to generate the T versus ppO_2 and T versus composition phase diagrams at a constant ppO_2, whereas the predom module is used to obtain the ppH_2O/ppH_2 versus T phase diagrams. The operating conditions chosen for evaluating the oxygen carriers are a ppO_2 of 0.01 atm in the reducer, an operating temperature range of 800–1000 °C for the chemical looping system, and a total pressure of 1 atm. These operating conditions correspond in general with CLPO processes. In this section, the phase diagrams for oxygen carriers which exhibit oxygen uncoupling behavior for applications to CLPO processes are given. Both single and bimetallic oxygen carriers are considered, with Fe a common component in all bimetallic oxygen carriers. The phase diagrams for the FeTi bimetallic system, which is characterized by an ionic diffusion mechanism described in Section 2.4.1, are discussed.[20]

T versus ppO_2 Phase Diagram

The *T* versus ppO_2 phase diagram provides information on the amount of oxygen that can be released in the reducer by the oxygen carrier and also on the conditions in which it can be re-oxidized in the combustor. This diagram shows the oxygen uncoupling behavior of oxygen carriers under reducing conditions and identifies the potential for their use in CLOU processes. It is desirable for a CLOU oxygen carrier to provide a large amount of oxygen release per gram of oxygen carrier in the reducer reactor, so that the solids inventory and the solids circulation rate can be at a minimum for complete fuel conversion in the reducer. It is noted that most of the oxygen carriers used in the CLOU process are concerned with combustion applications. For CLPO applications, however, an additional factor regarding the syngas selectivity needs to be considered, which is reflected in ppH_2O/ppH_2 versus *T* phase diagrams.

T versus Composition Phase Diagram at Constant ppO_2

Depending on their composition and the operating conditions, binary metal oxygen carriers form considerably more phases than single metal oxygen carriers. The *T* versus composition phase diagram at a ppO_2 of 0.01 atm provides a composition of oxygen carrier which has a sharp phase transition from its oxidized to reduced phase within the desired temperature range. A sharp phase transition implies a small two-phase region that is of an intermediate phase made up of a mixture of both oxidized and reduced phases. In reduction from the oxidized to intermediate phases, low O_2 is released that leads to a low solids conversion. Thus, the intermediate phase is undesired in the metal oxide reduction process. However, when an intermediate phase is present in an oxygen carrier composition range, it is preferable to be of a small region. It is over this specific composition range that the selectivity of the reduced phase is analyzed, based on the ppH_2O/ppH_2 versus *T* phase diagram.

ppH_2O/ppH_2 versus *T* Phase Diagram

The ppH_2O/ppH_2 and $ppCO_2/ppCO$ versus *T* phase diagrams determine the maximum syngas conversion that can be achieved for a certain oxygen carrier in a syngas CLC system.[4] For CLC applications, the reduced phase of the oxygen carrier is desired to be in equilibrium with high values of ppH_2O/ppH_2, as well as $ppCO_2/ppCO$ ratios at the reducer outlet. In contrast, low ppH_2O/ppH_2 and $ppCO_2/ppCO$ equilibrium ratios are preferred for reforming applications. In this section, only the ppH_2O/ppH_2 versus *T* phase diagram is considered, since the $ppCO_2/ppCO$ ratio follows a similar trend

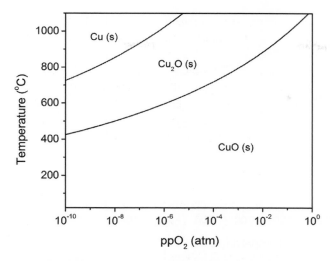

Figure 2.63 T versus ppO_2 phase diagram for $Cu-O_2$ system at 1 atm pressure, generated using FactSage 7.0 software.

with respect to temperature. This phase diagram is also used to select an appropriate gas–solid contact pattern for a given chemical looping application.

It is noted that this phase diagram is part of an analysis in the oxygen carrier selection for CLPO processes. The gas composition in a reducer reactor is a mixture of CO_2, CO, H_2, H_2O, CH_4, and other gases based on the type of fuel used. Thus, the reduced oxygen carrier has ideally to be in equilibrium with the syngas in the gas mixture. In the following discussion, only H_2 and H_2O are considered in qualitatively examining the selectivity of the oxygen carrier towards syngas. Further, as certain phases of an oxygen carrier do not form at all the conditions considered for these phase diagrams, for a system with only H, O, and metal as elements, only a limited number of phases of an oxygen carrier can exist.

2.5.1 Single Metal Oxygen Carriers

Copper

CuO has good oxygen uncoupling properties compared to all other single metal oxygen carriers.[124] As seen from Figure 2.63, between 800 and 1000 °C, the reduction from CuO to Cu_2O occurs at 900 °C for a ppO_2 of 0.01 atm, by equation (2.5.1),

$$CuO \rightarrow \tfrac{1}{2}Cu_2O + \tfrac{1}{4}O_2. \qquad (2.5.1)$$

Based on the above equation, oxygen release capacity is found to be equivalent to 0.1 g O_2/g CuO. The O_2 release is facilitated by a lower ppO_2 and therefore occurs at relatively lower temperatures. However, the reduction kinetics at these lower temperatures is slow for all practical purposes. After assessing the oxygen release from CuO, the lowest equilibrium ppH_2O/ppH_2 ratio for Cu_2O is evaluated. As seen from Figure 2.64,

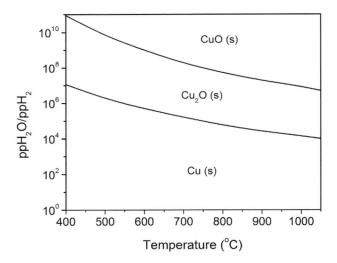

Figure 2.64 ppH_2O/ppH_2 versus T phase diagram for Cu–H_2–H_2O system at 1 atm pressure, generated using FactSage 7.0 software.

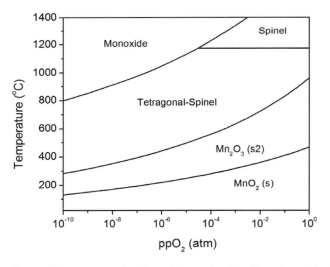

Figure 2.65 T versus ppO_2 phase diagram for Mn–O_2 system at 1 atm pressure, generated using FactSage 7.0 software.

the minimum ppH_2O/ppH_2 at which the CuO to Cu_2O transition takes place in the 800–1000 °C range is 5×10^6. The large ppH_2O/ppH_2 ratio suggests that CuO over-oxidizes H_2. Additionally, Cu_2O exists in equilibrium with the ppH_2O/ppH_2 ratios from 10^4 to 10^5, which are considerably high to achieve a good syngas selectivity. Therefore, Cu may not be a desired choice as an oxygen carrier for CLPO processes. The relatively low melting point of Cu_2O and high cost of Cu oxides, in general, also adds to the disadvantages for this process application.

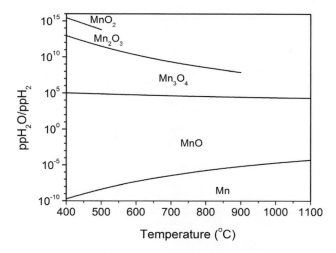

Figure 2.66 ppH_2O/ppH_2 versus T phase diagram for Mn–H_2–H_2O system at 1 atm pressure, generated using FactSage 7.0 software.

Manganese

From Figure 2.65, it is seen that for a 0.01 atm ppO_2, the Mn_2O_3 to Mn_3O_4 (tetragonal-spinel) reduction takes place at 720 °C, via equation (2.5.2),

$$Mn_2O_3 \rightarrow \tfrac{2}{3}Mn_3O_4 + \tfrac{1}{6}O_2. \qquad (2.5.2)$$

This reaction releases about 0.03 g O_2/g Mn_2O_3, which is relatively less compared with CuO. For a very low ppO_2, reduction to MnO (monoxide) is feasible, which releases 0.1 g O_2/g Mn_2O_3, comparable to CuO reduction, based on equation (2.5.3),

$$Mn_2O_3 \rightarrow 2MnO + \tfrac{1}{2}O_2. \qquad (2.5.3)$$

From Figure 2.66, Mn_3O_4 to Mn_2O_3 transition occurs at a ppH_2O/ppH_2 ratio of approximately 10^8 and Mn_3O_4 is in equilibrium with high ppH_2O/ppH_2 values. On the other hand, MnO is stable for ppH_2O/ppH_2 ratios in the range 10^{-6} to 10^{-5} for temperatures between 800 and 1000 °C. A high syngas selectivity is therefore possible with MnO as the reduced phase at the outlet of the reducer reactor. One disadvantage, however, of using Mn_2O_3 as an oxygen carrier is its inability to re-oxidize in typical combustor conditions. This is seen from Figure 2.64, where it is only for lower temperatures (200–600 °C) that Mn_2O_3 exists at a ppO_2 of 0.05 atm, which is typically assumed to be a fluidized bed combustor outlet condition.

Cobalt

The reduction from Co_3O_4 (spinel) to CoO takes place at a temperature of about 800 °C at 0.01 atm ppO_2, as seen in Figure 2.67, based on equation (2.5.4),

$$Co_3O_4 \rightarrow 3CoO + \tfrac{1}{2}O_2. \qquad (2.5.4)$$

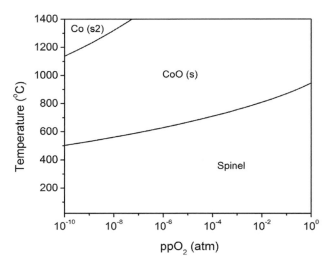

Figure 2.67 T versus ppO$_2$ phase diagram for Co–O$_2$ system at 1 atm pressure, generated using FactSage 7.0 software.

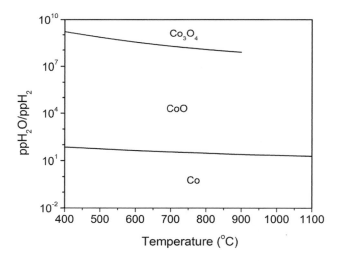

Figure 2.68 ppH$_2$O/ppH$_2$ versus T phase diagram for Co–H$_2$–H$_2$O system at 1 atm pressure, generated using FactSage 7.0 software.

This amounts to 0.07 g O$_2$/g Co$_3$O$_4$, which is comparable to the oxygen release from CuO. It is only for temperatures higher than 1100 °C and low ppO$_2$ that Co$_3$O$_4$ reduction to Co is possible. From Figure 2.68, the lowest ppH$_2$O/ppH$_2$ ratio that CoO is in equilibrium with is of the order of 10, which represents a good syngas selectivity.

Therefore, Co$_3$O$_4$ can be considered as a potential oxygen carrier for CLPO processes. However, the high cost, toxicity, and endothermic nature of reactions with fuels for Co$_3$O$_4$/CoO prohibit their use for large-scale chemical looping applications.[125]

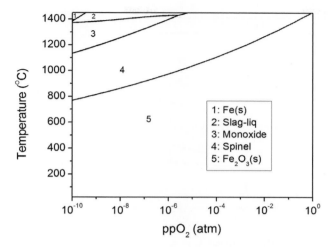

Figure 2.69 T versus ppO_2 phase diagram for Fe–O_2 system at 1 atm pressure, generated using FactSage 7.0 software.

Iron

The T versus ppO_2 phase diagram for Fe is very different from all other metal oxides that have been analyzed here. As seen from Figure 2.69, the reduction from Fe_2O_3 to Fe_3O_4 (spinel), given in equation (2.5.5), occurs at very high temperatures (>1100 °C) for a ppO_2 of 0.01 atm:

$$Fe_2O_3 \rightarrow \tfrac{2}{3}Fe_3O_4 + \tfrac{1}{6}O_2. \qquad (2.5.5)$$

Although at lower ppO_2, this transition temperature decreases, it is still high compared to Cu, Mn, and Co metal oxides. Also, the O_2 released in this process is relatively low at 0.03 g O_2/g Fe_2O_3. Fe oxides are therefore not known for their oxygen uncoupling properties and their primary mechanism for reaction is through ionic diffusion, as detailed in Section 2.3. From Figure 2.70, Fe_3O_4 is in equilibrium with a ppH_2O/ppH_2 ratio of 2 between 800 and 1000 °C, which is the lowest among all other single metal oxides considered here. Hence, Fe based oxygen carriers are considered to be selective towards syngas production. Fe oxides are relatively of low cost and non-hazardous compared to other metal oxides which make them a strong candidate for CLPO processes.

2.5.2   Binary Metal Oxygen Carriers

For chemical looping combustion and reforming applications, binary metals or bimetallic oxygen carriers exhibit a superior performance compared to single metal oxide oxygen carriers.[2] As seen in Section 2.3, single metal oxides begin to sinter and lose activity over a period of redox operation. Metal oxides are therefore either supported on an inert material, like TiO_2, or are mixed with other metal oxides to

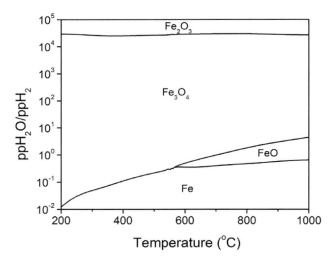

Figure 2.70 ppH_2O/ppH_2 versus T phase diagram for Fe–H_2–H_2O system at 1 atm pressure, generated using FactSage 7.0 software.

increase their sintering resistance. The drawbacks of low melting points, inability to oxidize at higher temperatures, high cost, and attrition resistance, previously mentioned, can also be mitigated through the addition of an inert or active support. Copper oxide has been tested with the use of such supports as $MgAl_2O_4$, Al_2O_3, SiO_2, ZrO_2, and TiO_2 to improve its sintering and attrition resistance.[127] Similarly, manganese oxide has been doped with metals, like Ca, Fe, Mg, Ni, to improve its ability to re-oxidize at the desired temperatures.[125] In this section, iron oxide is chosen as the secondary metal in bimetallic oxygen carriers because of the high selectivity towards syngas of its Fe_3O_4 phase.

The T versus composition at 0.01 atm ppO_2 and ppH_2O/ppH_2 versus T phase diagrams are discussed for all of the binary metal oxygen carriers considered here. The phase diagrams, however, do not take into account the rate of formation of different phases. Thus, a particular phase depicted in the phase diagram may have slow formation rates and cannot be formed in the actual system. Additionally, for the oxygen uncoupling materials, it is possible that the kinetics of oxygen release might be slow even though the oxygen release amount is high. It is observed that bimetallic system phase diagrams evaluated at conditions with only O_2 contain different phases compared to those that contain both H_2 and H_2O. This is one of the limitations of using these phase diagrams, which under a limited set of conditions are not representative of the actual system.

It is noted that T versus composition phase diagrams in the predom module of FactSage software have the different phases labeled as spinel, monoxide, bixbyite, corundum, and rutile in some of the bimetallic systems. The FactSage software lists several possible options for each of these phases. For example, a spinel phase in a FeMn system could represent Fe_3O_4, Mn_3O_4, $FeMn_2O_4$, or $MnFe_2O_4$. The exact chemical

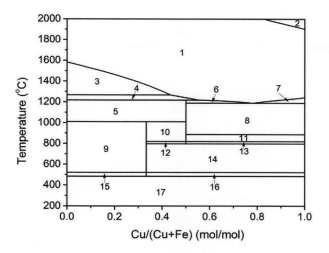

1: Slag-liq
3: Slag-liq+Spinel
5: $Fe_2O_3(s)+(Cu_2O)(Fe_2O_3)(s2)$
7: Slag-liq+$Cu_2O(s)$
9: $Fe_2O_3(s)+(CuO)(Fe_2O_3)(s3)$
11: $(Cu_2O)(Fe_2O_3)(s2)+CuO(s)$
13: $(Cu_2O)(Fe_2O_3)(s)+CuO(s)$
15: $Fe_2O_3(s)+(CuO)(Fe_2O_3)(s2)$
17: $Fe_2O_3(s)+(CuO)(s)$

2: Slag-liq+Cu(s)
4: Slag-liq+$Fe_2O_3(s)$
6: Slag-liq+$(Cu_2O)(Fe_2O_3)(s2)$
8: $(Cu_2O)(Fe_2O_3)(s2)+Cu_2O(s)$
10: $(CuO)(Fe_2O_3)(s3)+(Cu_2O)(Fe_2O_3)(s2)$
12: $(CuO)(Fe_2O_3)(s3)+(Cu_2O)(Fe_2O_3)(s)$
14: $(CuO)(Fe_2O_3)(s3)+CuO(s)$
16: $(CuO)(Fe_2O_3)(s2)+CuO(s)$

Figure 2.71 T versus ppO_2 phase diagram for Cu–Fe–O_2 system at 1 atm pressure, generated using FactSage 7.0 software.

formula of the spinel phase can be estimated by using the lever rule along a tie line or through experimental verification. For this discussion, an assumption about the actual description of a particular phase has been made, based on the composition of the oxygen carrier and ppH_2O/ppH_2 versus T phase diagrams, as will be illustrated later.

FeCu Bimetallic System

From Figure 2.71, a composition of Cu/(Cu + Fe) >0.5 has a transition from CuO to Cu_2O between 800 and 1000 °C, which translates to a high O_2 release in the reducer. However, the ability of CuO to release O_2 at low temperatures (400–600 °C) is eliminated upon addition of Fe. Figure 2.72 shows the $(Cu_2O)(Fe_2O_3)$ phase in equilibrium with a ppH_2O/ppH_2 ratio in the range 10^6 to 10^8 between 800 and 1000 °C. The order of magnitude for this ratio is similar to that for Cu_2O, as seen in Section 2.5.1. Therefore, the addition of Fe does not favorably affect the selectivity of Cu towards syngas, but instead decreases its oxygen uncoupling ability. Overall, the FeCu oxygen carrier does not seem a good fit for syngas production.

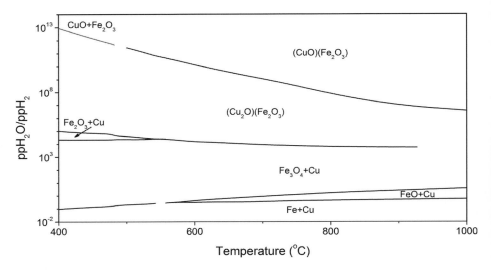

Figure 2.72 ppH_2O/ppH_2 versus T phase diagram for Cu–Fe–H$_2$–H$_2$O ($0.5 < $ Cu/(Cu + Fe) $ < 1$) system at 1 atm pressure, generated using FactSage 7.0 software.

FeMn Bimetallic System

Figure 2.73 shows that the preferred composition (Mn/(Mn + Fe)) for the FeMn oxygen carrier is from 0.6 to 0.8, as a sharp transition from bixbyite to spinel phase is observed. Table 2.4 lists the phases predicted by the predom module of FactSage software.

Figure 2.74 is generated for Mn/(Mn + Fe) between 0.3333 and 1, to include the high Mn composition range. It is observed that the $(MnO)(Fe_2O_3) + Mn_3O_4$ mixture of spinel phases is in equilibrium with a ppH_2O/ppH_2 ratio of the order of 10^5. Also, there is an intermediate $Mn_2O_3 + Fe_3O_4$ (bixbyite + spinel) phase, between the oxidized and reduced phases, whose formation involves low O_2 release. Apart from the low O_2 release for this intermediate phase, the ppH_2O/ppH_2 value for the FeMn oxygen carrier in general is too high to be used effectively for CLPO processes.

FeCo Bimetallic System

The spinel phase suggested by Factsage software could be Fe_3O_4, Co_3O_4, $(CoO)(Fe_2O_3)$ and $(FeO)(Co_2O_3)$, or a combination of the above phases, whereas the monoxide could consist of FeO and CoO. From Figure 2.75, the transition from spinel to spinel + monoxide occurs in the desired temperature range of 800–1000 °C. This transition occurs at around 900 °C for a composition of Co/(Co + Fe) from 0.3333 to 1. Again, as is observed for FeCu and FeMn systems, the oxygen uncoupling ability of Co_3O_4 at lower temperatures is reduced due to the addition of Fe.

From Figure 2.76, $(CoO)(Fe_2O_3) + CoO$ (spinel + monoxide) is seen to be stable for ppH_2O/ppH_2 ratios ranging from 10^2 to 10^8, which is very high for the FeCo oxygen carrier to be selective towards syngas production.

Table 2.4 Description of phases listed by the FactSage 7.0 software.

Phase	Description
Bixbyite	Mn_2O_3 with Fe_2O_3 in dilute amounts
Spinel	Fe_3O_4, Mn_3O_4, $MnFe_2O_4$, $FeMn_2O_4$
Corundum	Fe_2O_3 with Mn_2O_3 in dilute amounts
Monoxide	FeO, MnO with Mn_2O_3, Fe_2O_3 in dilute amounts
Tetragonal-Spinel	Mn_3O_4 with Fe_2O_3

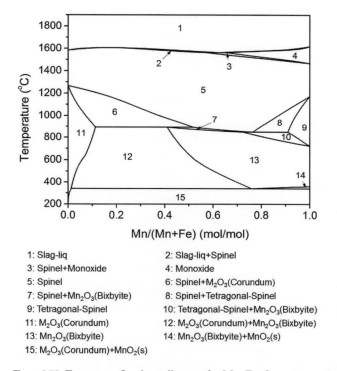

1: Slag-liq
2: Slag-liq+Spinel
3: Spinel+Monoxide
4: Monoxide
5: Spinel
6: Spinel+M$_2$O$_3$(Corundum)
7: Spinel+Mn$_2$O$_3$(Bixbyite)
8: Spinel+Tetragonal-Spinel
9: Tetragonal-Spinel
10: Tetragonal-Spinel+Mn$_2$O$_3$(Bixbyite)
11: M$_2$O$_3$(Corundum)
12: M$_2$O$_3$(Corundum)+Mn$_2$O$_3$(Bixbyite)
13: Mn$_2$O$_3$(Bixbyite)
14: Mn$_2$O$_3$(Bixbyite)+MnO$_2$(s)
15: M$_2$O$_3$(Corundum)+MnO$_2$(s)

Figure 2.73 T versus ppO_2 phase diagram for Mn–Fe–O_2 system at 1 atm pressure, generated using FactSage 7.0 software.

FeTi Bimetallic System

The Fe–Ti binary metal oxygen carrier is evaluated for applications in CLPO processes. This oxygen carrier, as mentioned earlier, has experimentally demonstrated applications for syngas production.[39]

FactSage software describes the corundum (M_2O_3) phase to be Fe_2O_3 and the rutile phase as TiO_2 with Ti_2O_3 in dilute amounts. It is observed from Figure 2.77 that there is no transition from oxidized to reduced phase between 800 and 1000 °C. The lack of transition points to the fact that FeTi is not an oxygen uncoupling material. Ionic diffusion is the predominant reduction mechanism for FeTi, as described experimentally in Section 2.3.2 and supported by the DFT calculation in Section 2.4.1. The

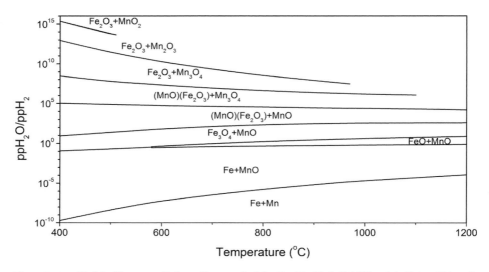

Figure 2.74 ppH_2O/ppH_2 versus T phase diagram for Mn–Fe–H$_2$–H$_2$O ($0.3333 <$ Mn/(Mn + Fe) < 1) system at 1 atm pressure, generated using FactSage 7.0 software.

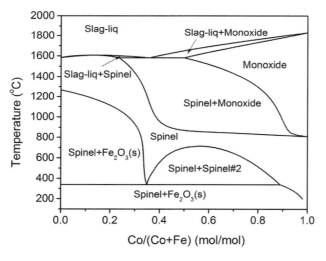

Figure 2.75 T versus ppO_2 phase diagram for Co–Fe–O$_2$ system at 1 atm pressure, generated using FactSage 7.0 software.

ppH_2O/ppH_2 versus T phase diagram is shown in Figure 1.40, where the lowest ppH_2O/ppH_2 ratio with which the reduced phase, FeO_3TiO_2, is in equilibrium is of order of magnitude 10^{-2} for the temperature range 400–1000 °C. This demonstrates good syngas selectivity and hence indicates FeTi to be a suitable oxygen carrier for syngas production.

In summary, phase diagrams are an important tool for evaluating single metal and binary metal oxygen carriers for CLPO processes. The T versus ppO_2 and T versus

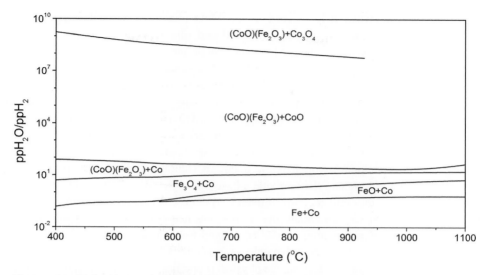

Figure 2.76 ppH_2O/ppH_2 versus T phase diagram for Co–Fe–H_2–H_2O ($0.3333 <$ Co/(Co + Fe) $<$ 1) system at 1 atm pressure, generated using FactSage 7.0 software.

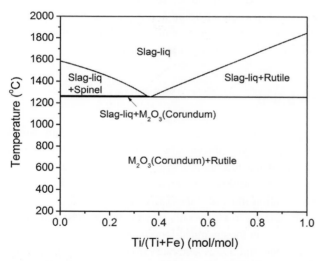

Figure 2.77 T versus ppO_2 phase diagram for Ti–Fe–O_2 system at 1 atm pressure, generated using FactSage 7.0 software.

composition at 0.01 atm ppO_2 phase diagrams identify the oxidized and reduced phases of oxygen carriers in the desired temperature range. For binary metal oxygen carriers, these diagrams are also used to decide the composition of the oxygen carrier which has a sharp transition between the two phases under reducing conditions. The ppH_2O/ppH_2 versus T phase diagrams estimate the syngas selectivity of the reduced phase of an oxygen carrier. All the single metal oxygen uncoupling materials have a high oxygen release capacity but a poor syngas selectivity, except for Mn oxygen carrier when it is reduced to the MnO phase. Fe has a low ppH_2O/ppH_2 ratio in equilibrium with Fe_3O_4 as its reduced phase, which indicates a higher selectivity towards syngas compared to other single metal oxygen carriers considered. In binary metal oxygen carriers, Fe is added to each of the single metal oxygen carriers examined. The addition of Fe decreases the oxygen uncoupling behavior of the oxygen carrier and there is no major change observed in selectivity towards syngas. Finally, the suitability of FeTi for CLPO processes is confirmed using the phase diagrams. It is essential to note that there are limitations to using this approach. In spite of the limitations of these phase diagrams, they are still a useful source of information. Although phase diagrams cannot fully simulate oxygen carriers within a CLPO system, they provide a basis for estimating the behavior of an oxygen carrier under the influence of different system conditions. Experiments will always be required to substantiate the results obtained from phase diagrams, but the range of experimental conditions can be gauged from these diagrams.

Besides the thermodynamic assessment discussed above that is of importance to oxygen carrier selection, the physical properties of the oxygen carriers, such as crushing strength and attrition resistance, are also critical, especially when applying them to continuous operation of a chemical looping system for syngas production. Clearly, the amount of attrition of the oxygen carrier in the chemical looping system would determine the rate of fresh particle make-up and hence the economic feasibility of the CLPO process. In the following section, issues related to the mechanical strength of oxygen carriers and tests for quantifying it are discussed.

2.6   Oxygen Carrier Attrition

There are three simple and well-defined mechanical stresses on bulk solids material or particles, namely, compression, impact, and shear. Compression stress represents a simple characterization of the mechanical stress property of the solid material that is conducted under the static condition. The impact and shear stresses are measured under flow conditions. This section illustrates these two stresses.

Attrition of oxygen carrier particles occurs when they are subjected to flows involved in handling and processing in the chemical looping system. Attrition can cause numerous difficulties in unit operation. Fines generated from attrition will not only lead to loss of valuable oxygen carrier material but also increase the burden of fine removal devices. The expensive operational costs of solids make-up and fine removal devices may make the chemical looping process economically unfavorable. Furthermore, attrition changes the overall physical properties of oxygen carrier particles, i.e. particle size distribution and surface area. Hydrodynamic characteristics of reactors, heat transfer characteristics of the reactor bed, and kinetics of the oxygen carrier particles are all influenced by particle size distribution. Bridging of the oxygen carrier particles in the standpipe of a chemical looping system may also occur due to a decrease in particle size and occurrence of fines, causing particle agglomeration and local overheat.[131] Therefore, attrition induced changes can affect the quality of the chemical product and normal operation of the system.

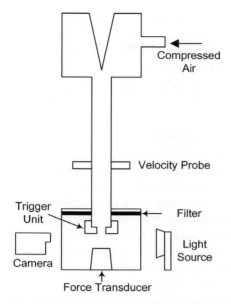

Figure 2.78 Schematic diagram of vertical impact test rig.[133,135]

There are two different modes of attrition, namely, fragmentation and abrasion. Abrasion generates fines from the surface of the particles, resulting in two distinct groups of particles: one group with diameter close to the original particles and the other with a much smaller diameter. Fragmentation destroys the particles and produces a number of particles all of which are distinctly smaller than the original.[132] The size distribution of the particles after fragmentation is broader with a smaller mean diameter than the initial batch of particles. Both of these two modes occur in most industrial processes but to varying extents.

2.6.1   Types of Attrition Tests

Numerous test methods have been developed to quantify impact and shear stresses. Some test methods are described below.

The Impact Test

Impact stress very often occurs whenever particles come into contact with the walls of process equipment or other particles. Attrition can also occur upon impact of particles with grid jets, in cyclones and bends, or in the bed.[132] There are a variety of tests for single particle impact which are used to delineate the intrinsic behavior of the material properties and forces or stresses applied under controlled environments.[132-134] However, the more relevant impact tests are directly associated with pneumatic acceleration of particles impinged onto a target.[135]

The single particle impact test equipment is broadly used in industrial applications as friability tests. The impact test apparatus, as proposed by Yuregir et al.,[133,135] is shown schematically in Figure 2.78. The particles are accelerated to the required velocity in a

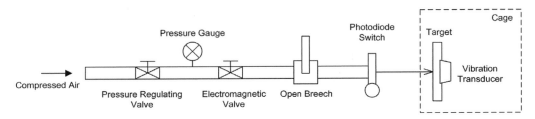

Figure 2.79 A schematic diagram of the continuous flow gas gun.[138]

vertically downward direction using a concentric air eductor, where compressed air at 0.5 MPa is allowed to expand after passing through a nozzle. The particles are introduced individually into the air stream and impacted onto an instrumented target plate in random orientation. The vertical tube has a 3.9 mm ID and a 1 m length. The particle velocity can be determined from the flight time of the particle between a pair of photoelectric sensors of known distance located close to the target plate. The target assembly is equipped with a filter to capture the debris generated from particle impact.[133] The impact force is measured by a force transducer. The various stages of impact are recorded photographically using a high speed camera. Cleaver et al.[136] used a slightly modified device for determining attrition rates of a batch of particles with weight ranging from 5 to 10 g and high speed imaging of the impact process of single particles.

A single particle impact test in the horizontal direction has been conducted. A gas gun was developed for accelerating solids particles.[137] In a set-up shown in Figure 2.79,[138] particles are introduced individually into the open breech after a steady gas flow is established. The particles are accelerated with a driving pressure up to 2 bar along an 8 mm diameter barrel with a length of 0.3 m. Velocities up to about 35 m/s could be reproducibly obtained for particles with size range 2.8–7.6 mm. The impact velocity is determined based on the time needed to travel through a distance interval. This type of impact test with gas gun is quick to operate, allowing multiple duplications and probability studies of particle breakage in a short time.

Figure 2.80 shows the set-up used by Davuluri and Knowlton.[139] The amount of sample required for one test is about 50 to 100 g. The test time is short, about 1 min. In this test, solids particles are accelerated down a long, vertical tube by a high-velocity air stream and impacted on a flat rigid plate. Since the solids are accelerated to a very high velocity, even plastics could be degraded in this apparatus. The velocity when complete shattering of the material occurs is defined as the threshold velocity. For most materials, this velocity is greater than 85 m/s. When the particle velocity exceeds the threshold velocity, particles will be shattered, and therefore no attrition information will be obtained. The benefit of such an impact test is that it can be applied to a wide range of particle properties, represented by Geldart's group C to group D.[132]

Batch impact experiments under high temperature conditions were also conducted with horizontal acceleration apparatus.[140] The experiments were conducted at 25–580 °C with 100 g of limestone particles whose initial harmonic mean diameter was 689

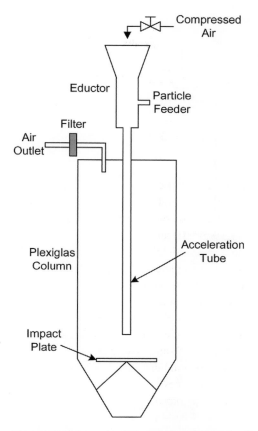

Figure 2.80 Schematic of Davuluri and Knowlton's impact test device.[139]

μm. Based on the particle size distributions before and after the impact test, it was concluded that impacts changed the particle size distribution and mean particle diameter significantly with a conveying gas velocity in the range 20–100 m/s. An increase in temperature decreases the attrition due to a decrease in particle impact velocity and an increase in the threshold particle impact velocity. Over the range of temperatures tested, the threshold particle velocities for breaking limestone particles varied from ~8.5 to 13.5 m/s.

Air Jet Fluidized Bed

The air jet fluidized bed can be used for testing attrition of oxygen carrier particles due to both fragmentation and abrasion mechanisms. Most fluidized bed tests are so-called submerged jet tests, where high speed gas jets are submerged in a fluidized bed, generating high attrition rates of particles in a relatively short period of time. Forsythe and Hertwig[141] initiated the fluidized bed attrition test for characterizing attrition performance of fluid catalytic cracking (FCC) catalysts. The device, as shown in Figure 2.81a, consists of a 2.54 cm ID glass pipe of length 1.52 m with a single jet

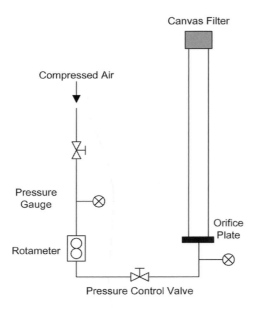

(a)

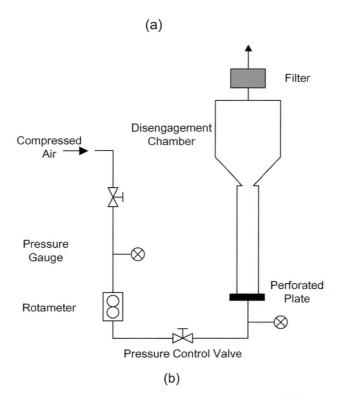

(b)

Figure 2.81 Schematics of the air jet fluidized bed test.[141,144]

orifice 0.39 mm in diameter at the bottom. A canvas filter is installed at the top of the pipe to keep all the particles inside. Small batches of FCC particles are placed in the pipe for test. During the test, the FCC particles are fluidized with air flowing through the orifice, with the orifice gas velocity approaching the speed of sound. The Forsythe and Hertwig test is the basis for standard tests like ISO 5937, which specifies an orifice diameter of 0.4 mm, air flow rate 7 lpm, and about 300 kPa back pressure, thus giving rise to choking conditions and a presumably sonic air jet velocity of about 440 m/s. The Forsythe and Hertwig test apparatus has also been used by other researchers.[142,143] Contractor et al.[143] concluded that relative attrition rates of different particles can only be obtained from the same attrition test method and that accelerated attrition rates in the attrition test apparatus may be misleading as the mechanism of attrition in the test apparatus can be different from that in the reactor. Cairati et al.[142] found that the Forsythe and Hertwig test may require times far exceeding one hour for particles to reach an equilibrium attrition rate. It is thus very difficult to compare two material types with different equilibrium periods. Also, since the fines generated from attrition remain in the system, the accumulated fines will affect the attritability of the particles over time. The results from the test are thus time integrated and can only be assessed with the aid of an initial particle size distribution. To make comparison of different samples tested in the apparatus meaningful, the samples should have similar initial particle size distribution and identical pretreatment.

To overcome the shortcomings of Forsythe and Hertwig's apparatus, Gwyn modified the set-up by adding an enlarged disengagement chamber on top of the straight column of the fluidized bed[137,144], as shown in Figure 2.81b. The attrition-produced fines, whose size was less than the cut-off size determined by the gas velocity in the disengagement chamber, would be immediately entrained out of the system by the gas flow. Three same-sized orifices in a triangular pitch are used instead of the single orifice gas inlet in the original apparatus. The ASTM D 5757–95 test follows the Gwyn device, with an orifice gas velocity of 310 m/s.

There still are some drawbacks of this technique. First, this test requires a large batch of test samples of weight about 450 g. Second, Geldart group B and D particles slug in the column. And the required attrition time for plastic type materials may be up to 20 hr. The long testing time makes it impossible for this method to be a standard test method.

Jet Cup Test

The Grace Davison jet cup attrition test is another common method of testing particle attrition for circulating fluidized bed applications.[145] It is a standard test procedure in the petroleum industry for measuring the relative attrition tendency of a FCC catalyst. The Grace Davison jet cup, as shown in Figure 2.82a, consists of a cup 2.54 cm in diameter. The cup, which is attached to the bottom of a large disengagement chamber, has a tangential gas inlet. A specific amount of the test material, of order 5 to 10 g, is placed into the Grace Davison jet cup. High speed gas at a velocity of 144 m/s is supplied to the cup through the gas inlet. Fines generated in the cup due to attrition enter the disengagement section, where the larger particles are refluxed back

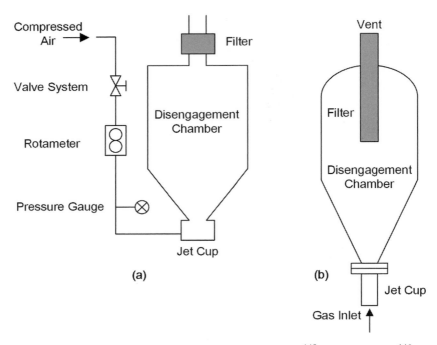

Figure 2.82 Schematic diagrams of jet cup (a) Grace Davison[145]; (b) Cocco et al.[146]

into the jet cup and very fine particles are entrained out of the chamber and collected by the filter. After an hour, the particle size distribution is analyzed, and the attrition loss is determined.

The Grace Davison jet cup works extremely well for small particles with a diameter of less than 200 μm. However, larger particles tend to slug in the small cup. With samples less than 10 g, collecting all the materials and keeping a mass balance of the particles is very difficult. Davuluri and Knowlton[139] expanded on this concept by using a jet cup 7.6 cm in diameter so that the test can process larger amounts of particles to improve the mass balance and minimize slugging phenomena for larger particles. The large jet cup can accommodate ~100 g of particles. The larger sample size reduces measuring error and results in a 95% mass recovery. The larger jet cup size also allows attrition testing of larger Geldart group B or D materials with the gas inlet velocity kept constant at 144 m/s by changing the orifice diameter of the gas inlet.

Cocco et al.[146] studied the underlying hydrodynamics responsible for particle attrition in the jet cup device using both CFD simulation and a cold flow experiment. Results showed that the standard cylindrical jet cup design, for both 2.54 cm and 7.6 cm jet cups, is insufficient for moving all the particles. A portion of the bed remains stagnant unless the gas jet velocity is unrealistically high for a practical attrition test. A conical jet cup, shown in Figure 2.82b, which expands from a 3.8 cm diameter at the bottom to a 7.6 cm diameter at the top, was designed. The new jet cup is able to achieve better particle mobility and higher attrition rates.

In the jet cup test, attrition takes place mainly due to the impact of sample particles against the walls of the jet cup chamber. The main attrition mechanism is a fracturing process rather than abrasion. The particle attrition is proportional to the average probability of particle fracture, resulting in a constant attrition rate while the number of particles impinging on the jet cup walls remains constant. Since the prevailing stress mechanisms in the jet cup test differ from those in real processes, the jet cup test can only provide relative comparisons among different particles; while the quantitative attrition rate in commercial processes can't be predicted. In spite of this limitation, jet cups are still commonly used for ranking various materials using the attrition index.

2.6.2 Attrition Studies on Different Oxygen Carriers

A considerable amount of work has been devoted in recent years to developing suitable oxygen carrier particles for chemical looping processes. Compared to oxygen carrying capacity and the reactivity of particles, much less attention has been focused on particle strength, i.e. its resistance to attrition in the chemical looping process.

Gayan and his co-workers tested several kinds of oxygen carrier particles made of different metals with a fluidized bed operated under cyclic redox conditions.[147-153] The fluidized bed, as shown in Figure 2.83, has an inner diameter of 55 mm and a height of 500 mm, and is electrically heated. The elutriation data were collected over time or number of redox cycles by measuring solids trapped in the two hot filters placed after the fluidized bed.

Copper based oxygen carriers of different compositions with alumina as a support and prepared using both wet impregnation and dry impregnation methods have been tested.[149] Commercial γ-alumina particles of 100–320 μm were used. The samples, prepared with both wet impregnation and dry impregnation, were calcined in an air atmosphere at 500 °C for 30 minutes before being calcined for one hour at 550, 850, or 950 °C in an air atmosphere. In addition, a commercial alumina-supported catalyst with 13 wt% CuO, Cu13Al-C, was analyzed for comparison purposes.

One hundred redox cycles were performed in the batch fluidized bed at 800 °C to analyze the attrition rate and reactivity of the oxygen carrier particles which did not agglomerate during the experiments. Figure 2.84 shows the attrition rates of these oxygen carriers. A high attrition rate was observed for the first few cycles due to the rounding effects and detachment of fines stuck during oxygen carrier preparation. The attrition rate then decreased with cycles and reached a near steady attrition rate after 50 cycles. The oxygen carriers showed low attrition rates with a mean attrition rate of less than 0.01 wt%/cycle (<0.02%/h in the experiments), which leads to a lifetime of the particles of over 10,000 cycles. Increasing the calcination temperature decreases the attrition rate. Preparation method has little effect on attrition rate, the wet impregnated particles having a slightly higher rate than the dry impregnated particles. The attrition behavior of oxygen carriers Cu17Al-WI-550 and Cu15Al-DI-550 was similar to that of commercial particles.

Nickel based and Ni/Cu-based oxygen carrier particles with both γ-Al_2O_3 and α-Al_2O_3 as supporting materials prepared by dry impregnation have also been tested.[154]

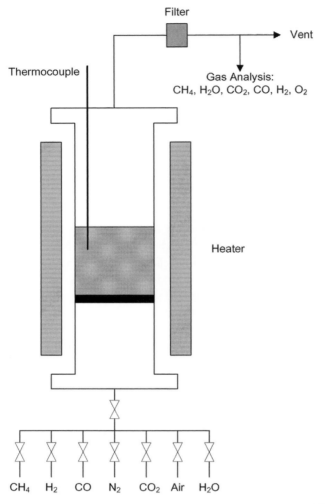

Figure 2.83 Fluidized bed reactor for multiple redox cyclic tests.[149]

For Ni-based oxygen carrier particles, the NiO weight content in the oxygen carrier particles was in the range 12–30 wt% for impregnation on γ-Al$_2$O$_3$ and 6–16 wt% for impregnation on α-Al$_2$O$_3$. Ni/Cu-based oxygen carrier particles were prepared with NiO/CuO weight ratios from 0.06 to 4. For comparison of Cu-based oxygen carriers with others, a 15% CuO, supported on γ-Al$_2$O$_3$, Cu15-γAl, was prepared by dry impregnation.

The tests were conducted at 950 °C with a superficial gas velocity to the reactor of 0.1 m/s and with 100 reduction/oxidation cycles for the oxygen carriers. As shown in Figure 2.85, variation of the attrition rate of oxygen carriers with the number of reduction–oxidation cycles indicates that the attrition rates are higher in the first few cycles, and later are lower. The initial higher rate is caused by the rounding effects of

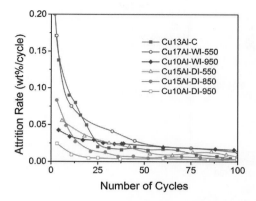

Figure 2.84 Attrition rates of some Cu-based oxygen carriers (legend shows the copper content, impregnation method and calcination temperatures).[149] A black and white version of this figure will appear in some formats. For the color version, please refer to the plate section.

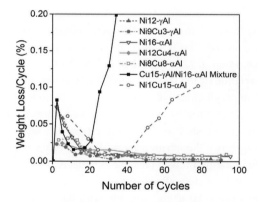

Figure 2.85 Attrition rates of several Ni-based and Ni/Cu-based oxygen carriers prepared by impregnation (legend shows Ni/Cu content and support material).[153] A black and white version of this figure will appear in some formats. For the color version, please refer to the plate section.

the particles and by entrapment of the fine materials on the particles during their production. Later, a decrease in the internal structure variation during the redox process leads to a decrease in the attrition rate. The test results indicate that attrition rates from redox cycles 30 to 100 were progressing at a rate of 0.01%/cycle for the oxygen carriers tested. Based on this attrition rate, it can be estimated that the oxygen carrier particle may sustain for 10,000 cycles or even more. This estimated attrition rate, if it can be realized in a commercial system, represents a desired attrition property for oxygen carrier particles. Figure 2.85 also shows the oxygen carriers with high attrition rates after multi-cycle testing.

In the fluidized bed rig described above, the effects of supports, such as γ-alumina, α-alumina, magnesium aluminate, calcium aluminate, as well as γ-alumina with thermal

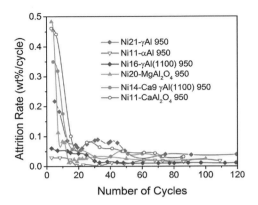

Figure 2.86 Attrition rates of Ni-based oxygen carriers with redox cycles in a batch fluidized bed (legend with Ni content, support material, calcination temperature).[150] A black and white version of this figure will appear in some formats. For the color version, please refer to the plate section.

and chemical pretreatments, on the performance of Ni-based oxygen carriers prepared by dry impregnation were also tested.[150] Oxygen carriers were prepared by slowly adding a volume of a saturated solution (20 °C, 4.2 M) of $Ni(NO_3)_2 \cdot 6H_2O$ (>99.5% purity) equal to the total pore volume of the support particles, with thorough stirring at room temperature. Successive impregnations were applied to the particles followed by calcination at 550 °C in air atmosphere for 30 minutes, and sintered for one hour at the desired calcination temperature. The tested oxygen carrier particles were in the size range 100–500 μm. Figure 2.86 shows the attrition rates of these oxygen carriers. The attrition rates were usually high in the first few cycles but then decreased to a low value, after 20–40 cycles, which remained constant afterwards. The lowest attrition rates were demonstrated by oxygen carriers prepared using supports with the highest mechanical strengths, i.e. $\alpha\text{-}Al_2O_3$ and $\gamma\text{-}Al_2O_3$ (1100), using thermal treatment, and $MgAl_2O_4$. Alternatively, oxygen carriers prepared using $\gamma\text{-}Al_2O_3$ and $CaAl_2O_4$ as support had the highest attrition rates due to their low mechanical strength. The effect of NiO content on the performance of Ni-based oxygen carriers prepared by incipient wet impregnation, at ambient (AI) and hot conditions (HI), and by deposition–precipitation (DP) methods using $\gamma\text{-}Al_2O_3$ and $\alpha\text{-}Al_2O_3$ as supports, was also examined.[151] The experiments were carried out at 950 °C, with an inlet superficial gas velocity into the reactor of 0.1 m/s, and using 300 g of oxygen carrier particles.

Figure 2.87 shows the attrition rates measured with the different oxygen carriers. After 20–40 cycles of reaction, the attrition rates of the oxygen carriers decreased compared to the initial redox cycles and then became almost constant. The higher strength of $\alpha\text{-}Al_2O_3$ compared to $\gamma\text{-}Al_2O_3$ resulted in oxygen carriers prepared using the former as support which also had a lower attrition rate. The preparation method for oxygen carriers had an effect on their attrition rate in the first few cycles. Oxygen carriers synthesized by the DP method had higher attrition rates because of the processing steps involved in precipitation and filtration for this method. Additionally,

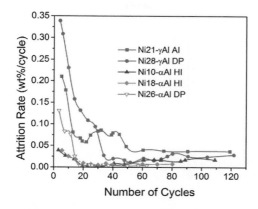

Figure 2.87 Attrition rates of Ni-based oxygen carriers prepared with different methods and supports (legend shows Ni content, support material, and preparation method).[151] A black and white version of this figure will appear in some formats. For the color version, please refer to the plate section.

the precipitation, filtration, and drying steps in this method were time intensive. The oxygen carriers prepared by the different methods had similar attrition rates after the first 20–40 redox cycles. The attrition rates, in general, were low after the first few cycles indicating a reduced effect of redox cycles on the mechanical properties of the oxygen carriers. SEM images of fresh and used oxygen carriers did not show any holes or cracks on the surface. Based on these experimental results, oxygen carriers synthesized using AI, HI, or DP methods on α-Al_2O_3 support were found to be suitable for use in the CLC process. However, among the preparation methods, HI was preferred as it produced particles of a high Ni content with reduced tendency towards agglomeration in fluidized beds, and thus the HI method appears to be more desirable for producing low cost oxygen carriers. The HI method using α-Al_2O_3 as support allowed, with relative ease, the production of oxygen carriers with a high active metal content, high reactivity, and high selectivity towards CO_2 and H_2O. Moreover, they had low attrition rates with no particle agglomeration observed when operated in a fluidized bed system.

The attrition rate for an ilmenite oxygen carrier under redox cycles in a fluidized bed reactor was also tested.[152] The oxygen carrier particles were made of ilmenite concentrated from a natural ore and had a particle diameter between 100 and 300 μm with particle density 4100 kg/m^3. Fresh particles were calcined in air at 950 °C for 24 h to improve their performance as an oxygen carrier. The main components of the calcined ilmenite particles were 55.5 wt% of Fe_2TiO_5, 10.6 wt% of Fe_2O_3, 28.4 wt% of TiO_2, and 5.5 wt% of other inert compounds. The crushing strength of fresh particles was 2.2 N, measured using a Shimpo FGN-5 crushing strength apparatus. The temperature of the fluidized bed reactor was maintained at 900 °C during the test. The reducing gas consisted of H_2, CO, CO_2, H_2O, and N_2, and had a gas velocity of 0.30 m/s at the inlet of the fluidized bed. The test lasted for 56 h with a total of 100 redox cycles performed. Particles elutriated from the fluidized bed reactor were

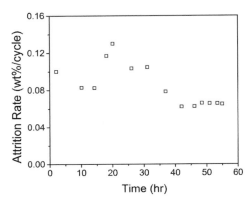

Figure 2.88 Attrition rate of ilmenite particles in a fluidized bed reactor under redox cycles.[152]

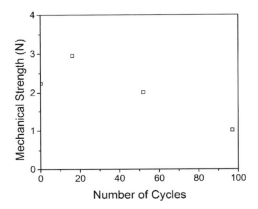

Figure 2.89 Mechanical strength of ilmenite oxygen carrier particles over 100 redox cycles.[152]

retained in a filter, and were taken every ten cycles. Fines of oxygen carrier particles were defined as those of diameter less than 40 μm. The particles showed a low attrition rate as a function of fluidization time, as shown in Figure 2.88. The initial attrition rate was about 0.08–0.12%/h, mainly due to the rounding off of initial particles. After 40 h of operation, the attrition rate stabilized to 0.066 %/h, mainly attributed to the surface attrition rather than particle breakage. After a number of redox cycles, the particles formed a core–shell structure with Fe_2O_3 outside and TiO_2 inside. The XRD analysis of fines showed only iron oxide with no titanium oxide or iron titanates observed.

The fresh ilmenite particles maintained a relatively high crushing strength for the first 20 redox cycles, which was between 2.2 and 2.9 N.[152] The crushing strength did not vary between the fully oxidized and reduced ilmenite particle after activation with CO, and was equal to 2.9 N. After 50 redox cycles, the crushing strength deteriorated and reached a value of 1 N after 100 redox cycles, as shown in Figure 2.89. This variation in crushing strength is crucial for deciding on the feasibility of using ilmenite particles in circulating fluidized bed CLC systems.

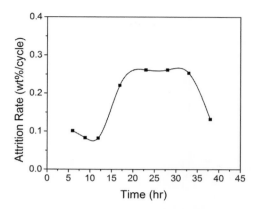

Figure 2.90 Attrition rate of activated bauxite waste samples under redox cycles in batch mode fluidized bed.[148]

Although the crushing strength value to start with was acceptable for use in circulating fluidized bed CLC systems, after 100 cycles, it reduced to a value which was on the borderline, close to the acceptable limit for use in these systems. The lower crushing strength after 100 redox cycles was attributed to the Fe_2O_3 layer formed on the surface of the ilmenite particles and was not representative of the particle as a whole.

A bauxite waste from alumina production, supplied by Alcoa Europe-Alúmina Española S.A., which contains 71 wt% of Fe_2O_3 with β-Al_2O_3 as the major constituent, has also been tested in the fluidized bed system shown in Figure 2.83.[147] The bauxite waste sample was dried at room temperature for 72 h before being sieved to the desired particle size of 150–300 μm. The dried sample was calcined at 1473 K for 18 h to ensure complete oxidation and to increase the resistance to attrition. Batch samples of 300 g oxygen carrier particles were placed in the fluidized bed for tests. The oxygen carrier particles were exposed to alternating reducing and oxidizing conditions at a temperature in the range 1100–1223 K for 40 h. During the reducing period, gas mixtures of H_2, CO, CH_4, CO_2, and H_2O with N_2 with different fractions were supplied with a gas velocity of 0.3 m/s in the fluidized bed. The experiments were operated with redox cycles of the particles between Fe_2O_3 and Fe_3O_4. The attrition rates at different times during operation, defined as the mass percentage of particles of size less than 40 μm generated per hour of operation, are shown in Figure 2.90. The averaged attrition rate value during the 40 h operation is about 0.2%/h, which corresponds to an average lifetime of the particles of 500 h.

Brown et al.[154] investigated the attrition rates of mechanically mixed unsupported iron oxide and copper oxide impregnated on γ-alumina in a fluidized bed under both inert and redox cycle conditions, as well as the impact test. The fluidized bed, similar to that shown in Figure 2.83, was made of a stainless steel tube with 40 mm inner diameter and a height of 1 m from the gas distributor to the top exit. The distributor was a perforated plate with 55 holes of 0.5 mm diameter drilled on a triangular pitch. The elutriated fines were collected from two sintered brass filters at the outlet of the fluidized

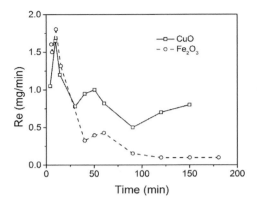

Figure 2.91 Elutriation rates of unsupported iron oxide particles and copper based oxygen carrier particles in a fluidized bed without reaction.[154]

bed. A three-way valve was installed at the exit before the two filters so that the two filters could be used alternately and thus the fines could be collected over a specific time period. The apparatus was operated at atmospheric pressure and electrically heated to different temperatures. Iron oxide particles were prepared by spraying Fe_2O_3 powder (Sigma Aldrich, <10 μm diameter and >99 wt% purity) with reverse osmosis water and mixing them manually, followed by roasting in a muffle furnace at 950 °C for 3 h. The particles were sieved to a size range of 400–600 μm. The copper oxide particles were prepared by impregnating γ-alumina particles in a size range of 600–900 μm with a saturated solution of copper nitrate and aluminum nitrate. The mixture was thoroughly stirred and stood for 18 h at 60 °C before being dried on a tray overnight at 150 °C, and roasted in a fluidized bed at 825 °C. The obtained copper based oxygen carrier particles had ~24 wt% CuO. The attrition rates of both oxygen carrier particles without reaction, as shown in Figure 2.91, were first tested by putting 100 g of oxygen carrier particles in a fluidized bed at an operating temperature of 850 °C and a fluidizing velocity of 0.8 m/s. Both oxygen carriers showed a decreasing attrition rate with time at the initial stage and a steady attrition rate after about two hours. The initially high attrition rate is probably due to the initial rounding of the fresh particles as surface irregularities are broken off. However, under redox cycle conditions, the mechanical strength of iron oxide particles deteriorated significantly after repeated cycles of oxidation and reduction, due to reaction induced structural changes. In comparison, the CuO based particles could withstand repeated reactions, with no obvious signs of increase in elutriation or fragmentation during the experiment. These results highlight the importance of support on the mechanical strength of the oxygen carrier particles. The experimental results also indicated that it is insufficient to screen oxygen carrier particles by only examining the initial strength of fresh particles. It is essential to do a full characterization of oxygen carrier particles under long-term repeated redox cycle conditions.

To measure the attrition resistance of different oxygen carriers, Lyngfelt and his co-workers at Chalmers University of Technology developed a customized jet cup test rig based on PSRI's conical jet cup design.[155–157] It was sized to hold a 5 g sample and consisted of a conical cup 39 mm high with an inner diameter of 13 mm at the bottom and 25 mm at the top. Air with a velocity of approximately 100 m/s was injected tangentially through a nozzle of diameter 1.5 mm. Using this jet cup, Rydén et al.[155] compared the attrition resistance of fresh oxygen carrier particles with those obtained from chemical looping combustion experiments in circulating fluidized beds; carried out at Chalmers University of Technology. A total of 25 different oxygen carrier particles were tested where the active metal oxide of the oxygen carrier was either nickel, iron, manganese, or copper and the supports used included $NiAl_2O_4$, $MgAl_2O_4$, Al_2O_3, ZrO_2, Mg-ZrO_2, and TiO_2. Oxygen carriers made of composite materials were also tested, which included iron/manganese based metal oxides and calcium/manganese based perovskite structures. All the oxygen carrier particles that were tested were divided into four groups, A–D, based on their attrition rate in the circulating fluidized beds for the chemical looping combustion process. Group A particles had been operated at the 10 kWth scale with a low attrition rate and were generating fines between 0.02 and 0.002 wt%/h, where fines were particles smaller than 45 μm. Oxygen carriers in groups B and C were operated in a 300 Wth chemical looping combustion system and group D particles demonstrated the highest attrition rate among all other groups. Group B particles had a low attrition rate, generating fines below 0.05 wt%/h. Group C particles had a higher attrition rate compared to group B particles, generating fines in the range 0.05–0.5 wt%/h. Group D particles generated fines at a rate greater than 0.5 wt%/h. Table 2.5 summarizes the results of crush strength and jet cup attrition tests for both fresh and used oxygen carrier particles. In Table 2.5, A_{tot} represents the total amount of fines produced over the 1 hr test period, whereas A_i is the rate of fines produced in the last 30 min of the test. Nickel and iron based metal oxides with Al_2O_3, $NiAl_2O_4$, or $MgAl_2O_4$ as the support and the calcium/manganese based perovskite structures had the highest attrition resistance. Iron/manganese based composite metal oxides, copper based metal oxides, and oxygen carriers with zirconia also had high attrition resistance. Overall, no strong correlation between crushing strength and attrition resistance measured using jet cup could be established, but the jet cup tests showed a better correlation to the attrition behavior of particles in actual circulating fluidized bed operations for chemical looping.

Azimi et al.[156] investigated the attrition resistance of Mn–Fe oxygen carriers with the addition of Al_2O_3 as support. Spray-dried oxygen carrier particles with Mn:Fe molar ratios of 80:20 and 33:67, together with different amounts of Al_2O_3 as supporting material were prepared. Oxygen carriers were calcined for 4 h at different temperatures of 950 °C, 1100 °C, or 1200 °C. It was concluded that addition of Al_2O_3 to materials with a Mn:Fe molar ratio of 80:20 was not advantageous. The attrition resistance of Al_2O_3 supported oxygen carriers with a Mn:Fe ratio of 80:20 was affected by the calcination temperature. For a calcination temperature of 950 °C, the oxygen carriers demonstrated low attrition resistance, whereas after calcining at 1200 °C, they exhibited a high attrition resistance which came at the cost of poor gas conversion. Al_2O_3-supported oxygen carriers with a Mn:Fe molar ratio of 33:67, in general, demonstrated a high attrition resistance with an attrition index (A_i) of 0.45–3.7 wt%/h. This value of

Table 2.5 Attrition performance of oxygen carriers used in Chalmers' chemical looping systems.[155]

Composition (wt.%)	Production method	Calcination	Crush strength (N)	Group	Status	A_{tot} (wt.%/ h)	A_i (wt. %/h)	Attrition behavior
40/60 NiO/NiAl$_2$O$_4$	Spray drying	4 h at 1450 °C	2.3	A	Fresh	0.7	0.48	Logarithmic
					Used	0.08	0.08	Linear
40/60 NiO/MgAl$_2$O$_4$	Spray drying	4 h at 1400 °C	2.3	A	Fresh	0.95	0.64	Logarithmic
					Used	0.08	0.08	Linear
60/40 Fe$_2$O$_3$/ MgAl$_2$O$_4$	Freeze granulation	6 h at 1100 °C	0.7	A	Fresh	2.23	1.28	Logarithmic
					Used	0.32	0.12	Logarithmic
60/40 NiO/NiAl$_2$O$_4$	Spin flash drying	Not applicable	5.1	A	Fresh	0.58	0.16	Logarithmic
					Used	0.16	0.16	Linear
CaMn$_{0.9}$Mg$_{0.1}$O$_{3-\delta}$	Spray drying	4 h at 1300 °C	2.6	A	Fresh	1.03	0.6	Logarithmic
					Used	0.38	0.32	≈Linear
40/60 NiO/NiAl$_2$O$_4$	Freeze granulation	6 h at 1600 °C	2.5	A	Fresh	–	–	Unavailable
					Used	0.61	0.57	≈Linear
20/80 NiO/MgAl$_2$O$_4$	Freeze granulation	6 h at 1400 °C	2.5	B	Fresh	0.22	0.16	Logarithmic
					Used	0.12	0.12	Linear
18/82 NiO/α-Al$_2$O$_3$	Wet impregnation	1 h at 950 °C	4.1	B	Fresh	2.96	1.44	Logarithmic
					Used	0.5	0.52	Linear
21/79 NiO/γ-Al$_2$O$_3$	Wet impregnation	1 h at 950 °C	2.6	B	Fresh	3.16	2.61	Logarithmic
					Used	0.54	0.56	Linear
60/40 Fe$_2$O$_3$/Al$_2$O$_3$	Freeze granulation	6 h at 1100 °C	1.3	B	Fresh	8.41	7.25	≈Linear
					Used	2.67	2.67	Linear
≈ FeTiO$_x$	Crushing, beneficiation	None	3.7	B	Fresh	0.75	0.46	Logarithmic
					Used	6.56	5.68	≈Linear
60/40 Fe$_2$O$_3$/ MgAl$_2$O$_4$	Freeze granulation	6 h at 1150 °C	0.6	B	Fresh	10.95	9.65	≈Linear
					Used	0.52	0.4	≈Linear
CaMn$_{0.875}$Ti$_{0.125}$O$_{3-\delta}$	Freeze granulation	3 h at 1200 °C	1.2	B	Fresh	20.12	8.52	Logarithmic
					Used	8.63	6.72	≈Linear
≈ Fe$_2$O$_3$	Crushing	24 h at 950 °C	8.3	C	Fresh	1.66	1.48	≈Linear
					Used	0.99	0.83	≈Linear
60/40 NiO/Mg-ZrO$_2$	Freeze granulation	6 h at 1400 °C	1.7	C	Fresh	21.91	18.97	Logarithmic
					Used	13.9	10.6	Logarithmic
20/80 CuO/ZrO$_2$	Spray drying	4 h at 1150 °C	0.4	C	Fresh	19.5	20.22	Linear
					Used	27.12	27.22	Linear
40/60 Mn$_3$O$_4$/Mg-ZrO$_2$	Spray drying	4 h at 1200 °C	n/a	C	Fresh	3.22	3.26	Linear
					Used	4.69c		Destroyed
40/60 Mn$_3$O$_4$/Mg-ZrO$_2$	Freeze granulation	6 h at 1150 °C	0.7	C	Fresh	41.02c		Destroyed
					Used	4.05c		Destroyed
≈ Mn$_2$O$_3$ (Eastern European Ore)	Crushing	None	n/a	D	Fresh	–	–	Unavailable
					Used	5.79	3.94	Logarithmic
≈ Mn$_2$O$_3$ (Brazilian Ore)	Crushing	None	1.3	D	Fresh	5.08	2.01	Logarithmic
					Used	6.16	3.01	Logarithmic
40/60 CuO/ZrO$_2$	Spray drying	4 h at 950 °C	3.4	D	Fresh	5.96	3.47	Logarithmic
					Used	7.92	6.78	≈Linear
40/60 CuO/CeO$_2$	Spray drying	4 h at 950 °C	1	D	Fresh	16.37c		Destroyed
					Used	10.36c		Destroyed

Table 2.5 (*cont.*)

Composition (wt.%)	Production method	Calcination	Crush strength (N)	Group	Status	A_{tot} (wt.%/h)	A_i (wt.%/h)	Attrition behavior
66.8/32.2 (Fe/Mn)$_2$O$_3$	Spray drying	4 h at 1100 °C	2.7	D	Fresh	32.86	33.12	Linear
					Used	8.41c		Destroyed
20.7/79.3 (Fe/Mn)$_2$O$_3$	Spray drying	4 h at 950 °C	0.7	D	Fresh	18.40c		Destroyed
					Used	1.47c		Destroyed
60/40 Fe$_2$O$_3$/Mg-ZrO$_2$	Freeze granulation	6 h at 1100 °C	1.1	D	Fresh	50.71c		Destroyed
					Used	23.75c		Destroyed

A_i was significantly better than that of unsupported oxygen carriers with the same Mn: Fe ratio, having an A_i of 30 wt%/h.[158]

Arjmand et al.[157] tested the attrition resistance of oxygen carriers with 11 different contents of Mn, Fe, and Si, in the jet cup test rig as well as the crushing test. Oxygen carriers were produced by spray drying and then calcined at two different temperatures of 1100 °C and 1200 °C for 4 h. The particles are sieved to a size range of 125–180 μm. As shown in Table 2.6, the crushing strength for oxygen carriers in general is lower than those given in Table 2.5 and it is seen that the crushing strength did not correlate linearly with the attrition resistance in the jet cup test. For most materials, the attrition index, A_i (wt%/h), which was defined as the slope of the accumulated attrition of the oxygen carriers in the last 30 min of the test, was substantially lower when sintered at 1200 °C, with the exceptions of MFS5, MFS6, and MFS7, which exhibited low attrition rates at both sintering temperatures. Material composition had a significant effect on the attrition resistance of the oxygen carriers sintered at 1100 °C. The attrition index of oxygen carriers with a silica content of ~30% was considerably lower than others. However, the trend was not clear for oxygen carrier particles sintered at 1200 °C. Similar attrition tests for different oxygen carrier particles were conducted by Baek et al.[159–161] using a modified three-hole air jet fluidized bed attrition tester based on ASTM D 5757–95, which follows the Gwyn device, mentioned earlier.

NiO contents of 60, 70, and 80 wt% mixed with γ-Al$_2$O$_3$ using a spray drying method were prepared for test.[159] To examine the effect of pseudoboehmite on the performance of the oxygen carriers, 5 and 10 wt% contents of pseudoboehmite were also used as raw support materials to mix with 70 wt% of NiO and γ-Al$_2$O$_3$. The oxygen carriers were sintered at different temperatures, 1000, 1100, 1200, and 1300 °C to investigate the effect of sintering temperature on the performance of oxygen carriers. The attrition performances of different oxygen carriers are shown in Figure 2.92. It is shown that oxygen carriers with 70 wt% NiO have higher attrition resistance than other oxygen carriers, and are stronger than a commercial FCC catalyst, Akzo as a reference, which has an attrition index of 22.5%. The calcination temperature required to obtain sufficient mechanical strength was lowered to 1000 °C. With the increase of the pseudoboehmite content, the mechanical strength decreased slightly.

Table 2.6 Attrition performance of tested Mn–Fe–Si oxygen carriers[157]

Name	Synthesis composition Mn_3O_4: SiO_2: Fe_2O_3 (wt%)	Sintering temperature (°C)	A_i (wt %/h)	Crushing strength (N)
MFS1	44.6:8.8:46.6	1100	18.5	0.3
		1200	0.7	0.6
MFS2	67.5:8.9:23.6	1100	29.5	0.4
		1200	4	1.5
MFS3	46:18:36	1100	27.6	0.5
		1200	3.9	0.5
MFS4	57.7:18.1:24.2	1100	43.3	0.4
		1200	1.4	0.8
MFS5	23.4:27.6:49.0	1100	0.8	1
		1200	0.8	1
MFS6	35.5:27.8:36.9	1100	0.4	1.2
		1200	0.4	0.9
MFS7	47.3:27.9:24.8	1100	1.5	1
		1200	1.2	0.5
MFS8	59.5:28.1:12.4	1100	25.4	0.7
		1200	11.5	0.7
MFS9	36.3:38.2:25.5	1100	43.2	0.3
		1200	5.2	0.6
MFS10	48.8:38.4:12.8	1100	23.9	0.5
		1200	1.6	0.3
MFS11	25.0:48.9:26.1	1100	57.1	0.4
		1200	25.7	0.8

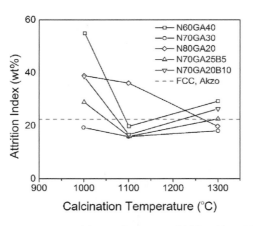

Figure 2.92 Attrition performance of NiO with γ-Al_2O_3 and pseudoboehmite (legend shows Ni content, alumina content, and pseudoboehmite content).[159]

The effect of MgO on the physical properties of spray dried NiO oxygen carriers was also investigated.[159,161] Oxygen carriers were prepared with 70 wt% NiO content, 0, 4.2, and 8.4 wt% MgO balanced with α-Al_2O_3 or γ-Al_2O_3. Table 2.7 shows the attrition performance of the oxygen carrier tested air jet fluidized bed device shown in

Table 2.7 Attrition index of different MgO- and Al$_2$O$_3$-supported NiO oxygen carriers.[159,161]

Oxygen carrier	Support content MgO:Al$_2$O$_3$ (wt%)	Calcination temperature (°C)	Attrition index (wt%)
ΥAlMg0	0:30	1300 (5h)	18.2
		1400 (5h)	1.1
ΥAlMg4	4.2:25.8	1300 (5h)	68.3
		1350 (5h)	16.7
		1350 (10h)	6.1
		1400 (5h)	0.4
ΥAlMg8	8.4:21.6	1200 (5h)	47.2
		1300 (5h)	17.9
		1400 (5h)	1.7
αAlMg0	0:30	1300 (5h)	61.7
		1400 (5h)	9
		14500 (5h)	2.8
		1500 (5h)	0.5
αAlMg4	4.2:25.8	1400 (5h)	73.6
		1500 (5h)	7.4
αAlMg8	8.4:21.6	1400 (5h)	44.1
		14500 (5h)	22.5
		1450 (10h)	9
		1500 (5h)	2

Figure 2.81. It is seen that the attrition resistance is strongly dependent on the temperature and time of calcination. Oxygen carriers with γ-Al$_2$O$_3$ as supporting materials had higher mechanical strength than others calcined at the same temperature. To be comparable to the attrition performance of reference particles of a commercial FCC catalyst, the minimum required calcination temperature was 1300 °C for γ-Al$_2$O$_3$-supported oxygen carriers, while it was 1400 °C for α-Al$_2$O$_3$-supported ones. Adding MgO as supporting material decreased the attrition resistance of the oxygen carrier particles. However, the oxygen carriers with 8.4 wt% MgO contents had better attrition resistance than those with 4.2 wt% MgO, although the attrition performances for both cases were worse than those without MgO content. Increasing calcination temperature and calcination time could increase the attrition strength of the oxygen carrier particles, mainly due to the stronger bond of grains with higher temperature and longer time.

2.7   Enhancement Techniques for Oxygen Carrier Performance

Enhanced morphological properties like surface area, pore volume, and pore size distribution help reduce oxygen carrier susceptibility to sintering at higher temperatures. Ionic conductivity reflects on the ionic diffusion in the oxygen carrier particle. Thus, a good solid-state ionic conductivity can compensate for the sintering caused loss in surface area and pore volume. Furthermore, better control over lattice oxygen diffusion is desirable for higher product selectivity in partial oxidation processes. The following section discusses different methods for improving oxygen carrier morphology and solid-state ionic diffusion, which have a direct correlation with their reactivity, recyclability, and product selectivity in partial oxidation processes. Section 2.5 discusses attrition behavior that reflects techniques for improving the strength and attrition resistance of the oxygen carrier particles. Some additional physical property enhancement techniques are discussed in Section 2.7.1. Section 2.7.2 describes additional techniques for improving chemical properties of materials, such as reactivity and recyclability.

2.7.1   Other Physical Property Enhancement Techniques

Wet chemical synthesis of oxygen carrier particles includes a variety of processes such as the sol–gel process, hydrothermal synthesis, Pechini method, spray drying, aerosol spray pyrolysis, and cryochemical synthesis. It uses a liquid phase at one of the process stages and produces much smaller grains (crystallites) compared to conventional solid-state methods. The wet chemical route has the advantages of ease in synthesis, inexpensive equipment facility, low working temperature, and high percentages of pure products. Consequently, wet chemical synthesis is largely used for small particle growth with desirable characteristics, e.g. high specific surface area, fine grain size and size distribution, and low particle agglomeration. For example, the precursors can be prepared from an aqueous solution of metal nitrates, with the precipitate formed by mixing with a solution of alkali, at a constant pH at room temperature. The precipitate is then aged and washed with deionized water, and then filtered and dried in air. The as-dried precursor is then subject to calcination in air to obtain the desired metal oxide particles with improved surface area and pore volume.

Wet chemical synthesis has been applied to various materials, including CuO, FeO, and perovskite materials such as $LaMn_{1-x}Co_xO_{3-\delta}$, $LaFeO_{3-\delta}$, $LaCoO_{3-\delta}$, and $LaFe_{1-x}Co_xO_{3-\delta}$.[162] In spite of the uniform nanoparticle distribution and much improved surface area, agglomeration can often be observed when the concentration of oxygen carrier material is high. For example, a heavy sintering effect is observed when the oxygen carrier is made up of 82.5 wt% CuO nanoparticles and cycled at temperatures above 900 °C. When the CuO loading is reduced to 60 wt%, the sintering effect can be significantly reduced with redox cycles at 800–1000 °C. The stable dispersion of CuO nanocrystals on the solid support contributes to a reduction in the sintering effect.[163] Similarly in perovskite materials, a maximum concentration of 60 wt% oxygen carrier is currently used regardless of the size of particles.

2.7.2   Other Chemical Property Enhancement Techniques

Controlled Physical/Chemical Thin Film Deposition

Physical deposition is used to produce high quality thin films as inert supports using mechanical, electromechanical, or thermodynamic methods. Physical deposition systems require a low-pressure vapor environment for correct functioning. Under high vacuum, the target material is subject to high-energy treatment so that it can be vaporized and deposits as a solid film on the solid material surface. Examples of physical deposition include atomic layer deposition (ALD), laser molecular beam epitaxy (MBE), and sputtering. Chemical vapor deposition (CVD) is somewhat similar to physical vapor deposition. There are different types of CVDs, such as laser CVD, photochemical CVD, low pressure CVD, and metal organic CVD. To coat a layer of target material using CVD, the target material is first converted into a chemical compound, which is then vaporized in a reaction chamber at a certain temperature. In the reaction chamber, the gaseous chemical compound reacts with the solid substrate material or decomposes to form a layer of the target material on the solid substrate material. In a CVD synthesis apparatus, there should be a gas delivery system, reaction chamber, substrate loading mechanism, and an energy supplier. In addition, ultra-high vacuum is essential to ensure that there is no other gas contamination in the reacting gas in the chamber. The substrate temperature is critical for vapor deposition. Thus, accurate control over the temperature and pressure inside the vacuum chamber is of high importance. The initial chemical component needs to be both volatile and stable, in order to be vaporized without decomposition before being coated onto the substrate. One issue in the synthesis of oxygen carrier particles using CVD is the undesired chemical reaction of target oxygen carrier material with the substrate. For example, in an effort to eliminate sintering issues that slow down the gas splitting reaction, ferrites deposited on porous support materials such as SiO_2, Al_2O_3, ZrO_2, and yttria-stabilized zirconia (YSZ) are fabricated using a CVD method.[164] However, these supports are not always chemically inert to the chemical vapor, which tends to form an undesirable spinel structure and consequently slows down the cyclic redox reaction.

Core–Shell Structures

As discussed in Sections 2.3.2 and 2.3.3, the formation of core–shell structures over multiple redox cycles could negatively affect the oxygen carrier performance. However, the core–shell concept can also be used to design oxygen carriers with better structural stability and improved ionic diffusion through the lattice. Shafiefarhood et al.[98] have proposed the use of oxygen carriers with a perovskite shell and a Fe_2O_3 core for the partial oxidation of methane to syngas. Pure and inert supported Fe_2O_3 based oxygen carriers are found to have low activity and syngas selectivity from methane oxidation. The reactivity and syngas selectivity can be enhanced by using tailored support materials like the perovskite-structured mixed ionic–electronic conductive support, lanthanum strontium ferrite (LSF). LSF-supported Fe_2O_3 has higher activity and syngas selectivity from methane oxidation due to enhanced O^{2-} and electron transport facilitated by the support. When designed in the core–shell form, with the Fe_2O_3 as the core and the LSF as the shell, it is believed that the structural stability and sintering resistance are improved, along with improved resistance towards carbon deposition. It is shown that the relatively well-defined surface of the core–shell structure compared to supported composites provides better syngas selectivity. The core–shell particles have shown 10–200 times higher activity towards methane partial oxidation compared to inert supported oxygen carriers like Al_2O_3 and $MgAl_2O_4$. They have even shown better reactivity, product selectivity after multiple redox cycles, and higher carbon resistance when compared with composite Fe_2O_3- and LSF-based oxygen carriers. The core–shell structure has been found to be stable even after 100 redox cycles at 900 °C. The better performance is attributed to the steady and controlled diffusion of lattice oxygen from the Fe_2O_3 core into the mixed conductive LSF shell. Also, the absence of metallic Fe on the surface in a reducing environment improves the resistance for carbon formation. XRD analysis of reacted core–shell oxygen carriers has shown that the perovskite structure of LSF is maintained while the core is completely reduced to metallic Fe. This means that the core–shell structure has the ability to preferentially donate lattice oxygen from the iron oxide core. Since lattice oxygen is believed to be more stable in the perovskite structure than iron oxide, the LSF shell acts in the controlled transport of O^{2-} ions between the core and the surface thereby giving a high activity for methane partial oxidation, high product selectivity, and good recyclability.

Neal et al.[100] investigated the mechanism of syngas formation from methane using a core–shell Fe_2O_3–LSF catalyst. They found that the reduction of the catalyst by methane could be divided into four regions: (i) deep or unselective oxidation to CO_2; (ii) a transition region between unselective and selective oxidation; (iii) a selective oxidation region; and (iv) a coking region. The different regions corresponded to the change in the oxidation states of iron oxide present in the core of the particles. The particles underwent changes in the surface oxygen available, lattice oxygen concentration, and the degree of reduced iron on the surface. The high selectivity towards formation of syngas was attributed to a relatively oxygen depleted surface which had a higher concentration of metallic iron as compared to the particles in regions (i) and (ii). Transient pulse studies conducted by Shafiefarwood et al.[98] showed that there is also a change in the mechanism of methane conversion from Eley-Rideal in region (i) to a Langmuir–Hinshelwood-like mechanism in region (iii). They concluded, based on their observations, that it was the amount and type of surface oxygen available that determines the mechanism of methane oxidation.

Natural and Artificial Doping

After extensive studies on the morphological enhancement of oxygen carrier particles, attention has gradually moved to the optimization of solid-phase ionic conductive capacity. As described in Section 2.3, solid-phase ionic diffusion determines how fast the solid product can be transported inward into the solid-grain phase and/or how fast the solid reactant can be transported outward to the solid-grain surface. Thus, a good solid-phase ionic conductivity can compensate the unavoidable loss in surface area and pore volume, thereby maintaining relatively stable reactivity and recyclability.[15] Oxygen carrier particles with high ionic conductivity can be grouped into two major categories: (1) artificially doped materials and (2) naturally doped minerals.

Because the ionic diffusion in a solid crystal structure is determined by the diffusion energy barrier and diffusion pathways which are usually constrained by the crystal structure, most pure metals/metal oxides (e.g. Fe and Fe_2O_3) have low intrinsic ionic conductive capability due to relatively well-organized ion packing.[80] However, when exotic ions are introduced into the well-organized crystal structures, the rigid crystal lattice can possibly be deformed by the hetero-ions, thereby creating additional diffusion pathways (vacancies and interstitial sites) and lowering the diffusion energy barrier. As a result, it is possible that the overall ionic conductivity is significantly enhanced.

A commonly performed doping or mixing procedure is to simply sinter a solids mixture, e.g. NiO/YSZ, and Fe_2O_3/TiO_2.[54,165,166] The mixing process is to ensure homogeneous distribution of the solid components, while a sintering process can make one component diffuse into the other, or the two components mutually diffuse into each other's crystal lattice. If the amount of dopant is significantly smaller than the primary component, they tend to form just one solid phase. The solid–solid interface would still exist if the concentration of the doped component exceeds its maximum solubility in the other solid. When multiple solid phases exist, there are usually structural defects at the solid–solid interface, which can possibly induce enhanced ionic conductivity along the boundary. Alternatively, materials with proven oxygen anion conductivity, such as YSZ, can be directly mixed with primary oxygen carrier material(s). Such materials are expected to provide stable ionic diffusion pathways in the mixture particle, through which the oxygen anions can enter and leave the particle with less resistance.[165,166] However, this approach does not always guarantee better ionic conductivity for a binary solid system, especially when the oxygen carrier material and support form a chemically inert species. It has been reported that a mixture of NiO and Al_2O_3 can form $NiAl_2O_4$, which is a chemically stable and low-reactivity component for chemical looping applications.[166,167]

Another typical artificially formed composite material is perovskite-type materials, such as $CaMn_{0.875}Ti_{0.125}O_3$.[168] The idea behind the synthesis of these materials is possibly based on the notion that materials with a perovskite structure usually have high levels of oxygen vacancies, resulting in enhanced oxygen anion conductivity. Materials of this unique structure are widely used in solid oxide fuel cells and other similar areas which require solid materials with a high oxygen conductive capability.[169] Hence, their application as chemical looping oxygen carrier materials looks to be promising.

It is known that some naturally occurring minerals and industrial by-products contain multiple metal components in a composite form, such as ilmenite ($FeTiO_3$), chromite ($FeCrO_4$), etc. Some multi-metal crystal structures can be regarded as a primary material "naturally" doped with another component, such as $FeO + TiO_2 \rightarrow FeTiO_3$, thus generating a high level of intrinsic defects due to a similar mechanism to that described above. Hence, some naturally occurring materials are able to perform as well as artificially doped composite particles. Li et al.[16] studied the ionic diffusion process in crystal structures of ilmenite ($FeTiO_3$) and pure FeO using DFT calculations, and found that the naturally doped ilmenite has a significantly lower diffusion energy barrier for oxygen anions than pure Fe. Fossdal et al.[170] studied the possibility of using some Norwegian industrial tailings and by-products as oxygen carriers for chemical looping combustion. Among all the tested samples, one manganese ore showed relatively good redox stability, but no detailed explanation was provided. Similar particle screening studies were also reported for other Fe-based and Mn-based materials. State of the art studies generally lack an effective approach or well-established theories to accurately predict the ionic diffusion behavior in naturally doped materials, especially those with complex crystal structures and compositions; current selection of these materials still follows the conventional trial and error strategy.

2.8   Concluding Remarks

Oxygen carrier particles in chemical looping play an important role in increasing process efficiencies compared to conventional fossil fuel conversion processes, especially when CO_2 capture is considered. As one of the most important factors towards their future commercial success, oxygen carrier particles need to maintain stable and satisfactory reactivity and recyclability. The most important feature that distinguishes oxygen carrier particles from catalysts and "one-time" consumable solid reactants is that oxygen carrier particles undergo cyclic gas–solid reactions through defects and ionic diffusion. During cyclic gas–solid reactions, the solid reactants are subject to cyclic changes in morphology and crystal structure and when coupled with the attrition induced during flows, they pose considerable challenges to maintenance of the physical integrity of the solid reactants.

As metal oxides commonly used in oxygen carrier particles undergo reaction induced morphological changes, these changes can influence the reactivity and recyclability of the oxygen carrier particles over their lifetime in the system. Although the surface area and pore volume of metal oxides may increase or decrease over a certain time period during the cyclic redox process, in general, however, physical properties of oxygen carriers tend to deteriorate at the end of many cycles in redox reactions. Most previous work attributes this loss to high-temperature sintering, which has led research efforts towards optimization of the initial solid morphological properties and anti-sintering properties. These are valid perspectives and countermeasures in catalytic reactions where the catalysts are not chemically converted, as well as "one-time" gas–solid reactions where only the initial solid morphology matters. As previously mentioned, the cyclic reaction per se is found to be an independent deteriorating factor towards the solid morphology, which implies that deterioration of the initial solid morphology is unavoidable. That is, even though the initial solid surface and pore volume can be significantly increased using some sophisticated techniques and be stabilized at a higher level, the long-term stability of the morphology for oxygen carriers should also be accounted for in the design of metal oxides.

In another aspect, it requires intense attention to the ionic conductive capability of the solid. A good ionic conductivity can sustain its reactivity at the chemical reaction controlled step, and the key to redox reactions is defect chemistry, including the creation and consumption of oxygen vacancies through the optimal diffusion pathways. Increased ionic conductivity can compensate the reactivity loss due to morphological deterioration. Because the ionic diffusive property is much less susceptible to sintering and morphological deterioration, effective enhancement of this property becomes important. For an oxygen carrier material, it is necessary to examine systematically the effects of various chemical and physical characteristics of the crystal structure on the overall ionic diffusion process, with a particular focus on the diffusion energy barrier and diffusion pathway. Understanding the mechanism can provide both theoretical guidance to artificial particle synthesis and predict particle performance with a minimum of trial and error screening. However, because of a lack of such understanding, the trial and error screening strategy still dominates the current approach to ionic conductivity optimization. It is noted that since most state of the art metal oxide materials that exhibit high intrinsic ionic conductivity also exhibit low physical strength, a novel design of metal oxide oxygen carriers thus often seeks a compromise between these two properties.

Mass and ionic diffusion, change of lattice structure during redox reactions, heat stress, impurities, and shear and impact stresses during the flow all contribute to attrition of the metal oxide materials in a chemical looping system. To screen metal oxide materials for their mechanical properties, it is desirable to first screen their crushing/compression stresses during non-flow redox conditions prior to examining their attrition properties using elaborative flow reactors with and without redox reactions. There are correlations for certain metal oxide materials between crushing stresses and attrition property in the reactor when the crushing stress resistance is high. Various attrition test devices that can be used to evaluate the impact and shear effects of metal oxide materials with and without reactions are discussed in this chapter. Various techniques that can be used to enhance the metal oxide performance are also presented.

References

1. Fan, L.-S., "Metal Oxide Reaction Engineering and Particle Technology Science – A Gateway to Novel Energy Conversion Systems," AIChE 67th Institute Lecture, Salt Lake City, UT, November 11 (2015).
2. Li, F., H. R. Kim, D. Sridhar, F. Wang, L. Zeng, J. Chen, and L.-S. Fan, "Syngas Chemical Looping Gasification Process: Oxygen Carrier Particle Selection and Performance," *Energy & Fuels*, 23, 4182–4189 (2009).
3. Adánez, J., L. F. de Diego, F. García-Labiano, P. Gayán, A. Abad, and J. M. Palacios, "Selection of Oxygen Carriers for Chemical-Looping Combustion," *Energy & Fuels*, 18, 371–377 (2004).
4. Fan, L.-S., *Chemical Looping Systems for Fossil Energy Conversions*, John Wiley & Sons, Hoboken, NJ (2010).
5. de Diego, L. F., F. García-Labiano, J. Adánez, P. Gayán, A. Abad, B. M. Corbella, and J. María Palacios, "Development of Cu-Based Oxygen Carriers for Chemical-Looping Combustion," *Fuel*, 83, 1749–1757 (2004).
6. Lee, J.-B., C.-S. Park, S.-I. Choi, Y.-W. Song, Y.-H. Kim, and H.-S. Yang, "Redox Characteristics of Various Kinds of Oxygen Carriers for Hydrogen Fueled Chemical-Looping Combustion," *Journal of Industrial and Engineering Chemistry*, 11, 96–102 (2005).

7. Yu, F.-C., N. Phalak, Z. Sun, and L.-S. Fan, "Activation Strategies for Calcium-Based Sorbents for CO_2 Capture: A Perspective," *Industrial & Engineering Chemistry Research*, 51, 2133–2142 (2012).

8. Mattisson T., M. Johansson, and A. Lyngfelt, "Multicycle Reduction and Oxidation of Different Types of Iron Oxide Particles Application to Chemical-Looping Combustion," *Energy & Fuels*, 18, 628–637 (2004).

9. Johansson M., T. Mattisson, and A. Lyngfelt, "Investigation of Fe_2O_3 with $MgAl_2O_4$ for Chemical-Looping Combustion," *Industrial & Engineering Chemistry Research*, 43, 6978–6987 (2004).

10. Sun, Z. "Morphological Property Variation and Ionic Transfer Behaviors of Solid Reactants in Fe-based and CaO-based Chemical Looping Processes," PhD Dissertation, The Ohio State University, Columbus, OH (2012).

11. Rubel A., K. Liu, J. Neathery, and D. Taulbee, "Oxygen Carriers for Chemical Looping Combustion of Solid Fuels," *Fuel*, 88, 876–884 (2009).

12. Qin, L., Z. Cheng, M. Guo, M. Xu, J. A. Fan, and L.-S. Fan, "Impact of 1% Lanthanum Dopant on Carbonaceous Fuel Redox Reactions with an Iron-Based Oxygen Carrier in Chemical Looping Processes," *ACS Energy Letters*, 2, 70–74 (2017).

13. Luo, S., L. Zeng, and L.-S. Fan, "Chemical Looping Technology: Oxygen Carrier Characteristics," *Annual Review of Chemical and Biomolecular Engineering*, 6, 53–75 (2015).

14. Paul, A., "Effect of Thermal Stabilization on Redox Equilibria and Colour of Glass," *Journal of Non-Crystalline Solids* 71, 269–278 (1985).

15. Zeng, L., M. V. Kathe, E. Y. Chung, and L.-S. Fan, "Some Remarks on Direct Solid Fuel Combustion Using Chemical Looping Processes," *Current Opinion in Chemical Engineering*, 1, 290–295 (2012).

16. Li, F., S. Luo, Z. Sun, X. Bao, and L.-S. Fan, "Role of Metal Oxide Support in Redox Reactions of Iron Oxide for Chemical Looping Applications: Experiments and Density Functional Theory Calculations," *Energy & Environmental Science*, 4, 3661–3667 (2011).

17. Arjmand, M., A.-M. Azad, H. Leion, A. Lyngfelt, and T. Mattisson, "Prospects of Al_2O_3 and $MgAl_2O_4$-Supported Cuo Oxygen Carriers in Chemical-Looping Combustion (CLC) and Chemical-Looping with Oxygen Uncoupling (CLOU)," *Energy & Fuels*, 25, 5493–5502 (2011).

18. de Diego, L. F., M. Ortiz, F. García-Labiano, J. Adánez, A. Abad, and P. Gayán, "Hydrogen Production by Chemical-Looping Reforming in a Circulating Fluidized Bed Reactor Using Ni-Based Oxygen Carriers," *Journal of Power Sources*, 192, 27–34 (2009).

19. Otsuka, K., Y. Wang, E. Sunada, and I. Yamanaka, "Direct Partial Oxidation of Methane to Synthesis Gas by Cerium Oxide," *Journal of Catalysis*, 175, 152–160 (1998).

20. Luo, S., L. Zeng, D. Xu, M. Kathe, E. Chung, N. Deshpande, L. Qin, A. Majumder, T.-L. Hsieh, A. Tong, Z. Sun, and L.-S. Fan, "Shale Gas-to-Syngas Chemical Looping Process for Stable Shale Gas Conversion to High Purity Syngas with a H_2:CO Ratio of 2:1," *Energy & Environmental Science*, 7, 4104–4117 (2014).

21. Otsuka, K., E. Sunada, T. Ushiyama, and I. Yamanaka, "The Production of Synthesis Gas by the Redox of Cerium Oxide," in *Studies in Surface Science and Catalysis*, de Pontes, M., R. L. E. C. P. Nicolaides, J. H. Scholtz and M. S. Scurrell, eds., 107, 531–536, Elsevier, Amsterdam, Netherlands (1997).

22. Gupta, A., M. S. Hegde, K. R. Priolkar, U. V. Waghmare, P. R. Sarode, and S. Emura, "Structural Investigation of Activated Lattice Oxygen in $Ce_{1-x}Sn_xO_2$ and $Ce_{1-x-y}Sn_xPd_yO_{2-\delta}$ by EXAFS and DFT Calculation," *Chemistry of Materials*, 21, 5836–5847 (2009).

23. Wang, S., T. Kobayashi, M. Dokiya, and T. Hashimoto, "Electrical and Ionic Conductivity of Gd-Doped Ceria," *Journal of the Electrochemical Society*, 147, 3606–3609 (2000).

24. Salazar-Villalpando, M. D., D. A. Berry, and A. Cugini, "Role of Lattice Oxygen in the Partial Oxidation of Methane over Rh/Zirconia-Doped Ceria. Isotopic Studies," *International Journal of Hydrogen Energy*, 35, 1998–2003 (2010).

25. Wu, Q., J. Chen, and J. Zhang, "Effect of Yttrium and Praseodymium on Properties of $Ce_{0.75}Zr_{0.25}O_2$ Solid Solution for CH_4–CO_2 Reforming," *Fuel Processing Technology*, 89, 993–999 (2008).

26. Kang, Z. C., and L. Eyring, "Lattice Oxygen Transfer in Fluorite-Type Oxides Containing Ce, Pr, and/or Tb," *Journal of Solid State Chemistry*, 155, 129–137 (2000).

27. Sadykov, V. A., T. G. Kuznetsova, G. M. Alikina, Y. V. Frolova, A. I. Lukashevich, Y. V. Potapova, V. S. Muzykantov, V. A. Rogov, V. V. Kriventsov, D. I. Kochubei, E. M. Moroz, D. I. Zyuzin, V. I. Zaikovskii, V. N. Kolomiichuk, E. A. Paukshtis, E. B. Burgina, V. V. Zyryanov, N. F. Uvarov, S. Neophytides, and E. Kemnitz, "Ceria-based Fluorite-like Oxide Solid Solutions as Catalysts of Methane Selective Oxidation into Syngas by the Lattice Oxygen: Synthesis, Characterization and Performance," *Catalysis Today*, 93–95, 45–53 (2004).

28. Jalibert, J. C., M. Fathi, O. A. Rokstad, and A. Holmen, "Synthesis Gas Production by Partial Oxidation of Methane from the Cyclic Gas-Solid Reaction Using Promoted Cerium Oxide," in Spivey, J. J., E. Iglesia, and T.H. Fleisch, eds., *Studies in Surface Science and Catalysis, 136, 301–306*, Elsevier, Amsterdam, Netherlands (2001).

29. Li, R. J., C. C. Yu, X. P. Dai, and S. K. Shen, "Partial Oxidation of Methane to Synthesis Gas Using Lattice Oxygen instead of Molecular Oxygen," *Chinese Journal of Catlysis*, 23, 381–387 (2002).

30. Jeong, H. H., J. H. Kwak, G. Y. Han, and K. J. Yoon, "Stepwise Production of Syngas and Hydrogen through Methane Reforming and Water Splitting by Using a Cerium Oxide Redox System," *International Journal of Hydrogen Energy*, 36, 15221–15230 (2011).

31. Chen, J., C. Yao, Y. Zhao, and P. Jia, "Synthesis Gas Production from Dry Reforming of Methane over $Ce_{0.75}Zr_{0.25}O_2$-Supported Ru Catalysts," *International Journal of Hydrogen Energy*, 35, 1630–1642 (2010).

32. Johansson, M., T. Mattisson, A. Lyngfelt, and A. Abad, "Using Continuous and Pulse Experiments to Compare Two Promising Nickel-Based Oxygen Carriers for Use in Chemical-Looping Technologies," *Fuel*, 87, 988–1001 (2008).

33. de Diego, L. F., M. Ortiz, J. Adánez, F. García-Labiano, A. Abad, and P. Gayán, "Synthesis Gas Generation by Chemical-Looping Reforming in a Batch Fluidized Bed Reactor Using Ni-Based Oxygen Carriers," *Chemical Engineering Journal*, 144, 289–298 (2008).

34. Jerndal, E., T. Mattisson, and A. Lyngfelt, "Investigation of Different $NiO/NiAl_2O_4$ Particles as Oxygen Carriers for Chemical-Looping Combustion," *Energy & Fuels*, 23, 665–676 (2008).

35. Zafar, Q., T. Mattisson, and B. Gevert, "Redox Investigation of Some Oxides of Transition-State Metals Ni, Cu, Fe, and Mn Supported on SiO_2 and $MgAl_2O_4$," *Energy & Fuels*, 20, 34–44 (2006).

36. Adánez, J., A. Abad, F. García-Labiano, P. Gayán, and L. F. de Diego, "Progress in Chemical-Looping Combustion and Reforming Technologies," *Progress in Energy and Combustion Science*, 38, 215–282 (2012).

37. Rydén, M., A. Lyngfelt, and T. Mattisson, "Synthesis Gas Generation by Chemical-Looping Reforming in a Continuously Operating Laboratory Reactor," *Fuel*, 85, 1631–1641 (2006).

38. Pröll, T., J. Bolhàr-Nordenkampf, P. Kolbitsch, and H. Hofbauer, "Syngas and a Separate Nitrogen/Argon Stream via Chemical Looping Reforming – A 140 kW Pilot Plant Study," *Fuel*, 89, 1249–1256 (2010).

39. Azadi, P., J. Otomo, H. Hatano, Y. Oshima, and R. Farnood, "Interactions of Supported Nickel and Nickel Oxide Catalysts with Methane and Steam at High Temperatures," *Chemical Engineering Science*, 66, 4196–4202 (2011).

40. Nakayama, O., N. Ikenaga, T. Miyake, E. Yagasaki, and T. Suzuki, "Partial Oxidation of CH_4 with Air to Produce Pure Hydrogen and Syngas," *Catalysis Today*, 138, 141–146 (2008).

41. Steinfeld, A., P. Kuhn, and J. Karni, "High-Temperature Solar Thermochemistry: Production of Iron and Synthesis Gas by Fe_3O_4-Reduction with Methane," *Energy*, 18, 239–249 (1993).

42. Thomas, T. J., L.-S. Fan, P. Gupta, and L. G. Velazquez-Vargas, "Combustion Looping using Composite Oxygen Carriers," U.S. Patent 7,767,191 (2010).

43. Li, F., L. Zeng, L. G. Velazquez-Vargas, Z. Yoscovits, and L.-S. Fan, "Syngas Chemical Looping Gasification Process: Bench-Scale Studies and Reactor Simulations," *AIChE Journal*, 56, 2186–2199 (2010).

44. Sridhar, D., A. Tong, H. Kim, L. Zeng, F. Li, and L.-S. Fan, "Syngas Chemical Looping Process: Design and Construction of a 25 kWth Subpilot Unit," *Energy & Fuels*, 26, 2292–2302 (2012).

45. Tong, A., D. Sridhar, Z. Sun, H. R. Kim, L. Zeng, F. Wang, D. Wang, M. V. Kathe, S. Luo, Y. Sun, and L.-S. Fan, "Continuous High Purity Hydrogen Generation from a Syngas Chemical Looping 25 kWth Sub-Pilot Unit with 100% Carbon Capture," *Fuel*, 103, 495–505 (2013).

46. Kim, H. R., D. Wang, L. Zeng, S. Bayham, A. Tong, E. Chung, M. V. Kathe, S. Luo, O. McGiveron, A. Wang, Z. Sun, D. Chen, and L.-S. Fan, "Coal Direct Chemical Looping Combustion Process: Design and Operation of a 25-kWth Sub-Pilot Unit," *Fuel*, 108, 370–384 (2013).

47. Bayham, S. C., H. R. Kim, D. Wang, A. Tong, L. Zeng, O. McGiveron, M. V. Kathe, E. Chung, W. Wang, A. Wang, A. Majumder, and L.-S. Fan, "Iron-Based Coal Direct Chemical Looping Combustion Process: 200-h Continuous Operation of a 25-kWth Sub-pilot Unit," *Energy & Fuels*, 27, 1347–1356 (2013).

48. Tong, A., L. Zeng, M. V. Kathe, D. Sridhar, and L.-S. Fan, "Application of the Moving-Bed Chemical Looping Process for High Methane Conversion," *Energy & Fuels*, 27, 4119–4128 (2013).

49. Pineau, A., N. Kanari, and I. Gaballah, "Kinetics of Reduction of Iron Oxides by H_2: Part II. Low Temperature Reduction of Magnetite," *Thermochimica Acta*, 456, 75–88 (2007).

50. Pineau, A., N. Kanari, and I. Gaballah, "Kinetics of Reduction of Iron Oxides by H_2: Part I: Low Temperature Reduction of Hematite," *Thermochimica Acta*, 447, 89–100 (2006).

51. Li, S., S. Krishnamoorthy, A. Li, G. D. Meitzner, and E. Iglesia, "Promoted Iron-Based Catalysts for the Fischer–Tropsch Synthesis: Design, Synthesis, Site Densities, and Catalytic Properties," *Journal of Catalysis*, 206, 202–217 (2002).

52. Chen, S., Q. Shi, Z. Xue, X. Sun, and W. Xiang, "Experimental Investigation of Chemical-Looping Hydrogen Generation Using Al_2O_3 or TiO_2-Supported Iron Oxides in a Batch Fluidized Bed," *International Journal of Hydrogen Energy*, 36, 8915–8926 (2011).

53. Liu, L. and M. R. Zachariah, "Enhanced Performance of Alkali Metal Doped Fe_2O_3 and Fe_2O_3/Al_2O_3 Composites As Oxygen Carrier Material in Chemical Looping Combustion," *Energy & Fuels*, 27, 4977–4983 (2013).

54. Li, F., Z. Sun, S. Luo, and L.-S. Fan, "Ionic Diffusion in the Oxidation of Iron—Effect of Support and its Implications to Chemical Looping Applications," *Energy & Environmental Science*, 4, 876–880 (2011).

55. Galinsky, N. L., Y. Huang, A. Shafiefarhood, and F. Li, "Iron Oxide with Facilitated O^{2-} Transport for Facile Fuel Oxidation and CO_2 Capture in a Chemical Looping Scheme", *ACS Sustainable Chemistry & Engineering*, 1, 364–373 (2013).

56. Galinsky, N. L., A. Shafiefarhood, Y. Chen, L. Neal, and F. Li, "Effect of Support on Redox Stability of Iron Oxide for Chemical Looping Conversion of Methane," *Applied Catalysis B: Environmental*, 164, 371–379 (2015).

57. Kodama, T., H. Ohtake, S. Matsumoto, et al., "Thermochemical Methane Reforming Using a Reactive WO_3/W Redox System," *Energy*, 25, 411–425 (2000).

58. Kodama, T., T. Shimizu, T. Satoh, and K.-I. Shimizu, "Stepwise Production of CO-Rich Syngas and Hydrogen via Methane Reforming by a WO_3-Redox Catalyst," *Energy*, 28, 1055–1068 (2003).

59. Steinfeld, A., M. Brack, A. Meier, A. Weidenkaff, and D. Wuillemin, "A Solar Chemical Reactor for Co-Production of Zinc and Synthesis Gas," *Energy*, 23, 803–814 (1998).

60. Aoki, A., T. Shimizu, Y. Kitayama, and T. Kodama, "A Two-Step Thermochemical Conversion of CH_4 to CO, H_2 and C_2-Hydrocarbons below 1173 K," *Journal de Physique IV*, 09, 337–342 (1999).

61. Halmann, M., A. Frei, and A. Steinfeld, "Thermo-Neutral Production of Metals and Hydrogen or Methanol by the Combined Reduction of the Oxides of Zinc or Iron with Partial Oxidation of Hydrocarbons," *Energy*, 27, 1069–1084 (2002).

62. Wei, H. J., Y. Cao, W. J. Ji, and C. T. Au, "Lattice Oxygen of $La_{1-x}Sr_xMO_3$ (M = Mn, Ni) and $LaMnO_{3-\alpha}F_\beta$ Perovskite Oxides for the Partial Oxidation of Methane to Synthesis Gas," *Catalysis Communications*, 9, 2509–2514 (2008).

63. Dai, X. P., R. J. Li, C. C. Yu, and Z. P. Hao, "Unsteady-State Direct Partial Oxidation of Methane to Synthesis Gas in a Fixed-Bed Reactor Using $AFeO_3$ (A = La, Nd, Eu) Perovskite-Type Oxides as Oxygen Storage," *Journal of Physical Chemistry B*, 110, 22525–22531 (2006).

64. Wang, Y., J. Zhu, L. Zhang, X. Yang, L. Lu, and X. Wang, "Preparation and Characterization of Perovskite $LaFeO_3$ Nanocrystals," *Materials Letters*, 60, 1767–1770 (2006).

65. Mihai, O., D. Chen, and A. Holmen, "Catalytic Consequence of Oxygen of Lanthanum Ferrite Perovskite in Chemical Looping Reforming of Methane," *Industrial & Engineering Chemistry Research*, 50, 2613–2621 (2011).

66. Nalbandian, L., A. Evdou, and V. Zaspalis, "$La_{1-x}Sr_xM_yFe_{1-y}O_{3-\delta}$ Perovskites as Oxygen-Carrier Materials for Chemical-Looping Reforming," *International Journal of Hydrogen Energy*, 36, 6657–6670 (2011).

67. Kodama T., T. Shimizu, T. Satoh, M. Nakata, and K.-I. Shimizu, "Stepwise Production of CO-Rich Syngas and Hydrogen via Solar Methane Reforming by Using a Ni(II)-Ferrite Redox System," *Solar Energy*, 73, 363–374 (2002).

68. Sturzenegger, M., L. D'Souza, R. P. W. J. Struis, and S. Stucki, "Oxygen Transfer and Catalytic Properties of Nickel Iron Oxides for Steam Reforming of Methane," *Fuel*, 85, 1599–1602 (2006).

69. Cha, K.-S., B.-K. Yoo, H.-S. Kim, T.-G. Ryu, K.-S. Kang, C.-S. Park, and Y.-H. Kim, "A Study on Improving Reactivity of Cu-Ferrite/ZrO$_2$ Medium for Syngas and Hydrogen Production from Two-Step Thermochemical Methane Reforming," *International Journal of Energy Research*, 34, 422–430 (2010).

70. Cha, K.-S., H.-S. Kim, B.-K. Yoo, Y.-S. Lee, K.-S. Kang, C.-S. Park, and Y.-H. Kim, "Reaction Characteristics of Two-Step Methane Reforming over a Cu-Ferrite/Ce–ZrO$_2$ Medium," *International Journal of Hydrogen Energy*, 34, 1801–1808 (2009).

71. Nakayama, O., N. Ikenaga, T. Miyake, E. Yagasaki, and T. Suzuki, "Production of Synthesis Gas from Methane Using Lattice Oxygen of NiO–Cr$_2$O$_3$–MgO Complex Oxide," *Industrial & Engineering Chemistry Research*, 49, 526–534 (2010).

72. Deshpande, N., A. Majumder, L. Qin, and L.-S. Fan, "High-Pressure Redox Behavior of Iron-Oxide-Based Oxygen Carriers for Syngas Generation from Methane," *Energy & Fuels*, 29, 1469–1478 (2015).

73. Rydén, M., A. Lyngfelt, and T. Mattisson, "Chemical-Looping Combustion and Chemical-Looping Reforming in a Circulating Fluidized-Bed Reactor Using Ni-Based Oxygen Carriers," *Energy & Fuels*, 22, 2585–2597 (2008).

74. Chen, L., Y. Lu, Q. Hong, J. Lin, and F. M. Dautzenberg, "Catalytic Partial Oxidation of Methane to Syngas over Ca-Decorated-Al$_2$O$_3$-Supported Ni and NiB Catalysts," *Applied Catalysis A: General*, 292, 295–304 (2005).

75. Kobayashi, Y., J. Horiguchi, S. Kobayashi, Y. Yamazaki, K. Omata, D. Nagao, M. Konno, and M. Yamada, "Effect Of NiO Content in Mesoporous NiO–Al$_2$O$_3$ Catalysts for High Pressure Partial Oxidation of Methane to Syngas," *Applied Catalysis A: General*, 395, 129–137 (2011).

76. Abad, A., F. García-Labiano, L. F. de Diego, P. Gayán, and J. Adánez, "Reduction Kinetics of Cu-, Ni-, and Fe-Based Oxygen Carriers Using Syngas (CO + H$_2$) for Chemical-Looping Combustion," *Energy & Fuels*, 21, 1843–1853 (2007).

77. Go, K. S., S. R. Son, and S. D. Kim, "Reaction Kinetics of Reduction and Oxidation of Metal Oxides for Hydrogen Production," *International Journal of Hydrogen Energy*, 33, 5986–5995 (2008).

78. García-Labiano, F., J. Adánez, L. F. de Diego, P. Gayán, and A. Abad, "Effect of Pressure on the Behavior of Copper-, Iron-, and Nickel-Based Oxygen Carriers for Chemical-Looping Combustion," *Energy & Fuels*, 20, 26–33 (2006).

79. Hsia, C., G. R. S. Pierre, and L.-S. Fan, "Isotope Study on Diffusion in CaSO$_4$ Formed During Sorbent-Flue-Gas Reaction," *AIChE Journal*, 41, 2337–2340 (1995).

80. Chiang, Y.-M., D. P. Birnie, and W. D. Kingery, *Physical Ceramics: Principles for Ceramic Science and Engineering*, John Wiley & Sons, New York, NY (1997).

81. Sun, Z., S. Luo, and L.-S. Fan, "Ionic Transfer Mechanism of COS Reaction with CaO: Inert Marker Experiment and Density Functional Theory (DFT) Calculation," *AIChE Journal*, 58, 2617–2620 (2012).

82. Sun, Z., Q. Zhou, and L.-S. Fan, "Reactive Solid Surface Morphology Variation via Ionic Diffusion," *Langmuir*, 28, 11827–11833 (2012).

83. Day, R. J. and M. A. Frisch, "Chromium Chromate as an Inert Marker in Copper Oxidation," *Surface and Interface Analysis*, 8, 33–36 (1986).

84. Petit, F. S. and J. B. Wagner Jr., "Oxidation of Cobalt in CO-CO$_2$ Mixtures in the Temperature Range 920° C–1200° C," *Acta Metallurgica*, 12, 41–47 (1964).

85. Virkar, A. V., J. L. Huang, and R. A. Cutler, "Strengthening of Oxide Ceramics by Transformation-Induced Stress," *Journal of the American Ceramic Society*, 70, 164–170 (1987).

86. Sun, Z., H. Chi, and L.-S. Fan, "Physical and Chemical Mechanism for Increased Surface Area and Pore Volume of CaO in Water Hydration," *Industrial & Engineering Chemistry Research*, 51, 10793–10799 (2012).

87. Sun, Z., Q. Zhou, and L.-S. Fan, "Formation of Core–Shell Structured Composite Microparticles via Cyclic Gas–Solid Reactions," *Langmuir*, 29, 12520–12529 (2013).

88. Qin, L., A. Majumder, J. A. Fan, D. Kopechek, and L.-S. Fan, "Evolution of Nanoscale Morphology in Single and Binary Metal Oxide Microparticles During Reduction and Oxidation Processes," *Journal of Materials Chemistry A*, 2, 17511–17520 (2014).

89. Fu, Y., J. Chen, and H. Zhang, "Synthesis of Fe_2O_3 Nanowires by Oxidation of Iron," *Chemical Physics Letters*, 350, 491–494 (2001).

90. Wen, X., S. Wang, Y. Ding, Z. L. Wang, and S. Yang, "Controlled Growth of Large-Area, Uniform, Vertically Aligned Arrays of α-Fe_2O_3 Nanobelts and Nanowires," *Journal of Physical Chemistry B*, 109, 215–220 (2005).

91. Dang, H. Y., J. Wang, and S. S. Fan, "The Synthesis of Metal Oxide Nanowires by Directly Heating Metal Samples in Appropriate Oxygen Atmospheres," *Nanotechnology*, 14, 738 (2003).

92. Farbod, M., N. Meamar Ghaffari, and I. Kazeminezhad, "Fabrication of Single Phase CuO Nanowires and Effect of Electric Field on their Growth and Investigation of their Photocatalytic Properties," *Ceramics International*, 40, 517–521 (2014).

93. Fang, H. Z., S. L. Shang, Y. Wang, Z. K. Liu, D. Alfonso, D. E. Alman, Y. K. Shin, C. Y. Zou, A. C. T. van Duin, Y. K. Lei, and G. F. Wang, "First-Principles Studies on Vacancy-Modified Interstitial Diffusion Mechanism of Oxygen in Nickel, Associated with Large-Scale Atomic Simulation Techniques," *Journal of Applied Physics*, 115, 043501 (2014).

94. Lloyd, G. J. and J. W. Martin, "The Diffusivity of Oxygen in Nickel Determined by Internal Oxidation of Dilute Ni–Be Alloys," *Metal Science Journal*, 6, 7–11 (1972).

95. Liu, H., B. Wang, M. Fan, et al., "Study on Carbon Deposition Associated with Catalytic CH_4 Reforming by Using Density Functional Theory," *Fuel*, 113, 712–718 (2013).

96. Huang, J., Z. Huang, S. Yi, Y. Liu, M. Fang, and S. Zhang, "Fe-Catalyzed Growth of One-Dimensional α-Si_3N_4 Nanostructures and their Cathodoluminescence Properties," *Scientific Reports*, 3, (2013).

97. Qin, L., Z. Cheng, J. A. Fan, D. Kopechek, D. Xu, N. Deshpande, and L.-S. Fan, "Nanostructure Formation Mechanism and Ion Diffusion in Iron–Titanium Composite Materials with Chemical Looping Redox Reactions," *Journal of Materials Chemistry A*, 3, 11302–11312 (2015).

98. Shafiefarhood, A., N. Galinsky, Y. Huang, Y. Chen, and F. Li, "Fe_2O_3@$La_xSr_{1-x}FeO_3$ Core-Shell Redox Catalyst for Methane Partial Oxidation," *ChemCatChem*, 6, 790–799 (2014).

99. Shafiefarhood, A., J. C. Hamill, L. M. Neal, and F. Li, "Methane Partial Oxidation Using FeO_x@$La_{0.8}Sr_{0.2}FeO_{3-\delta}$ Core–Shell Catalyst – Transient Pulse Studies," *Physical Chemistry Chemical Physics*, 17, 31–39 (2015).

100. Neal, L., A. Shafiefarhood, and F. Li, "Dynamic Methane Partial Oxidation Using a Fe_2O_3@$La_{0.8}Sr_{0.2}FeO_{3-\delta}$ Core-Shell Redox Catalyst in the Absence of Gaseous Oxygen," *ACS Catalysis*, 4, 3560–3569 (2014).

101. Chen, W. K. and N. L. Peterson, "Effect of the Deviation from Stoichiometry on Cation Self-Diffusion and Isotope Effect in Wüstite, $Fe_{1-x}O$," *Journal of Physics and Chemistry of Solids*, 36, 1097–1103 (1975).

102. Nakamura, R., D. Tokozakura, H. Nakajima, J.-G. Lee, and H. Mori, "Hollow Oxide Formation by Oxidation of Al and Cu Nanoparticles," *Journal of Applied Physics*, 101, 074303 (2007).

103. Gemelli, E. and N. H. A. Camargo, "Oxidation Kinetics of Commercially Pure Titanium," *Matéria (Rio de Janeiro)*, 12, 525–531 (2007).

104. Fellows, R. A., A. R. Lennie, D. J. Vaughan, and G. Thornton, "A LEED Study of the $FeTiO_3(0001)$ Surface Following Annealing in O_2 Partial Pressures," *Surface Science*, 383, 50–56 (1997).

105. Wilson, N. C., J. Muscat, D. Mkhonto, P. E. Ngoepe, and N. M. Harrison, "Structure and Properties of Ilmenite from First Principles," *Physical Review B*, 71, 075202 (2005).

106. Ando, K. and Y. Oishi, "Self-Diffusion Coefficients of Oxygen Ion in Single Crystals of $MgO \cdot nAl_2O_3$ Spinels," *Journal of Chemical Physics*, 61, 625–629 (1974).

107. Oishi, Y. and K. Ando, "Self-Diffusion of Oxygen in Polycrystalline $MgAl_2O_4$," *Journal of Chemical Physics*, 63, 376–378 (1975).

108. Reddy, K. P. R. and A. R. Cooper, "Oxygen Diffusion in MgO and $\alpha\text{-}Fe_2O_3$", *Journal of the American Ceramic Society*, 66, 664–666 (1983).

109. Uberuaga, B. P., D. Bacorisen, R. Smith, J. A. Ball, R. W. Grimes, A. F. Voter, and K. E. Sickafus, "Defect Kinetics in Spinels: Long-Time Simulations of $MgAl_2O_4$, $MgGa_2O_4$, and $MgIn_2O_4$," *Physical Review B*, 75, 104116 (2007).

110. Erhart, P. and K. Albe, "First-principles Study of Migration Mechanisms and Diffusion of Oxygen in Zinc Oxide," *Physical Review B*, 73, 115207 (2006).

111. Vineyard, G. H., "Frequency Factors and Isotope Effects in Solid State Rate Processes," *Journal of Physics and Chemistry of Solids*, 3, 121–127 (1957).

112. Cheng, Z., L. Qin, M. Guo, J. A. Fan, D. Xu, and L.-S. Fan, "Methane Adsorption and Dissociation on Iron Oxide Oxygen Carriers: The Role of Oxygen Vacancies," *Physical Chemistry Chemical Physics*, 18, 16423–16435 (2016).

113. Hara, Y., "Mathematical Model of the Shaft Furnace for Reducing Iron Ore Pellets," *Tetsu-to-Hagane*, 62, 315–323 (1976).

114. Hara, Y., M. Tsuchiya, and S. Kondo, "Intraparticle Temperature of Iron-Oxide Pellet During the Reduction," *Tetsu-to-Hagane*, 60, 1261–1270 (1974).

115. Turkdogan, E. T. and J. V. Vinters, "Gaseous Reduction of Iron Oxides: Part I. Reduction of Hematite in Hydrogen," *Metallurgical and Materials Transactions B*, 2, 3175–3188 (1971).

116. Szekely, J. and J. W. Evans, "A Structural Model for Gas—Solid Reactions with a Moving Boundary," *Chemical Engineering Science*, 25, 1091–1107 (1970).

117. Szekely, J. and J. W. Evans, "A Structural Model for Gas-Solid Reactions with a Moving Boundary-II: The Effect of Grain Size, Porosity and Temperature on the Reaction of Porous Pellets," *Chemical Engineering Science*, 26, 1901–1913 (1971).

118. Sohn, H. Y. and J. Szekely, "A Structural Model for Gas-Solid Reactions with a Moving Boundary—III: A General Dimensionless Representation of the Irreversible Reaction between a Porous Solid and a Reactant Gas," *Chemical Engineering Science*, 27, 763–778 (1972).

119. Sohn, H. Y. and J. Szekely, "A Structural Model for Gas—Solid Reactions with a Moving Boundary—IV. Langmuir—Hinshelwood Kinetics," *Chemical Engineering Science*, 28, 1169–1177 (1973).

120. Szekely, J., C. I. Lin, and H. Y. Sohn, "A Structural Model for Gas—Solid Reactions with a Moving Boundary—V an Experimental Study of the Reduction of Porous Nickel-Oxide pellets with Hydrogen," *Chemical Engineering Science*, 28, 1975–1989 (1973).

121. Szekely, J. and M. Propster, "A Structural Model for Gas Solid Reactions with a Moving Boundary—VI: The Effect of Grain Size Distribution on the Conversion of Porous Solids," *Chemical Engineering Science*, 30, 1049–1055 (1975).

122. Ishida, M. and C. Y. Wen, "Comparison of Zone-Reaction Model and Unreacted-Core Shrinking Model in Solid—Gas Reactions—I Isothermal Analysis," *Chemical Engineering Science*, 26, 1031–1041 (1971).

123. Vargas, L. G. V., "*Development of Chemical Looping Gasification Processes for the Production of Hydrogen from Coal*," PhD Dissertation, The Ohio State University, Columbus, OH (2007).

124. Zeng, L., "*Multiscale Study of Chemical Looping Technology and Its Applications for Low Carbon Energy Conversions*," PhD Dissertation, The Ohio State University, Columbus, OH (2012).

125. Huang, J., T. Furukawa, and K. Aoto, "High Temperature Behavior of Na–Fe Oxides in H_2O+CO_2 Atmosphere," *Journal of Physics and Chemistry of Solids*, 66, 388–391 (2005).

126. Mattisson, T., A. Lyngfelt, and H. Leion, "Chemical-Looping with Oxygen Uncoupling For Combustion of Solid Fuels," *International Journal of Greenhouse Gas Control*, 3, 11–19 (2009).

127. Frick, V., M. Rydén, H. Leion, T. Mattisson, and A. Lyngfelt, "Screening of Supported and Unsupported Mn-Si Oxygen Carriers for CLOU (Chemical-Looping with Oxygen Uncoupling)," *Energy*, 93, 544–554 (2015).

128. Leion, H., T. Mattisson, and A. Lyngfelt, "Using Chemical-Looping with Oxygen Uncoupling (CLOU) for Combustion of Six Different Solid Fuels," *Energy Procedia*, 1, 447–453 (2009).

129. Shafiefarhood, A., A. Stewart, and F. Li, "Iron-Containing Mixed-Oxide Composites as Oxygen Carriers for Chemical Looping with Oxygen Uncoupling (CLOU)," *Fuel*, 139, 1–10 (2015).

130. Allendorf, M. D., R. B. Diver, N. P. Siegel, and J.E. Miller, "Two-Step Water Splitting Using Mixed-Metal Ferrites: Thermodynamic Analysis and Characterization of Synthesized Materials," *Energy & Fuels*, 22, 4115–4124 (2008).

131. Wang, D. and L.-S. Fan, "Bulk Coarse Particle Arching Phenomena in a Moving Bed with Fine Particle Presence," *AIChE Journal*, 60, 881–892 (2014).

132. Werther, J. and J. Reppenhagen, "Attrition in Fluidized Beds and Pneumatic Conveying Lines," Chapter 7 in Yang, W. C., ed., *Fluidization, Solids Handling, and Processing*, Noyes Publications, Westwood, NJ (1998).

133. Yuregir, K. R., M. Ghadiri, and R. Clift, "Observations on Impact Attrition of Granular Solids," *Powder Technology*, 49, 53–57 (1986).

134. Werther, J. and J. Reppenhagen, "Attrition," Chapter 8 in Yang, W. C., ed., *Handbook of Fluidization and Fluid-Particle Systems*, Marcel Dekker, New York, NY (2003).

135. Yuregir, K. R., M. Ghadiri, and R. Clift, "Impact Attrition of Sodium Chloride Crystals," *Chemical Engineering Science*, 42, 843–853 (1987).

136. Cleaver, J. A. S., M. Ghadiri, and N. Rolfe, "Impact Attrition of Sodium Carbonate Monohydrate Crystals," *Powder Technology*, 76, 15–22 (1993).

137. Hutchings, I. M. and R. E. Winter, "A Simple Small-Bore Laboratory Gas-Gun," *Journal of Physics E*, 8, 84 (1975).

138. Salman, A. D., D. A. Gorham, and A. Verba, "A Study of Solid Particle Failure under Normal and Oblique Impact," *Wear*, 186–187, Part 1, 92–98 (1995).

139. Davuluri, R. P. and T. M. Knowlton, "Development of a Standardized Attrition Test Procedure," Fluidization IX: Proceedings of the Ninth Engineering Foundation Conference on Fluidization, 333–340 (1998).

140. Chen, Z., C. Jim Lim, and J. R. Grace, "Study of Limestone Particle Impact Attrition," *Chemical Engineering Science*, 62, 867–877 (2007).

141. Forsythe, W. L. and W. R. Hertwig, "Attrition Characteristics of Fluid Cracking Catalysts," *Industrial & Engineering Chemistry Research*, 41, 1200–1206 (1949).

142. Cairati, L., L. Di Flore, P. Forzatti, I. Pasquon, and F. Trifiro, "Oxidation of Methanol in a Fluidized Bed. 1. Catalyst Attrition Resistance and Process Variable Study," *Industrial & Engineering Chemistry Process Design and Development*, 19, 561–565 (1980).

143. Contractor, R. M., H. E. Bergna, U. Chowdry, and A. W. Sleight, "Attrition Resistant Catalysts for Fluidized Bed Systems", Fluidization VI: Proceedings of the International Conference on Fluidization, 589–596 (1989).

144. Gwyn, J. E., "On the Particle Size Distribution Function and the Attrition of Cracking Catalysts," *AIChE Journal*, 15, 35–39 (1969).

145. Weeks, S. A. and P. Dumbill, "Method Speeds FCC Catalyst Attrition Resistance Determinations," *Oil & Gas Journal*, 88:16, (1990).

146. Cocco, R., Y. Arrington, R. Hays, J. Findlay, S. B. R. Karri, and T. M. Knowlton, "Jet Cup Attrition Testing," *Powder Technology*, 200, 224–233 (2010).

147. Mendiara, T., L. F. de Diego, F. García-Labiano, P. Gayán, A. Abad, and J. Adánez, "Behaviour of a Bauxite Waste Material as Oxygen Carrier in a 500 W_{th} CLC Unit with Coal," *International Journal of Greenhouse Gas Control*, 17, 170–182 (2013).

148. Mendiara, T., L. F. de Diego, F. García-Labiano, P. Gayán, A. Abad, and J. Adánez, "Use of an Fe-based Residue from Alumina Production as an Oxygen Carrier in Chemical-Looping Combustion," *Energy & Fuels*, 26, 1420–1431 (2012).

149. de Diego, L. F., P. Gayán, F. García-Labiano, J. Celaya, A. Abad, and J. Adánez, "Impregnated CuO/Al$_2$O$_3$ Oxygen Carriers for Chemical-Looping Combustion: Avoiding Fluidized Bed Agglomeration," *Energy & Fuels*, 19, 1850–1856 (2005).

150. Gayán, P., L. F. de Diego, F. García-Labiano, et al., "Effect of Support on Reactivity and Selectivity of Ni-Based Oxygen Carriers for Chemical-Looping Combustion," *Fuel*, 87, 2641–2650 (2008).

151. Gayán, P., C. Dueso, A. Abad, et al., "NiO/Al$_2$O$_3$ Oxygen Carriers for Chemical-Looping Combustion Prepared by Impregnation and Deposition–Precipitation Methods," *Fuel*, 88, 1016–1023 (2009).

152. Cuadrat, A., A. Abad, J. Adánez, L. F. de Diego, F. García-Labiano, and P. Gayán, "Behavior of Ilmenite as Oxygen Carrier in Chemical-Looping Combustion," *Fuel Processing Technology*, 94, 101–112 (2012).

153. Adánez, J., F. García-Labiano, L. F. de Diego, P. Gayán, J. Celaya, and A. Abad, "Nickel–Copper Oxygen Carriers to Reach Zero CO and H$_2$ Emissions in Chemical-Looping Combustion," *Industrial & Engineering Chemistry Research*, 45, 2617–2625 (2006).

154. Brown, T. A., F. Scala, S. A. Scott, J. S. Dennis, and P. Salatino, "The Attrition Behaviour of Oxygen-Carriers Under Inert and Reacting Conditions," *Chemical Engineering Science*, 71, 449–467 (2012).

155. Rydén, M., P. Moldenhauer, S. Lindqvist, T. Mattisson, and A. Lyngfelt, "Measuring Attrition Resistance of Oxygen Carrier Particles for Chemical Looping Combustion with a Customized Jet Cup," *Powder Technology*, 256, 75–86 (2014).

156. Azimi, G., T. Mattisson, H. Leion, M. Rydén, and A. Lyngfelt, "Comprehensive Study of Mn–Fe–Al Oxygen-Carriers for Chemical-Looping with Oxygen Uncoupling (CLOU)," *International Journal of Greenhouse Gas Control*, 34, 12–24 (2015).

157. Arjmand, M., V. Frick, M. Rydén, H. Leion, T. Mattisson, and A. Lyngfelt, "Screening of Combined Mn-Fe-Si Oxygen Carriers for Chemical Looping with Oxygen Uncoupling (CLOU)," *Energy & Fuels*, 29, 1868–1880 (2015).

158. Azimi, G., H. Lrion, M. Rydén, T. Mattisson, and A. Lyngfelt, "Investigation of Different Mn-Fe Oxides as Oxygen Carrier for Chemical-Looping with Oxygen Uncoupling (CLOU)," *Energy & Fuels*, 27, 367–377 (2013).

159. Baek, J.-I., J. Ryu, J. B. Lee, et al., "Highly Attrition Resistant Oxygen Carrier for Chemical Looping Combustion," *Energy Procedia*, 4, 349–355 (2011).

160. Baek, J.-I., C. K. Ryu, J. Ryu, J.-W. Kim, T. H. Eom, J. B. Lee, and J. Yi, "Performance Comparison of Spray-Dried Oxygen Carriers: The Effect of NiO and Pseudoboehmite Content in Raw Materials," *Energy & Fuels*, 24, 5757–5764 (2010).

161. Baek, J.-I., S.-R. Yang, T. H. Eom, J. B. Lee, and C. K. Ryu, "Effect of MgO Addition on the Physical Properties and Reactivity of the Spray-Dried Oxygen Carriers Prepared with a High Content of NiO And Al_2O_3," *Fuel*, 144, 317–326 (2015).

162. Pishahang, M., E. Bakken, S. Stølen, C. I. Thomas, and P. I. Dahl, "Oxygen Non-Stoichiometry, Redox Thermodynamics, and Structure of $LaFe_{1-x}Co_xO_{3-\delta}$," *Ionics*, 19, 869–878 (2012).

163. Song, Q., W. Liu, C. D. Bohn, R. N. Harper, E. Sivaniah, S. A. Scott, and J. S. Dennis, "A High Performance Oxygen Storage Material for Chemical Looping Processes with CO_2 Capture," *Energy & Environmental Science*, 6, 288–298 (2012).

164. Scheffe, J. R., M. D. Allendorf, E. N. Coker, B. W. Jacobs, A. H. McDaniel, and A. W. Weimer, "Hydrogen Production via Chemical Looping Redox Cycles Using Atomic Layer Deposition-Synthesized Iron Oxide and Cobalt Ferrites," *Chemistry of Materials*, 23, 2030–2038 (2011).

165. Jin, H., T. Okamoto, and M. Ishida, "Development of a Novel Chemical-Looping Combustion: Synthesis of a Looping Material with a Double Metal Oxide of CoO–NiO," *Energy & Fuels*, 12, 1272–1277 (1998).

166. Ishida, M. and H. Jin, "A Novel Combustor Based on Chemical-Looping Reactions and its Reaction Kinetics," *Journal of Chemical Engineering of Japan*, 27, 296–301 (1994).

167. Sedor, K. E., M. M. Hossain, and H. I. de Lasa, "Reactivity and Stability of Ni/Al_2O_3 Oxygen Carrier for Chemical-Looping Combustion (CLC)," *Chemical Engineering Science*, 63, 2994–3007 (2008).

168. Leion, H., Y. Larring, E. Bakken, R. Bredesen, T. Mattisson, and A. Lyngfelt, "Use of $CaMn_{0.875}Ti_{0.125}O_3$ as Oxygen Carrier in Chemical-Looping with Oxygen Uncoupling," *Energy & Fuels*, 23, 5276–5283 (2009).

169. Huang, Y.-H., R. I. Dass, Z.-L. Xing, and J. B. Goodenough, "Double Perovskites as Anode Materials for Solid-Oxide Fuel Cells," *Science*, 312(5771), 254–257 (2006).

170. Fossdal, A., E. Bakken, B. A. Øye, C. Schøning, I. Kaus, T. Mokkelbost, and Y. Larring, "Study of Inexpensive Oxygen Carriers for Chemical Looping Combustion," *International Journal of Greenhouse Gas Control*, 5, 483–488 (2011).

3   Oxidative Coupling of Methane

E. Y. Chung, W. Wang, S. Luo, S. Nadgouda, and L.-S. Fan

3.1   Introduction

The conversion of carbonaceous feedstocks into higher value products occurs through two major routes: (1) indirect oxidation; and (2) direct oxidation, as shown in Figure 3.1.[1–3] In the indirect oxidation approach, carbonaceous feedstock is first oxidized to syngas, which can then be further converted to value-added products. With respect to syngas generation, the thermodynamics and reaction characteristics of metal oxide oxygen carriers are discussed in Chapter 2; chemical looping reactor configurations are discussed in Chapter 4; and the techno-economic analyses of several chemical looping processes are discussed in Chapter 6. In the direct oxidation approach, the carbonaceous feedstock is directly upgraded to the desired product.[4] Conceptually, the direct route simplifies the overall process and reduces costs by removing processing steps, but no commercial-scale system exists.[4–6] Early research activity on direct oxidation processes, specifically for methane, can be traced back to the 1920s, but the utilization of methane in the chemical industry has been limited due to its high molecular stability.[7] Currently, direct methane utilization involves conversion to value-added products such as aromatics, oxygenates, olefins, and paraffins, of which the three major oxidative processes are partial oxidation to methanol, to formaldehyde, as given in Section 5.4, and to ethylene and ethane via the oxidative coupling of methane (OCM).[1,5,8]

Volatility in petroleum prices and limited petroleum reserves have made natural gas resources progressively more attractive as an energy source.[9,10] Recent discoveries of natural gas reserves, increased accessibility to shale gas, and low natural gas prices have propelled a resurgence in methane-to-chemicals research.[11–13] Monetizing cheap natural gas to obtain higher value-added products would provide a critical opportunity to the petroleum and chemical industry.[9] Research into one promising direct route, OCM, has exhibited peaks and troughs, as shown in Figure 3.2. To date, pilot-scale systems were constructed and tested at the Atlantic Richfield Company (ARCO) in the 1990s.[14,15] In 2012, Honeywell announced plans to scale up a proof-of-concept direct methane conversion process to ethylene.[16] More recently, in 2015 Siluria Technologies began operating a pilot-scale OCM demonstration unit to convert methane to ethylene or gasoline.[17]

In this chapter, the direct oxidation of methane to ethylene and higher hydrocarbons through OCM is presented. It introduces, in Section 3.2, the reactions, thermodynamics,

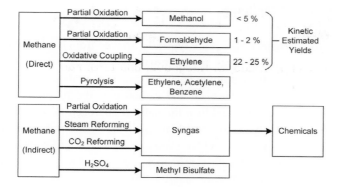

Figure 3.1 Various routes for direct and indirect methane conversion processes to value added products.[1]

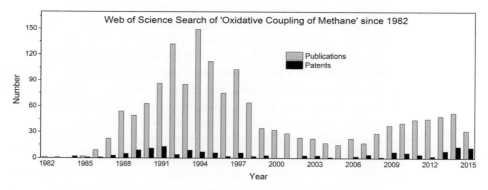

Figure 3.2 Citations for oxidative coupling of methane publications and patents.

and modes of OCM, including co-feed, where reactions between molecular oxygen and methane take place over a metal oxide catalysts, and redox, where reactions take place with catalytic metal oxides and methane. It also discusses the important features of the co-feed OCM reaction mechanism. Section 3.3 summarizes the common properties of co-feed metal oxide catalysts in OCM reactions, the characteristics for an effective metal oxide catalyst, and the developmental progress made since the pioneering work in the early 1980s. Section 3.4 evaluates the catalytic metal oxides for specific redox reactions for eventual use in a chemical looping system. Section 3.5 illustrates alternative types of catalyst, including halogen based, sulfur based, and complex metal oxides. Section 3.6 discusses the kinetics of OCM, from both a co-feed and redox perspective, since OCM is a kinetically dominated process. It also presents novel reactor configurations that can potentially increase product yields. Section 3.7 provides an overview of the processes that integrate the OCM technology and describes their importance in the purview of commercial applications.

3.2   Oxidative Coupling of Methane

Currently, light hydrocarbons (C_2–C_4) are typically derived from crude oil and natural gas.[18] Both natural gas and shale gas are gas mixtures that are typically composed of greater than 90% methane, the remaining balance being higher alkanes such as ethane, propane, and butane, as well as nitrogen and carbon dioxide. At present, the direct oxidation approach is less economical than the indirect approach as product yields are low, and it requires potentially costly separation methods.[1,19] Several issues, including the key one concerning low product yield, remain to be resolved for the direct oxidation approach. For example, using methane as feedstock, the yield of product is less than 5% for methanol, 1–2% for formaldehyde and 22–25% for ethylene and ethane (C_2) hydrocarbons, as shown in Figure 3.1.[1]

Of particular interest, ethylene is a critical intermediate in the petrochemical industry for products such as polyethylene, 1,2-dichloroethane, and ethylene oxide. Presently, ethylene is most commonly produced from the steam cracking of hydrocarbon derivatives such as naphtha (a distillation product of petroleum) or ethane (cryogenically distilled from natural gas resources) and uses 40% of the energy in the chemical and petrochemical industry.[18] The steam cracking of ethane is a high temperature process, 750–900 °C, that also forms undesirable coke deposits, which are tempered by the addition of steam, with a typical steam:ethane ratio of between 0.2 and 0.4 kg/kg.[18] The ethane cracking reaction is highly endothermic and energy intensive, which requires a large heat input, consuming approximately 25% of the total energy required in the steam cracking process.[18] As ethane steam cracking produces a wide product distribution, a fractionation train is necessary to obtain pure ethylene. The ethane steam cracking process requires upstream separation to obtain the ethane feed and downstream separation to obtain the ethylene. The importance of ethylene as a chemical, and its production process intensification feature, and higher yields from using OCM compared to other direct approaches, result in a considerable focus being placed on OCM development.[11,20]

Research into oxidative conversion from methane to ethylene and other value-added hydrocarbons has been reported since the early 1980s.[1,21–30] This research area was first named oxidative coupling of methane by the early pioneers Keller and Bhasin, and is also termed oxidative dimerization of methane.[22] Since that time, extensive OCM research and development has been conducted with a primary focus on developing optimal materials, understanding the fundamental mechanisms, and engineering the proper OCM reactor and process design. However, in recent years, research activity in these areas has reflected the price volatility involved with crude oil and natural gas.[9,10,31,32]

The Gibbs free energy for the non-oxidative coupling reaction of methane to ethane and hydrogen is +71.0 kJ/mol at 727 °C. Thus, methane conversion and C_2 yield are limited.[33] However, the Gibbs free energy for the partial oxidation of methane to ethane and water is −121.6 kJ/mol at 727 °C, meaning that the products are thermodynamically more stable than the reactants, so it is possible to produce C_2 with high yields.[33]

For this chapter, conversion, selectivity, and yield for methane reaction in the reactor are generally defined in equations (3.2.1) to (3.2.3):

$$\text{Conversion} = \frac{[\text{Carbon}]_{\text{Methane in}} - [\text{Carbon}]_{\text{Methane out}}}{[\text{Carbon}]_{\text{Methane in}}} \times 100\% \qquad (3.2.1)$$

$$\text{Selectivity} = \frac{[\text{Carbon}]_{\text{OCM hydrocarbon products}}}{[\text{Carbon}]_{\text{OCM products}}} \times 100\% \qquad (3.2.2)$$

$$\text{Yield} = \text{Conversion} \times \text{Selectivity}, \qquad (3.2.3)$$

where OCM hydrocarbon products are the desired products, such as C_2 and higher hydrocarbons, and OCM products include carbon monoxide and carbon dioxide in addition to the hydrocarbon products. [Carbon] represents the moles of carbon in a particular compound. It is noted that when the total carbon is balanced between the reactor inlet and outlet, the moles of total methane converted in the reactor are equal to the carbon moles in the products. However, in calculating the selectivity reported in the literature, the moles of OCM products that are referred to from the experimental measurements do not necessarily imply those that satisfy the total carbon balance.

The OCM reaction system has a challenging trade-off relationship between achieving high methane conversion and high C_2 selectivity. During high methane conversion, partial oxidation of methane thermodynamically favors undesirable and competing combustion products, such as carbon dioxide and carbon monoxide, which reduces the C_2 selectivity. It is projected that the industrial OCM process should have a C_2 yield of between 25 and 40% and C_2 selectivity exceeding 70% in order to be profitable.[11,22,34] Figure 3.3 shows the variation of the methane conversion and C_2 selectivity obtained experimentally for various materials from the OCM literature. The general trend for the variation from Figure 3.3 indicates that a high methane conversion gives rise to a low C_2 selectivity, while a low methane conversion gives rise to a high C_2 selectivity. Directly comparing the variations in the results is difficult due to the

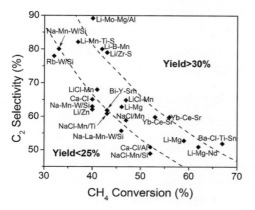

Figure 3.3 Compilation of methane conversion versus C_2 selectivity from the literature for various OCM materials.[11,19] (Reproduced from Zavyalova, U. et al., 2011, *ChemCatChem*, 3, 1935–1947, with the permission of John Wiley and Sons).

considerable differences involved in material synthesis methods, reactor types, reaction conditions, gas sampling, and data analyses. Further, there has been a lack of confirmation of mass balance between the reactants and the entire product slate, particularly with respect to carbon balance, which must include carbon deposition on the solid material. Although OCM is not currently economically profitable, this direct approach is attractive due to the conceptual simplicity of the process. With the recent rise in cheap and abundant shale gas and the rising price for olefins, the economics of OCM could improve significantly under this shale gas boom scenario.[9,12,13] With such prospects, the direct oxidation of methane to olefins via OCM is entering another revival peak.[15,35–37]

3.2.1   Oxidative Coupling of Methane Reactions

The reactions that comprise OCM can be divided into two categories: (1) reactions that form the desired product, C_2 (ethylene and ethane), higher hydrocarbons (C_{2+}); and (2) reactions that form undesired products, CO_x (carbon monoxide and carbon dioxide).[38] The oxidative coupling of methane to ethylene occurs through two sequential reactions involving the partial oxidation of methane followed by that of ethane, as given in reaction (3.2.4) and reaction (3.2.5). Reaction (3.2.6) is the net reaction:

$$2\,CH_4 + \tfrac{1}{2}\,O_2 \rightarrow C_2H_6 + H_2O \tag{3.2.4}$$

$$C_2H_6 + \tfrac{1}{2}\,O_2 \rightarrow C_2H_4 + H_2O \tag{3.2.5}$$

$$2\,CH_4 + O_2 \rightarrow C_2H_4 + 2\,H_2O. \tag{3.2.6}$$

The formation of carbon oxides (CO_x), such as CO and CO_2, is inevitable in the OCM process since they are thermodynamically more stable than both the reactants and the desired C_2 products. This results in CO_x formation from both gas-phase reactions and heterogeneous reactions on the catalyst surface.[39,40] The reactions forming CO_x, shown in reactions (3.2.7) to (3.2.9), are unwanted side reactions that have a negative impact on the OCM reaction by decreasing the C_2 yield through parallel oxidation reactions of reactant (methane) and consecutive oxidation reactions of desirable target (C_2) products:

$$CH_4 + O_2 \rightarrow CO/CO_2 \tag{3.2.7}$$

$$C_2H_6 + O_2 \rightarrow CO/CO_2 \tag{3.2.8}$$

$$C_2H_4 + O_2 \rightarrow CO/CO_2. \tag{3.2.9}$$

The source of oxygen for the OCM reactions can be either gaseous, molecular oxygen, or lattice oxygen, and defines the mode of operation for OCM.[22,33,41,42] If gaseous, molecular oxygen (or air) is the source of oxygen, then methane and oxygen are simultaneously reacted over a catalyst. This simultaneous[33,43] co-feeding of methane and oxygen (or air) approach is also classified as a purely catalytic OCM approach or co-feed approach. If lattice oxygen from a reducible catalytic metal oxide is the source of oxygen, then methane and oxygen are sequentially reacted to produce

Table 3.1 OCM classification

Oxygen source	Gaseous, molecular	Lattice
OCM operating mode	Co-feed	Redox/chemical looping
Methane/oxygen feed	Simultaneous	Sequential, cyclic, periodic, or alternating
Classification	Pure catalytic	Chemical looping
Solid phase role	Catalyst	Catalytic metal oxide

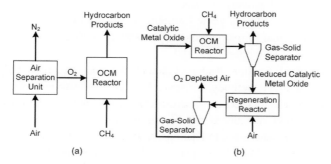

Figure 3.4 Typical approaches to OCM: (a) co-feed and (b) redox/chemical looping.

ethylene and to regenerate the reduced catalytic metal oxide, respectively. This sequential[33,43,44], cyclic, or periodic (or alternating) feeding of methane and oxygen individually from natural gas and air, respectively, is classified as reduction–oxidation (redox) or chemical looping OCM. Table 3.1 summarizes the oxygen source and OCM mode of operation.

The process schemes of these two operating modes are shown in Figure 3.4. In the co-feed approach, oxygen and methane are introduced into a single reactor where the OCM reactions occur, while in the redox/chemical looping approach, two reactors are required. In the first reactor, reducible catalytic metal oxides react with methane to yield gaseous products as well as a reduced catalytic metal oxide. In the second reactor, the reduced catalytic metal oxide is regenerated by air. In general, these two operating modes give similar product yields, with advantages and disadvantages for both. For all aspects of the co-feed approach, reaction mechanism and kinetics, reactor design, and process analysis have been extensively studied. This mode of operation requires a capital- and energy-intensive air separation unit (ASU), as pure oxygen is needed as the oxidant. The redox approach eliminates competing gas-phase reactions and the need for an ASU for oxygen generation.[41,45] Further, the oxidation step in the redox mode of operation burns off any undesirable coke or carbon deposited during methane oxidation.[41,45] However, the redox operating mode may have disproportionate residence times, where the lattice oxygen from the catalytic metal oxide is depleted quickly during the reduction step (~2–30 min), while the oxidation step requires a significantly longer time (~130 min).[46]

3.2.2   Co-feed Ethylene Formation Mechanism

The fundamental studies of the complex OCM reaction network have predominately focused on understanding the multifaceted mechanism of OCM, which occurs as heterogeneous reactions on the catalyst surface and homogeneous reactions in the gas phase.[5,35,43,47–49] It is generally accepted that there are two major OCM steps: (1) the heterogeneous reaction, where methane reacts with adsorbed oxygen on the metal oxide catalyst surface to form a gas-phase methyl radical ($CH_3\bullet$) and surface hydroxyl species; and (2) heterogeneous and homogeneous reactions of the methyl radicals on the surface and in the gas phase respectively to produce C_2 products.[33,48,50–52] The simplified overall reaction mechanism for ethylene formation is given in Figure 3.5.[53] It is noted that the hydroxyl group that forms on the catalytic surface during hydrogen abstraction reactions is not depicted in Figure 3.5.

For the first step, the formation of methyl radicals has been experimentally detected through direct and indirect methods for several OCM catalysts and catalytic metal oxides. However, studies focused on determining the mechanism of the C–H bond cleavage to form methyl radicals is not certain, with two prevailing mechanisms: (1) heterolytic cleavage where methyl anions react with oxygen to produce methyl radicals in two steps; or (2) homolytic cleavage, the more commonly accepted theory, where methyl radicals are directly formed from methane in one step, as shown in Figure 3.5.[54–56]

The formation of methyl radicals through heterolytic cleavage is akin to an acid–base reaction where the catalyst is a basic metal oxide functioning as the base and methane as the acid.[54,57–61] First, the methane (CH_4) dissociates into a methyl anion (CH_3^-) and a

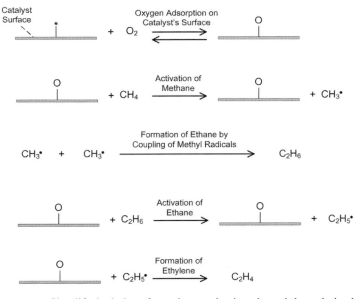

Figure 3.5 Simplified ethylene formation mechanism through homolytic cleavage, proposed by Lee et al.[53]

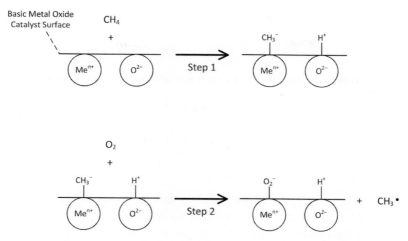

Figure 3.6 Two-step heterolytic cleavage of methane to methyl radical.[54,57,58]

hydrogen cation (H^+), where CH_3^- is bound to the positively charged metal (Me^{n+}) and the H^+ is bound to the negatively charged lattice oxygen (O^{2-}). Upon reaction with gaseous oxygen (O_2), substitution of CH_3^- by O_2 occurs, where the Me^{n+} becomes bound to O_2^-, and a methyl radical is released.[54,57,58] Figure 3.6 illustrates the two-step reaction for heterolytic cleavage of methane to form a methyl radical.[54,57,58]

The direct, homolytic conversion of methane to methyl radicals is more widely accepted and has been experimentally observed. Lunsford et al. detected surface derived gas-phase methyl radicals using two spectroscopic methods: (1) matrix isolation electron spin resonance (MIESR); and (2) resonance enhanced multiphoton ionization (REMPI).[62–66] Temporal analysis of products (TAP) has also been utilized by multiple groups to demonstrate the formation of methyl radicals in the gas phase.[67–69] The theory of homolytic C–H bond cleavage is also supported from experiments using hydrogen–deuterium (H–D) exchange.[54,70] However, these techniques are not direct measurements. More recently, Luo et al. employed synchrotron vacuum ultraviolet photoionization mass spectroscopy (SVUV–PIMS) for direct in situ detection of gas-phase methyl radicals; they further confirmed the generation of methyl radicals by methane activation on the surface.[71]

The two mechanisms for methyl radical formation are generally believed to occur, but with the homolytic mechanism being dominant as the more basic sites would be neutralized by carbon dioxide, leaving only the less basic sites that can only form methyl radicals through the homolytic mechanism.[54,66] There is still uncertainty as to the exact homolytic mechanism, either following a Eley–Rideal mechanism or a Mars–van Krevelen mechanism.[66,72,73] The methyl formation reaction following the Eley–Rideal mechanism is given in reaction (3.2.10) and reaction (3.2.11), and the Mars–van Krevelen mechanism is given in reaction (3.2.12) to reaction (3.2.14):[72,73]

$$O^{2-}_{(ads)} + *_{(ads)} + \tfrac{1}{2}O_{2(g)} \leftrightarrow 2O^-_{(ads)} \qquad (3.2.10)$$

$$O^-_{(ads)} + CH_{4(g)} \rightarrow CH_{3(g)} \bullet + \ OH^-_{(ads)} \qquad (3.2.11)$$

where * represents an oxygen vacancy.

$$CH_{4(g)} \leftrightarrow CH_{4(ads)} \qquad (3.2.12)$$

$$O_{2(g)} \leftrightarrow O_{2(ads)} \ \text{either} \ O_2^- \ \text{or} \ O_2^{2-} \qquad (3.2.13)$$

$$CH_{4(ads)} + O_{2(ads)} \rightarrow CH_3 \bullet + \ HO_2 \bullet \qquad (3.2.14)$$

In the second step, the heterogeneous and homogeneous reactions that form the desired C_2 products are more complex: a series of parallel and consecutive reactions occurring both on the surface and in the gas phase. It is generally accepted that ethane is formed from the combination of two methyl radicals. Ethylene is formed from the homogeneous or heterogeneous dehydrogenation of ethane.

The selectivity and yield for C_2 formation are governed by kinetics and not thermodynamics.[1] There is a trade-off between selective oxidation to C_2s and total oxidation to CO_x; for the same conversion, with an increase in over-oxidation to carbon oxides, there is a decrease in the yield of C_2 products. This reciprocal relationship between carbon oxide formation and C_2 product formation results in the need to reduce the oxygen concentration in the OCM reactor in order to decrease carbon oxide formation, but this also reduces the overall methane conversion. The general reaction trends for OCM with inclusion of CO_x formation are provided in Section 3.2.3.

3.2.3   Co-feed Carbon Oxide Formation Mechanism

The thermodynamically favorable CO_x product forms through several reaction pathways in a typical OCM reaction. The numerous heterogeneous gas–solid and homogeneous gas-phase reactions that occur make characterization of the OCM reaction network difficult to study, since neither traditional heterogeneous nor homogeneous reaction mechanisms can be applied directly without significant modifications to accommodate for both types of reaction.[74] The challenge is to account for intermediates that also interact with the surface of the catalyst. Parameters such as temperature, pressure, contact time, and methane:oxygen ratio dictate the conversion and selectivity for a particular metal oxide catalyst, and hence, play an important role in the amount of CO_x generated and the source of CO_x. While no two catalysts will respond identically to varying reaction conditions, general trends can be observed on a macro scale.

Temperature has a significant impact on the reactions, where increasing reaction temperature increases C_2 selectivity, or in other words, CO_x selectivity decreases. For example, increasing the reaction temperature from 550 °C to 750 °C increases the C_2

selectivity from 37% to 89%, while the CO_x selectivity decreases from 63% to 11% on a Sm_2O_3 catalyst.[75] O_2 and CH_4 pressures also affect product yields. Increasing the O_2 pressure increases CH_4 conversion and CO_x selectivity. Also, the methane:oxygen ratio is proportional to the ethane:ethylene ratio, where decreasing the methane:oxygen ratio decreases the ethane:ethylene ratio. To avoid over-oxidation of methane to CO_x, the methane:oxygen ratio is typically in the range 4:1 to 12:1, which is higher than the 2:1 stoichiometric ratio required for complete conversion of oxygen, as given in reaction (3.2.6).

For the OCM reaction network, at a low methane conversion, below 15%, the reactant methane is the major source of CO_x formation, whereas at higher methane conversions, the major source of CO_x is from the products.[39] At a low methane conversion, two popular reaction mechanisms are presented to explain the general trends of higher temperature and lower oxygen concentration, favoring high C_2 selectivity. Sinev et al. first investigated the OCM reaction mechanism at low methane conversions, which occurs through reactions (3.2.15)–(3.2.17):[76]

$$O_{(ads)} + CH_4 \xrightarrow{k_1} CH_3 \bullet + OH_{(ads)} \tag{3.2.15}$$

$$2\ CH_3 \bullet \xrightarrow{k_2} C_2H_6 \tag{3.2.16}$$

$$CH_3 \bullet + O_2 \underset{k_{-3}}{\overset{k_3}{\leftrightarrow}} CH_3O_2 \bullet. \tag{3.2.17}$$

Reaction (3.2.16) and reaction (3.2.17) are competing reactions and reaction (3.2.17) is an exothermic, reversible reaction, so reaction (3.2.16) would be favored at higher temperatures and form more C_2. Ito et al. suggested that $CH_3O_2\bullet$, formed in reaction (3.2.17), leads to the formation of carbon oxides.[77] The evolution of $CH_3O_2\bullet$ into CO_x products is shown in reaction (3.2.18) and reaction (3.2.19):[33,76,77]

$$2\ CH_3O_2 \bullet \rightarrow CH_2O + CH_3OH + O_2 \xrightarrow{k_4} CO/CO_2 \tag{3.2.18}$$

$$CH_3 \bullet + CH_3O_2 \bullet \xrightarrow{k_5} CO/CO_2. \tag{3.2.19}$$

Further, Sinev et al. derived equation (3.2.20) for the selectivity of ethane, based on the assumption that reaction (3.2.17) could reach equilibrium:[14,76]

$$\text{selectivity of ethane} = \left[1 + K_3^* P_{O_2} \left((K_{3^*} P_{O_2}^* k_4 + k_5)/k_2 \right) \right]^{-1}, \tag{3.2.20}$$

where K_3 is the equilibrium constant of reaction (3.2.17), P_{O_2} is the oxygen partial pressure, and k_2–k_5 are the rate constants of reactions. Equation (3.2.20) focuses only on the selectivity of methane to ethane in the gas phase and assumes no CO_x formation occurs at low conversions.

Determining the role of C_2 products in the formation of CO_x at higher conversions is difficult due to the complex nature of the OCM mechanism. Isotope studies with

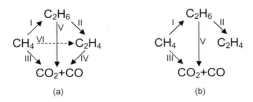

Figure 3.7 Potential reaction pathways for OCM.[75] (Reproduced from Otsuka et al., 1986, J. Catal., 100(2), 353–359, with permission from Elsevier.)

ethane or ethylene mixed with methane as the feed have been conducted to elucidate the mechanism. The distinction between the CO_x formed from C_2 products or from methane was determined by using different isotopes of carbon in the reactant feed. Ekstrom et al. were the first to use this method on Sm_2O_3 and found that the concentration of CO_x in the product increases sharply when the methane conversion increases above 15%.[78]

Temperature also plays an important role when it comes to the source of CO_x. Otsuka et al. have investigated the reaction framework to determine how to achieve high C_2 selectivity.[75] By using CH_4, C_2H_6, and C_2H_4 as reactants in three sets of experiments, they discovered that C_2H_4 formation is the result of stepwise reactions, and most of the carbon oxides are formed directly from methane.[75] They suggested that the OCM reaction network occurred through that shown in Figure 3.7b rather than Figure 3.7a, which could explain why increasing the temperature results in higher ethylene:ethane ratios. Furthermore, they calculated the reaction activation energies of these reactions, which have the order $E_{aII} > E_{aI} > E_{aV} > E_{aIII}$, confirming that increasing the temperature results in higher ethylene:ethane ratios. Ito et. al. suggested that at temperatures below 700 °C, reaction (3.2.17) leads to the formation of carbon oxides for the Li/MgO metal oxide catalyst, suggesting the source of CO_x as being methane. Nelson and Cant confirmed this hypothesis by determining the source of CO_x as C_2 products at $T > 740$ °C, and methane for $T < 700$ °C over the Li/MgO metal oxide catalyst.[79]

3.2.4   Role of Oxygen Species

Extensive studies have been performed on the type of oxygen species involved in OCM reactions. Generally, there is no definitive consensus on the type of oxygen species except for its homogeneous or heterogeneous nature. This lack of agreement is usually attributed to the difficulty in determining the active sites of the reactions due to variations in OCM materials and reaction conditions, such as temperature and reactant concentrations.[48]

The oxygen species participating in the reaction differ based on the types of materials.[15,64,80] For this chapter, OCM materials can be further broadly defined by the role of oxygen into three categories: (1) pure catalytic metal oxides, known as metal oxide catalysts; (2) multifunctional metal oxides that can operate as a catalyst and/or oxygen carrier, known as catalytic metal oxides; and (3) unique catalysts such as halogen-doped oxides, solid electrolytes, and non-oxide based OCM catalysts. These categories are not necessarily exclusive as there are OCM materials that qualify for multiple categories.

Regardless of the type of materials, in co-feed OCM, the first step is the surface adsorption of molecular oxygen at surface oxygen anion vacancies or interstitial sites.[15,47,48,53,80] Molecular oxygen adsorption can occur via two mechanisms: (1) dissociative adsorption; or (2) non-dissociative adsorption. One common experimental method in determining the mechanism of oxygen adsorption is the use of oxygen isotope exchange as reaction tracers. Takanabe et al. found $^{18}O-^{16}O$ on the surface of the $Na_2WO_4/Mn/SiO_2$ metal oxide catalyst using $^{18}O_2$ and $^{16}O_2$, which indicates a dissociative mechanism for O_2 adsorption.[80] Other oxygen isotope tracer results have studied the Li/MgO metal oxide catalyst surface, and are also supportive of surface dissociative adsorption over non-dissociative adsorption. For OCM processes using irreducible metal oxide catalysts, the presence of paramagnetic O^- plays a vital role.[48,74,77,82–84] Further it has been discovered that other charged oxygen species such as superoxide (O_2^-), peroxide (O_2^{2-}), and oxide (O^{2-}) are also found on the surface of OCM metal oxide catalysts.[48,72,84,85] During the metal oxide catalyst reaction, the materials are affected by electronic transfers that are facilitated by impurity defects, or intrinsic defects such as anion vacancies or interstitial sites, with the oxygen transfer mechanism as shown in Figure 3.8. In contrast, the redox OCM process with reducible catalytic metal oxide utilizes lattice oxygen, rather than the oxygen species adsorbed on the catalyst surface.[48,86] These redox catalytic metal oxides, unlike the irreducible catalysts, undergo changes in oxidation state where the electron transfer is to the oxide ion. Of the OCM material categories, these groupings of (1) OCM co-feed metal oxide catalysts, (2) OCM redox catalytic metal oxides, and (3) OCM with substituted and complex metal oxide catalysts, are further discussed in Sections 3.3, 3.4, and 3.5, respectively.

$$O_2 \rightleftharpoons O_{2\,ads} \qquad\qquad Null \rightleftharpoons h^+ + e^-$$
$$O_{2\,ads} + e^- \rightleftharpoons O_2^- \qquad O_{2\,ads} \rightleftharpoons O_2^- + h^+$$
$$O_2^- + e^- \rightleftharpoons O_2^{2-} \qquad\quad O_2^- \rightleftharpoons O_2^{2-} + h^+$$
$$O_2^{2-} \rightleftharpoons 2O^- \qquad\qquad O_2^{2-} \rightleftharpoons 2O^-$$
$$O^- + e^- \rightleftharpoons O^{2-} \qquad\quad O^- \rightleftharpoons O^{2-} + h^+$$

(a)   (b)

Figure 3.8 Molecular oxygen transformation mechanism to lattice oxygen for: (a) reducible catalytic metal oxides and (b) irreducible metal oxides where e^- represents electron and h^+ represents holes.[72]

3.3   Co-feed Metal Oxide Catalysts

Metal oxide catalysts play a key role in increasing the yield of olefins. For co-feed metal oxide catalysts, most researchers employ the co-feeding of methane and gaseous, molecular oxygen, since the cyclic feeding of methane and oxygen that is typically used for reducible catalytic metal oxides has demonstrated low activity.[48] The co-feed metal oxide catalyst materials are usually rare earth metal oxides and/or alkaline earth metals. The performance of the most commonly used metal oxide catalysts is shown in Table 3.2.[15]

Extensive studies have been conducted on screening for optimal metal oxide catalysts. These metal oxide catalysts are placed at the lower section of a fixed bed reactor. Methane, oxygen, and an inert gas are introduced from the bottom of the reactor where they react with a metal oxide catalyst to produce the product gases, which are then analyzed. Since the space velocity through the reactor is often small, it is expected that complete mixing of the gases will be achieved in the reactor. Nitrogen physisorption by the Brunett–Emmett–Teller (BET) method is often used to determine the surface area and pore volume for the catalyst, before and after reaction. Improved surface area and pore volume typically improves the reaction kinetics. Similar to other catalytic systems, morphological properties and physiochemical properties play a role in improving catalytic performance.

OCM materials have also been extensively studied for their physiochemical properties especially in terms of structural defects. Structural defects have been known to improve the electrical conductivity of catalysts. The electronic properties of a catalyst play a vital role in facilitating methane activation from the charged oxygen species.[87-90] Catalysts with p-type and oxygen-ion conductivity within the bandgap of 5–6 eV are the best performing OCM materials. Studies have shown that there is a correlation between electrical conductivity and C_2 yields, where metal oxides with low bandgaps, such as manganese oxides, were observed to have the best catalyst performance for OCM.[48,91,92]

Methane conversion and C_2 selectivity are the primary bases for the selection of OCM materials. Many types of oxides have been tested as catalysts for OCM, including, but not limited to, PbO, Bi_2O_3, SnO_2, Ga_2O_3, GeO_2, In_2O_3, ZnO, CdO, CaO, MgO, and Al_2O_3. Different dopant or support materials have also been added to improve performance. A general conclusion obtained from material screening is that rare earth metal oxides, particularly the lanthanide oxides (except Ce, Pr, Tb) tend to have high C_2 selectivity (>70%), while the acidic oxides, including Al_2O_3 and ZSM-5, often show poor C_2 selectivity.[59,72,85]

First reported by Kimble et al. and Driscoll et al., lithium (Li^+) doped MgO is an early OCM metal oxide catalyst that is still widely investigated.[62,93] The addition of Li^+ to MgO significantly increases the C_2 product selectivity and methane conversion, even when the concentration of Li^+ was as low as 0.4 wt%.[94-96] This phenomenon was difficult to understand initially since the addition of Li^+ decreases the surface area, which is contrary to the general notion of increased surface area promoting reaction kinetics.

Table 3.2 Specific activity data for oxidative coupling of methane with molecular oxygen using metal oxide catalysts[15]

Metal oxide catalyst	$CH_4/$ O_2 mole ratio	Temp (°C)	Methane feed rate (mol $gcat^{-1}$ hr^{-1})	10^4 Specific activity (mol product m^{-2} hr^{-1})						
				H_2	CO_2	CO	C_2H_4	C_2H_6	Total C_2	Total C^*
5%Li/MgO	3	710	0.0616^*	7.7	16.1	1.3	2.5	8.0	10.5	38.4
MgO	3	710	0.0616^*	3.7	4.5	1.3	0.8	0.8	1.6	9.0
5%Li/La_2O_3	3	710	0.0616^*	20	24	5.1	1.8	8.9	10.7	50.5
La_2O_3	3	710	0.0616^*	11	17	1.8	4.2	4.9	9.1	37.0
5%Li/Sm_2O_3	3	710	0.0616^*	16	97	3.9	29.0	42.0	71.0	242.9
Sm_2O_3	3	710	0.0616^*	19	28	3.5	5.0	6.8	11.8	55.1
MgO	2.1	700	0.0148	—	—	—	0.02	0.06	0.08	1.4
7%Li/MgO	2.1	700	0.0185	—	—	—	3.0	3.0	6.0	16.5
20%Na/MgO	2.1	700	0.0185	—	—	—	0.21	0.43	0.64	4.2
30%K/MgO	2.1	700	0.0185	—	—	—	0	0.02	0.02	3.0
CaO	2.1	700	0.0185	—	—	—	0.05	0.54	0.59	7.9
5%Li/CaO	2.1	700	0.0185	—	—	—	2.8	3.9	6.7	16.7
15%Na/CaO	2.1	700	0.0185	—	—	—	4.9	5.5	10.4	30.8
23%K/CaO	2.1	700	0.0185	—	—	—	3.5	4.9	8.4	28.7
Y_2O_3	46	700	0.0122	—	—	—	0.20	0.46	0.66	1.7
La_2O_3	46	700	0.0122	—	—	—	0.15	0.28	0.43	1.0
Sm_2O_3	46	700	0.0122	—	—	—	1.1	2.3	3.4	7.3
Gd_2O_3	46	700	0.0122	—	—	—	0.37	0.83	1.2	3.0
Ho_2O_3	46	700	0.0122	—	—	—	0.24	1.16	1.4	3.3
Yb_2O_3	46	700	0.0122	—	—	—	0.1	0.5	0.6	11.4
PbO	46	700	0.0122	—	—	—	0.2	2.3	2.5	—
Bi_2O_3	46	700	0.0122	—	—	—	0.3	1.6	1.9	5.1
$LaAlO_3$	1	710	0.179^*	—	—	—	—	—	43.1	178
La_2O_3	1	710	0.179^*	—	—	—	—	—	8.7	37.8
Sm_2O_3	1	710	0.179^*	—	—	—	—	—	22.5	120
CeO_2	2	750	0.0298	—	—	—	1.5	0.2	1.7	26.8
Yb_2O_3	2	750	0.0298	—	—	—	2.2	2.2	4.4	19.7
$Ce_{0.95}Yb_{0.1}O_{1.95}$	2	750	0.0298	—	—	—	0	0.2	0.2	83.4
SrO	2	750	0.0298	—	—	—	5.6	2.0	7.6	24.4
$SrCeO_3$	2	750	0.0298	—	—	—	12.5	14.7	27.2	110.0
$SrCe_{0.9}Yb_{0.1}O_{2.95}$	2	750	0.0298	—	—	—	20.2	13.4	33.6	112.0

* Based on methane converted into carbon-containing product.

Identification of active sites for methane coupling is important since it is directly related to developing effective OCM metal oxide catalysts. Extensive research has been conducted since the 1980s to study active sites, and it is generally accepted that formation of the [LiO] center on the surface is responsible for improved methane conversion and C_2 selectivity. Methyl radicals are generated from the interaction of CH_4 with the [LiO] center, either by thermal treatment or by high intensity electron bombardment. There are direct correlations between methane conversions and [LiO] centers. The [LiO] center is formed by the replacement of Mg^{2+} by Li^+.[77,96,97]

The similarity in ionic radii could be a critical factor for effective ionic substitution. Ionic radii may explain why certain dopant materials such as Na^+ (99 pm) are not as effective as Li^+ (68 pm) for MgO (Mg^{2+}: 66 pm), but much more effective for CaO (Ca^{2+}: 97 pm). Furthermore, the primary effects of the alkali oxide [LiO] center in MgO or the [NaO] center in CaO have also been explored. It was found that the O^- in the [LiO] or [NaO] center is the primary reason for the high methane conversion and C_2 selectivity.[15] Similar effects of Li dopant have also been observed on Sm_2O_3. O^- is not just effective for methane activation as a monomer, its dimer form, O_2^{2-} has also been found to be effective. This has been observed directly from C_2 formation over solid peroxides, including Na_2O_2 and BaO_2. In addition to O^-, superoxide O_2^- has also been found to be effective in promoting formation of the methyl radical, which may be the active site for La_2O_3.[98]

Although low cost Li/MgO is effective, there are potential issues that prevent its practicability as a commercial catalyst. The Li/MgO metal oxide catalyst quickly deactivates at temperatures higher than 800 °C, and its stability remains an issue. This is because of the evaporation of Li. The concentration of Li cannot be higher than 0.03% in MgO; but in order to work as a desirable OCM material, the Li concentration should be higher than 0.4%. A constant loss of Li from the solid is observed; thus, Li/MgO cannot work as a stable metal oxide catalyst over long time periods. Many metal oxides have been added to the Li/MgO system, where improved selectivity and stability have been observed; however, long-term stability still remains an issue.[95,96] Typically, dopants are added through a trial and error process with some directional rationale for the OCM material synthesis. For example, the addition of ions with a 3+ charge, such as Fe^{3+}, at a 1:1 atomic ratio of Fe^{3+} to Li^+, improves the stability of the metal oxide catalyst by acting as a charge-compensating ion. While iron doping can increase methane conversion, it can also simultaneously decrease the C_2 selectivity.[94]

Mn/Na_2WO_4-based metal oxide catalysts, first reported by Fang et al., are another type of OCM material that demonstrates a C_2 yield of up to 32% in a fixed bed reactor with reasonable stability.[99,100] Ji et al. suggested that WO_4, with suitable geometry and energy-matching properties for extracting hydrogen from methane, were the active sites of metal oxide catalysts that contributed to high CH_4 conversion and C_2 selectivity.[101] Specifically, $Mn/Na_2WO_4/SiO_2$ has been widely tested as an effective OCM metal oxide catalyst.[53] WO_4 is believed to have the desirable chemical and physical structures for CH_4 activation, and thus, is important for high C_2 yield.

OCM catalysts are typically synthesized by the incipient wetness impregnation method, slurry method, or sol–gel method. Among these, the impregnation method is most widely used for $Mn/Na_2WO_4/SiO_2$ metal oxide catalysts due to the excellent distribution of the active components Na, W, and Mn. It is reported that the incipient wetness impregnation method produces OCM metal oxide catalysts with higher catalytic activity than those synthesized by the other two methods. Recent work has focused on novel synthesis methods to improve C_2 yields. For example, 5% Mn/ 10% Na_2WO_4/SiO_2 by solution combustion synthesis, which is a method that combines sol–gel and combustion synthesis, has shown promise, with C_2 yields of 27% with ethylene:ethane ratios of 3.6.[102,103] These improvements were attributed to additional energy provided by the fuel combustion that acts as a complexing agent.

Since the reaction mechanism of OCM is quite complicated and is not yet well understood, alternative stochastic approaches have also been applied more recently in order to find new catalysts. For example, Huang et al. found metal oxide catalysts that have 27.8% C_2 yield as a result of screening a combination of six metal components.[104] From statistical analysis based on the design of the experiment, Chua et al. demonstrated that $Mn/Na_2WO_4/SiO_2$ produced C_2 yields of 31–33%.[105,106] In general, current metal oxide catalysts achieve C_2 selectivities of 50–70% and methane conversions of 35–55%, which result in C_2 yields of 21–32%.[48,107]

3.4   Redox Catalytic Metal Oxides

Chemical looping uses the lattice oxygen from an oxygen carrier that is circulated between a reducer (or fuel reactor) and a combustor (or air reactor) to transfer oxygen and has been extensively studied for combustion applications.[108,109] Chemical looping avoids direct contact between gaseous oxygen and the fuel, such as methane, thereby eliminating the need for an air separation unit (ASU), reducing the risk from highly flammable methane–oxygen mixtures, and allowing for inherent separation of CO_2. Although chemical looping is still a relatively new term, the principle has been widely used in OCM, especially in pioneering OCM work.[14,22,23,25–30,41,73,110] The redox OCM mode of operation using reducible catalytic metal oxides is conceptually analogous to redox cycles and reducible metal oxides used in chemical looping combustion, gasification, and reforming. However, the controlling factors for the two differ as the products from chemical looping gasification and reforming, as discussed in Chapters 2 and 4, are dictated by thermodynamics while the product distribution from OCM is dictated by reaction kinetics. In this sense, the catalytic metal oxides from OCM behave more similarly to the functionality and mechanism of catalytic metal oxides in Sections 5.2, 5.4, 5.5, and 5.6.

The chemical looping approach for OCM has been historically referred to as the reducible catalytic metal oxide OCM approach or the OCM reduction–oxidation (redox) approach. In order to minimize non-selective methane oxidation, Keller and Bhasin fed methane and air cyclically over catalytic metal oxides, separated by a short inert gas purge period.[22] Essentially, the catalytic metal oxides used in this OCM mode act as oxygen carriers that deliver oxygen from air to methane, while undergoing redox reactions. The general reaction scheme is similar to the catalytic process, except that the oxygen is solely derived from the lattice oxygen of the oxygen carrier, as demonstrated in reactions (3.4.1) and (3.4.2):

$$nCH_4 + MO_{z+y} \rightarrow C_nH_{4n-2y} + yH_2O + MO_z \tag{3.4.1}$$

$$MO_z + (y/2)O_2 \rightarrow MO_{z+y}. \tag{3.4.2}$$

Both Union Carbide and ARCO have devoted major research and development efforts toward the commercialization of this technology.[25–30,41,110,111] The majority of the solid materials tested were reducible, catalytic metal oxides that produced higher hydrocarbons up to C_7, such as toluene. As a result of the oil price crash in the 1980s, these technologies were not scaled to commercial olefin production or commercial liquid fuel synthesis.

Similar to other chemical looping processes, catalytic metal oxides or oxygen carriers play an essential role in determining process efficiency, as discussed in Chapters 2, 4, and 6. There are many possible reactions other than methane partial oxidation to C_2, and the desirable products could be further oxidized to CO or CO_2. Selectivity to the desired products and reactant conversion are largely dependent on the catalytic metal oxides used in the process.

A series of metal oxides have been identified as oxygen carrier candidates, of which a supported manganese oxide was found to produce the most promising results.[41] However, among the reducible catalytic metal oxides, at 800 °C, manganese oxides produced favorable results with C_2 yields of 20%, which is short of the target yield for commercial OCM economic feasibility.[11,22,34] Thus, the redox mode may require innovative reactor design in order to maintain high yields. Previous research has compared the co-feed and chemical looping modes of manganese oxides, and demonstrated that for the same conversion, higher C_2 selectivity occurs with the chemical looping redox mode than in the co-feed mode, as is shown in Figure 3.9.[112]

Under a redox OCM scheme, research has focused on selection of a suitable reducible metal oxide.[41,45,112,113] Also, there is an indication that the support

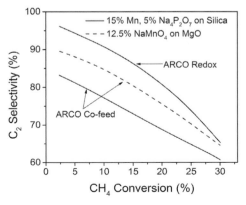

Figure 3.9 Comparison of OCM chemical looping (redox) to co-feed for Mn-based metal oxides.[112]

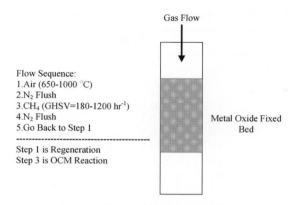

Flow Sequence:
1. Air (650-1000 °C)
2. N_2 Flush
3. CH_4 (GHSV=180-1200 hr^{-1})
4. N_2 Flush
5. Go Back to Step 1

Step 1 is Regeneration
Step 3 is OCM Reaction

Gas Flow

Metal Oxide Fixed
Bed

Figure 3.10 Chemical looping OCM reactor for catalytic metal oxide screening.[41]

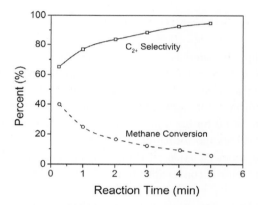

Figure 3.11 Time dependency of methane conversion and C_{2+} selectivity over 15% Mn, 5% $Na_4P_2O_7/SiO_2$.[41]

material should be low in acidity in order to obtain high selectivity. Silica is often used as the support material for this reason. The procedure for this screening process is shown in Figure 3.10. In addition, under a redox operating mode, because of the low oxygen carrying capacity of the reducible catalytic metal oxides, the lattice oxygen is quickly depleted during methane oxidation, rendering it difficult to improve yields. In general, methane conversion decreases and C_2 selectivity increases over time, and available lattice oxygen is depleted within 30 minutes. One example of the time dependence of methane conversion and C_2 selectivity is illustrated in Figure 3.11 for a 15% Mn, 5% $Na_4P_2O_7/SiO_2$ catalytic metal oxide.

The screening results in these early tests are given in Table 3.3. Manganese based catalytic metal oxides demonstrate high methane conversion, which has provided the basis for further material optimization. When pure Mn_2O_3 was tested, C_2 selectivity did

Table 3.3 Sample of chemical looping OCM results with various catalytic metal oxides at 800 °C and 860 hr^{-1} GHSV.[41]

Catalyst	(%) CH$_4$ convn	Selectivity (%)								
		C$_2$H$_4$	C$_2$H$_6$	C$_3$	C$_4$	C$_5$	Benzene	Toluene	CO	CO$_2$
5% Mn/Al$_2$O$_3$	11.0	45.0^a								55.0
5% Mn/SiO$_2$	13.3	21.8	15.7	2.2	0.8	0.1	0.6	0.1	16.5	42.2
15% Mn/SiO$_2$	26.0	36.3	14.3	3.8	2.3	0.2	1.8	0.3	14.9	26.1
5% Bi/SiO$_2$	5.1	23.1	30.7	1.9	1.2	0.2	0.8	0.3	24.5	17.3
5% Ge/SiO$_2$	1.4	38.5	24.6	3.5	2.3	0.2	1.7	0.2	27.0	2.0
5% In/SiO$_2$	7.3	26.7	19.8	2.9	1.1	0.1	1.4	0.2	17.3	30.5
5% Pb/SiO$_2$	2.5	32.6	5.4	3.5	2.0	0.2	1.9	0.2	26.9	27.3
5% Sb/SiO$_2$	1.2	31.6	27.6	1.1	0.5	0.1	0.9	0.1	38.1	0.0
5% Sn/SiO$_2$	3.2	26.8	21.2	1.7	1.2	0.1	0.4	0.1	18.2	30.3
α-Al$_2$O$_3$	0.3	50.2	42.2	4.7	1.7	0.1	1.0	0.1	0.0	0.0

^{a} Combined C$_2$H$_4$ and C$_2$H$_6$.

not go beyond 20%. When support materials such as silica or alumina were used, the selectivity could be improved by three times. It was observed that braunite, Mn$_7$SiO$_{12}$, forms in the silica-supported manganese oxides, and this may be one of the reasons for the improved reaction performance. Furthermore, the addition of alkali components such as sodium pyrophosphate could further improve selectivity by about 10%. It was hypothesized that sodium promotes the formation of braunite, thus producing more selective and more active catalytic metal oxides. The metal oxide catalyst Mn/Na$_2$WO$_4$/ SiO$_2$ has been observed, with methane conversions from 12–31% and C$_{2+}$ selectivities from 60–80%.[114–117]

The general reaction mechanism for chemical looping OCM is shown in reaction (3.4.1) and reaction (3.4.2). For chemical looping OCM, there are two main aspects: product formation, C$_{2+}$ products, and the catalytic metal oxide redox reaction. The mechanism of higher hydrocarbon formation is relatively easy to study, and the product distribution provides useful information on the reaction pathway. For a given methane conversion, the product distribution over different catalytic metal oxides was found to be similar. This observation indicated that the mechanism behind the formation of hydrocarbons is likely to be conducted in the gas phase rather than on the catalytic metal oxide surface. Further, methane/ethylene as a feedstock produced propylene, while a methane/propylene mixture produced butylene. These results support that the mechanism of the redox OCM reaction occurs through gas-phase methyl radicals. However, the catalytic metal oxide redox reaction mechanism, particularly the reduction of catalytic metal oxide by methane, remains unclear.[41]

Results from The Ohio State University confirm the formation of higher hydrocarbons, with the product distribution provided in Table 3.4, with a C$_{2+}$ selectivity of 63.24% at a methane conversion of 36.7% for a C$_{2+}$ yield of 23.2%.[118–120] The selectivity and yield were calculated based on equations (3.2.2) and (3.2.3),

Table 3.4 Chemical looping product distribution from fixed bed experiments.

Component	Mole %[*]
Methane	70.70
Ethylene	8.99
Ethane	2.41
Propylene	0.46
Propane	0.07
Propadiene	0.01
Propyne	0.03
1-Butene	0.03
1,3-Butadiene	0.20
C_5	0.02
Benzene	0.06
C_7	0.01
Carbon dioxide	15.06
Hydrogen	1.96

[*] Total does not sum to 100.0 due to independent rounding.

respectively, while the methane conversion for OCM production was calculated specifically based on moles of carbon in the product, as given in equation (3.4.3):

$$\text{Conversion for OCM Reaction} = \frac{[\text{Carbon}]_{\text{OCM products}}}{[\text{Carbon}]_{\text{Methane out}} + [\text{Carbon}]_{\text{OCM products}}} \times 100\%.$$

(3.4.3)

Hydrocarbons up to C_7 were observed but the C_5 and C_7 compounds could not be analytically identified. The chemical looping OCM experiments were conducted in a fixed bed at 840 °C using a manganese oxide based catalytic metal oxide with alternating methane and oxygen flows at a gas hourly space velocity (GHSV) of 2400 hr^{-1} of methane.

For the redox OCM reaction, oxygen plays a critical role. Oxygen diffusion in the catalytic metal oxide could take place via various mechanisms, such as diffusion along the interstices, exchange of vacancies and ions, or simultaneous cyclic replacement of atoms.[121] If the oxygen diffusion rate is too slow, the oxygen in the bulk phase will have a minimal effect on the reaction. Various chemical composition and defective structures have been tested to control the change in oxygen mobility, and hence, the reaction rate of the OCM reaction.[122–124]

3.5 Substituted and Complex Metal Oxide Catalysts

OCM materials are not only limited to metal oxides; studies have utilized non-oxides or combined metal oxides with other materials. Research into increasingly complex catalysts to facilitate the coupling of methane has been performed.[48]

3.5.1   Halogen-Based Metal Catalysts

The addition of non-oxygen transfer based compounds has been extensively explored. Halogen-doped metal oxides have demonstrated high C_2 selectivity. A higher selectivity towards ethylene rather than ethane has been shown by catalysts containing chloride ions, or for reaction conditions where a chlorine-containing gaseous compound is introduced along with the reactants.[125] C_{2+} selectivity as high as 90% with hydrocarbons up to C_4 have been reported by Wohlfahrt et al., who explored the catalytic behavior for various halides in OCM reactions.[126] The higher selectivities of ethylene for halogen systems prompted multiple studies on pure halide catalysts. Research was conducted particularly on unsupported chloride catalysts. Baldwin et al. compared the performance of pure $BaCl_2$, $MgCl_2$, $CaCl_2$, $SrCl_2$, and $MnCl_2$, of these chloride catalysts supported by pumice, and of alkali chloride promoted chloride catalysts.[125] The pure chloride catalysts had lower activity and selectivity compared to other sets of catalysts. Pulse studies over these catalysts point to the fact that the active sites for the OCM reaction may be at the metal oxide/chloride interface. Also the presence of alkali chloride promoters was found to increase the stability of the catalyst by retention of chloride ions on the catalyst surface.

Favorable effects with the presence of alkali chlorides on catalytic behavior were also observed for pure metal oxide catalysts. For example, Otsuka et al. demonstrated 31% C_2 yield with a manganese oxide with lithium chloride under co-feed conditions.[127] Alkaline earth bromides and fluorides were found to be less effective in increasing catalyst activity and selectivity as compared to the performance of alkaline earth chloride promoters.[128] For chlorine based catalysts the chlorine radicals were believed to take part in the de-hydrogenation of ethane to ethylene, thereby enhancing catalytic activity, and also playing a role in suppressing total oxidation.[127,129–131]

3.5.2   Solid Electrolyte Metal Catalysts

As a result of their ability to conduct oxygen ions while maintaining good mechanical strength and chemical resistance at high temperatures, solid electrolytes, like ZrO_2, ThO_2, and CeO_2 are used in OCM as materials in membrane reactors. Based on the specific properties of these solid electrolytes they can be used as electrochemical oxygen pumps, solid fuel cells, and in mixed oxygen ion–electron conduction. In the aforementioned applications, the generalized mechanism for the OCM reaction involves permeation of oxygen through the solid electrolyte, conversion of methane to methyl radicals via consumption of the transported oxygen species, coupling of methyl radicals to form C_2 products, or the total oxidation of methyl radicals or C_2 species to carbon oxides. Considering the above mechanism, solid electrolytes with good catalytic properties in terms of methane activation and high oxygen ion conductivity to match the methane activation rate would be most suited for OCM reaction applications. Solid electrolytes, such as perovskite materials, have also been used in OCM because of their favorable properties for OCM reactions, and these materials are described in more detail in the paragraphs below.[87,132–138] These perovskite materials have been used to study OCM chemical looping kinetics, as described in Section 3.6.1 and are also used in membrane reactors, which are further explained in Section 3.6.2.

Perovskite type oxides, i.e. ABO_3, where A and B are two different types of cations (also see Section 2.2.3), have sparked an interest in catalytic applications. The partial substitution of either A or B type cations by other cations of different oxidation states allows for tuning of their structural and functional properties and affects their catalytic behavior. The substitution of A type cations by lower valency cations has been found to alter the oxidation states of the B type cations and create oxidation defects. This effect was strongly evident in the $La_{(1-x)}Sr_xMO_{3-\delta}$ (where M = Co, Cr, Fe, Mn) perovskites where substitution of La by Sr resulted in multivalent oxidation states of M and also created oxygen vacancies. Both these modifications increased catalytic activity of these perovskites towards total oxidation of ammonia, CO, and hydrocarbons.[139]

Perovskites, because of their oxygen vacancies and strong redox abilities, are known for their use in the total oxidation of hydrocarbons and carbon monoxide. However, for OCM, the strong redox abilities of perovskites are more useful in increasing the selectivity of reaction towards ethylene and ethane, whereas the oxygen vacancies would increase the selectivity towards total oxidation to CO_2.[140] Hence, balancing these two properties of perovskites would enable optimal catalytic activity and selectivity in the OCM reaction. Some examples of perovskites that have been used for OCM applications are $La_{1-x}Ba_xCo_{1-y}Fe_yO_{3-\delta}$, $Ca_{1-x}Sr_xTi_{1-y}Fe_yO_{3-\delta}$, and $SrCe_{1-y}Yb_yO_{3-\delta}$.

The redox ability, especially of the B site cations, and the oxygen vacancies influence the type of oxygen species that are adsorbed onto the perovskite materials. The type of adsorbed oxygen species is a significant factor for OCM reactions as it has been found that divalent oxygen species, such as O^{2-} and O_2^{2-}, are more selective to ethane or C_{2+} products, whereas the monovalent oxygen species such as O^- are responsible for total oxidation of ethane. Temperature-programmed desorption of oxygen (O_2-TPD) experiments have shown that the oxygen vacancies promote the dissociative adsorption of O_2 resulting in formation of O^- species, also known as α oxygen.[139] In comparison, the O^{2-} and O_2^{2-} species, also known as β oxygen, are associated with the partially reduced B site cations.[139] Therefore, to increase the suitability of the perovskite material for use in OCM reactions, the density of different oxidation states of B site cations needs to be controlled and the number of oxygen vacancies has to be reduced. Both these tasks have been accomplished by doping halide anions into the perovskite structures and have resulted in an increase in ethylene selectivity for ethane oxidation reactions.[139–141] The halide anions, after being doped in the perovskites, either substitute the oxygen anions or occupy the oxygen vacancies. Dai et al. have schematically represented the halide ($X^- = Cl^-$, F^-) substitution in $La_{1-x}Sr_xFeO_{3-\delta}$ type perovskites, as shown in Figure 3.12.[139]

If the halide ion substitutes O^{2-}, it would then result in a decrease in oxidation state of Fe from Fe^{4+} to Fe^{3+}. In comparison, the oxidation state would increase from Fe^{3+} to Fe^{4+} when the halide ions occupy the oxygen vacancies. Therefore, halide doping was found to vary both the valence of B site cations and to reduce the oxygen vacancies to a desired amount resulting in an increase in selectivity towards ethylene by suppressing

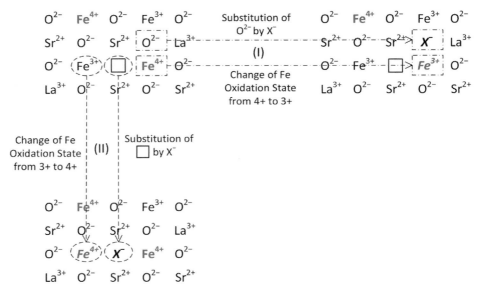

Figure 3.12 Models of halide substitution of the defects in the perovskite $La_{1-x}Sr_xFeO_{3-\delta}X_\sigma$ (where $X^- = F^-$ or Cl^-), such that (I) X^- replaces a lattice O^{2-} ion, or (II) X^- occupies an oxygen vacancy ($\square$).[139]

the total oxidation of ethane. Additionally, halides such as Cl^- and F^-, which are more electronegative than O^{2-}, on inclusion into the perovskite catalyst decrease the electron density of O^{2-} and weaken the cation–oxygen bonding. This makes the lattice oxygen more active in the reaction and increases the catalytic activity of the perovskite.[139]

Many studies have been carried out on the effect of halogen doping on the selective oxidation of ethane to ethylene. Dai et al. compared the halide-doped and undoped catalysts for $SrFeO_{3-\delta}Cl_\sigma$, $SrCoO_{3-\delta}Cl_\sigma$, $La_{1-x}Sr_xFeO_{3-\delta}X_\sigma$, $YBa_2Cu_3O_{7-\delta}X_\sigma$, $La_{1.85}Sr_{0.15}CuO_{4-\delta}X_\sigma$, and $Nd_{1.85}Ce_{0.15}CuO_{4-\delta}X_\sigma$ perovskites, where X is Cl or F.[141] In all cases halogen doping has been found to significantly increase the activity and selectivity of the catalyst for oxidative dehydrogenation of ethane. The factors that make halide-doped perovskites highly effective for selective oxidation of ethane could also be beneficial for OCM reactions. However, few studies have investigated the effect of halide-doped perovskites on the selectivity of an OCM reaction towards ethylene, which could present an avenue for future research.

3.5.3   Soft Oxidant Based Metal Catalysts

Recently, non-oxygen based OCM reactions have been of particular interest. The problems of over-oxidation of methane by O_2 and the highly exothermic nature of reactions that cause hotspots in the reactors have made researchers look towards different types of oxidants. Disulfur (S_2) has been found to be a molecule that is analogous to dioxygen. The tendency for methane to be over-oxidized by S_2 is far less thermodynamically favorable as compared to O_2. Zhu et al. have explored the use of elemental sulfur as a "soft" oxidant in the presence of transition metal sulfide catalysts for conversion of methane to ethylene.[142] Known as soft oxidation, metal sulfides and gaseous sulfur have demonstrated high yields of C_2 for the OCM reaction.[142]

The factor determining the activity and selectivity of the metal sulfide catalysts was found to be the catalyst metal–sulfur bond strength. It was observed both by experiments and by DFT calculations that weakly bonded sulfur atoms on the catalyst surface promote activation of methane via H abstraction, resulting in higher conversion of methane. On the other hand, the strongly bonded S atoms favorably affected the selectivity of the catalyst towards ethylene formation. The effects of the catalyst metal–sulfur bond strength were seen in experiments performed with MoS_2, RuS_2, PdS, and TiS_2 catalysts (arranged in increasing metal–S bond energy). For the above catalysts, it was also found that a high CH_4/S feed ratio, higher weight hourly space velocity (WHSV), and provision of supports, increased their selectivity towards ethylene production from methane. Supported PdS catalysts were found to be the best performing catalysts, which showed a C_2 selectivity of 20% and methane conversions of 16% at 1050 °C. In general, metal sulfides demonstrated promise for methane activation at the weakly bonded sulfur atom sites, and C_2 selectivity was correlated to the strongly bonded sulfur surface sites. This provides a motivation for looking into other options for "soft" oxidants that may allow for better control over the selectivity and reactor temperature for OCM reactions.

3.6 Reaction Engineering

While much work has focused on the development and optimization of OCM metal oxides materials, reaction engineering, specifically reaction kinetics and catalytic reactor characteristics, also play an important role in improving C_{2+} yields. While typical reactors for OCM operate in fixed beds and fluidized beds, they are limited in C_{2+} yields. Reactor systems have the potential to increase C_{2+} yields using existing catalysts through improved reactant mixing and/or product separations. This section provides an overview of the OCM reaction network, reaction kinetic models, and kinetic model effects on reactor performance in the formation of C_{2+} products for the co-feed reaction system. The redox reaction kinetics are also illustrated. Reactor designs for both co-feed and redox reaction systems are discussed.

3.6.1 Co-feed Elementary Kinetic Model

Developing an accurate kinetic model from elementary steps requires careful design. The inclusion of a large number of elementary reactions is computationally demanding

Table 3.5 Specific network of reactions for oxidative coupling of methane.[53]

		Reactions	
Adsorption reactions	3.6.1:	$O_2 + 2s \overset{K_{O_2}}{\leftrightarrow} 2O(ads)$	$\left(K_{O_2} = k_{O_2}/k_{O_2}^{rev} \right)$
	3.6.2:	$2OH(ads) \overset{K_{H_2O}}{\leftrightarrow} H_2O + O(ads) + s$	$\left(K_{H_2O} = k_{H_2O}/k_{H_2O}^{rev} \right)$
Catalytic surface reactions	3.6.3:	$CH_4 + O(ads) \overset{k_1}{\rightarrow} CH_3\bullet + OH(ads)$	
	3.6.4:	$C_2H_6 + O(ads) \overset{k_2}{\rightarrow} C_2H_5\bullet + OH(ads)$	
	3.6.5:	$C_2H_5\bullet + O(ads) \overset{k_3}{\rightarrow} C_2H_4 + OH(ads)$	
	3.6.6:	$CH_3\bullet + 3O(ads) \overset{k_4}{\rightarrow} CHO(ads) + 2OH(ads)$	
	3.6.7:	$CHO(ads) + O(ads) \overset{k_5}{\rightarrow} CO + OH(ads) + s$	
	3.6.8:	$CO + O(ads) \overset{k_6}{\rightarrow} CO_2 + s$	
	3.6.9:	$C_2H_4 + O(ads) \overset{k_7}{\rightarrow} C_2H_3\bullet + OH(ads)$	
Gas phase reactions	3.6.10:	$2CH_3\bullet \overset{k_8}{\rightarrow} C_2H_6$	
	3.6.11:	$CH_3\bullet + O_2 \overset{k_9}{\rightarrow} CHO\bullet + H_2O$	
	3.6.12:	$C_2H_3\bullet + O_2 + OH\bullet \overset{k_{10}}{\rightarrow} 2CHO\bullet + H_2O$	
	3.6.13:	$CHO\bullet + O_2 \overset{k_{11}}{\rightarrow} CO + HO_2\bullet$	
	3.6.14:	$CO + HO_2\bullet \overset{k_{12}}{\rightarrow} CO_2 + OH\bullet$	
	3.6.15:	$C_2H_6 \overset{k_{13}}{\rightarrow} C_2H_5\bullet + H\bullet$	
	3.6.16:	$C_2H_5\bullet + H\bullet \overset{k_{14}}{\rightarrow} C_2H_4 + H_2$	

and can potentially result in inaccurate parameter values due to a lack of experimental data; yet relying solely on global kinetic data has limited applicability, as will be shown in Section 3.6.3.[53] A reduction in the total number of reactions can be achieved in two ways: (1) by assuming the main reaction steps are the elementary steps in equilibrium; or (2) focusing only on the key elementary steps responsible for the main products (C_2s and CO_x). However, the first way cannot accurately explain the rate-limiting step, and the second way relies on experimental data with key parameters estimated from empirical regression.

One example of a reaction mechanism developed from the elementary steps, introduced in Section 3.2.1, is proposed by Lee et al.[53] The first step is the adsorption of oxygen molecules on the surface. As previously discussed in Section 3.2.4, oxygen isotope tracer experiments support surface dissociative adsorption over the non-dissociative adsorption mechanism; reaction (3.6.1) in Table 3.5 from Lee et al.[53] In reaction (3.6.1), s represents a surface adsorption site where the gaseous oxygen adsorbs on the metal oxide catalyst. The surface O combines with methane in the gas phase, generating methyl radicals and an OH group: reaction (3.6.3).[53] It is assumed that OH radicals reach equilibrium concentrations on the metal oxide catalyst surface, and two OH groups lead to the formation of a surface O, water, and a vacancy: reaction (3.6.2). Methane adsorption follows the Eley–Rideal mechanism and is considered to be irreversible since the $CH_{4x}D_x$ does not increase significantly with CH_4 conversion in a CD_4/CH_4 isotopic scan from an H–D exchange.

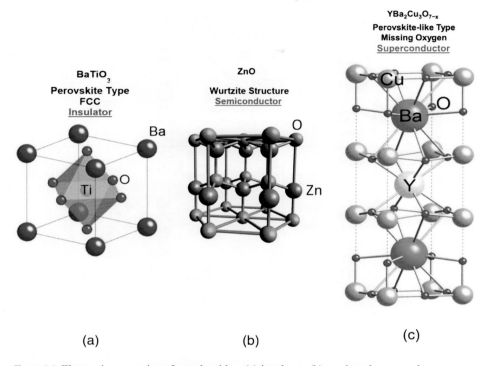

Figure 1.4 Electronic properties of metal oxides: (a) insulator; (b) semiconductor; and (c) superconductor.

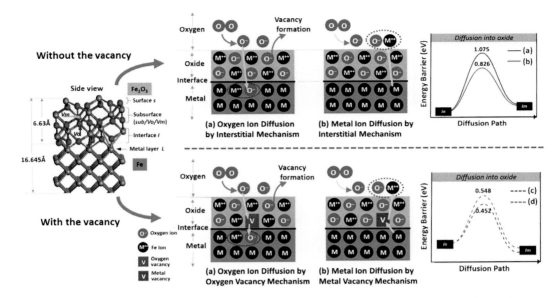

Figure 1.9 Ionic diffusion mechanism with diffusion activation energies obtained from DFT calculations.

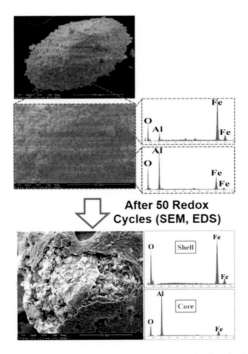

Figure 1.10 SEM and EDS spectral analysis of mixed iron oxide–aluminum oxide at the beginning of the experiment and at the completion of 50 reduction–oxidation cycles.

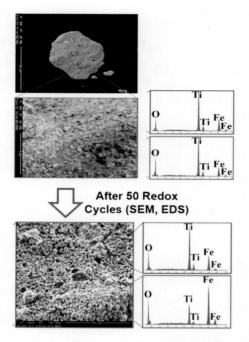

Figure 1.11 SEM and EDS spectral analysis of mixed iron oxide–titanium dioxide at the beginning of the experiment and at the completion of 50 reduction–oxidation cycles.

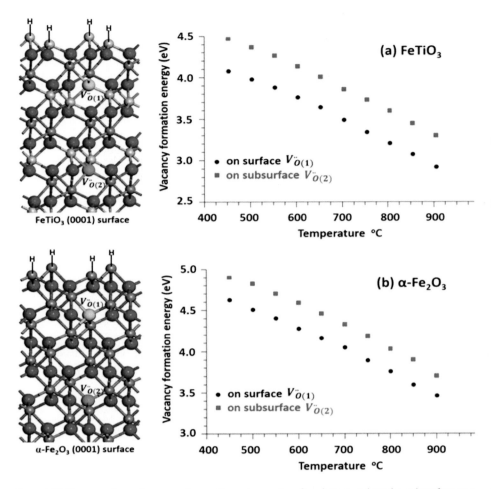

Figure 1.15 Vacancy formation energies on the outermost surface layers and on the subsurface as a function of temperature at pressure 1 atm for (a) $FeTiO_3$ and (b) α-Fe_2O_3 (Qin et al. *J. Mater. Chem. A*, 2015, 3, 11302–11312 – reproduced by permission of The Royal Society of Chemistry).

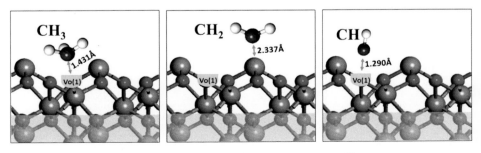

Figure 1.17 CH_x fragments adsorption. Vo(1) denotes oxygen vacancy.[34] (Cheng Z. et al. *Phys. Chem. Chem. Phys.*, 2016, 18, 16423–16435 – reproduced by permission of The Royal Society of Chemistry).

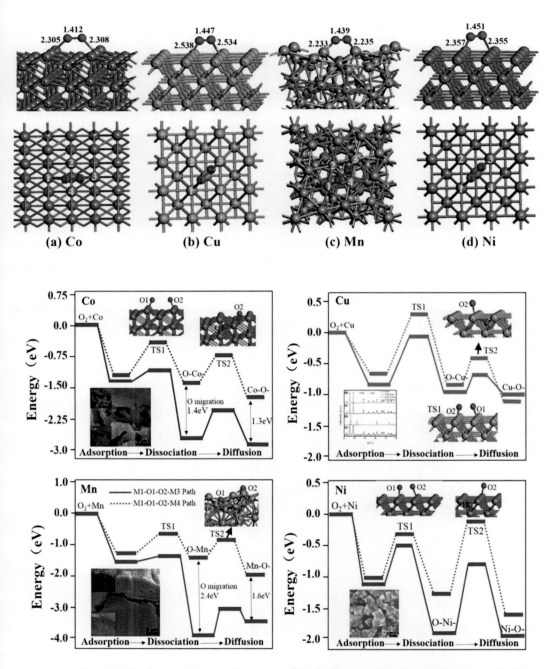

Figure 1.19 Surface structures and energy profiles for the oxidation of transition metals Co, Cu, Mn, and Ni (top: the most stable adsorption structures of O_2 on Co, Cu, Mn, and Ni. The distances of the O–O bond and the O–metal bond are indicated (Å). Bottom: calculated reaction coordinates of O_2 dissociation and diffusion on Co, Cu, Mn, and Ni (100) surfaces. Inset: SEM images of oxidized Co, Mn, and Ni and XRD spectra of CuO during redox reactions).

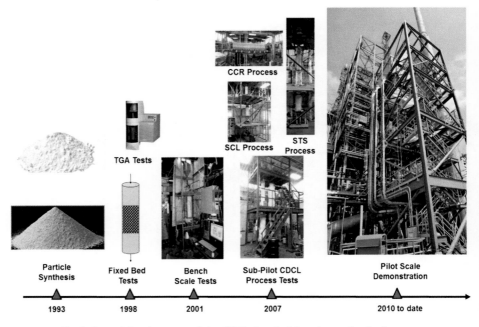

Figure 1.28 Evolution of development of the OSU chemical looping technologies.

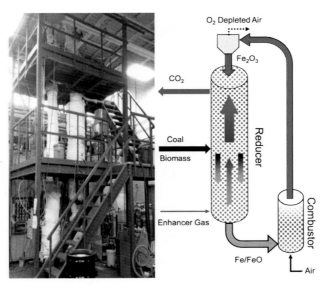

Figure 1.30 Sub-pilot scale demonstration unit for the CDCL process (left); flow schematic of the CDCL process for power production (right).

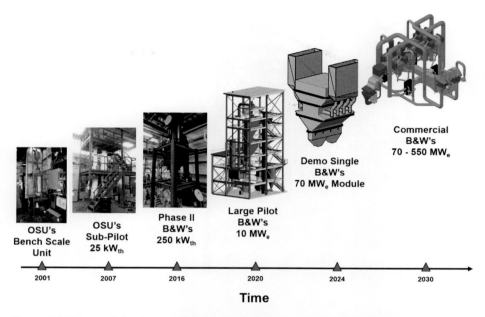

Figure 1.34 Commercialization pathway for CDCL technology by OSU and B&W Power Generation Group.

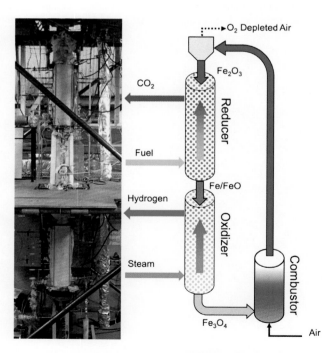

Figure 1.35 Sub-pilot scale demonstration unit for the SCL process for H_2 production (left); flow schematic of the SCL process unit for H_2 production (right).

Figure 1.36 SCL pilot-scale demonstration unit at the National Carbon Capture Center, Wilsonville, AL (reprinted from Fan, L.-S. et al., 2015, *AIChE J.,* 61: 2–22 with permission from John Wiley and Sons).

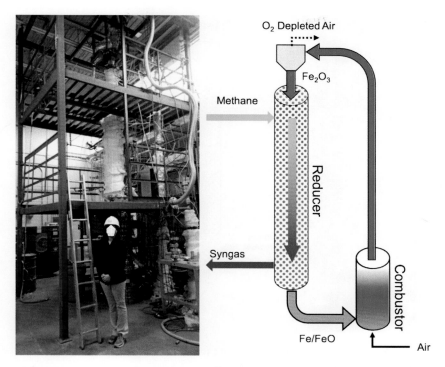

Figure 1.43 Sub-pilot scale demonstration unit for the STS process (left); flow schematic of the STS process unit (right).

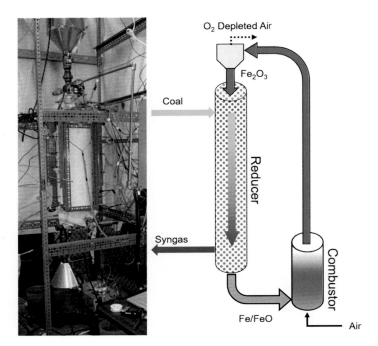

Figure 1.46 Bench-scale demonstration unit for the reducer (left); flow schematic of the CTS process unit (right).

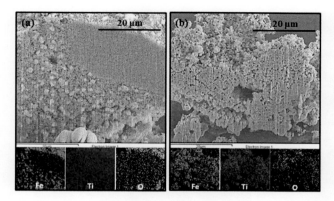

Figure 2.20 SEM and EDS elemental mapping of cross-sections of reduced ITCMO particles. Samples reduced under H_2, $Y_{H_2} = 0.5$ and T = 900 °C. (a) 1 atm and (b) 10 atm. Reprinted with permission from Deshpande et al. *Energy & Fuels*, **2015**, 29(3), 1469–1478. Copyright 2015 American Chemical Society.

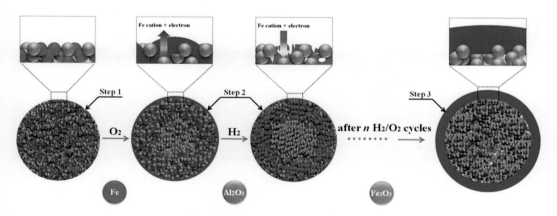

Figure 2.23 Formation of a core–shell structured composite of Fe_2O_3–Al_2O_3 micro-particles via cyclic redox reactions. Reprinted with permission from Sun et al. *Langmuir*, **2013**, 29(40), 12520–12529. Copyright 2013 American Chemical Society.

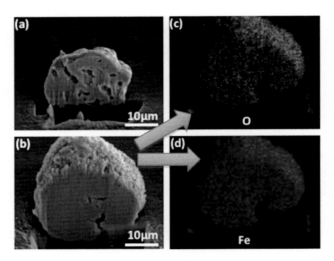

Figure 2.26 Fe micro-particle. (a) SEM image of cross-section of fresh micro-particle; (b) SEM image of cross-section of Fe_2O_3 micro-particle after Fe oxidation at 700 °C; (c) EDS mapping of O from (b); (d) EDS mapping of Fe from (b). Qin et al. *J. Mater. Chem. A*, **2014**, 2(41), 17511–17520. Reproduced by permission of The Royal Society of Chemistry.

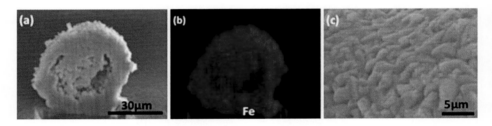

Figure 2.28 Fe micro-particle after one oxidation–reduction cycle. (a) SEM image of cross-section; (b) EDS mapping of Fe; (c) SEM image of surface. Qin et al. *J. Mater. Chem. A*, **2014**, 2(41), 17511–17520. Reproduced by permission of The Royal Society of Chemistry.

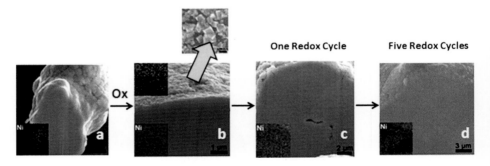

Figure 2.29 Cross-sectional SEM image of a Ni micro-particle subjected to redox cycles at 700 °C in the presence of H_2 and O_2. (a) Fresh Ni, (b) after oxidation, (c) after oxidation and reduction, single cycle, (d) after five redox cycles.

One Redox Cycle   Five Redox Cycles

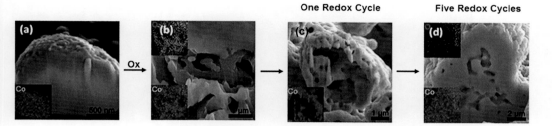

Figure 2.31 Cross-sectional SEM image of the Co micro-particle subjected to redox cycles at 700 °C in the presence of H_2 and O_2: (a) fresh Co, (b) after oxidation, (c) after oxidation and reduction, single cycle, (d) after five redox cycles.

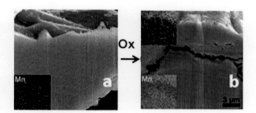

Figure 2.32 Cross-sectional SEM image analysis of the Mn micro-particle: (a) fresh Mn, (b) after oxidation at 700 °C in the presence of O_2.

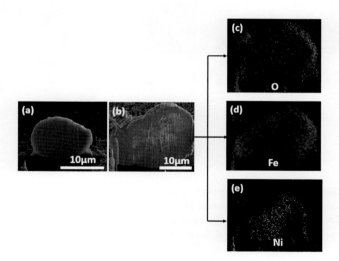

Figure 2.34 FeNi micro-particle. SEM images of the cross-section: (a) fresh micro-particle; (b) oxidized particle after oxidation at 700 °C for 0.5 h; EDS mapping of oxidized micro-particles: (c) O; (d) Fe; and (e) Ni. Qin et al. *J. Mater. Chem. A*, **2014**, 2(41), 17511–17520. Reproduced by permission of The Royal Society of Chemistry.

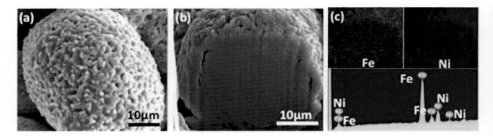

Figure 2.36 FeNi micro-particle after one oxidation–reduction cycle: (a) SEM image of surface; (b) SEM image of cross-section; (c) EDS mapping of cross-section and EDS spectrum of surface. Qin et al. *J. Mater. Chem. A*, **2014**, 2(41), 17511–17520. Reproduced by permission of The Royal Society of Chemistry.

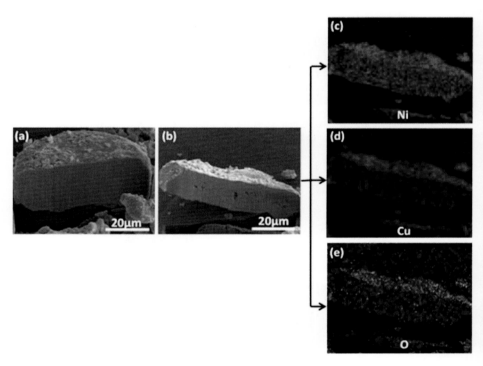

Figure 2.37 CuNi micro-particle. SEM images of the cross-section: (a) fresh micro-particle; (b) oxidized micro-particle after oxidation at 700 °C for 0.5 hour; oxidized micro-particle EDS mapping: of (c) Ni; (d) Cu; and (e) O. Qin et al. *J. Mater. Chem. A*, **2014**, 2(41), 17511–17520. Reproduced by permission of The Royal Society of Chemistry.

Figure 2.40 FeTi micro-particle: (a) SEM image of cross-section; (b) EDS mapping of Fe and Ti, and EDS spectrum of the surface. Qin et al. *J. Mater. Chem. A*, **2015**, 3(21), 11302–11312. Reproduced by permission of The Royal Society of Chemistry.

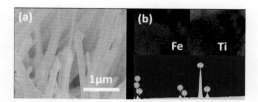

Figure 2.41 Oxidized FeTi micro-particles at 700 °C: (a) SEM image of the surface at higher magnification; (b) EDS mapping of the cross-section and EDS spectrum of the surface. Qin et al. *J. Mater. Chem. A*, **2015**, 3(21), 11302–11312. Reproduced by permission of The Royal Society of Chemistry.

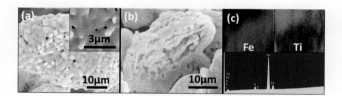

Figure 2.42 FeTi micro-particle after one oxidation–reduction cycle at 700 °C: (a) SEM image of surface; inset: higher magnification SEM image; (b) SEM image of cross-section; and (c) EDS mapping and spectrum of (b). Qin et al. *J. Mater. Chem. A*, **2015**, 3(21), 11302–11312. Reproduced by permission of The Royal Society of Chemistry.

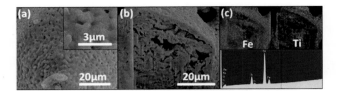

Figure 2.43 FeTi micro-particle after five oxidation–reduction cycles at 700 °C: (a) SEM image of surface; inset: higher magnification SEM image; (b) SEM image of cross-section; and (c) EDS mapping and EDS spectrum of (b). Qin et al. *J. Mater. Chem. A*, **2015**, 3(21), 11302–11312. Reproduced by permission of The Royal Society of Chemistry.

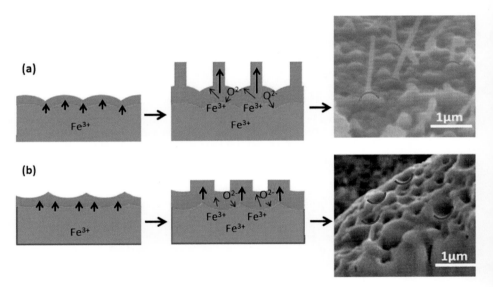

Figure 2.46 Growth mechanisms of (a) iron oxide nanowires and (b) iron oxide nanopores. Qin et al. *J. Mater. Chem. A*, **2014**, 2(41), 17511–17520. Reproduced by permission of The Royal Society of Chemistry.

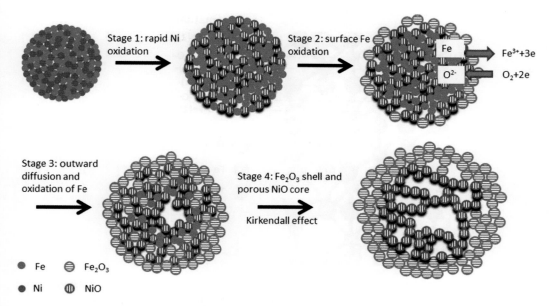

Figure 2.47 Core–shell structure formations in oxidized FeNi. Qin et al. *J. Mater. Chem. A*, **2014**, 2 (41), 17511–17520. Reproduced by permission of The Royal Society of Chemistry.

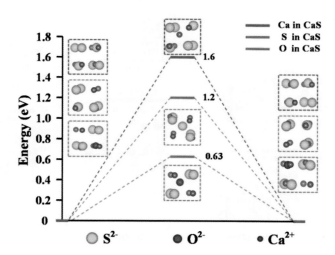

Figure 2.48 The DFT calculated energy barriers of three key diffusing ions in the reaction of COS with CaO. Reproduced with permission from Sun et al. *AIChE J.*, **2012**, 58(8), 2617–2620. Copyright©2011 American Institute of Chemical Engineers (AIChE).

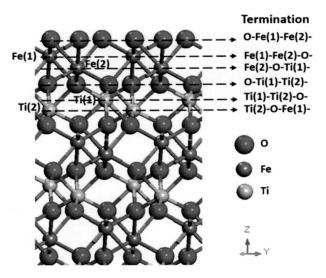

Figure 2.49 FeTiO$_3$ (0001) surfaces and possible surface termination. Qin et al. *J. Mater. Chem. A*, **2015**, 3(21), 11302–11312. Reproduced by permission of The Royal Society of Chemistry.

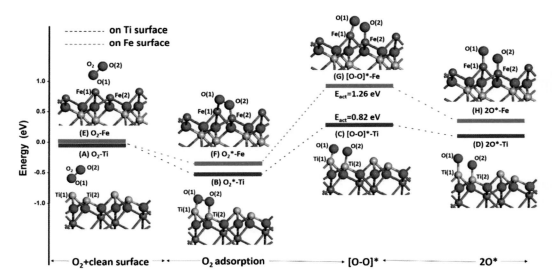

Figure 2.50 O$_2$ molecule adsorption for Ti–Ti–O and Fe–Fe–O surface terminations. Qin et al. *J. Mater. Chem. A*, **2015**, 3(21), 11302–11312. Reproduced by permission of The Royal Society of Chemistry.

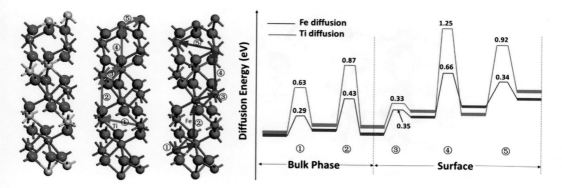

Figure 2.51 Ti and Fe diffusion paths and the associated energies. Qin et al. *J. Mater. Chem. A,* **2015**, 3(21), 11302–11312. Reproduced by permission of The Royal Society of Chemistry.

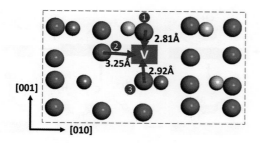

Figure 2.53 Cross-section through $FeTiO_3$ bulk unit cell in the (100) plane, where red spheres: O anions, purple spheres: Fe cations, gray spheres: Fe cations.

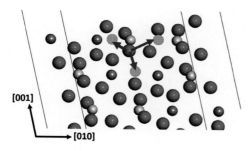

Figure 2.55 Schematic illustration of oxygen interstitial diffusion in $FeTiO_3$ bulk, where red spheres: O anions, purple spheres: Fe cations, gray spheres: Fe cations, dotted red spheres: Fe–Ti interstitial site.

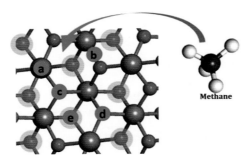

Figure 2.56 Methane adsorption sites on the surface layer of Fe_2O_3 system (top view).

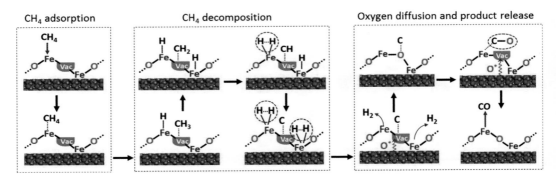

Figure 2.57 Proposed mechanism of CH_4 partial oxidation to syngas on Fe_2O_3 system based on DFT calculations.

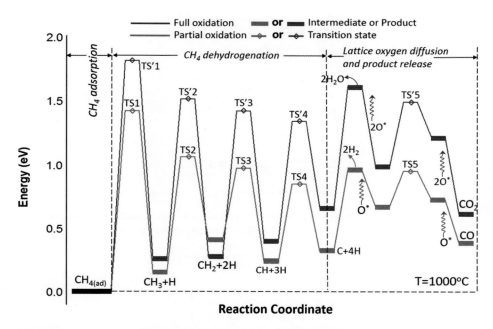

Figure 2.58 CH_4 oxidation energy profile through full oxidation (pathway 1) and partial oxidation (pathway 2).

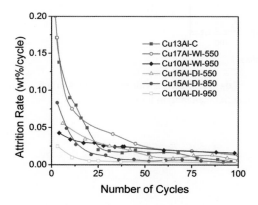

Figure 2.84 Attrition rates of some Cu-based oxygen carriers (legend shows the copper content, impregnation method and calcination temperatures).

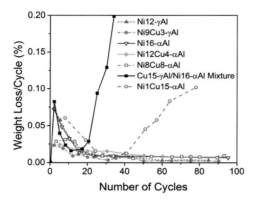

Figure 2.85 Attrition rates of several Ni-based and Ni/Cu-based oxygen carriers prepared by impregnation (legend shows Ni/Cu content and support material).

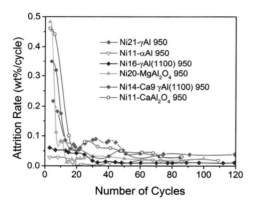

Figure 2.86 Attrition rates of Ni-based oxygen carriers with redox cycles in a batch fluidized bed (legend with Ni content, support material, calcination temperature).

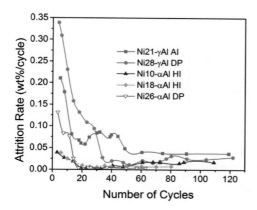

Figure 2.87 Attrition rates of Ni-based oxygen carriers prepared with different methods and supports (legend shows Ni content, support material, and preparation method).

Figure 4.7 Shell partial oxidation technology plants: (left: Bintulu); (right: Pearl).

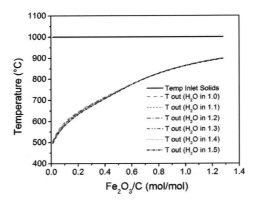

Figure 6.56 Temperature of ITCMO at reducer outlet as a function of Fe_2O_3:C ratio.

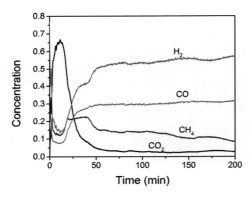

Figure 6.67 Syngas generation from natural gas in a fluidized bed reducer using ITCMO.

In the OCM reaction, the generated methyl radicals could either dimerize to produce ethane in the gas phase: reaction (3.6.10); or the methyl radicals can yield oxygen-containing components such as CO and CO_2; reactions (3.6.6) to (3.6.8). It is believed that ethane is formed from methyl radicals in the gas phase, since the residence time of ethane is observed to be close to that of an inert tracer.[80] The formation of ethane and oxygen containing components occurs in two competing reactions. It is observed that the latter takes place more easily on the surface of metal oxides that have multiple oxidation states.[81,143-145]

Ethylene is the product of dehydrogenation of ethane by surface oxygen, as seen in reactions (3.6.4) and (3.6.5). This has been observed with electron paramagnetic resonance (EPR).[144] This is consistent with the observation that an increase in ethylene yield often coincides with a decrease in ethane yield for similar methane conversions.[146] CO and CO_2 can form on the metal oxide catalyst surface through reactions (3.6.7) and (3.6.8) or in the gas phase through reactions (3.6.13) and (3.6.14). Chen et al. stated that the majority of CO and CO_2 formed is a result of the surface reaction between $CH_3\bullet$ radicals and atomic oxygen.[145] However, CO could also form in the gas phase from the reaction between $CH_3\bullet$ and $C_2H_3\bullet$ radicals with oxygen molecules.[147] Furthermore, it is assumed that the intermediate $CHO\bullet$ is in a quasi-steady state on the metal oxide catalyst surface. In order to develop a kinetic model with reasonable accuracy, the dehydrogenation of ethyl radicals also needs to be considered. Assuming that O_2 and H_2O adsorption reach equilibrium quickly and the remaining reactions are the rate determining steps, then the reaction mechanism can be described through rate equations (3.6.17)–(3.6.21):

$$r_{CH_4} = -k_1\theta_O P_{CH_4} \tag{3.6.17}$$

$$r_{C_2H_6} = k_8 P_{CH_3\bullet}^2 - (k_2\theta_O + k_{13})P_{C_2H_6} \tag{3.6.18}$$

$$r_{C_2H_4} = (k_3\theta_O + k_{14}P_{H\bullet})P_{C_2H_5\bullet} - k_7\theta_O P_{C_2H_4} \tag{3.6.19}$$

$$r_{CO} = k_5\theta_{CHO} + k_{11}P_{CHO\bullet}P_{O_2} - (k_6\theta_O + k_{12}P_{HO_2\bullet})P_{CO} \tag{3.6.20}$$

$$r_{CO_2} = (k_6\theta_O + k_{12}P_{HO_2\bullet})P_{CO}, \tag{3.6.21}$$

where P_x is the partial pressure of specie "x",

$$\theta_O = \left(\sqrt{K_{O_2}P_{O_2}}\right)\theta_S \ , \ \theta_{CHO} = \alpha\theta_S^2, \ \theta_S = \left(-\beta + \sqrt{\beta^2 + 4\alpha}\right)/(2\alpha)$$

$$\alpha = \frac{k_4 K_{O_2}}{k_5}P_{O_2}P_{CH_3\bullet}, \beta = 1 + \sqrt{K_{O_2}P_{O_2}} + \sqrt{\frac{P_{H_2O}\sqrt{K_{O_2}P_{O_2}}}{K_{H_2O}}}.$$

Table 3.6 Calculated results from Lee et al.[53]

Reaction rate constants	Temperature (°C)				$k_{pre,i}$	E_a (kJ/mol)	R^2
	760	790	820	850			
$k_{1(r\ 3.6.3)}$	4.17E−09	6.08E−09	1.08E−08	1.67E−08	1.44E−01	149.1	1.000
$k_{2(r\ 3.6.4)}$	1.14E−09	2.71E−09	6.12E−09	1.33E−08	2.18E+04	262.8	0.960
$k_{3(r\ 3.6.5)}$	2.51E−05	4.80E−05	8.84E−05	1.58E−04	2.24E+05	196.9	0.736
$k_{4(r\ 3.6.6)}$	2.05E−03	5.20E−03	1.25E−02	2.88E−02	4.23E+11	283.2	0.984
$k_{5(r\ 3.6.7)}$	4.78E−05	9.58E−05	1.85E−04	3.44E−04	2.46E+06	211.7	0.903
$k_{6(r\ 3.6.8)}$	3.45E−08	5.59E−08	8.81E−08	1.36E−07	9.03E−01	146.7	0.998
$k_{7(r\ 3.6.9)}$[a]	2.64E−08	2.64E−08	2.64E−08	2.64E−08	2.64E−08	—	—
$k_{8(r\ 3.6.10)}$	4.92E−08	5.74E−08	6.65E−08	7.64E−08	1.22E−05	47.2	0.342
$k_{9(r\ 3.6.11)}$[a]	7.72E−12	7.72E−12	7.72E−12	7.72E−12	7.72E−12	—	—
$k_{10(r\ 3.6.12)}$	5.51E−10	7.99E−10	1.13E−09	1.58E−09	2.87E−04	113.1	0.872
$k_{11(r\ 3.6.13)}$	9.39E−05	2.09E−04	4.44E−04	9.07E−04	1.84E−08	243.1	0.902
$k_{12(r\ 3.6.14)}$	7.28E−10	1.07E−09	1.53E−09	2.16E−09	5.53E−04	116.4	0.999
$k_{13(r\ 3.6.15)}$	2.00E−09	2.56E−09	3.23E−09	4.03E−09	1.25E−05	75.0	1.000
$k_{14(r\ 3.6.16)}$	3.73E−10	5.15E−10	6.99E−10	9.33E−10	3.42E−05	98.3	0.787
Adsorption equilibrium Constants					$K_{pre,i}$	ΔH (kJ/mol)	
K_{O2}	3.47E−06	1.93E−06	1.11E−06	6.61E−07	3.63E−15	−177.6	0.6388
K_{H2O}	2.56E+02	2.24E+02	1.97E+02	1.75E+02	2.23E+00	−40.6	0.3271

[a] These parameters were assumed to be independent of temperature and averaged values were used, since energies were calculated to be close to zero.

Reaction constants and adsorption equilibrium constants are determined by fitting the experimental data to the Arrhenius and Van't Hoff equations at various reaction temperatures, respectively. The calculated results are shown in Table 3.6, from Lee et al.[53]

Of special importance is the temperature range in which the OCM reactions are applicable, where excessively high temperatures result in the deactivation of metal oxide catalysts and too low temperatures result in low reactivity. Within the temperature range where the OCM reaction kinetics are reasonable, the kinetic model properly explains the effects of temperature and methane:oxygen ratio.

Methane conversion increases with increasing temperature. For a fixed methane: oxygen ratio, higher temperatures improve ethylene production and simultaneously decrease ethane production. Since reaction (3.6.4) is the reaction for the ethane consumption, reaction (3.6.10) is the reaction for ethane production, and the activation energy of reaction (3.6.4) is higher than that of reaction (3.6.10), a temperature increase has a more significant impact on reaction (3.6.4) than on reaction (3.6.10), resulting in a decrease in ethane production.

When all reaction conditions are fixed, a higher methane:oxygen ratio could generate more ethylene and ethane at the expense of lower methane conversion. This is explained by the decreased reaction rate of CO and CO_2 generation. Unfortunately, although the qualitative trend is reasonable, the average errors are still significant between calculated results and quantitative experimental values, particularly in terms of methane conversion and CO selectivity. This results from both inaccuracy of experimental measurement and the imperfection of the kinetics models. For example, side reactions may occur, including the formation of methanol. Further, there are many experimental factors that have not been incorporated in this model. For instance, the particle size and geometric shape of the metal oxide catalyst could have an impact on reaction results.[53]

The equations incorporated in the rate equations result from empirical selection in order to fit the experimental results. For example, the Langmuir–Hinshelwood mechanism has been widely used. Although the rate equations could generate results with reasonable accuracy, there is a lack of sound physical or chemical rationale. It is assumed that adsorption–desorption equilibrium is important in the Langmuir–Hinshelwood type of rate equations, but it is actually impossible for methane or ethane to achieve substantial coverage on the oxide surfaces at typical OCM temperatures.

Kinetic models of this catalytic process remain difficult to develop, as there are still limitations. Even if the rate equations accurately describe OCM reactivity in lab-scale kinetic experiments, there are still obvious differences between small-scale kinetic experiments and realistic commercial-scale industrial reactors. For example, temperature gradients could be an important factor in industrial reactors, which is often overlooked in lab-scale kinetic experiments.

3.6.2   Co-feed Global Kinetic Models

As mentioned in Section 3.3, one of the most promising catalysts for OCM is a sodium tungstate, manganese catalyst supported on silica, $Mn/Na_2WO_4/SiO_2$. The reaction kinetics with $Mn/Na_2WO_4/SiO_2$ as the catalyst has been heavily researched, with several reaction mechanisms, reaction networks, and kinetic models developed, all of which vary widely.[148-151] From experimental results obtained from micro-catalytic fixed bed reactors (micro-reactors), kinetic models are obtained to describe the OCM reaction network by first selecting the reactions to be included in the kinetic model, then selecting an appropriate kinetic expression for each of the reactions, and finally data fitting the experimental results to obtain the parameters for each kinetic expression in the proposed kinetic model.

In this section and Section 3.6.3, three OCM kinetic models, denoted as R1, R2, and R3 in Table 3.7[148-151], are selected from the literature for detailed analysis and discussion. These models are collectively represented by 13 separate reactions, as given in the table, i.e. reactions (3.6.22) to (3.6.34), with their kinetic expressions also given. The number of reactions considered for each of the three kinetic models varies between five and ten, with only three reactions that are common to all three kinetic models. These three reactions are reaction (3.6.22), reaction (3.6.23), and reaction (3.6.25) representing selective oxidation of methane to ethane, selective oxidation of ethane to ethylene, and methane full combustion, respectively. Two of the models have three additional reactions in common. These three reactions are reaction (3.6.30), reaction (3.6.33), and reaction (3.6.34), representing ethylene partial combustion, water–gas shift, and

Table 3.7 Global reaction kinetics for the reactions considered in the three kinetic models. [145-148]

Reaction		Equation	Pre-exponential factor	E_a	Parameters		Ref. (Model)
Ethane formation model: R1, R2, R3	$2CH_4 + 0.5O_2 \rightarrow C_2H_6 + H_2O$ (3.6.22)	$r_1 = \dfrac{k_{01}e^{-E_{a,1}/RT}\left(K_{0,O_2}e^{-\Delta H_{adO_2}/RT}P_{O_2}\right)^{n_1}P_{CH_4}^{m_1}}{\left[1+\left(K_{0,O_2}e^{-\Delta H_{adO_2}/RT}P_{O_2}\right)^{n_1}\right]^2}$	$k_{01} = 29.4$ kmol/(kg s p^{m+n})	$E_{a,1} = 212.6$ kJ/mol	$m_1 = 1$ $\Delta H_{adO_2} = -121.9$ kJ/mol	$n_1 = 0.75$ $K_{0,O_2} = 4.39 \times 10^{-11}$ Pa^{-1}	149,150 (R3)
		$r_1 = \dfrac{k_{0,1}e^{-E_{a,1}/RT}P_{CH_4}^{m_1}P_{O_2}^{n_1}}{\left(1+K_{1,CH_4}e^{-\Delta H_{ad,1,CH_4}/RT}P_{CH_4}+K_{1,O_2}e^{-\Delta H_{ad,1,O_2}/RT}P_{O_2}\right)^2}$	$k_{0,1} = 1.066\times10^{-3}$ mol/(g s Pa^{m+n})	$E_{a,1} = 133$ kJ/mol	$m_1 = 0.501$ $K_{1,CH_4} = 5.5\times10^{-14}$ Pa^{-1} $\Delta H_{ad,1,CH_4} = -126$ kJ/mol	$n_1 = 0.504$ $K_{1,O_2} = 1.96 \times 10^{-13}$ Pa^{-1} $\Delta H_{ad,1,O_2} = -125$ kJ/mol	148 (R2)
		$r_1 = k_1 P_{CH_4}^{m_1} P_{O_2}^{n_1}$	$k_1 = 1.18\times10^{-8}$ mol/(kg s Pa^{m+n})		$m_1 = 1$	$n_1 = 0.36$	151 (R1)
Ethylene formation model: R1, R2, R3	$C_2H_6 + 0.5O_2 \rightarrow C_2H_4 + H_2O$ (3.6.23)	$r_5 = k_{05}e^{-E_5/RT}P_{C_2H_6}^{m_5}P_{O_2}^{n_5}$	$k_{05} = 2.70\times10^{-3}$ kmol/(kg s p^{m+n})	$E_{a,5} = 153.5$ kJ/mol	$m_5 = 0.91$	$n_5 = 0.5$	149,150 (R3)
		$r_4 = \dfrac{k_{0,4}e^{-E_{a,4}/RT}P_{C_2H6}^{m_4}P_{O_2}^{n_4}}{\left(1+K_{4,C_2H_6}e^{-\Delta H_{ad,4,C_2H_6}/RT}P_{C_2H_6}^{m_4}+K_{4,O_2}e^{-\Delta H_{ad,4,O_2}/RT}P_{O_2}^{n_4}\right)^2}$	$k_{0,4} = 1.26\times10^{-5}$ mol/(g s Pa^{m+n})	$E_{a,4} = 230$ kJ/mol	$m_4 = 0.295$ $K_{4,C_2H_6} = 1.98\times10^{-13}$ Pa^{-1} $\Delta H_{ad,4,C_2H_6} = -147$ kJ/mol	$n_4 = 1.22$ $K_{4,O_2} = 7.86 \times 10^{-14}$ Pa^{-1} $\Delta H_{ad,4,O_2} = -99$ kJ/mol	148 (R2)
		$r_3 = k_3 P_{C_2H_6}^{m_3} P_{O_2}^{n_3}$	$k_3 = 2.008\times10^{-7}$ mol/(kg s Pa^{m+n})		$m_3 = 1$	$n_3 = 0.58$	151 (R1)

Model	Reaction	Eq.	Rate expression	k	E_a	m	n	Ref
Ethylene formation/ ethane dehydrogenation model: R3	$C_2H_6 \rightarrow C_2H_4 + H_2$	(3.6.24)	$r_8 = k_{08} e^{-E_{a,8}/RT} p_{C_2H_6}^{m_8}$	$k_{08} = 1.08\times10^7$ mol/(m^3 s P^{m+n})	$E_{a,8} = 291.9$ kJ/mol	$m_8 = 0.88$	$n_8 = 0$	149,150 (R3)
Methane full combustion model: R1, R2, R3	$CH_4 + 2O_2 \rightarrow CO_2 + 2H_2O$	(3.6.25)	$r_2 = k_{02} e^{-E_{a,2}/RT} p_{CH_4}^{m_2} p_{O_2}^{n_2}$	$k_{02} = 3.07\times10^{-7}$ kmol/(kg s P^{m+n})	$E_{a,2} = 98.54$ kJ/mol	$m_2 = 0.85$	$n_2 = 0.5$	149,150 (R3)
			$r_3 = \dfrac{k_{0,3} e^{-E_{a,3}/RT} p_{CH_4}^{m_3} p_{O_2}^{n_3}}{\left(1 + K_{3,CH_4} e^{-\Delta H_{ad,3,CH_4}/RT} p_{CH_4}^{m_3} + K_{3,O_2} e^{-\Delta H_{ad,3,O_2}/RT} p_{O_2}^{n_3}\right)^2}$	$k_{0,3} = 1.36\times10^{-9}$ mol/(g s Pa^{m+n})	$E_{a,3} = 24.5$ kJ/mol	$m_3 = 0.875$; $K_{3,CH_4} = 2.54\times10^{-14}$ Pa^{-1}; $\Delta H_{ad,3,CH_4} = -99.8$ kJ/mol	$n_3 = 0.047$; $K_{3,O_2} = 3.79\times10^{-11}$ Pa^{-1}; $\Delta H_{ad,3,O_2} = -99.8$ kJ/mol	148 (R2)
			$r_2 = k_2 p_{CH_4}^{m_2} p_{O_2}^{n_2}$	$k_2 = 7.02\times10^{-9}$ mol/(kg s Pa^{m+n})		$m_2 = 0.59$	$n_2 = 1$	151 (R1)
Methane partial combustion model: R2(3.6.27), R3(3.6.26)	$CH_4 + O_2 \rightarrow CO + H_2O + H_2$	(3.6.26)	$r_3 = k_{03} e^{-E_{a,3}/RT} p_{CH_4}^{m_3} p_{O_2}^{n_3}$	$k_{03} = 6.65\times10^{-8}$ kmol/(kg s P^{m+n})	$E_{a,3} = 146.8$ kJ/mol	$m_3 = 0.5$	$n_3 = 1.57$	149,150 (R3)
	$CH_4 + 3/2\,O_2 \rightarrow CO + 2H_2O$	(3.6.27)	$r_2 = \dfrac{k_{0,2} e^{-E_{a,2}/RT} p_{CH_4}^{m_2} p_{O_2}^{n_2}}{\left(1 + K_{2,CH_4} e^{-\Delta H_{ad,2,CH_4}/RT} p_{CH_4}^{m_2} + K_{2,O_2} e^{-\Delta H_{ad,2,O_2}/RT} p_{O_2}^{n_2}\right)^2}$	$k_{0,2} = 6.82\times10^{-9}$ mol/(g s Pa^{m+n})	$E_{a,2} = 30$ kJ/mol	$m_2 = 0.604$; $K_{2,CH_4} = 7.6\times10^{-14}$ Pa^{-1}; $\Delta H_{ad,2,CH_4} = -93$ kJ/mol	$n_2 = 0.297$; $K_{2,O_2} = 1.1\times10^{-14}$ Pa^{-1}; $\Delta H_{ad,2,O_2} = -156$ kJ/mol	148 (R2)
CO Full Combustion model: R3	$CO + 0.5O_2 \rightarrow CO_2$	(3.6.28)	$r_4 = k_{04} e^{-E_{a,4}/RT} p_{CH_4}^{m_4} p_{O_2}^{n_4}$	$k_{04} = 5.26\times10^{-4}$ kmol/(kg s P^{m+n})	$E_{a,4} = 114.6$ kJ/mol	$m_4 = 0.5$	$n_4 = 0.5$	149,150 (R3)
Ethylene full combustion model: R1	$C_2H_4 + 3O_2 \rightarrow 2CO_2 + 2H_2O$	(3.6.29)	$r_4 = k_4 p_{C_2H_4}^{m_4} p_{O_2}^{n_4}$	$k_4 = 5.2\times10^{-8}$ mol/(kg s Pa^{m+n})		$m_4 = 1$	$n_4 = 1$	151 (R1)

Table 3.7 (cont.)

Reaction		Equation	Pre-exponential factor	E_a	Parameters		Ref. (Model)
Ethylene partial combustion model: R2, R3	$C_2H_4 + 2O_2 \rightarrow 2CO + 2H_2O$	(3.6.30)	$r_6 = k_{06}e^{-E_{a,6}/RT}p_{CH_4}^{m_6} p_{O_2}^{n_6}$	$k_{06} = 0.181$ kmol/(kg s P^{m+n})	$E_{a,6} = 174.4$ kJ/mol	$m_6 = 0.72$	$n_6 = 0.4$	149,150 (R3)
			$$r_5 = \frac{k_{0,5}e^{-E_{a,5}/RT}p_{C_2H_4}^{m_5} p_{O_2}^{n_5}}{\left(1+K_{5,C_2H_4}e^{-\Delta H_{ad,5,C2H4}/RT}p_{C_2H_4}^{m_5}+K_{5,O_2}e^{-\Delta H_{ad,5,O_2}/RT}p_{O_2}^{n_5}\right)^2}$$	$k_{05} = 4.78\times 10^{-4}$ mol/(g s Pa^{m+n})	$E_{a,5} = 110$ kJ/mol	$K_{5,C_2H_4} = 3.46\times 10^{-13}$ Pa^{-1} — $\Delta H_{ad,5,C_2H_4} = -167$ kJ/mol — $m_5 = 0.42$	$K_{5,O_2} = 6.87\times 10^{-9}$ Pa^{-1} — $\Delta H_{ad,5,O_2} = -135$ kJ/mol — $n_5 = 0.31$	148 (R2)
Ethylene partial combustion model: R3	$C_2H_4 + 2H_2O \rightarrow 2CO + 4H_2$	(3.6.31)	$r_7 = k_{07}e^{-E_{a,7}/RT}p_{C_2H_4}^{m_7} p_{H_2O}^{n_7}$	$k_{07} = 4.61\times 10^2$ kmol/(kg s P^{m+n})	$E_{a,7} = 394.2$ kJ/mol	$m_7 = 1.62$	$n_7 = 0.71$	149,150 (R3)
Ethane full combustion model: R1	$C_2H_6 + 7/2O_2 \rightarrow 2CO_2 + 3H_2O$	(3.6.32)	$r_5 = k_5 p_{C_2H_6}^{m_5} p_{O_2}^{n_5}$	$k_5 = 3.31\times 10^{-8}$ mol/(kg s Pa^{m+n})		$m_5 = 1$	$n_5 = 1$	151 (R1)
Water–gas shift Model: R2, R3	$CO + H_2O \rightarrow CO_2 + H_2$	(3.6.33)	$r_{10} = k_{010}e^{-E_{a,10}/RT}p_{CO}^{m_{10}} p_{H_2O}^{n_{10}}$	$k_{010} = 5.24\times 10^{-6}$ kmol/(kg s P^{m+n})	$E_{a,10} = 131.3$ kJ/mol	$m_{10} = 1$	$n_{10} = 1$	149,150 (R3)
			$r_6 = k_{0,6}e^{-E_{a,6}/RT}p_{CO}^{m_6} p_{H_2O}^{n_6}$	$k_{06} = 5.27\times 10^{-7}$ mol/(g s Pa^{m+n})	$E_{a,6} = 53.8$ kJ/mol	$m_6 = 0.5$	$n_6 = 0.5$	148 (R2)
Reverse water–gas shift model: R2, R3	$CO_2 + H_2 \rightarrow CO + H_2O$	(3.6.34)	$r_9 = k_{09}e^{-E_{a,9}/RT}p_{CO_2}^{m_9} p_{H_2}^{n_9}$	$k_{09} = 5.77\times 10^{-3}$ kmol/(kg s P^{m+n})	$E_{a,9} = 158$ kJ/mol	$m_9 = 1$	$n_9 = 1$	149,150 (R3)
			$r_7 = k_{0,7}e^{-E_{a,7}/RT}p_{CO_2}^{m_7} p_{H_2}^{n_7}$	$k_{0,7} = 3.9\times 10^{-4}$ mol/(g s Pa^{m+n})	$E_{a,7} = 99$ kJ/mol	$m_7 = 0.5$	$n_7 = 0.5$	148 (R2)

Table 3.8 Catalyst synthesis and composition.[148–151]

	Shahri and Alavi	Daneshpayeh et al.	Tiemersma et al.
Catalyst	2% Mn/5%Na_2WO_4/SiO_2	4% Mn/5% Na_2WO_4/SiO_2	2% Mn/5% Na_2WO_4/SiO_2
Synthesis method	Incipient wet impregnation	Two-step incipient wet impregnation	Two-step incipient wet impregnation
Drying	100 °C overnight	130 °C for 24 hr	n/a
Calcine	850 °C for 8 h	800 °C for 8 hr	850 °C or 900 °C
Size (μm)	250–400	150–250	300–500

Table 3.9 Reaction conditions for kinetic parameter estimation.[148-151]

	Shahri and Alavi	Daneshpayeh et al.	Tiemersma et al.
Reactor material	Quartz	Quartz	Quartz
ID (mm)	4 or 5	7	5
Sample size (g)	0.2	0.1	0.25–0.50
Temperature (°C)	800–900	750–875	800–900
Total pressure (bar)	1	0.876	2
CH_4/O_2	4–5	4–7.5	3–14
M_{cat}/V_{STP} (kg s/m^3)	85–345	30–160	60–120
Carbon balance (%)	>95	>95	>99
Data fitting	Non-linear fitting	Decimal genetic algorithm	Least squares minimization
Average relative error (%)	<15	>9	n/a

reverse water–gas shift, respectively. The kinetic expressions also vary widely, with Daneshpayeh et al. predominantly using power law, Shahri and Alavi mainly using Langmuir–Hinshelwood–Hougen–Watson, and Tiemersma et al. using pure power law.[148–151] To maintain consistency among all three references, m_j and n_j are reversed from the original reference for Tiemersma et al.[151] It is to be noted that the partial pressure of specie x is denoted as P_x for Section 3.6.2.

Even though the kinetic expressions derived from micro-reactor experiments are all obtained from catalysts of same or similar compositions, variations in catalyst synthesis, range of reaction parameters, product analysis, and kinetic parameter estimation method result in a wide variation of reactions involved and kinetic expressions obtained. Table 3.8 provides the catalyst synthesis conditions for the three kinetic models, while Table 3.9 provides the range of reaction parameters used to develop the kinetic models. The methods for gas analyses for all three groups use gas chromatography but differ in the columns and detectors used. Shahri and Alavi used a gas chromatograph (GC) with a thermal conductivity detector, Porapak-Q column, and a 5 Å molecular sieve column, while Daneshpayeh et al. used a Carle 400 AGC with a Porapak-Q column, methanizer, and FID, and Tiemersma et al. used a micro-GC with a thermal conductivity detector, PoraPLOT Q column, and two 5 Å molecular sieve columns.[148–151]

While each of the kinetic models developed fits its own research group's experimental results well, this does not guarantee that the kinetic models will provide similar reactor performance under a *ceteris paribus* assumption. This is evidenced by Shahri and Alavi, who conclude that largely neglecting undesired gas-phase reactions in their kinetic model "may lead to significant errors on the reaction engineering simulations of catalytic reactor performance"; further confirmed by Daneshpayeh et al., who conclude that "oxidative and thermal dehydrogenation of ethane and total oxidation of hydrocarbons have significant effects on OCM reaction."[148-150] Since the OCM reaction is catalytic, kinetic expressions must be relied upon for OCM reactor modeling in order to obtain the product composition, because the formation of carbon oxides is always thermodynamically favored. The results of kinetic reactor modeling are essential for two reasons. First, if OCM is to ever become a truly viable, competitive alternative to steam cracking for ethylene production, a model of the OCM reactor performance based on specifics such as catalyst loading, reactor aspect ratio, temperature, pressure, and methane:oxygen ratio is essential for sizing, costing, and process design. Second, scale-up from a micro-reactor to lab- and bench-scale reactor and eventually pilot and commercial scale requires an accurate, predictive model for reactor design and sensitivity.

### 3.6.3	Co-feed Fixed Bed Reactor Behavior

Using Aspen Plus®, the reactor effects, as defined by methane conversion and product distribution, of the three kinetic models described in the previous section are analyzed and compared by maintaining equivalent process simulation reactor conditions, and only varying the kinetic model.

In order to translate a proposed kinetic model into an Aspen Plus® reactor performance simulation, the components of the proposed kinetic model, given in Table 3.7, are used as the inputs into the Aspen Plus® plug-flow reactor module, RPlug. General is selected for the kinetic expressions as it allows for mixed kinetic expressions. Peng-Robinson is selected as the property method, since it accurately estimates properties of hydrocarbon mixtures. For determining the reactor performance of the kinetic model, the stream class is conventional since the solid-phase catalyst does not participate directly in the reaction. For a given kinetic model, the catalyst only participates indirectly in the process simulation where reaction kinetics parameters are used as the input into the plug-flow reactor. The chemical compounds included in the databank for the simulations are shown in Table 3.10.

Table 3.10 List of selected compounds for co-feed OCM in Aspen.

Component name	Component ID	Type
Methane	CH_4	Conventional
Ethane	C_2H_6	Conventional
Ethylene	C_2H_4	Conventional
Carbon monoxide	CO	Conventional
Carbon dioxide	CO_2	Conventional
Hydrogen	H_2	Conventional
Water	H_2O	Conventional
Oxygen	O_2	Conventional

Reactants and Products

The inlet natural gas composition is simplified to pure methane. Higher hydrocarbons and other impurities typically found in natural gas are excluded due to a lack of data, as OCM kinetic studies focus on methane reaction kinetics and not higher hydrocarbons.

Results

The effects of the three kinetic models are compared under the following static, identical conditions. The reactor temperature is 850 °C, reactor pressure is 1 atm, methane flow is 1000 kmol/hr, and oxygen flow is 250 kmol/hr. To allow for a comparison, three separate RPlug reactors are modeled with each RPlug reactor corresponding to a kinetic model. R1, R2, and R3 model the kinetic expressions used in Tiemersma et al., Shahri and Alavi, and Daneshpayeh et al., respectively.[148-151] Figure 3.13 shows the general set-up for the simulation.

A systematic approach to comparing reactor performance results is conducted. In Aspen, it is possible to selectively isolate individual reactions that occur within each reactor. For example, each kinetic model has several reactions. Using the RPlug module, it is possible to select which reactions occur. This feature can be used to observe the effect of specific reactions on the overall reactor performance. First, the three reactions common to all three kinetic models are compared, followed by the three reactions common to two models, and then finally all reactions are modeled to obtain the final results. Methane conversions for reaction (3.6.22) range from 18% to 100%. When reactions (3.6.23) and (3.6.25) occur, methane conversion decreases slightly, with a range 17.3% to 99.6%. Even though methane conversions have a wide range, the ethylene formation in model R1 matches closely with model R2 at a production rate of 60 kmol/hr, but model R3 only produces 1.8 kmol/hr, reactions (3.6.22) and (3.6.23) are responsible for the formation of ethylene through methane partial oxidation, whereas reactions (3.6.25) to (3.6.34) are side reactions that do not lead to ethylene formation. When reaction (3.6.25) is included in the reaction set, no trends across the three kinetic models exist. For model R1, methane conversion increases, ethane formation increases, and ethylene formation decreases. For model R2, methane conversion decreases, ethane formation decreases, and ethylene formation decreases. However, carbon dioxide formation is nearly identical for the model, by both models R1 and R2. The model proposed by Daneshpayeh et al. has little change with the addition of reactions (3.6.23) and (3.6.24), with nearly complete conversion of methane to ethane for all

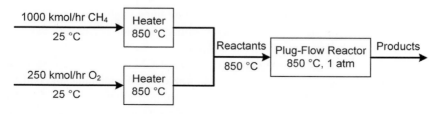

Figure 3.13 General set-up for simulation of a co-feed OCM reactor.

Table 3.11 Methane conversion as a function of reaction.

	Model R1	Model R2	Model R3
Methane conversion (%) with reaction (3.6.22)	17.9	90.5	100
Methane conversion (%) with reaction (3.6.22), (3.6.23)	17.3	84.5	99.6
Methane conversion (%) with reaction (3.6.22), (3.6.23), (3.6.25)	19.8	64.4	99.6
Methane conversion (%) with reaction (3.6.22), (3.6.23), (3.6.25), (3.6.30)		58.8	99.6

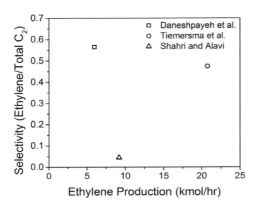

Figure 3.14 Ethylene selectivity versus ethylene production rate for three kinetic models.

cases. With the addition of reaction (3.6.30), the methane conversion in model R3 changes insignificantly while model R2 decreases in all aspects. When all reactions are applied to obtain the final result for the OCM reactor, the final methane conversions range from 17.5% to 50% and an ethylene selectivity range, defined as ethylene/total C_2, between 0.04 and 0.56. Table 3.11 and Figure 3.14 provide the details of the results.

The effect of the three kinetic models obtained from micro-reactor experiments using a sodium–tungstate–manganese catalyst supported on silica synthesized via incipient wet impregnation under identical reactor conditions yielded widely varying results, which could be due to a number of factors. First, the reactor conditions selected best encompassed an overlap in all three kinetic models. However, given the experimental range and conditions used for determination of all three kinetic models, there was no single, unique condition that encompassed the range of all three micro-reactor experiments. Thus, a broad variation in the results of the process simulation reactor performance using one single set of conditions for all three models is expected. The results do not devalue kinetic plug-flow reactor simulations nor negate the validity of kinetic models. Rather, the results from the micro-reactor kinetic models and the effect on reactor results highlight the complexity of the OCM reaction mechanism, the difficulty of elucidating a kinetic model that is applicable over a wide range of conditions, and the need for additional research in closely coupling kinetic reactor data with process simulations to further advance the field of OCM and reduce the error that exists in developing a more robust, predictive model.

3.6.4 Redox Reaction Kinetics

The redox approach to OCM differs from the co-feed approach since oxygen diffusion within the catalytic metal oxide is critical to its performance. This additional factor increases the complexity of the OCM redox kinetics since diffusion of the lattice oxygen from the bulk to the surface must be considered. Reshetnikov et al. derived a relationship between oxygen diffusion and the OCM reaction for a $K_{0.125}Na_{0.125}Sr_{0.75}CoO_{3-x}$ perovskite.[152] The catalytic metal oxide perovskite was assumed to be comprised of catalytic metal oxide centers with adsorbed oxygen, catalytically active metal oxide centers, and reduced catalytic metal oxide centers, denoted by ZO, Z, and Z_R, respectively.

Reaction (3.6.35) is the reaction between adsorbed oxygen on the catalytic metal oxide surface (ZO) with methane (A) to form product (P) and a catalytically active metal oxide center (Z). The reaction rate for reaction (3.6.35) is given in equation (3.6.36). Once the adsorbed oxygen is depleted, lattice oxygen dominates the oxygen supply for methane oxidation. The reaction of the active catalytic metal oxide center (Z) with methane is given in reaction (3.6.37) and its corresponding reaction kinetics is given in equation (3.6.38). Re-oxidation of the reduced catalytic metal oxide (Z_R) occurs from lattice oxygen within the catalytic metal oxide and is shown in reaction (3.6.39) with corresponding kinetics given in equation (3.6.40):

$$A + n\,ZO \rightarrow P + n\,Z \tag{3.6.35}$$

$$r_2 = k_2 C\theta_{ZO} \tag{3.6.36}$$

$$A + Z \rightarrow Z_R \tag{3.6.37}$$

$$r_3 = k_3 C\theta_Z \tag{3.6.38}$$

$$[O]_s + Z_R \rightarrow Z \tag{3.6.39}$$

$$r_5 = k_5 \theta_{Z_R} \sigma_s, \tag{3.6.40}$$

where r_i is the rate of reaction; k_i is the rate constant of the rate equation; C is the mole fraction of methane; θ_{ZO}, θ_Z, and θ_{Z_R} are the fractions of ZO, Z, and Z_R; and σ_s is the oxidizability of the catalytic metal oxide surface, where oxidizability of the catalytic metal oxide (σ) is defined as the fraction of oxygen in the lattice with respect to its maximum oxygen concentration.

Equation (3.6.41) and equation (3.6.42) express the rate of change of methane concentration and fraction of catalytic centers as a function of time, respectively, assuming the reactions occurred in a continuous stirred tank reactor (CSTR). Equation (3.6.43) provides the relationship for the conservation of the total number of catalytic metal oxide centers. Equation (3.6.44) provides the initial condition necessary for solving equations (3.6.41) and (3.6.42):

$$\frac{dC}{dt} = \frac{1}{\tau_g}\left(C^f - C\right) + a\sum_{k=1}^{N_r} v_{ki}r_k(C,\theta) \tag{3.6.41}$$

$$\frac{d\theta_j}{dt} = \sum_{k=1}^{N_r} v_{kj}\, r_k(C, \theta) \tag{3.6.42}$$

$$\theta_Z + \theta_{ZO} + \theta_{Z_R} = 1 \tag{3.6.43}$$

$$C = C^0,\ \theta_j = \theta_j^0 \quad \text{at} \quad t = 0, \tag{3.6.44}$$

where τ_g is the contact time; a is a parameter for the total amount of active centers; N_r is the number of reactions; v_{kj} are the stoichiometric coefficients; and j = ZO, Z, Z_R.

From experiments with $K_{0.125}Na_{0.125}Sr_{0.75}CoO_{3-x}$ perovskite as the catalytic metal oxide, the calculated coefficients were $k_2 = 1\ s^{-1}$, $k_3 = 1.1 \times 10^{-2}\ s^{-1}$, $k_5 = 1.2 \times 10^{-3}\ s^{-1}$.[152]

At the surface of the catalytic metal oxide, the extent of oxidizability is a function of oxygen diffusion within the crystal lattice to the surface and the rate of reduction of the surface by reaction with methane. The oxidizability of the catalytic metal oxide is then calculated using equation (3.6.45), where the left-hand side represents the rate of change in oxidizability of the catalytic metal oxide and the right-hand side represents the oxygen diffusion within the lattice:

$$\frac{\partial \sigma}{\partial \tau} = \frac{1}{\varphi^2} \frac{\partial^2 \sigma}{\partial \xi^2}. \tag{3.6.45}$$

The boundary conditions for equation (3.6.45) are

$$\text{for } \xi = 0 : \frac{d\sigma}{d\xi}\bigg|_{\xi=0} = 0$$

$$\text{for } \xi = 1 : \frac{d\sigma}{d\xi}\bigg|_{\xi=1} = -\varphi^2 \theta_{Z_R} \sigma_s,$$

where $\varphi^2 = L^2 k_5/D$ is the Thiele parameter; D is the effective coefficient of volumetric oxygen diffusion in the catalytic metal oxide; L is the characteristic dimension of the oxide crystallite; $\tau = k_5 t$ is the dimensionless time; and $\xi = l/L$ is a dimensionless coordinate in the crystallite.

In this model, the Thiele parameter was found to be the most important factor to evaluate the effect of oxygen diffusivity on the OCM reaction dynamics. A small Thiele parameter corresponded to a high rate of lattice oxygen diffusivity and hence a faster rate of surface oxygen regeneration. With small values of φ, the lattice oxygen concentration was found to be uniform within the crystallite, but with large values of φ, a lattice oxygen gradient was observed in the crystallite since the oxygen diffusivity was insufficient to re-oxidize the reduced catalytic metal oxide surface. Hence, with increasing values of φ, the methane conversion decreases with time due to a decrease in oxidizability of the surface. Reshetnikov et al. found that as long as φ is less than 7, the surface reaction will not be limited by oxygen diffusion. Also, based on the fact that L for mixed metal oxide crystallites falls in the range 2–30 nm, the value for D was found to be in the range 10^{-18} to 10^{-16} cm^2/s, which is around the typical value for the effective coefficient of oxygen diffusion in metal oxides.[152]

3.6.5   Co-feed Reactor Design

In recent years, reactor design has increasingly become the focus of OCM studies.[143] The C_2 yield is limited by catalyst selectivity and the reactions producing undesired products, CO_x. The ideal co-feed reactor would be designed such that there is a uniform concentration of oxygen throughout the reactor and ethylene is obtained through continuous removal. By controlling the oxygen concentration, a low partial pressure of oxygen can be maintained, which reduces the reactions resulting in CO_x.[153–155] The continuous removal of ethylene simplifies downstream separation and purification steps. Several reactor designs have attempted to incorporate these two concepts and are discussed.

Membrane reactors offer a possible way to provide a uniform oxygen concentration throughout the reactor.[156] The membrane wall thickness is one of the most important parameters affecing its performance. A thicker membrane wall improves oxygen distribution, which results in a higher ethylene yield. Jašo et al. compared the performance of OCM in three types of reactors: a fixed bed reactor (FBR), a conventional packed bed membrane reactor (CPBMR), and a novel proposed packed bed membrane reactor (PPBMR).[146,157] The configurations of these reactors are shown in Figure 3.15.

In the FBR, methane and oxygen are co-fed into the reactor bed. The structure of the FBR is the basis for the CPBMR and PPBMR. The difference lies in the distribution of the oxygen or CH_4 stream. Lu et al. found that C_2 selectivity could be significantly increased through optimal distribution of the oxygen feed.[156]

In the CPBMR, the catalytic material fills the tube side, and oxygen is supplied from the shell side through a porous, tubular membrane. Since it has been observed in the FBR that a higher methane:oxygen ratio is desirable to improve C_2 selectivity, injection of oxygen on the shell side of the reactor reduces the formation of CO_x by controlling the local oxygen concentration.

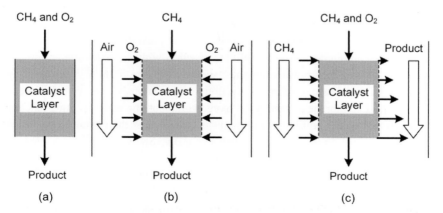

Figure 3.15 Different reactor configurations: (a) fixed bed reactor (FBR), (b) conventional packed bed membrane reactor (CPBMR), (c) proposed packed bed membrane reactor (PPBMR).[146,157]

In the PPBMR, the tube side is still filled with catalytic materials. In addition to the co-feeding of methane and oxygen from the tube side, additional methane is introduced from the shell side. Similar to the CPBMR, the rationale of increasing C_2 selectivity by increasing the local methane:oxygen ratio does not change. Instead of decreasing the O_2 concentration in the reaction zone, an increasing methane:oxygen ratio is achieved by increasing the CH_4 concentration in the CPBMR. In doing so, the products are continuously removed from the other side of the reactor to avoid further oxidation of the products.[146] Jašo et al. presented simulation results showing that it is feasible to obtain a relatively high methane conversion in the FBR, while a relatively high selectivity could be achieved in the CPBMR since the oxygen supply is more controlled. One key drawback of the FBR is the occurrence of severe hotspots since isothermal operation is impossible in the FBR, even in a dilute environment. Instead, membrane reactors provide a better approach to managing the exothermic heat of reaction through distributed feeding of the reactant as compared to fixed bed reactors. In addition to the enhanced product yield, it is also possible to operate membrane reactors under isothermal conditions, although hotspots with 100 K difference still occur. The CPBMR has the potential to produce desirable product at the highest yield, but methane back-permeation remains an issue. Furthermore, the CPBMR requires long contact times in order to obtain a desirable product yield. The PPBMR is more flexible than the FBR with respect to operating conditions and does not suffer from methane back-permeation. However, due to its oxygen feeding process, the highest achievable yield in the PPBMR is almost identical to the FBR.

In summary, membrane reactors have many advantages, but in order to be used at an industrial level, the slow reaction rate caused by low oxygen concentration in the OCM reaction side must be overcome.[146] The C_2 yield has yet to exceed 35%. In general, the metal oxide catalyst and the reactor are two intertwined factors. Catalytic activity has minimal effects on the performance of the PPBMR and the FBR but is important for the reaction performance of the CPBMR system. For example, at intermediate temperatures and membrane thicknesses, the La_2O_3/CaO metal oxide catalyst performs better in the CPBMR while the $Mn/Na_2WO_4/SiO_2$ metal oxide catalyst has a better performance in the PPBMR reactor.

Increasingly complex reactors have recently been developed. For example, Cao et al. studied combining the water splitting reaction with the oxidative coupling of methane using oxygen transport membranes, shown in Figure 3.16.[158] Since the oxygen generated from the water splitting reaction is continuously removed through the oxygen transport membrane driven by the oxygen gradient, the water splitting reaction is favored even though the equilibrium constant is quite low at the operating conditions. With an oxygen transport membrane consisting of a $Ba_{0.5}Sr_{0.5}Co_{0.8}Fe_{0.2}O_{3-x}$ (BSCF) perovskite material and a $Mn/Na_2WO_4/SiO_2$ catalyst, a C_2 yield of 6.5% and water conversion of 9% were obtained at 950 °C. A hydrogen generation rate of 3.3 $cm^3/(min\ cm^2)$ was achieved from the water splitting reaction.[158]

Kyriakou et al. paired steam electrolysis with methane coupling in a solid electrolyte cell.[136] In this study, they examined different materials at the anode of the electrochemical cell; Ag and perovskite ($SrZr_{0.95}Y_{0.05}O_{3-a}$) were tested, where C_2 yields were

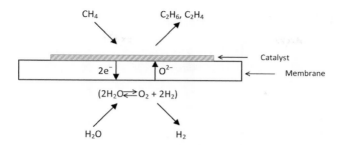

Figure 3.16 OCM reaction scheme of oxygen generated from a water splitting reaction in an oxygen transport membrane.[158]

7.6%. At the cathode (Pt), results demonstrated production of up to 65% of H_2. Recently, Kondratenko and Rodemerck proposed a dual reactor system of utilizing OCM with selective hydrogenation of the carbon oxides, which demonstrated C_2 yields of 34%.[159] Thus, dual-purpose reactor systems present a new avenue for OCM research as there is the potential to diversify the desired value-added products.

So far, the reactors discussed have mainly focused on reactant separation to produce an evenly distributed feed throughout the reactor. However, as discussed in Section 3.2.1, one of the major issues with OCM is the over oxidation of desired C_2 products to CO_x. Instead of reactant separation prior to reaction, immediate separation of reactants and desired products in the reactor would be beneficial by increasing C_2 yields and simplifying downstream processing. A countercurrent moving bed chromatographic reactor (CMBCR) has been proposed, where a chromatographic medium, which is a solid composed of material that is typical of the stationary phase of a chromatography column, selectively separates the C_2s from the unreacted methane in the product gas stream through physical adsorption. The chromatographic medium selected must have a higher affinity towards C_2 than methane. Ideally, the solid phase OCM catalyst and the chromatographic medium flow countercurrent to the gaseous reactants, methane and oxygen. Since the chromatographic medium loses its affinity towards hydrocarbons at the high reaction temperatures of OCM, an actual CMBCR would require the chromatographic medium to be separate from the OCM reactor and to operate at a lower temperature.[160] To incorporate the chromatographic medium into OCM, several studies have been conducted using a simulated countercurrent moving bed chromatographic reactor (SCMCR).[160–164]

The SCMCR operation simulates a countercurrent moving bed reactor by moving the inlet feed port of the reactant gas along the length of a fixed bed reactor.[160] Since the CMBCR would have to be divided into two sections, the OCM reactor and a column filled with chromatographic medium for separations, the SCMCR is also divided into those two sections. Throughout the literature, a four-reactor set-up connected in series is most common for SCMCR. As shown in Figure 3.17, each OCM reactor is followed by a column filled with the chromatographic medium to separate the C_2s from the unreacted methane. First, the methane and oxygen feed enter the OCM reactor. In the top of the separations column, the C_2s are strongly adsorbed to the

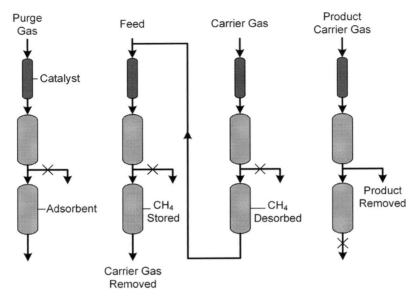

Figure 3.17 Experimental set-up for SCMCR.[164]

chromatographic medium while weakly adsorbed methane is at the bottom of the column. All gas ports simultaneously shift to the left, where carrier gas enters the reactor/column vacated by the feed thereby desorbing the methane and oxygen and carrying it into the new feed column. The C_2s are desorbed after two cycles through the use of product carrier gas, which is at a higher flow rate than the carrier gas used for desorption of methane and oxygen from the chromatographic medium. These stages are cycled between the four reactors to simulate a countercurrent moving bed chromatographic reactor.

Several studies using different co-feed catalysts have been investigated. Most notably, samarium oxide produced a yield as high as 50% with the SCMCR.[160,164] Of the chromatographic mediums tested, hydrophobic carbon molecular sieves were best suited for OCM conditions.[161] Several optimization studies of SCMCR have also been reported where parameters such as cycle time, feed switching time, methane to oxygen ratio, make-up feed rate, column length, and flow rates in different sections were varied.[163] A three-section model was also examined through numerical simulation and optimization studies.[162]

3.6.6   Redox Reactor Design

A typical continuous redox system would have two reactors, where one reactor performs the OCM reaction (reducer) and one reactor regenerates the catalyst (combustor) while circulating the solid catalytic metal oxide between the two reactors, as shown in Figure 3.4. A reactor system that allows for continuous circulation of solids

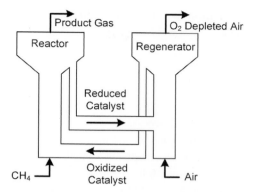

Figure 3.18 ARCO redox dual circulating fluidized bed set-up.[41]

will always provide superior performance compared to a periodically operated fixed bed redox design, where the oxidation state of the catalytic metal oxide is dynamic with time, resulting in a product distribution that varies with time. Fluidized bed reactors and moving bed reactors are two examples of reactors that can continuously circulate solids.

ARCO Dual Circulating Fluidized Beds

Circulating fluidized beds possess excellent heat and mass transfer properties, which is crucial to the success of OCM in the redox approach. The hydrodynamics of circulating fluidized beds are also well known, which allows the gas and solid flows in the circulating systems to be well controlled. ARCO has conducted OCM in the redox approach using a circulating fluidized bed for each reactor, as shown in Figure 3.18.[41] Using this set-up, ARCO tested its catalytic metal oxide for 30,000 redox cycles over a period of six months, where the catalytic metal oxide maintained its reactivity and selectivity as well as its fluidization properties.[165]

Moving Bed Reactor Design

So far, a moving bed reactor has been discussed for physical separation of hydrocarbon products and methane in a CMBCR for co-feed OCM operation. The use of moving bed reactors, however, can also be applied to chemical reactions. It has been suggested that a moving bed reactor design would be beneficial for redox OCM applications.[166,167]

Similar to the co-feed CMBCR system, a countercurrent moving bed reducer reactor in the chemical looping scheme can be considered. However, operating a countercurrent moving bed reducer without a chromatographic medium present for C_2 separations would lead to lower selectivity, since the desired products would be in contact with the highest oxidation state of the catalytic metal oxide, resulting in over-oxidation to CO_x. Thus, when OCM is to be operated in a chemical looping scheme, a moving bed in a cocurrent mode is more desirable. This design of a cocurrent moving bed reducer reactor, as discussed in Section 4.5, would have several advantages over ARCO's fluidized bed reducer. Also, as discussed in Section 1.6, a moving bed is preferred over a fluidized bed for use as a reducer from the viewpoint of the chemical and physical aspects of metal oxide performance.

3.7   Process Engineering

To achieve the threshold yield of 25–40% necessary for economic profitability, as mentioned in Section 3.2.1, OCM research has focused primarily on catalyst development and reactor design.[48,146,166] Process development, which considers the entire OCM process from inlet feed to purified product, has received less attention but cannot be overlooked as heat integration, separations, and process design all have a significant impact on process energetics and economics.

3.7.1   Process Design Perspective

The classical approach to process design divides a process into sections based on function, and each section is developed independently.[169–171] The OCM process can be divided into two sections: (1) product generation; and (2) product processing. Product generation occurs through either the co-feed or redox approach and product generation encompasses the reactants, OCM catalyst, and reactor design to generate the OCM products. The product processing section includes downstream separations and upgrading of OCM reactor products from the product generation section. The downstream processing of OCM reactor products is applicable to products formed from either the co-feed or redox approach, as product distribution is similar for both. From a process design standpoint, it is important to consider both product generation and product processing sections as the inlet feed and reactor design influences the product composition, which then determines the most appropriate arrangement for downstream processing.[172] With respect to general process equipment composing each section, the product generation section is independent of the product processing section. The independence of the two sections allows for independent analysis of each.

Within the product generation section, research for the co-feed and redox approach has focused mainly on OCM catalyst and reactor design. For a specific OCM catalyst formulation, methane conversion and product distribution are controlled by the reaction kinetics, which are dictated by the reactor operating conditions and reactor design. It is important to note that the results from reactor performance and optimization studies are typically applicable to only one specific catalyst formulation.[106,146,147,153,168,173–189] From a process design perspective, regardless of the OCM approach, one of the most important considerations is heat integration as the OCM reactions are highly exothermic. Heat integration is explored further in the next section.

Once the OCM product distribution is defined, the product processing section can be developed and evaluated. The product processing section, which includes purification and upgrading, has only recently received attention, where process arrangements for improving OCM economics by reducing energy requirements for separation, upgrading to liquid fuels, and considering co-generation of chemicals and electricity have been investigated.[38,111,190–199]

Technische Universität Berlin (TU Berlin) hosts one of the more active centers in OCM research and has built a mini-plant facility with the ability to operate OCM in a co-feed mode using either a fixed, fluidized, or membrane reactor, and to remove carbon dioxide using either an amine based solvent or a membrane.[197] In order to reduce process development time, they use a concurrent engineering approach towards process development where reaction, separations, and simulations are investigated in parallel.[172,200] Experimentally, the results indicate that using a membrane reactor provides the highest selectivity and the use of a membrane for carbon dioxide removal requires the lowest energy.

In the absence of obtaining a 25%–40% yield, additional processes such as formaldehyde synthesis, methanol synthesis, methane reforming, and electricity generation have been coupled with the OCM process in an effort to improve the overall economics.[191,194,201] The resulting analysis concluded that under certain economic conditions, the co-generation of ethylene and methanol could be profitable. Formaldehyde synthesis was not profitable due to high purification costs. Co-production of electricity was less profitable than OCM alone, due to the low cost of electricity.

3.7.2   Redox Heat Integration

From a process standpoint, the redox (chemical looping) approach to OCM possesses several inherent advantages as compared with the co-feed approach. These include the ability to operate outside the flammability limit of methane, the elimination of gas-phase oxidative reactions, and the ability to manage the heat generated from the OCM reactions.[22,41,124,174] The importance of heat management cannot be understated due to the large heat release from OCM reactions. Without proper heat removal from the OCM reactor, the adiabatic temperature rise negatively impacts the C_{2+} yields, since higher temperatures thermodynamically favor CO_x formation.[47] For example, a 200 °C adiabatic temperature rise occurs at 10% methane conversion and increases exponentially to 1060 °C at 30% methane conversion, based on product selectivity at that methane conversion.[47] Even when neglecting additional CO_x formation at higher methane conversions, a 600 °C temperature rise occurs at 30% methane conversion.[47] Co-feed reactors have been mainly confined to a fixed bed design where heat removal is an external process, which is difficult to control at large heating rates.[174] The redox approach to OCM using a circulating loop of catalytic metal oxide solids provides multiple heat integration and heat removal options, since the two reactors will be of a fluidized or moving bed type where internal heat transfer tubes can be placed inside the reactor for heat removal. One chemical looping heat integration example is provided below with additional possibilities summarized.

Aspen Plus® Process Set-up
Aspen Plus® v8.8 was used to model chemical looping OCM reactors. The stream class specified was CISOLID since the solid actively participates as a reactant in the OCM

Table 3.12 List of selected compounds for chemical looping OCM in Aspen

Component name	Component ID	Type
Methane	CH4	Conventional
Ethane	C2H6	Conventional
Propane	C3H8	Conventional
Propadiene	C3H4-1	Conventional
Propyne	C3H4-2	Conventional
Ethylene	C2H4	Conventional
Propylene	C3H6	Conventional
1-Butene	C4H8	Conventional
1,3-Butadiene	C4H6	Conventional
Cyclopentadiene	C5H6	Conventional
Benzene	C6H6	Conventional
Toluene	C7H8	Conventional
Hydrogen	H2	Conventional
Water	H2O	Conventional
Carbon monoxide	CO	Conventional
Carbon dioxide	CO2	Conventional
Oxygen	O2	Conventional
Nitrogen	N2	Conventional
Argon	AR	Conventional
Carbon-graphite	C	Solid
Manganese dioxide	MNO2	Solid
Dimanganese trioxide	MN2O3	Solid
Trimanganese tetraoxide	MN3O4	Solid
Manganese oxide	MNO	Solid
Manganese	MN	Solid

reactions. Peng–Robinson was selected as the property method since it accurately predicts hydrocarbon mixture interactions. Table 3.12 provides the list of selected components.

To evaluate the heat integration possibilities, natural gas, simplified to pure methane, was used as the inlet feed to the OCM reactor (reducer). The catalytic metal oxide used in the OCM experiments is a complex metal compound not found in any Aspen Plus® database. As such, the metal oxide was simplified to its redox pair of Mn_2O_3 and Mn_3O_4. Essentially, the catalytic metal oxide is split into two parts in the process simulation: the Mn_2O_3–Mn_3O_4 redox pair functions as the oxygen carrier, while the reducer is used to specify the products, thus acting in the role of catalyst. The product distribution from the reducer ranges from C_1 (unconverted CH_4, CO, CO_2) to C_7 and is given in Table 3.4. Air, whose specifications are given in Table 3.13, was then used to re-oxidize the reduced metal oxide in the combustor and complete the cyclic loop.

To simulate the OCM reaction, two reactors, two gas–solid separators, heat exchangers, and an air compressor are used. The reactions occurring in the reducer are a combination of coupling and cracking reactions. From the fixed bed experimental results given in Table 3.4, the exact reaction and extent of each reaction to form each product is unknown, so an RYIELD reactor is used to model the reducer, where only the

Table 3.13 Composition of air.

Component	Ambient air, vol %
Oxygen	20.74
Argon	0.92
Nitrogen	77.32
Water vapor	0.99
Carbon dioxide	0.03

product yields are necessary as an input. The combustor re-oxidizes the reduced catalytic metal oxide and is modeled using an RGIBBS reactor. After each reactor, a gas–solid separator is required in order to continually cycle the solids between the two reactors which are modeled using SSplit. Only the efficiency of the gas–solid separation is required as an input for SSplit, and the exact physical equipment is not specified, but can represent cyclones, baghouses, electrostatic precipitators, and in general any gas–solid separation equipment. Heat exchangers are used to minimize the use of utilities, both hot and cold, through heat exchange with existing process streams and utilization of the heats of reaction. The last piece of equipment is the air compressor, which is used to slightly pressurize the inlet air into the combustor.

Simulation Results – Base Case

1000 kmol/hr methane is fed into the reducer. The methane reacts with Mn_2O_3 to yield the product distribution given in Table 3.4. The resulting Mn_3O_4 is then regenerated with air as the oxygen source to produce Mn_2O_3. Initially, heaters, which represent the use of utilities, are used for heating and cooling streams to identify heat integration possibilities by quantifying the amount of heat either required or liberated from each process stream and reactor. Figure 3.19 shows the general process flow diagram, while Table 3.14 provides details of the corresponding streams in Figure 3.19.

The delivery pressure of natural gas is 30 atm. The gaseous stream from the reducer is always cooled to 40 °C regardless of final processing, either separations or upgrading. If the OCM products are separated using a cryogenic distillation fractionation train, then the OCM products are compressed to 35 atm. If the OCM products are upgraded to liquid fuels through oligomerization, then the OCM products are compressed to a minimum of 20 atm. In either case, the OCM products must first be cooled prior to further compression. This cooling, along with interstage cooling between compression stages, is an absolute necessity as ethylene begins to polymerize at around 100 °C. The depleted air is cooled to 40 °C to maintain a temperature differential, necessary for heat exchange. From Figure 3.19, both the reducer and combustor are exothermic. The amount of available heat is greater than the amount of required heat, which makes the OCM chemical looping process favorable from an energy standpoint. This heat can be used for electricity generation or heat integration purposes, depending on which provides the maximum economic benefit.

Table 3.14 Specifications for process streams in Figure 3.19.

Stream number	Component	Temperature (°C)
1	Cold natural gas	25
1′	Hot natural gas	840
2	Hot OCM products	840
2′	Cold OCM products	40
3	Cold air	29
3′	Hot air	840
4	Hot depleted air	840
4′	Cold depleted air	40

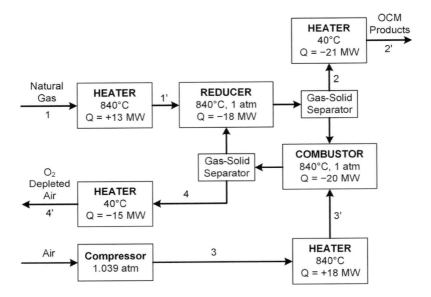

Figure 3.19 General process layout for chemical looping OCM.

From Figure 3.19, the simplest heat integration option would be to use stream 4 to heat stream 1, and stream 2 to heat stream 3. This option is shown in Figure 3.20.

From Figure 3.20, the incoming reactants are pre-heated using the product streams. The inlet methane to the reducer is pre-heated using the depleted air stream from the combustor. The methane is heated to 830 °C while cooling the depleted air to 145 °C. Air from the air fan is heated to 734 °C while cooling the OCM products stream to 75 °C. First, the OCM products can be compressed to 1.3 atm, which is the maximum pressure based on the constraint that the outlet temperature from the compressor must be less than 100 °C. Second, the OCM products stream can be cooled using utilities to 40 °C and then compressed to 2 atm. Since a heat exchanger using cooling water would be less costly than a compressor, and the achievable compression ratio increases from

Table 3.15 Heat exchange results for product–reactant heat exchange.

Heat exchanger	Hot stream	Cold stream
HEATX 1	4 → 4' (depleted air)	1 → 1' (natural gas)
	840 °C → 145 °C	25 °C → 830 °C
HEATX 2	2 → 2' (OCM products)	3 → 3' (air)
	850 °C → 42 °C	29 °C → 830 °C
HEATER	4b → 4c	Cold utility
	79 °C → 40 °C	

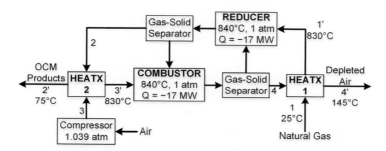

Figure 3.20 Reactant–product heat exchange.

1.3 to 2, it would be more advantageous to first cool the OCM products stream to 40 °C followed by compression, rather than direct compression, and this stream is further sent to product processing, which is explained later. Table 3.15 summarizes the heat exchange network when using reactants and products. To remove the excess heat from the reactors, steam can be generated to produce 9.7 MW$_e$ of gross electricity. The electricity can be used to operate auxiliary equipment with excess electricity sold. As mentioned earlier, internal heat removal can only occur in fluidized or moving bed reactors but not in fixed bed reactors.

While only one example is given, the heat integration options for the redox approach can occur through several means. The reactants can be partially heated either directly or indirectly using the exothermic heats of reaction. This would lower the initial pre-heat temperature while ensuring isothermal operations. Combinations of reactant–product and reactant–heats of reaction can also be performed. No matter the case, the chemical looping approach to OCM provides several possibilities for efficient heat removal and utilization.

3.7.3 Product Processing

Figure 3.21 shows the general process flow diagram for ethylene separation from OCM. The numerous stages, temperature differential between inlet gas and purified products, and overall complexity of the separations process leads to approximately half the capital cost being separations related. The product stream from the OCM reactor is cooled,

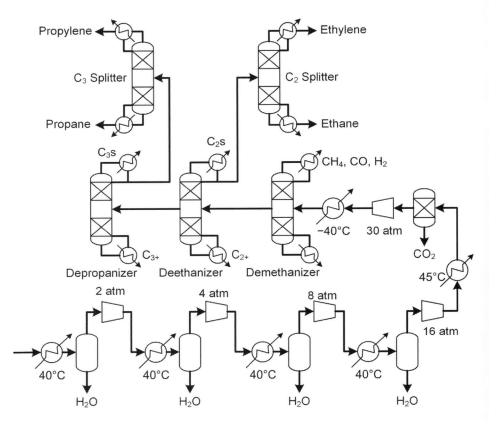

Figure 3.21 General set-up for ethylene purification.

compressed, removed of acid gases, dried, further compressed, and finally separated into individual components via cryogenic distillation.

Co-feed OCM is more complex than redox OCM since pure oxygen is impossible to obtain from an ASU. The lack of pure oxygen introduces gaseous impurities such as nitrogen, argon, and carbon dioxide that must also be separated in addition to the OCM products. The maximum concentration of ethylene in the product stream occurs when the oxygen source is pure, while decreasing the inlet oxygen purity further dilutes the ethylene concentration, which increases separation energy and cost requirements. The partial pressure of oxygen in the inlet feed also decreases with decreasing oxygen purity, which will affect reactions that are dependent upon the oxygen partial pressure.

The gas compression stage follows the guidelines set forth for CO_2 compression and sequestration.[202,203] Interstage cooling is essential and a refrigeration cycle instead of cooling water will be necessary.[204] The removal of acid gases occurs with a non-regenerative caustic soda wash, regenerative amine solvents, or regenerative Benfield process, with the selection dependent upon concentration and species of acid gases present.[204,205] Moisture removal can occur with either an ethylene glycol system or molecular sieves.[204] If an ethylene glycol system is employed, triethylene glycol is the solvent of choice.[206] The triethylene glycol absorbs the water from the gas stream, the gas stream leaves the absorber dry, and the rich triethylene glycol stream is regenerated in a stripper column.[206] The preferred method of gas dehydration uses molecular sieves located after the final compressor.[204] The remaining hydrocarbons are finally separated and purified using a series of cryogenic distillation columns. Beginning with the coldest temperature, the demethanizer separates methane from the C_{2+} components. The deethanizer separates the C_2s, ethane, and ethylene, from the C_{3+} components. The C_2s exiting the top of the deethanizer are separated into ethane and ethylene using a C_2 splitter while the deethanizer bottoms form the inlet for the depropanizer. The depropanizer and subsequent cryogenic distillation columns are an extension of the deethanizer–C_2 splitter combination for purification and are combined in series until all components are separated.

The complexity of the traditional ethylene separations process combined with the unique composition exiting the OCM reactor provides an opportunity to develop alternative separation schemes. The use of silver complexes to selectively bind with ethylene is one such promising idea. Complexation is used on the industrial scale, but so far no large-scale process for ethylene separation is in operation. Membranes have been researched for separation of a species from a gas mixture and can be applied for ethylene separation, either from typical pyrolysis gas or from OCM. Issues with cost, durability, and separation efficiency have limited the industrial practice of membranes for separations. One final idea is to directly consume or convert the hydrocarbons in the product stream as they are a rich blend of various hydrocarbons. For example, Mobil's olefin to gasoline and distillate (MOGD) process converts olefins into distillates and gasoline using a ZSM-5 catalyst. Coupling the OCM products with the MOGD process will convert C_{3+} olefins into distillates, ethylene into gasoline, and non-olefins remain inert. By directly converting or consuming the OCM reaction products, the possibility exists to directly transform the OCM reactor hydrocarbons into a useful commodity while completely bypassing the initial separations process.

From a process design standpoint, both co-feed and chemical looping OCM provide numerous advantages over traditional steam cracking for ethylene production. First, methane is directly upgraded to a high value product, ethylene, without intermediates. Second, the OCM reactions are highly exothermic whereas steam cracking is endothermic, reducing energy requirements. Chemical looping OCM is advantageous over co-feed OCM in this aspect since heat integration options are more flexible and easily applied. Finally, separations equipment is reduced using OCM since the separations require only a demethanizer, deethanizer, and C_2 splitter. Additional pathways for direct product upgrading, such as to gasoline through the MOGD process, are also possible without the need for separations.

3.8   Concluding Remarks

With the potential for diversifying fuel resources and utilizing ample natural gas reserves, there is the necessity for further research and development on efficient and cost-effective direct selective oxidation processes.[9] OCM provides an opportunity to take advantage of economical natural gas for value-added products such as ethylene to commercial consumer chemicals such as fine chemicals, synthetic fuels, and plastics.[2] However, continued efforts must be pursued in selective direct oxidation of methane in order to improve selectivity for desirable products and to decrease over-oxidation to CO_2. Since the 1980s, much work has focused on understanding the fundamental heterogeneous and homogeneous mechanisms behind the complex OCM reaction network. The OCM reaction network is a series of parallel and consecutive reactions occurring at both the catalytic material surface and the gas phase. It is generally accepted that ethane is formed from the recombination of two methyl radicals. Subsequently, ethylene is formed from the homogeneous or heterogeneous dehydrogenation of ethane. However, there is a trade-off between selective oxidation to desired products of ethane and ethylene and total oxidation to carbon oxides.

In order to develop the optimal OCM materials, extensive research is focused on developing metal oxide catalysts in the presence of gaseous oxygen, catalytic metal oxides for chemical looping, and alternative catalysts that utilize halogens or non-oxides. Of the metal oxide catalysts in the presence of gaseous oxygen, $Mn/Na_2WO_4/SiO_2$ has been widely tested as the most effective OCM metal oxide catalyst. Manganese based catalytic metal oxides have demonstrated the most promise as OCM chemical looping materials. Alternative catalysts such as non-oxygen transfer based compounds, halides and sulfides, have also been explored. The reaction mechanism and network are not fully understood, and OCM research is continuing. However, based on current fundamental knowledge, it is expected that catalysts alone cannot achieve C_2 yields beyond 30%. Thus, innovative reactor and process designs are necessary in order potentially to progress OCM.

Improving the C_2 yield from an OCM catalyst relies not only on catalyst development but also on reaction engineering. The exact reaction mechanism and network for an OCM catalyst or catalytic metal oxide remains unknown, but they can be explained using current catalysis mechanisms such as Langmuir–Hinshelwood and Eley–Rideal. In general, OCM catalysts behave similarly and it is important to understand them at the fundamental level in order to maximize the C_2 yield and minimize undesired products such as carbon oxides. Catalyst activity is a function of several parameters with temperature, pressure, and methane:oxygen ratio being the important ones. The general trends include: increasing the reaction operating temperature increases C_2 yield within a specific temperature range in which the catalyst is functional; increasing oxygen pressure increases methane conversion but decreases C_2 selectivity; increasing the methane pressure increases both the OCM reaction kinetics and C_2 selectivity; and decreasing the methane:oxygen ratio shifts the ethylene:ethane ratio towards ethylene. Since catalytic metal oxides also donate oxygen, the reaction mechanism and network for a catalytic metal oxide is more complex than a metal oxide catalyst, since oxygen diffusion from the bulk phase to the surface of the metal oxide becomes an additional factor.

By understanding the reaction parameters that influence the C_2 yields, reactors can be designed that exploit the OCM catalyst formulation and reaction kinetics to maximize C_2 yield. Membrane reactors have the potential to increase C_2 yield from the OCM reaction by either providing a distributed feed of O_2 or CH_4 with continuous removal of product. The advantages of distributing the O_2 or CH_4 feed through the reactor include increasing the local methane:oxygen ratio, reducing the formation of carbon oxides, and improving the heat management from the highly exothermic OCM reactions. Continuously removing product immediately as it is formed is advantageous as no consecutive gas-phase side reactions can consume the product and minimal additional purification is necessary. However, operational sustainability for reaction/separation reactors needs to be further ascertained.

Similar to the OCM reaction mechanism, OCM reaction kinetics cannot be perfectly modeled. From micro-kinetic experimental results using a sodium tungstate, manganese catalyst supported on silica, widely varying kinetic models have been developed. The reactor performance of three kinetic models ranging from 5–10 reactions and consisting of power law and Langmuir–Hinshelwood–Hougen–Watson expressions showed widely varying results when modeled in a plug-flow reactor in Aspen Plus® under otherwise static conditions. Kinetic model and reactor performance is a critical area that requires further refinement in order to develop a predictive model that allows for OCM catalyst co-feed scale-up. In terms of process development, the OCM chemical looping approach presents a promising reactor and process design that allows for heat management, which is crucial, as hotspot formation and adiabatic temperature rise in a co-feed packed bed reactor cannot be easily resolved. With the abundance of shale gas reserves worldwide, it is anticipated that OCM will enter another peak and progress into an economical, direct commercial-scale method of methane utilization.

References

1. Sinev, M. Y., Z. T. Fattakhova, V. I. Lomonosov, and Y. A. Gordienko, "Kinetics of Oxidative Coupling of Methane: Bridging the Gap between Comprehension and Description," *Journal of Natural Gas Chemistry*, 18(3), 273–287 (2009).
2. Zaman, J., "Oxidative Processes in Natural Gas Conversion," *Fuel Processing Technology*, 58, 61–81 (1999).
3. Ross, J. R. H., A. N. J. van Keulen, M. E. S. Hegarty, and K. Seshan, "The Catalytic Conversion of Natural Gas to Useful Products," *Catalysis Today*, 30(1–3), 193–199 (1996).
4. Gesser, H. D., N. R. Hunter, and C. B. Prakash, "The Direct Conversion of Methane to Methanol by Controlled Oxidation," *Chemical Reviews*, 85(4), 235–244 (1985).
5. Lunsford, J. H., "Catalytic Conversion of Methane to More Useful Chemicals and Fuels: a Challenge for the 21st Century," *Catalysis Today*, 63(2–4), 165–174 (2000).
6. Alvarez-Galvan, M. C., N. Mota, M. Ojeda, S. Rojas, R. M. Navarro, and J. L. G. Fierro, "Direct Methane Conversion Routes to Chemicals and Fuels," *Catalysis Today*, 171(1), 15–23 (2011).
7. Olivos-Suarez, A. I., À Szécsényi, E. J. M. Hensen, J. Ruiz-Martinez, E. A. Pidko, and J. Gascon, "Strategies for the Direct Catalytic Valorization of Methane Using Heterogeneous Catalysis: Challenges and Opportunities," *ACS Catalysis*, 6(5), 2965–2981 (2016).
8. Holmen, A., "Direct Conversion of Methane to Fuels and Chemicals," *Catalysis Today*, 142(1–2), 2–8 (2009).

9. Armor, J. N., "Emerging Importance of Shale Gas to Both the Energy & Chemicals Landscape," *Journal of Energy Chemistry*, 22(1), 21–26 (2013).

10. Armor, J. N., "Key Questions, Approaches, and Challenges to Energy Today," *Catalysis Today*, 236(Part B), 171–181 (2014).

11. Zavyalova, U., M. Holena, R. Schlögl, and M. Baerns, "Statistical Analysis of Past Catalytic Data on Oxidative Methane Coupling for New Insights into the Composition of High-Performance Catalysts," *ChemCatChem*, 3(12), 1935–1947 (2011).

12. Stangland, E. E., "The Shale Gas Revolution: A Methane-to-Organic Chemicals Renaissance," 2014 US Frontiers of Engineering Symposium, Irvine, CA, September 11–14 (2014).

13. Jenkins, S., "Shale Gas Ushers in Ethylene Feed Shifts," *Chemical Engineering*, October, 17–19 (2012).

14. Krylov, O. V., "Methods for Increasing the Efficiency of Catalysts for the Oxidative Condensation of Methane," *Russian Chemical Reviews*, 61(8), 851–858 (1992).

15. Hutchings, G. J., M. S. Scurrell, and J. R. Woodhouse, "Oxidative Coupling of Methane using Oxide Catalysts," *Chemical Society Reviews*, 18, 251–283 (1989).

16. Black, T. *Honeywell breakthrough seen transforming plastics industry*, http://www.bloomberg.com/news/articles/2012-04-25/honeywell-breakthrough-seen-transforming-plastics-industry-tech (Accessed on 25 April 2012).

17. Siluria, *Oxidative coupling of methane*, http://siluria.com/Technology/Oxidative_Coupling_of_Methane (Accessed on 2 September 2016).

18. Ren, T., M. Patel, and K. Blok, "Olefins from Conventional and Heavy Feedstocks: Energy Use in Steam Cracking and Alternative Processes," *Energy*, 31(4), 425–451 (2006).

19. Horn, R. and R. Schlögl, "Methane Activation by Heterogeneous Catalysis," *Catalysis Letters*, 145(1), 23–39 (2015).

20. Batiot, C. and B. K. Hodnett, "The Role of Reactant and Product Bond Energies in Determining Limitations to Selective Catalytic Oxidations," *Applied Catalysis A: General*, 137(1), 179–191 (1996).

21. Mitchell, H. L. and R. H. Waghorne, "Catalysts for the Conversion of Relatively Low Molecular Weight Hydrocarbons to Higher Molecular Weight Hydrocarbons and the Regeneration of the Catalysts," U.S. Patent 4,239,658A (1980).

22. Keller, G. E. and M. M. Bhasin, "Synthesis of Ethylene via Oxidative Coupling of Methane," *Journal of Catalysis*, 73(1), 9–19 (1982).

23. Hinsen, W. and M. Baerns, "Oxidative Coupling of Methane to C2-Hydrocarbons in the Presence of Different Catalysts," *Chemiker-Zeitung*, 107(7–8), 223–226 (1983).

24. Ito, T. and J. H. Lunsford, "Synthesis of Ethylene and Ethane by Partial Oxidation of Methane over Lithium-doped Magnesium Oxide," *Nature*, 314(6013), 721–722 (1985).

25. Jones, C. A., J. J. Leonard, and J. A. Sofranko, "Methane Conversion," U.S. Patent 4,443,644 (1984).

26. Jones, C. A., J. J. Leonard, and J. A. Sofranko, "Methane Conversion," U.S. Patent 4,443,645 (1984).

27. Jones, C. A., J. J. Leonard, and J. A. Sofranko, "Methane Conversion," U.S. Patent 4,443,646 (1984).

28. Jones, C. A., J. J. Leonard, and J. A. Sofranko, "Methane Conversion," U.S. Patent 4,443,647 (1984).

29. Jones, C. A., J. J. Leonard, and J. A. Sofranko, "Methane Conversion," U.S. Patent 4,443,648 (1984).

30. Jones, C. A., J. J. Leonard, and J. A. Sofranko, Methane Conversion, U.S. Patent 4,443,649 (1984).

31. Liss, W. E., "Impacts of Shale Gas Advancements on Natural Gas Utilization in the United States," *Energy Technology*, 2(12), 953–967 (2014).

32. Mitchell, S. F. and D. F. Shantz, "Future Feedstocks for the Chemical Industry—Where Will the Carbon Come From?," *AIChE Journal*, 61(8), 2374–2384 (2015).

33. Amenomiya, Y., V. I. Birss, M. Goledzinowski, J. Galuszka, and A. R. Sanger, "Conversion of Methane by Oxidative Coupling," *Catalysis Reviews*, 32(3), 163–227 (1990).

34. Su, S. Y., J. Y. Ying, and W. H. Green Jr., "Upper Bound on the Yield for Oxidative Coupling of Methane," *Journal of Catalysis*, 218(2), 321–333 (2003).

35. Guo, Z., B. Liu, Q. Zhang, W. Deng, Y. Wang, and Y. Yang, "Recent Advances in Heterogeneous Selective Oxidation Catalysis for Sustainable Chemistry," *Chemical Society Reviews*, 43(10), 3480–3524 (2014).

36. Ras, E. and G. Rothenberg, "Heterogeneous Catalyst Discovery using 21st Century Tools: A Tutorial," *RSC Advances*, 4(12), 5963–5974 (2014).

37. Tullo, A. H., "Ethylene from Methane," *Chemical & Engineering News*, 89(3), 20–21 (2011).

38. Choudhary, V. R. and B. S. Uphade, "Oxidative Conversion of Methane/Natural Gas into Higher Hydrocarbons," *Catalysis Surveys from Asia*, 8(1), 15–25 (2004).

39. Labinger, J. A., "Oxidative Coupling of Methane: An Inherent Limit to Selectivity?" *Catalysis Letters*, 1(11), 371–376 (1988).

40. Shi, C., M. P. Rosynek, and J. H. Lunsford, "Origin of Carbon Oxides during the Oxidative Coupling of Methane," *Journal of Physical Chemistry*, 98(34), 8371–8376 (1994).

41. Jones, C. A., J. J. Leonard, and J. A. Sofranko, "Fuels for the Future: Remote Gas Conversion," *Energy & Fuels*, 1(1), 12–16 (1987).

42. Fierro, J. L. G., "Catalysis in C_1 Chemistry: Future and Prospect," *Catalysis Letters*, 22(1), 67–91 (1993).

43. Lee, J. S. and S. T. Oyama, "Oxidative Coupling of Methane to Higher Hydrocarbons," *Catalysis Reviews*, 30(2), 249–280 (1988).

44. Hudgins, R. R., P. L. Silveston, C. Li, and A. A. Adesina, "Partial Oxidation and Dehydrogenation of Hydrocarbons," 4.2 in Silveston, P. L. and R. R. Hudgins, *Periodic Operation of Reactors*, 1st edn, Butterworth-Heinemann, Waltham, MA, pp. 79–122 (2013).

45. Sofranko, J. A., J. J. Leonard, and C. A. Jones, "The Oxidative Conversion of Methane to Higher Hydrocarbons," *Journal of Catalysis*, 103(2), 302–310 (1987).

46. Jones, C. A., J. J. Leonard, and J. A. Sofranko, "Methane Conversion," U.S. Patent 4,560,821 (1985).

47. Wolf, E. E., ed., *Methane Conversion by Oxidative Processes: Fundamental and Engineering Aspects*, Van Nostrand Reinhold, New York, NY (1992).

48. Kondratenko, E. V. and M. Baerns, "Oxidative Coupling of Methane," 13.17 in Ertl, G., H. Knozinger, F. Schuth, and J. Weitkamp, *Handbook of Heterogeneous Catalysis*, Wiley-VCH, Weinheim, Germany (2008).

49. Sinev, M. Y., L. Y. Margolis, V. Y. Bychkov, and V. N. Korchak, "Free Radicals as Intermediates in Oxidative Transformations of Lower Alkanes," *Studies in Surface Science and Catalysis*, 110, 327–335 (1997).

50. Driscoll, D. J. and J. H. Lunsford, "Gas-Phase Radical Formation during the Reactions of Methane, Ethane, Ethylene, and Propylene over Selected Oxide Catalysts," *Journal of Physical Chemistry*, 89(21), 4415–4418 (1985).

51. Tjatjopoulos, G. J. and I. A. Vasalos, "Reaction-Path Analysis of a Homogeneous Methane Oxidative Coupling Mechanism," *Applied Catalysis A: General*, 88(2), 213–230 (1992).
52. Campbell, K. D., E. Morales, and J. H. Lunsford, "Gas-Phase Coupling of Methyl Radicals during the Catalytic Partial Oxidation of Methane," *Journal of the American Chemical Society*, 109(25), 7900–7901 (1987).
53. Lee, M. R., M.-J. Park, W. Jeon, J.-W. Choi, Y.-W. Suh, and D. J. Suh, "A Kinetic Model for the Oxidative Coupling of Methane over $Na_2WO_4/Mn/SiO_2$," *Fuel Processing Technology*, 96, 175–182 (2012).
54. Dissanayake, D., J. H. Lunsford, and M. P. Rosynek, "Site Differentiation in Homolytic vs. Heterolytic Activation of Methane over Ba/MgO Catalysts," *Journal of Catalysis*, 146(2), 613–615 (1994).
55. Krylov, O. V., "Free-Radical Reactions Involved in C_1-C_3 Hydrocarbons Interactions with Oxide Catalysts," *Catalysis Today*, 13(4), 481–486 (1992).
56. Lunsford, J. H., "The Catalytic Oxidative Coupling of Methane," *Angewandte Chemie International Edition*, 34(9), 970–980 (1995).
57. Sokolovskii, V. D., S. M. Aliev, O. V. Buyevskaya, and A. A. Davydov, "Type of Hydrocarbon Activation and Nature of Active Sites of Base Catalysts in Methane Oxidative Dehydrodimerization," *Catalysis Today*, 4(3–4), 293–300 (1989).
58. Sokolovskii, V. D., S. M. Aliev, O. V. Buyevskaya, and A. A. Davydov, "Heterolytic Mechanism of Methane Activation in Oxidative Dehydrodimerization," *Studies in Surface Science and Catalysis*, 55, 437–446 (1990).
59. Choudhary, V. R. and V. H. Rane, "Acidity/Basicity of Rare-Earth Oxides and Their Catalytic Activity in Oxidative Coupling of Methane to C_2-Hydrocarbons," *Journal of Catalysis*, 130(2), 411–422 (1991).
60. Maitra, A. M., I. Campbell, and R. J. Tyler, "Influence of Basicity on the Catalytic Activity for Oxidative Coupling of Methane," *Applied Catalysis A: General*, 85(1), 27–46 (1992).
61. Lapszewicz, J. A. and X. Jiang, "Investigation of Reactivity and Selectivity of Methane Coupling Catalysts using Isotope Exchange Techniques," *Catalysis Letters*, 13(1–2), 103–115 (1992).
62. Driscoll, D. J., W. Martir, J.-X. Wang, and J. H. Lunsford, "Formation of Gas-phase Methyl Radicals over Magnesium Oxide," *Journal of the American Chemical Society*, 107(1), 58–63 (1985).
63. Lunsford, J. H., P. G. Hinson, M. P. Rosynek, C. Shi, M. Xu, and X. Yang, "The Effect of Chloride Ions on a Li-MgO Catalyst for the Oxidative Coupling of Methane," *Journal of Catalysis*, 147(1), 301–310 (1994).
64. Lunsford, J. H., "Formation and Reactions of Methyl Radicals over Metal Oxide Catalysts," Chapter 1 in Wolf, E. E., ed., *Methane Conversion by Oxidative Processes: Fundamental and Engineering Aspects*, Van Nostrand Reinhold, New York, NY, pp. 3–29 (1992).
65. Campbell, K. D. and J. H. Lunsford, "Contribution of Gas-Phase Radical Coupling in the Catalytic Oxidation of Methane," *Journal of Physical Chemistry*, 92(20), 5792–5796 (1988).
66. Lunsford, J. H., "The Catalytic Conversion of Methane to Higher Hydrocarbons," *Catalysis Today*, 6(3), 235–259 (1990).
67. Buyevskaya, O. V., M. Rothaemel, H. W. Zanthoff, and M. Baerns, "Transient Studies on Reaction Steps in the Oxidative Coupling of Methane over Catalytic Surfaces of MgO and Sm_2O_3," *Journal of Catalysis*, 146(2), 346–357 (1994).

68. Mallens, E., J. Hoebink, and G. Marin, "An Investigation of the Oxygen Pathways in the Oxidative Coupling of Methane over MgO-based Catalysts," *Journal of Catalysis*, 160(2), 222–234 (1996).

69. Beck, B., V. Fleischer, S. Arndt, M. G. Hevia, A. Urakawa, P. Hugo, and R. Schomäcker, "Oxidative Coupling of Methane – A Complex Surface/Gas Phase Mechanism with Strong Impact on the Reaction Engineering," *Catalysis Today*, 228, 212–218 (2014).

70. Nelson, P. F., C. A. Lukey, and N. W. Cant, "Isotopic Evidence for Direct Methyl Coupling and Ethane to Ethylene Conversion during Partial Oxidation of Methane over Lithium/Magnesium Oxide," *The Journal of Physical Chemistry*, 92(22), 6176–6179 (1988).

71. Luo, L., X. Tang, W. Wang, Y. Wang, S. Sun, F. Qi, and W. Huang, "Methyl Radicals in Oxidative Coupling of Methane Directly Confirmed by Synchrotron VUV Photoionization Mass Spectroscopy," *Scientific Reports*, 3(1625), 1–6 (2013).

72. Dubois, J.-L. and C. J. Cameron, "Common Features of Oxidative Coupling of Methane Cofeed Catalysts," *Applied Catalysis*, 67(1), 49–71 (1990).

73. Geerts, Johannes Wilhemus Maria Henricus, "Ethylene Synthesis by Direct Partial Oxidation of Methane," PhD Dissertation, Technische Universiteit Eindhoven, Eindhoven, Netherlands (1990).

74. Sinev, M. Y., V. N. Korshak, and O. V. Krylov, "The Mechanism of the Partial Oxidation of Methane," *Russian Chemical Reviews*, 58(1), 22–34 (1989).

75. Otsuka, K., K. Jinno, and A. Morikawa, "Active and Selective Catalysts for the Synthesis of C_2H_4 and C_2H_6 via Oxidative Coupling of Methane," *Journal of Catalysis*, 100(2), 353–359 (1986).

76. Sinev, M. Y., V. N. Korchak, and O. V. Krylov, "Kinetic Peculiarities of Oxidative Condensation of Methane on Oxide Catalysts in a Heterogeneous-Homogeneous Process," *Kinetic Catalysis*, 28(6), (1988).

77. Ito, T., J.-X. Wang, C. Lin, and J. H. Lunsford, "Oxidative Dimerization of Methane over a Lithium-promoted Magnesium Oxide Catalyst," *Journal of the American Chemical Society*, 107(18), 5062–5068 (1985).

78. Ekstrom, A., J. A. Lapszewicz, and I. Campbell, "Origin of the Low Limits in the Higher Hydrocarbon Yields in the Oxidative Coupling Reaction of Methane," *Applied Catalysis*, 56(1), L29-L34 (1989).

79. Nelson, P. F. and N. W. Cant, "Oxidation of C_2 Hydrocarbon Products during the Oxidative Coupling of Methane over a Lithium/Magnesia Catalyst," *Journal of Physical Chemistry*, 94(9), 3756–3761 (1990).

80. Takanabe, K. and E. Iglesia, "Mechanistic Aspects and Reaction Pathways for Oxidative Coupling of Methane on $Mn/Na_2WO_4/SiO_2$ Catalysts," *The Journal of Physical Chemistry C*, 113(23), 10131–10145 (2009).

81. Hewett, K. B., L. C. Anderson, M. P. Rosynek, and J. H. Lunsford, "Formation of Hydroxyl Radicals from the Reaction of Water and Oxygen over Basic Metal Oxides," *Journal of the American Chemical Society*, 118(29), 6992–6997 (1996).

82. Iwamoto, M. and J. H. Lunsford, "Surface Reactions of Oxygen Ions. 5. Oxidation of Alkanes and Alkenes by O_2^- on Magnesium Oxide," *Journal of Physical Chemistry*, 84(23), 3079–3084 (1980).

83. Carriero, J. A. S. P. and M. Baerns, "Catalytic Conversion of Methane by Oxidative Coupling to C_{2+} Hydrocarbons," *Reaction Kinetics and Catalysis Letters*, 35(1–2), 349–360 (1987).

84. Wu, J., S. Li, J. Niu, and X. Fang, "Mechanistic Study of Oxidative Coupling of Methane over Mn_2O_3-Na_2WO_4/SiO_2 Catalyst," *Applied Catalysis A: General*, 124(1), 9–18 (1995).

85. Maitra, A. M., "Critical Performance Evaluation of Catalysts and Mechanistic Implications for Oxidative Coupling of Methane," *Applied Catalysis A: General*, 104(1), 11–59 (1993).

86. Lee, J. S. and S. T. Oyama, "Oxidative Coupling of Methane to Higher Hydrocarbons," *Catalysis Reviews Science and Engineering*, 30(2), 249–280 (1988).

87. Liu, S., X. Tan, K. Li, and R. Hughes, "Methane Coupling Using Catalytic Membrane Reactors," *Catalysis Reviews*, 43(1–2), 147–198 (2001).

88. Gayko, G., D. Wolf, E. V. Kondratenko, and M. Baerns, "Interaction of Oxygen with Pure and SrO-Doped Nd_2O_3 Catalysts for the Oxidative Coupling of Methane: Study of Work Function Changes," *Journal of Catalysis*, 178(2), 441–449 (1998).

89. Kondratenko, E. V., D. Wolf, and M. Baerns, "Influence of Electronic Properties of Na_2O/CaO Catalysts on Their Catalytic Characteristics for the Oxidative Coupling of Methane," *Catalysis Letters*, 58(4), 217–223 (1999).

90. Borchert, H. and M. Baerns, "The Effect of Oxygen-Anion Conductivity of Metal–Oxide Doped Lanthanum Oxide Catalysts on Hydrocarbon Selectivity in the Oxidative Coupling of Methane," *Journal of Catalysis*, 168(2), 315–320 (1997).

91. Voskresenskaya, E. N., V. G. Roguleva, and A. G. Anshits, "Oxidant Activation over Structural Defects of Oxide Catalysts in Oxidative Methane Coupling," *Catalysis Reviews*, 37(1), 101–143 (1995).

92. Malekzadeh, A., A. Khodadadi, M. Abedini, M. Amini, A. Bahramian, and A. K. Dalai, "Correlation of Electrical Properties and Performance of OCM MOx/Na_2WO_4/SiO_2 Catalysts," *Catalysis Communications*, 2(8), 241–247 (2001).

93. Kimble, J. B. and J. H. Kolts, "Methane Conversion," U.S. Patent 4,654,460 (1987).

94. Simon, U., S. Arndt, T. Otremba, T. Schlingmann, O. Görke, K.-P. Dinse, R. Schomäcker, and H. Schubert, "Li/MgO with Spin Sensors as Catalyst for the Oxidative Coupling of Methane," *Catalysis Communications*, 18, 132–136 (2012).

95. Arndt, S., U. Simon, S. Heitz, A. Berthold, B. Beck, O. Görke, J.-D. Epping, T. Otremba, Y. Aksu, and E. Irran, "Li-doped MgO from Different Preparative Routes for the Oxidative Coupling of Methane," *Topics in Catalysis*, 54(16–18), 1266–1285 (2011).

96. Arndt, S., G. Laugel, S. Levchenko, R. Horn, M. Baerns, M. Scheffler, R. Schlögl, and R. Schomäcker, "A Critical Assessment of Li/MgO-based Catalysts for the Oxidative Coupling of Methane," *Catalysis Reviews*, 53(4), 424–514 (2011).

97. Johnson, M. A., E. V. Stefanovich, and T. N. Truong, "An *ab initio* Study on the Oxidative Coupling of Methane over a Lithium-doped MgO Catalyst: Surface Defects and Mechanism," *Journal of Physical Chemistry B*, 101(16), 3196–3201 (1997).

98. Lin, C.-H., K. D. Campbell, J.-X. Wang, and J. H. Lunsford, "Oxidative Dimerization of Methane over Lanthanum Oxide," *Journal of Physical Chemistry*, 90, 534–537 (1986).

99. Arndt, S., T. Otremba, U. Simon, M. Yildiz, H. Schubert, and R. Schomäcker, "Mn–Na_2WO_4/SiO_2 as Catalyst for the Oxidative Coupling of Methane. What is Really Known?," *Applied Catalysis A: General*, 425–426, 53–61 (2012).

100. Xueping, F., L. Shuben, L. Jingzhi, and C. Yanlai, "Oxidative Coupling of Methane on W-Mn Catalysts," *Journal of Molecular Catalysis*, 6, 003 (1992).

101. Ji, S., T. Xiao, S. Li, L. Chou, B. Zhang, C. Xu, R. Hou, A. P. E. York, and M. L. H. Green, "Surface WO_4 Tetrahedron: the Essence of the Oxidative Coupling of Methane over M-W-Mn/SiO_2 Catalysts," *Journal of Catalysis*, 220(1), 47–56 (2003).

102. Ghose, R., H. T. Hwang, and A. Varma, "Oxidative Coupling of Methane Using Catalysts Synthesized by Solution Combustion Method," *Applied Catalysis A: General*, 452, 147–154 (2013).

103. Ghose, R., H. T. Hwang, and A. Varma, "Oxidative Coupling of Methane Using Catalysts Synthesized by Solution Combustion Method: Catalyst Optimization and Kinetic Studies," *Applied Catalysis A: General*, 472, 39–46 (2014).

104. Huang, K., X. Zhan, F. Chen, and D. Lü, "Catalyst Design for Methane Oxidative Coupling by Using Artificial Neural Network and Hybrid Genetic Algorithm," *Chemical Engineering Science*, 58(1), 81–87 (2003).

105. Chua, Y. T., A. R. Mohamed, and S. Bhatia, "Oxidative Coupling of Methane for the Production of Ethylene over Sodium-Tungsten-Manganese-Supported-Silica Catalyst (Na-W-Mn/SiO$_2$)," *Applied Catalysis A: General*, 343(1–2), 142–148 (2008).

106. Thien, C. Y., A. R. Mohamed, and S. Bhatia, "Process Optimization of Oxidative Coupling of Methane for Ethylene Production Using Response Surface Methodology," *Journal of Chemical Technology & Biotechnology*, 82(1), 81–91 (2007).

107. Sahebdelfar, S., M. T. Ravanchi, M. Gharibi, and M. Hamidzadeh, "Rule of 100: An Inherent Limitation or Performance Measure in Oxidative Coupling of Methane?," *Journal of Natural Gas Chemistry*, 21(3), 308–313 (2012).

108. Fan, L.-S., L. Zeng, and S. Luo, "Chemical-Looping Technology Platform," *AIChE Journal*, 61(1), 2–22 (2015).

109. Fan, L.-S., *Chemical Looping Systems for Fossil Energy Conversions*, John Wiley & Sons, Hoboken, NJ (2010).

110. Jones, C. A., J. J. Leonard, and J. A. Sofranko, Methane Conversion, U.S. Patent 4,444,984 (1984).

111. Matherne, J. L. and G. L. Culp, "Direct Conversion of Methane to C$_2$'s and Liquid Fuels: Process Economics," Chapter 14 in Wolf, E. E., ed., *Methane Conversion by Oxidative Processes: Fundamental and Engineering Aspects*, Van Nostrand Reinhold, New York, NY, pp. 463–482 (1992).

112. Sofranko, J. A., J. J. Leonard, C. A. Jones, A. M. Gaffney, and H. P. Withers, "Catalytic Oxidative Coupling of Methane over Sodium-Promoted Mn/SiO$_2$ and Mn/MgO," *Catalysis Today*, 3(2–3), 127–135 (1988).

113. Jones, C. A., J. J. Leonard, and J. A. Sofranko, "The Oxidative Conversion of Methane to Higher Hydrocarbons over Alkali-Promoted Mn/SiO$_2$," *Journal of Catalysis*, 103(2), 311–319 (1987).

114. Salehoun, V., A. Khodadadi, Y. Mortazavi, and A. Talebizadeh, "Dynamics of Mn/Na$_2$WO$_4$/SiO$_2$ Catalyst in Oxidative Coupling of Methane," *Chemical Engineering Science*, 63(20), 4910–4916 (2008).

115. Wang, D., M. P. Rosynek, and J. H. Lunsford, "Oxidative Coupling of Methane Over Oxide-Supported Sodium-Manganese Catalysts," *Journal of Catalysis*, 155(2), 390–402 (1995).

116. Li, S.-B., "Oxidative Coupling of Methane over W-Mn/SiO$_2$ Catalyst," *Chinese Journal of Chemistry*, 19(1), 16–21 (2001).

117. Talebizadeh, A., Y. Mortazavi, and A. A. Khodadadi, "Comparative Study of the Two-Zone Fluidized-Bed Reactor and the Fluidized-Bed Reactor for Oxidative Coupling of Methane over Mn/Na$_2$WO$_4$/SiO$_2$ Catalyst," *Fuel Processing Technology*, 90(10), 1319–1325 (2009).

118. Chung, E. Y., W. K. Wang, H. Alkhatib, et al., "Process Development of Manganese-Based Oxygen Carriers for Oxidative Coupling of Methane in a Pressurized Chemical Looping

System," Presented at 2015 AIChE Spring Meeting and 11th Global Congress on Progress Safety, Austin, TX, April 26–30 (2015).

119. Chung, E. Y., W. K. Wang, H. Alkhatib, et al., "Examination of Oxidative Coupling of Methane by Traditional Catalysis and Chemical Looping with Manganese-Based Oxides," Presented at 2015 AIChE Fall Meeting, Salt Lake City, UT, November 8–13 (2015).

120. Chung, E. Y., "Investigation of Chemical Looping Oxygen Carriers and Processes for Hydrocarbon Oxidation and Selective Alkane Oxidation to Chemicals," PhD Dissertation, The Ohio State University, Columbus, OH (2016).

121. Kofstad, P., *Nonstoichiometry, Diffusion and Electrical Conductivity in Binary Metal Oxides*, John Wiley & Sons, New York, NY (1972).

122. Greish, A. A., L. M. Glukhov, E. D. Finashina, L. M. Kustov, J. Sung, K. Choo, and T. Kim, "Oxidative Coupling of Methane in the Redox Cyclic Mode over the Catalysts on the Basis of CeO_2 and La_2O_3," *Mendeleev Communications*, 20(1), 28–30 (2010).

123. Nagy, A. J., G. Mestl, and R. Schlögl, "The Role of Subsurface Oxygen in the Silver-Catalyzed, Oxidative Coupling of Methane," *Journal of Catalysis*, 188(1), 56–68 (1999).

124. Sung, J. S., K. Y. Choo, T. H. Kim, A. Greish, L. Glukhov, E. Finashina, and L. Kustov, "Peculiarities of Oxidative Coupling of Methane in Redox Cyclic Mode over $Ag-La_2O_3/SiO_2$ Catalysts," *Applied Catalysis A: General*, 380(1–2), 28–32 (2010).

125. Baldwin, T. R., R. Burch, E. M. Crabb, G. D. Squire, and S. C. Tsang, "Oxidative Coupling of Methane over Chloride Catalysts," *Applied Catalysis*, 56(1), 219–229 (1989).

126. Wolfahrt, K., M. Bergfeld, and H. Zengel, A Process for the Preparation of Ethylene–Ethane Mixtures, German Patent 3,503,664 (1986).

127. Otsuka, K., Q. Liu, M. Hatano, and A. Morikawa, "Synthesis of Ethylene by Partial Oxidation of Methane over the Oxides of Transition Elements with LiCl," *Chemistry Letters*, 15(6), 903–906 (1986).

128. Fujimoto, K., S. Hashimoto, K. Asami, and H. Tominaga, "Oxidative Coupling of Methane with Akaline Earth Halide Catalysts Supported on Alkaline Earth Oxides," *Chemistry Letters*, 16(11), 2157–2160 (1987).

129. Shigapov, A. N., M. A. Novoshilova, S. N. Vereshchagin, A. G. Anshits, and V. D. Sokolovskii, "Peculiarities in Oxidative Conversion of Methane to C_2 Hydrocarbons over $CaO-CaCl_2$ Catalysts," *Reaction Kinetics and Catalysis Letters*, 37(2), 397–402 (1988).

130. Burch, R., G. D. Squire, and S. C. Tsang, "Comparative Study of Catalysts for the Oxidative Coupling of Methane," *Applied Catalysis*, 43(1), 105–116 (1988).

131. Warren, B. K., "The Role of Chlorine in Chlorine-Promoted Methane Coupling Catalysts," *Catalysis Today*, 13(2–3), 311–320 (1992).

132. Machida, K. and M. Enyo, "Oxidative Dimerization of Methane over Cerium Mixed Oxides and Its Relation with Their Ion-Conducting Characteristics," *Journal of the Chemical Society, Chemical Communications*, (21), 1639–1640 (1987).

133. Stoukides, M., "Solid-Electrolyte Membrane Reactors: Current Experience and Future Outlook," *Catalysis Reviews*, 42(1–2), 1–70 (2000).

134. Omata, K., O. Yamazaki, K. Tomita, and K. Fujimoto, "Oxidative Coupling of Methane on an ABO_3 type Oxide with Mixed Conductivity," *Journal of the Chemical Society, Chemical Communications*, (14), 1647–1648 (1994).

135. Vermeiren, W. J. M., I. D. M. L. Lenotte, J. A. Martens, and P. A. Jacobs, "Perovskite-Type Complex Oxides as Catalysts for the Oxidative Coupling of Methane," *Studies in Surface Science and Catalysis*, 61, 33–40 (1991).

136. Kyriakou, V., I. Garagounis, and M. Stoukides, "Steam Electrolysis with Simultaneous Production of C_2 Hydrocarbons in a Solid Electrolyte Cell," *International Journal of Hydrogen Energy*, 39(2), 675–683 (2014).

137. Nomura, K., T. Hayakawa, K. Takehira, and Y. Ujihira, "Oxidative Coupling of Methane on Perovskite Oxides, (Ba,Ca)(Co,Fe)$O_{3-\delta}$," *Applied Catalysis A: General*, 101(1), 63–72 (1993).

138. Elshof, J. E. T., H. J. M. Bouwmeester, and H. Verweij, "Oxidative Coupling of Methane in a Mixed-Conducting Perovskite Membrane Reactor," *Applied Catalysis A: General*, 130(2), 195–212 (1995).

139. Dai, H. X., C. F. Ng, and C. T. Au, "Perovskite-Type Halo-Oxide $La_{1-x}Sr_xFeO_{3-\delta}X_\sigma$ (X= F, Cl) Catalysts Selective for the Oxidation of Ethane to Ethene," *Journal of Catalysis*, 189(1), 52–62 (2000).

140. Dai, H. X., C. F. Ng, and C. T. Au, "$YBa_2Cu_3O_{7-\delta}X_\sigma$ (X= F and Cl): Highly Active and Durable Catalysts for the Selective Oxidation of Ethane to Ethene," *Journal of Catalysis*, 193(1), 65–79 (2000).

141. Dai, H. X., C. T. Au, Y. Chan, K. C. Hui, and Y. L. Leung, "Halide-Doped Perovskite-Type $AMn_{1-x}Cu_xO_{3-\delta}$ (A = $La_{0.8}Ba_{0.2}$) Catalysts for Ethane-Selective Oxidation to Ethene," *Applied Catalysis A: General*, 213(1), 91–102 (2001).

142. Zhu, Q., S. L. Wegener, C. Xie, O. Uche, M. Neurock, and T. J. Marks, "Sulfur as a Selective 'Soft' Oxidant for Catalytic Methane Conversion Probed by Experiment and Theory," *Nature Chemistry*, 5(2), 104–109 (2013).

143. Nibbelke, R. H., J. Scheerova, M. H. J. M. Decroon, and G. B. Marin, "The Oxidative Coupling of Methane over MgO-Based Catalysts: A Steady-State Isotope Transient Kinetic Analysis," *Journal of Catalysis*, 156(1), 106–119 (1995).

144. Paganini, M. C., M. Chiesa, P. Martino, E. Giamello, and E. Garrone, "EPR Study of the Surface Basicity of Calcium Oxide. 2: The Interaction with Alkanes," *Journal of Physical Chemistry B*, 107(11), 2575–2580 (2003).

145. Chen, Q., J. H. B. J. Hoebink, and G. B. Marin, "Kinetics of the Oxidative Coupling of Methane at Atmospheric Pressure in the Absence of Catalyst," *Industrial & Engineering Chemistry Research*, 30(9), 2088–2097 (1991).

146. Jašo, S., H. R. Godini, H. Arellano-Garcia, M. Omidkhah, and G. Wozny, "Analysis of Attainable Reactor Performance for the Oxidative Methane Coupling Process," *Chemical Engineering Science*, 65(24), 6341–6352 (2010).

147. Kiatkittipong, W., T. Tagawa, S. Goto, S. Assabumrungrat, K. Silpasup, and P. Praserthdam, "Comparative Study of Oxidative Coupling of Methane Modeling in Various Types of Reactor," *Chemical Engineering Journal*, 115(1/2), 63–71 (2005).

148. Shahri, S. M. K., and S. M. Alavi, "Kinetic Studies of the Oxidative Coupling of Methane over the $Mn/Na_2WO_4/SiO_2$ Catalyst," *Journal of Natural Gas Chemistry*, 18(1), 25–34 (2009).

149. Daneshpayeh, M., A. Khodadadi, N. Mostoufi, Y. Mortazavi, R. Sotudeh-Gharebagh, and A. Talebizadeh, "Kinetic Modeling of Oxidative Coupling of Methane over $Mn/Na_2WO_4/SiO_2$ Catalyst," *Fuel Processing Technology*, 90(3), 403–410 (2009).

150. Daneshpayeh, M., A. Khodadadi, N. Mostoufi, Y. Mortazavi, R. Sotudeh-Gharebagh, and A. Talebizadeh, "Corrigendum to "Kinetic Modeling of Oxidative Coupling of Methane over $Mn/Na_2WO_4/SiO_2$ Catalyst," *Fuel Processing Technology*, 90(9), 1192 (2009).

151. Tiemersma, T. P., M. J. Tuinier, F. Gallucci, J. A. M. Kuipers, and M. van Sint Annaland, "A Kinetics Study for the Oxidative Coupling of Methane on a $Mn/Na_2WO_4/SiO_2$ Catalyst," *Applied Catalysis A: General*, 433–434, 96–108 (2012).

152. Reshetnikov, S. I., Y. I. Pyatnitskii, and L. Y. Dolgikh, "Effect of the Mobility of Oxygen in Perovskite Catalyst on the Dynamics of Oxidative Coupling of Methane," *Theoretical and Experimental Chemistry*, 47(1), 49–54 (2011).

153. Kao, Y. K., L. Lei, and Y. S. Lin, "A Comparative Simulation Study on Oxidative Coupling of Methane in Fixed-Bed and Membrane Reactors," *Industrial & Engineering Chemistry Research*, 36(9), 3583–3593 (1997).

154. Bhatia, S., C. Y. Thien, and A. R. Mohamed, "Oxidative Coupling of Methane (OCM) in a Catalytic Membrane Reactor and Comparison of Its Performance with Other Catalytic Reactors," *Chemical Engineering Journal*, 148(2–3), 525–532 (2009).

155. Vamvakeros, A., S. D. M. Jacques, V. Middelkoop, M. D. Michiel, C. K. Egan, I. Z. Ismagilov, G. B. M. Vaughan, F. Gallucci, M. van Sint Annaland, P. R. Shearing, R. J. Cernik, and A. M. Beale, "Real Time Chemical Imaging of a Working Catalytic Membrane Reactor during Oxidative Coupling of Methane," *Chemical Communications*, 51(64), 12752–12755 (2015).

156. Lu, Y., A. G. Dixon, W. R. Moser, and Y. H. Ma, "Oxidative Coupling of Methane in a Modified γ-Alumina Membrane Reactor," *Chemical Engineering Science*, 55(21), 4901–4912 (2000).

157. Godini, H. R., S. Jašo, S. Xiao, H. Arellano-Garcia, M. Omidkhah, and G. Wozny, "Methane Oxidative Coupling: Synthesis of Membrane Reactor Networks," *Industrial & Engineering Chemistry Research*, 51(22), 7747–7761 (2012).

158. Cao, Z., H. Jiang, H. Luo, S. Baumann, W. A. Meulenberg, H. Voss, and J. Caro, "Simultaneous Overcome of the Equilibrium Limitations in BSCF Oxygen-Permeable Membrane Reactors: Water Splitting and Methane Coupling," *Catalysis Today*, 193(1), 2–7 (2012).

159. Kondratenko, E. V. and U. Rodemerck, "A Dual-Reactor Concept for the High-Yielding Conversion of Methane into Higher Hydrocarbons," *Chem Cat Chem*, 5(3), 697–700 (2013).

160. Tonkovich, A. L., R. W. Carr, and R. Aris, "Enhanced C_2 Yields from Methane Oxidative Coupling by Means of a Separative Chemical Reactor," *Science*, 262(5131), 221–223 (1993).

161. Kruglov, A. V., M. C. Bjorklund, and R. W. Carr, "Optimization of the Simulated Counter-current Moving-Bed Chromatographic Reactor for the Oxidative Coupling of Methane," *Chemical Engineering Science*, 51(11), 2945–2950 (1996).

162. Kundu, P. K., A. K. Ray, and A. Elkamel, "Numerical Simulation and Optimization of Unconventional Three-Section Simulated Countercurrent Moving Bed Chromatographic Reactor for Oxidative Coupling of Methane Reaction," *Canadian Journal of Chemical Engineering*, 90(6), 1502–1513 (2012).

163. Kundu, P. K., Y. Zhang, and A. K. Ray, "Multi-Objective Optimization of Simulated Countercurrent Moving Bed Chromatographic Reactor for Oxidative Coupling of Methane," *Chemical Engineering Science*, 64(19), 4137–4149 (2009).

164. Tonkovich, A. L. and R. W. Carr, "A Simulated Countercurrent Moving-Bed Chromato-graphic Reactor for the Oxidative Coupling of Methane: Experimental Results," *Chemical Engineering Science*, 49(24A), 4647–4656 (1994).

165. Sofranko, J. A. and J. C. Jubin, "Natural Gas to Gasoline: the ARCO GTG Process," in *Proc. Symposium on Methane Activation, Conversion, and Utilization. International Congress of Pacific Basin Societies*. Honolulu, HI, December 17–19 (1989).

166. Iglesia, E., "Challenges and Progress in the Conversion of Natural Gas to Fuels and Chemicals," *ACS Preprints-Fuel Chemistry Division*, 47(1), 128–131 (2002).

167. Jones, C. A., J. J. Leonard, and J. A. Sofranko, Methane Conversion, U.S. Patent 4,665,260 (1987).

168. Mleczko, L. and M. Baerns, "Catalytic Oxidative Coupling of Methane - Reaction Engineering Aspects and Process Schemes," *Fuel Processing Technology*, 42(2–3), 217–248 (1995).

169. Sinnott, R. K., J. F. Richardson, and J. M. Coulson, *Coulson & Richardson's Chemical Engineering*, 4th edn, Butterworth-Heinemann, Burlington, MA (2005).

170. Seider, W. D., J. D. Seader, and D. R. Lewin, *Product and Process Design Principles: Synthesis, Analysis, and Evaluation*, 2nd edn, Wiley, New York (2004).

171. Smith, R., *Chemical Process Design and Integration*, Wiley, West Sussex, United Kingdom (2005).

172. Godini, H. R., S. Jašo, W. Martini, S. Stünkel, D. Salerno, S. N. Xuan, S. Song, S. Setarehalsadat, H. Trivedi, H. Arellano-Garcia, and G. Wozny, "Concurrent Reactor Engineering, Separation Enhancement and Process Intensification; Comprehensive UniCat Approach for Oxidative Coupling of Methane (OCM)," *Czasopismo Techniczne Mechanika*, 109(1-M), 63–74 (2012).

173. Edwards, J. H., R. J. Tyler, and S. D. White, "Oxidative Coupling of Methane over Lithium-Promoted Magnesium Oxide Catalysts in Fixed-bed and Fluidized-bed Reactors," *Energy & Fuels*, 4(1), 85–93 (1990).

174. Dautzenberg, F. M., J. C. Schlatter, J. M. Fox, J. R. Rostrup-Nielsen, and L. J. Christiansen, "Catalyst and Reactor Requirements for the Oxidative Coupling of Methane," *Catalysis Today*, 13(4), 503–509 (1992).

175. Jašo, S., H. R. Godini, H. Arellano-Garcia, and G. Wozny, "Oxidative Coupling of Methane: Reactor Performance and Operating Conditions," *Computer Aided Chemical Engineering*, 28, 781–786 (2010).

176. Santamaria, J. M., E. E. Miro, and E. E. Wolf, "Reactor Simulation Studies of Methane Oxidative Coupling on a Sodium/Nickel-Titanium Oxide (NiTiO$_3$) Catalyst," *Industrial & Engineering Chemistry Research*, 30(6), 1157–1165 (1991).

177. Schweer, D., L. Mleczko, and M. Baerns, "OCM in a Fixed-Bed Reactor: Limits and Perspectives," *Catalysis Today*, 21(2–3), 357–369 (1994).

178. Hoebink, J. H. B. J., P. M. Couwenberg, and G. B. Marin, "Fixed Bed Reactor Design for Gas Phase Chain Reactions Catalysed by Solids: the Oxidative Coupling of Methane," *Chemical Engineering Science*, 49(24), 5453–5463 (1994).

179. Yaghobi, N. and M. H. R. Ghoreishy, "Oxidative Coupling of Methane in a Fixed Bed Reactor over Perovskite Catalyst: A Simulation Study using Experimental Kinetic Model," *Journal of Natural Gas Chemistry*, 17(1), 8–16 (2008).

180. Andorf, R. and M. Baerns, "Oxidative Coupling of Methane in Fluidized- and Packed-Fluidized-Bed Reactors," *Catalysis Today*, 6(4), 445–452 (1990).

181. Pannek, U. and L. Mleczko, "Comprehensive Model of Oxidative Coupling of Methane in a Fluidized-Bed Reactor," *Chemical Engineering Science*, 51(14), 3575–3590 (1996).

182. Mleczko, L. and K.-J. Marschall, "Performance of an Internally Circulating Fluidized-Bed Reactor for the Catalytic Oxidative Coupling of Methane," *Canadian Journal of Chemical Engineering*, 75(3), 610–619 (1997).

183. Pannek, U. and L. Mleczko, "Reaction Engineering Simulations of Oxidative Coupling of Methane in a Circulating Fluidized-Bed Reactor," *Chemical Engineering & Technology*, 21(10), 811–821 (1998).

184. Al-Zahrani, S. M., "The Effects of Kinetics, Hydrodynamics and Feed Conditions on Methane Coupling Using Fluidized Bed Reactor," *Catalysis Today*, 64(3–4), 217–225 (2001).

185. Daneshpayeh, M., N. Mostoufi, A. Khodadadi, R. Sotudeh-Gharebagh, and Y. Mortazavi, "Modeling of Stagewise Feeding in Fluidized Bed Reactor of Oxidative Coupling of Methane," *Energy and Fuels*, 23(7), 3745–3752 (2009).

186. Jašo, S., H. Arellano-Garcia, and G. Wozny, "Oxidative Coupling of Methane in a Fluidized Bed Reactor: Influence of Feeding Policy, Hydrodynamics, and Reactor Geometry," *Chemical Engineering Journal*, 171(1), 255–271 (2011).

187. Jašo, S., "Modeling and Design of the Fluidized Bed Reactor for the Oxidative Coupling of Methane," PhD Dissertation, Technische Universität Berlin, Berlin, Germany (2012).

188. Eghbal-Ahmadi, M.-H., M. Zaerpour, M. Daneshpayeh, and N. Mostoufi, "Optimization of Fluidized Bed Reactor of Oxidative Coupling of Methane," *International Journal of Chemical Reactor Engineering*, 10(1), 1–21 (2012).

189. Edwards, J. H., K. T. Do, and R. J. Tyler, "Reaction Engineering Studies of Methane Coupling in Fluidised-Bed Reactors," *Catalysis Today*, 6(4), 435–444 (1990).

190. Suzuki, S., T. Sasaki, T. Kojima, M. Yamamura, and T. Yoshinari, "New Process Development of Natural Gas Conversion Technology to Liquid Fuels via OCM Reaction," *Energy & Fuels*, 10(3), 531–536 (1996).

191. Hugill, J. A., F. W. A. Tillemans, J. W. Dijkstra, and S. Spoelstra, "Feasibility Study on the Co-generation of Ethylene and Electricity through Oxidative Coupling of Methane," *Applied Thermal Engineering*, 25(8–9), 1259–1271 (2005).

192. Skutil, K. and M. Taniewski, "Some Technological Aspects of Methane Aromatization (Direct and via Oxidative Coupling)," *Fuel Processing Technology*, 87(6), 511–521 (2006).

193. Karimi, A., R. Ahmadi, H. R. Bozorg Zadeh, A. Jebreili Jolodar, and A. Barkhordarion, "Catalytic Oxidative Coupling of Methane-Experimental Investigation and Optimization of Operational Conditions," *Petroleum & Coal*, 49(3), 36–40 (2007).

194. Graf, P. O., "Combining Oxidative Coupling and Reforming of Methane: Vision or Utopia?," PhD Dissertation, University of Twente, Enschede, Netherlands (2009).

195. Stünkel, S., J.-U. Repke, and G. Wozny, "Ethylen Production via Oxidative Coupling of Methane (OCM) - Investigation of Alternative Separation Processes," *Czasopismo Techniczne Mechanika*, 6(5-M), 285–291 (2008).

196. Graf, P. O., and L. Lefferts, "Reactive Separation of Ethylene from the Effluent Gas of Methane Oxidative Coupling via Alkylation of Benzene to Ethylbenzene on ZSM-5," *Chemical Engineering Science*, 64(12), 2773–2780 (2009).

197. Stünkel, S., H. Trivedi, H.-R. Godini, S. Jašo, N. Holst, S. Arndt, J. Steinbach, and R. Schomäcker, "Oxidative Coupling of Methane: Process Design, Development and Operation in a Mini-Plant Scale," *Chemie Ingenieur Technik*, 84(11), 1989–1996 (2012).

198. Salerno-Paredes, D., "Optimal Synthesis of Downstream Processes using the Oxidative Coupling of Methane Reaction," PhD Dissertation, Technische Universität Berlin, Berlin, Germany (2013).

199. Ghareghashi, A., S. Ghader, H. Hashemipour, and H. R. Moghadam, "A Comparison of Co-current and Counter-current Modes for Fischer–Tropsch Synthesis in Two Consecutive Reactors of Oxidative Coupling of Methane and Fischer–Tropsch," *Journal of Natural Gas Science and Engineering*, 14, 1–16 (2013).

200. Stünkel, S., D. Illmer, A. Drescher, R. Schomäcker, and G. Wozny, "On the Design, Development and Operation of an Energy Efficient CO_2 Removal for the Oxidative

Coupling of Methane in a Miniplant Scale," *Applied Thermal Engineering*, 43, 141–147 (2012).

201. Salerno, D., H. Arellano-Garcia, and G. Wozny, "Techno-Economic Analysis for Ethylene and Oxygenates Products from the Oxidative Coupling of Methane Process," *Chemical Engineering Transactions*, 24, 1507–1512 (2011).

202. United States Department of Energy/National Energy Technology Laboratory, "Quality Guidelines for Energy System Studies: Process Modeling Design Parameters," United States Department of Energy /NETL, DOE/NETL-341/081911, Pittsburgh, PA (2012).

203. Black, J., "Cost and Performance Baseline for Fossil Energy Plants. Volume 1: Bituminous Coal and Natural Gas to Electricity," United States Department of Energy /NETL, DOE/NETL-2010/1397, Pittsburgh, PA (2010).

204. Kniel, L., O. Winter, and K. Stork, eds., *Ethylene-Keystone to the Petrochemical Industry*, Marcel Dekker, New York, NY (1980).

205. Peschel, A., A. Jörke, K. Sundmacher, and H. Freund, "Optimal Reaction Concept and Plant Wide Optimization of the Ethylene Oxide Process," *Chemical Engineering Journal*, 207, 656–674 (2012).

206. Mohamadbeigy, K., "Studying of the Effectiveness Parameters on Gas Dehydration Plant," *Petroleum & Coal*, 50(2), 47–51 (2008).

4 Syngas Generation

D. Xu, S. Luo, D. Wang, W. Wang, M. Kathe, and L.-S. Fan

4.1 Introduction

Syngas, predominantly a mixture of H_2 and CO, is an important feedstock for the production of many valuable commodity chemicals such as methanol, synthetic liquid fuels, ammonia, and hydrogen. Since syngas forms the backbone of many chemical industries, its purity, H_2:CO ratio, and cost have a major effect on the operation and economics of downstream chemical processing systems. Over the years, the process of syngas production has evolved from simply passing steam over hot coke to the use of large-scale solid circulating units that process a variety of hydrocarbon feedstocks. Some syngas production processes have been used at the commercial scale while others are still being developed at the lab and bench scale. Considerable knowledge has been gained from large-scale operations, and these have been valuable in paving the way for further innovations and new technologies development.

Gasification and reforming are the two processes that form the basis of all syngas production technologies. Syngas production technologies involve the use of oxidative reagents like oxygen, steam, CO_2, and more recently metal oxide oxygen carriers, to selectively convert various hydrocarbon feedstocks to syngas. The term "selective" is significant since the formation of full combustion products, CO_2 and H_2O, is undesirable. Therefore, careful consideration is required when selecting operating conditions like temperature, pressure, catalyst, and/or metal oxide oxygen carriers. The ultimate goal for generating syngas is a process that is simple in operation and low in energy and cost requirements. For operation with metal oxides, as in chemical looping processes, combustion applications have been studied extensively. Relatively little has been probed on chemical looping gasification and reforming. There is a distinct difference in operating strategy between chemical looping gasification/chemical looping reforming (CLG/CLR) and chemical looping combustion (CLC) processes. The CLC processes can achieve a complete reactant gas oxidation with a low reactant gas to oxygen carrier ratio. The CLG/CLR processes, on the other hand, can achieve a partial or selective reactant fuel oxidation with a high reactant fuel to oxygen carrier ratio.

In this chapter, syngas generation technologies based on conventional processes and chemical looping processes are discussed. Specifically, Section 4.2 provides a historical viewpoint on the development of current commercial-scale syngas generation processes. An account of each of the technologies followed by examples of their application in commercial operation is given. Solid fuel based technologies, including the use of coal and biomass, are illustrated along with gaseous fuel based technologies such as steam methane reforming (SMR), partial oxidation (POX), catalytic partial oxidation (CPOX), and autothermal reforming (ATR). Section 4.3 presents the thermodynamic rationale for metal oxide selection for chemical looping syngas generation. In Section 4.4, the use of fluidized bed reducers for chemical looping syngas generation is introduced. It discusses the integration of chemical looping combustion with steam methane reforming (SMR–CLC) and experimental results for CLG/CLR. Section 4.5 reviews CLG/CLR processes based on the moving bed reducer operation. Extensive experimental results that substantiate the design principles for the moving bed reducer reactor in these CLG/CLR systems, along with those of other reactors and auxiliary components such as gas sealing devices and solids flow control devices, are also described. Section 4.6 describes a novel CLG/CLR process that includes the utilization of CO_2, not only that recycled from downstream of the syngas product, but also intake of fresh CO_2 as feedstock, yielding a zero CO_2 emission or even negative CO_2 emission chemical looping process. The effects of CO_2 as feedstock on the H_2:CO ratio and a reduced modularization strategy that maximizes the syngas yield are discussed.

4.2   Conventional Syngas Generation Processes

Syngas as an intermediate feedstock can be used to produce various valuable chemicals or fuels. Figure 4.1 shows some examples of these chemicals and fuels, such as hydrogen, ammonia, methanol, and liquid fuels.[1]

Syngas is commonly produced from the partial oxidation of carbonaceous fuels through processes like gasification and reforming. Because of the versatility of syngas, gasification and reforming processes have been topics of interest for many decades. In 1780, Felice Fontana observed the production of a combustible gas when passing steam over red-hot coke; which was essentially syngas generation from coke gasification.[2] Syngas can be used for hydrogen generation via the water–gas shift (WGS) reaction, and subsequent CO_2 separation, power generation via the integrated gasification combined cycle (IGCC), and liquid fuel production via the Fischer–Tropsch (F–T) reaction. The H_2:CO ratio of syngas generated varies significantly depending on the feedstock used, thermodynamics under the operating conditions, and specific operating mode of the reactor, and is tailored to the desired downstream applications. For example, a H_2:CO molar ratio of 2:1 is usually used for the synthesis of methanol and liquid fuels such as gasoline and diesel; a ratio of 1:1 is suitable for production of acetic acid, acetic anhydride, and formaldehyde; and a ratio higher than 3:1 is usually used in hydrogen production via the WGS reaction.

4.2.1   Coal Gasification

Coal gasification produces gaseous product from coal and may be broadly classified into three distinct reactions, i.e. partial oxidation, pyrolysis, and hydrogenation. Coal

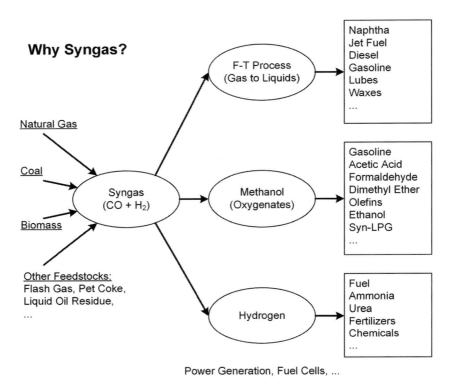

Figure 4.1 Syngas as an intermediate building block for value-added products from fossil fuels.[1]

gasification using partial oxidation utilizes an oxidative agent like oxygen and/or water to produce a syngas stream. Coal gasification by pyrolysis is the process of heating coal in the absence of air or any other reactants to produce a combustible gas mixture. Coal hydrogenation is the process of generating a combustible gas mixture that contains a significantly higher amount of H_2 per mole of carbon than that in coal. Coal gasification as a commercial technology has been in practice since 1812, when coal was gasified by pyrolysis in a batch mode to produce town gas for street lighting in London.[3,4] The use of coal gasification for producing town gas was phased out towards the end of the eighteenth century as more efficient alternatives were developed, including electric bulbs and cheap natural gas for space heating and cooking.

Currently, the dominant coal gasification technology is partial oxidation, which is used in most of the world's commercial coal gasification processes. Coal gasification using partial oxidation employs an oxidizing agent like air or oxygen and/or water to produce syngas. The earliest partial oxidation techniques for coal gasification were the water–gas and the producer gas processes.[5–8] The water–gas process was a batch process that was operated in two separate stages. The first stage involved blowing air through a fixed coal bed to heat the bed to ~1300 °C. The second stage involved a reaction with steam, wherein the steam was injected sequentially from the top and the bottom of the coal bed to optimize the heat transfer required for the endothermic syngas production reaction between water and coal. The producer gas process was one of the earliest continuous processes for coal gasification and it used air as the oxidant. The continuous nature of the process was maintained by a moving bed system, where humidified air was reacted countercurrently with lumps of coal, which were transported through the gasifier by gravity with a rotary grate at the bottom. In the late 1800s, the water–gas process was the primary method for producing syngas, as the syngas from the competing producer gas process that used air for the gasification reaction contained a large amount of N_2 and, hence, the syngas generated had a low heating value (~6500 KJ/m^3). The process using the Winkler fluidized bed for coal gasification was developed in the early 1920s, with air and steam as the oxidants.[9,10] This process was used to circumvent the clogging issues due to the presence of fines in moving coal bed systems. Both the moving bed and Winkler fluidized bed processes required voluminous reactor sizes for large capacity operation. Linde's development of cryogenic air separation for oxygen production provided an impetus to the development of the modern form of continuous gasification processes, with pure oxygen as the oxidant.[11,12] Using these gasification processes, high quality syngas can be produced. The Winkler fluidized bed gasification process (1926) and the Lurgi moving bed gasification process (1927) were the first continuously operating designs of large-scale partial oxidation gasification processes operated under pressurized conditions to be proposed.[13] The Koppers–Totzek process was developed in the early 1940s to operate in an entrained flow regime, which made it flexible for the type of coal used. Parallel to the development of improved gasifiers for partial oxidation of coal, several attempts were made at underground coal gasification.[14] These included experiments in England prior to World War I, pilot-scale operations in the Soviet Union from the 1940s to the late 1970s, and exploratory work in the United States in the 1980s and in Australia between 1999 and 2002.[3] The concept of underground coal gasification can eliminate the dangers associated with coal mining. However, concerns related to gas leakage into underground water and the prohibitive cost of developing effective oxygen/air supply systems that can be adapted to different coal seams have hampered any significant commercial development efforts. There was considerable activity in the 1960s and 1970s towards developing coal gasification based technologies for synthetic/substitute natural gas (SNG) production. Termed hydrogenation, these technologies were developed to produce substitute natural gas in regions where the natural gas supply was sparse and prices were high. Two notable attempts at producing synthetic natural gas using coal hydrogenation were the CO_2 acceptor process and the IGT HYGAS process.[15] The CO_2 acceptor process was developed by the Consolidation Coal Company, and later the Conoco Coal Development Company began with coal gasification using the reaction between coal and steam to produce H_2 by the addition of CaO; CaO in the gasifier removes CO_2 to form $CaCO_3$; CaO is regenerated by calcination that takes place in a separate reactor. It should be noted that the calcination and carbonation reactions for CaO and $CaCO_3$ underlie the calcium looping reaction. The HYGAS process was developed by IGT to hydrogenate coal for natural gas production by the methanation reaction. The required hydrogen for the methanation reaction was produced from the steam–iron reaction, where iron was obtained from iron oxide reduction by syngas produced from coal gasification.[16] It is noted that the reduction and oxidation of iron–iron oxide reaction underlies the iron-based chemical looping process. Both the CO_2 acceptor process and the IGT HYGAS process were demonstrated at pilot scale but were not commercialized, in part due to the low market price of natural gas and the high capital cost requirements resulting from only moderate efficiency of the process. The majority of coal gasification technologies today are partial oxidation based and can be classified by the type of gasifier used: (1) moving bed processes, (2) fluidized bed processes, and (3) entrained bed processes. These are elaborated below.

Moving Bed Coal Gasification Using Partial Oxidation

The Lurgi dry-ash gasifier and the British Gas Lurgi slagging (BGLS) gasifier are the two gasifier designs for moving bed coal gasification processes.[17,18] The Lurgi dry-ash gasifier is a packed moving bed in which a mixture of steam and pure oxygen is reacted countercurrently to the downward moving coal particles. The coal is distributed across the gasifier cross-section using a rotating mechanical device. The coal injected into the moving bed Lurgi gasifier undergoes drying and devolatilization in the top section of the gasifier while char is gasified in the bottom section of the gasifier. The reaction temperature is highest in the bottom half of the gasifier bed, where a high oxygen partial pressure promotes the exothermic combustion reaction over the endothermic gasification reactions. As the mixture of oxygen and steam moves upwards through the moving bed gasifier, endothermic gasification reactions dominate, leading to a temperature drop, which can be seen from the temperature profile in Figure 4.2.

The gas exiting the top of the gasifier flows to a direct quench cooler before being sent to downstream sulfur and ammonia clean-up units. The ash and the solids are transported to a collection device using a rotary grate located at the bottom of the reactor. An important design condition for the Lurgi dry-ash gasifier is that the

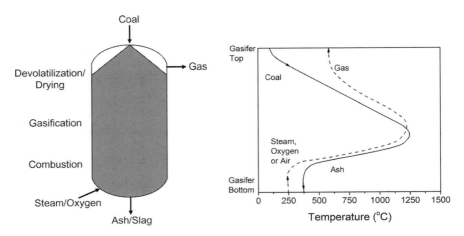

Figure 4.2 Temperature profile of moving bed gasifier.[18]

temperature of the ash exiting the gasifier is maintained below its softening temperature. This gasifier can only process mildly caking coals like lignite and sub-bituminous. The Lurgi type dry-ash gasifier is widely used commercially with plant operations in South Africa, USA, Germany, and China. The BGLS gasifier was developed in the early 1980s by British Gas and Lurgi as an improvement over the Lurgi dry-ash gasifier. A key design difference in the BGLS gasifier is in the temperature profile in which the ash is ensured to be in slag form, unlike the dry-ash Lurgi gasifier. Operating in the ash slagging regime in the BGLS allows for a significantly higher operating temperature than in the Lurgi gasifier, leading to a higher carbon conversion efficiency and a higher CO and H_2 concentration in the syngas.

Fluidized Bed Coal Gasification Using Partial Oxidation

Fluidized bed coal gasifiers can be classified into three major types depending on the fluidization regime in which they operate. These are: (1) bubbling/turbulent fluidized bed, (2) circulating fluidized bed, and (3) transport reactor.[3] Fluidized bed reactors are typically operated below the ash softening temperatures and because of the fluidized nature, can achieve a constant temperature profile in the gasifier, as shown in Figure 4.3.

The Winkler coal gasifiers discussed earlier, utilize a bubbling/turbulent fluidized bed to convert coal with the oxidants, steam, and oxygen, serving as the fluidizing medium. An increase in operating temperature and pressure for the Winkler coal gasifier can increase the carbon conversion efficiency and decrease reactor volume. Several high temperature Winkler gasifiers are in operation today, including Wesseling pilot plant in Germany, Versova 400 MW_e IGCC plant in the Czech Republic, and the Cobra IGCC plant in Germany.[19,20]

Circulating fluidized bed (CFB) coal gasifiers use a gas velocity higher than that in the bubbling/turbulent fluidized bed gasifiers, with the solid circulation through an

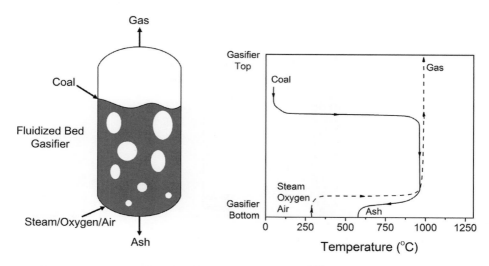

Figure 4.3 Fluidized bed coal gasification temperature profile.[3,18]

external loop. The high gas flow provides high heat transfer in a CFB, thereby reducing the tendency for tar formation and improving the carbon conversion efficiency of the gasifier over a bubbling/turbulent fluidized bed gasifier. The CFB coal gasifiers are capable of operating with a flexible fuel feedstock, including biomass based feedstocks. They are marketed commercially by both Foster-Wheeler and Lurgi. The latest developments in CFB coal gasification technologies include KBR's transport gasifier systems.[21,22] These use the high end of the gas velocity to transport the solids in a CFB, with significantly improved mass transfer and heat transfer efficiencies for the gasification system. The KBR gasifier uses air as an oxidant and processes low grade coals. It has been demonstrated on a pilot scale in Alabama, USA and is being used in the Kemper County IGCC commercial plant in Mississippi, USA and the gasification based plant in Dongguan, China.

Entrained Bed Coal Gasification Using Partial Oxidation

Entrained bed gasifiers operate in a cocurrent contact mode between coal and the oxidant.[23,24] Entrained bed gasifiers require fine coal powder (<100 microns) to promote mass transfer as the residence time for coal fines in a once through pass is short: on the order of seconds. The gasifiers operated at high carbon conversion efficiencies (>98%) can handle any kind of coal feedstock and operate in temperatures above the ash slagging temperature. The injection of coal can be dry fed using a carrier gas such as N_2 or wet fed using water as the carrier. These gasifiers are the most well developed and are the most widespread in use of all gasification technologies. Two examples of the entrained bed gasifier are GE and E-Gas gasifiers. The GE entrained flow gasifier exhibits a uniform gas temperature, as shown in Figure 4.4. It uses a single stage, cocurrent downward flow system and can use either radiant cooling or quench cooling for operation. The GE gasifier is used in over

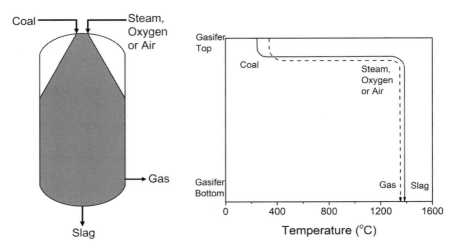

Figure 4.4 Downward flow entrained bed coal gasification temperature profile.[18]

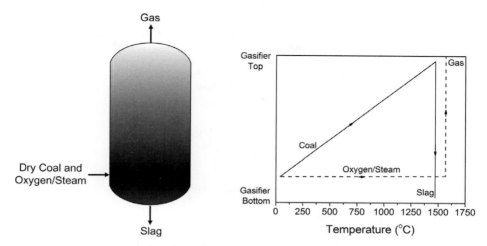

Figure 4.5 Upward flow entrained bed coal gasification temperature profile.[3]

65 gasification facilities worldwide, including Polk River IGCC in Polk County, Florida (USA), Eastman Chemicals complex in Kingston, Tennessee (USA), and Sarlux IGCC in Sardinia (Italy).[25,26]

The Shell, Prenflo, Koppers–Totzek, and Krupp–Koppers all use a single stage upflow gasifier with the coal being side-fired, as shown in Figure 4.5. These dry-fired entrained flow gasifiers utilize a lower amount of oxygen as compared to the water-slurry fed ones. Examples of this type of gasifier in use include the Buggenum IGCC plant, 3000 tpd Prenflo plant in Spain, and over 25 gasifiers in China.

The E-Gas gasifier uses a two-stage contact to improve the thermal efficiency by operating the first stage under ash non-slagging condition and the second stage under slagging condition. A staged operation reduces the syngas cooler duty and eliminates the requirement for a direct syngas quench.[27–29] The E-Gas technology is used at the Wabash River IGCC plant in Terre Haute, Indiana (USA), a polygen plant in Jamnagar (India), a H_2 production plant in Huizhou (China), and an SNG plant in Gwangyang (Korea). Table 4.1 summarizes the various major coal gasification based technologies in commercial operation today.

4.2.2 Natural Gas Reforming Technologies

Many groups have been developing natural gas reforming technologies for syngas production since the early 1900s.[30–36] The use of nickel based catalyst material which showed promise for methane conversion in the presence of steam was patented in 1912 by BASF.[30] Fischer and Tropsch investigated the conversion of methane into syngas suitable for F–T synthesis in the early 1920s.[31] The early development of natural gas reforming technologies was driven by a demand for H_2 to be used in petroleum refining.[34] In the late 1920s, Standard Oil established commercial H_2 production from a natural gas plant at their refinery in Bayway, New Jersey, and Baton Rouge, Louisiana,

Table 4.1 Characteristics of different types of coal gasifiers.

Moving bed coal gasification

Gasifier name	Stages	Oxidant flow direction	Oxidant	Coal feed condition	Coal flow direction
Lurgi dry ash	Single	Upward	Steam/O_2	Dry	Downward
Slagging BGL	Single	Upward	Steam/O_2	Dry	Downward

Fluidized bed coal gasification

Gasifier name	Stages	Oxidant flow direction	Oxidant	Feed	Coal flow direction
Winkler, HTW	Single	Upward	Steam/O_2	Dry	Downward

Entrained bed coal gasification

Gasifier name	Stages	Oxidant flow direction	Oxidant	Feed	Coal flow direction
Koppers–Totzek	Single	Upward	O_2	Dry	Upward
E-Gas	Double	Upward	O_2	Slurry	Upward
Texaco	Single	Downward	O_2	Slurry	Downward
Shell SCGP	Single	Upward	O_2	Dry	Upward

for petroleum refining. These reforming plants used a two-stage reactor, in which the first reactor performed steam reactions at a high temperature (870 °C), while the second reactor resembling a WGS type application was operated at a lower temperature (450 °C). The development of natural gas reforming catalysts has followed a parallel path to the development of steam reforming of light petroleum naphthas. In addition to producing H_2 for petroleum refining operations, reforming catalysts have found uses in a wide range of applications, including catalytic converters in automobiles, and providing H_2 for ammonia synthesis and town gas type applications.[37,38]

Steam Methane Reforming

Steam methane reforming is one of the earliest reforming technologies that has been widely used for the production of syngas from natural gas. The primary reaction in SMR is shown by equation (4.2.1):

$$CH_4 + H_2O \overset{Ni}{\leftrightarrow} CO + 3H_2 \qquad \Delta H = 225.5 \text{ kJ/mol at } 850\,°C. \qquad (4.2.1)$$

The typical operating conditions for a steam reformer ranges between pressures and temperatures of 14 and 20 atm and 800 and 1000 °C, respectively.[39] Since the SMR reaction is highly endothermic, it is often carried out in heat exchange reactors, also known as tubular reformer reactors. A typical steam methane reformer consists of multiple fixed bed reactor tubes filled with alumina-supported nickel catalyst. The heat required for the reactions occurring in the reformer tubes is provided by external fuel combustion. Figure 4.6 shows a typical SMR process. A pre-reformer is sometimes installed upstream of the reformer to partially convert the feed in order to reduce the residence time requirement, and hence the size, of the reformer. To maximize the methane conversion in the reversible reaction, excess steam with $H_2O:CH_4$ molar ratio ranging from 2 : 1 to 6 : 1 is generally used to drive the reforming reaction.[40,41] The

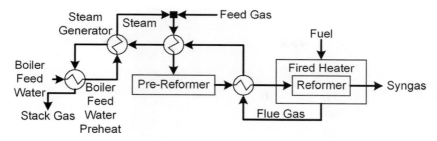

Figure 4.6 Typical SMR process.

WGS reaction also takes place as a side reaction in the reformer, producing H_2 and CO_2 by reaction (4.2.2):

$$CO + H_2O \leftrightarrow CO_2 + H_2 . \qquad (4.2.2)$$

A typical composition in vol% of the gas product from the steam methane reformer, on a dry basis, is 66% H_2, 12% CO, 12% CH_4, and 10% CO_2.[39]

The SMR process typically produces a hydrogen-rich syngas with a H_2:CO ratio above 3:1, which can be further processed to produce hydrogen using a WGS reactor. However, if methanol or liquid fuel is the desired end product, a secondary reforming stage can be employed where oxygen or CO_2 is used as an oxidant to decrease the H_2:CO ratio in the product gas.[42] In addition to the need for a secondary reforming stage, which adds to the capital cost, one of the major drawbacks of the SMR process is the energy penalty associated with generating steam for use in the WGS reactor, which reduces the thermal efficiency of the process. Nevertheless, SMR is still widely employed in commercial plants. For example, Methanex announced plans to expand the methanol production capacity of its Medicine Hat plant in Canada from 0.5 million t/yr to 1.3 million t/yr (5000 to 13,000 bbl/d), with syngas generated from SMR.[43]

Partial Oxidation

The POX technology produces syngas from methane without the use of a catalyst. In POX, oxygen is used to oxidize methane, as shown in reaction (4.2.3):

$$2CH_4 + O_2 \rightarrow 2CO + 4H_2 \qquad \Delta H = -51.5 \text{ kJ/mol at } 1300\,°C. \qquad (4.2.3)$$

Thus, an ASU is needed to obtain oxygen in POX. Compared to SMR, the advantages of POX technology include almost complete methane conversion and high H_2 and CO selectivity over H_2O and CO_2. The syngas produced by POX has an H_2:CO ratio close to 2:1, which is suitable for liquid fuel synthesis. As a result, the downstream processing of syngas is simplified before it enters the liquid fuel synthesis section. Energy-intensive steam generation in SMR is also avoided in the POX process.

One of the main challenges of POX processes comes with the extreme operating conditions. The exothermic reaction of methane partial oxidation brings the operating temperatures to above 1300 °C (2372 °F), which requires costly reactor materials.

Figure 4.7 Shell partial oxidation technology plants: (left: Bintulu); (right: Pearl).[44,49] A black and white version of this figure will appear in some formats. For the color version, please refer to the plate section.

Operating at an elevated pressure is preferred due to the processing conditions of the downstream units. Apart from the high temperature and pressure operating conditions, the reactor material of construction needs to be corrosion resistant under a reducing gas environment. The POX process is conducted with pure O_2 that requires energy intensive air separation units (ASUs) for O_2 separation. Further, a heat recovery system is needed for the POX process which can further increase process costs.[44-46]

The POX process is a well-developed industrial technology for syngas generation, especially for the gas-to-liquids (GTL) industry. One important example of the application of a POX-GTL process is in the Shell middle distillate synthesis (SMDS) plant located in Bintulu, Malaysia, which started operation in 1993. With an initial investment of US$850 million, the SMDS plant produced liquid fuel from natural gas via syngas at a design capacity of 12,500 bbl/d using the POX process.[44] In 2003, the plant underwent a US$50 million renovation for debottlenecking to further increase the capacity to 14,700 bbl/d.[44] Based on the extensive experience gained from the Bintulu SMDS plant, Shell scaled up its POX-GTL process in the Pearl GTL project located in Qatar. The Pearl GTL project is the largest GTL plant in the world with a designed capacity of 140,000 bbl/d, consuming 1600 MMSCFD wellhead gas and 860,000 m^3/hr (about 30,000 t/d) of air.[47,48] After four years of construction, the plant began operations in 2011 and reached full production capacity by 2013 (Figure 4.7).

Catalytic Partial Oxidation (CPOX)

The POX process for syngas production bears the advantages of high methane conversion, high syngas selectivity, and a desirable H_2:CO ratio. However, the POX reactor needs to be operated at a very high temperature and has some drawbacks, as indicated earlier. One approach to reducing the reaction temperature, while maintaining a desired syngas yield, is through the use of a catalyst for the reaction. This type of POX process is called the catalytic partial oxidation (CPOX) process.[50-52] Schmidt et al. studied the catalytic oxidation of methane and observed very high (>90%) selectivity towards both H_2 and CO, low CO_2 and H_2O content, and almost complete CH_4 conversion with a contact time of about 10^{-3} s.[51] Although the result was encouraging at the lab scale, the engineering scale-up of the CPOX process could be challenging. Active catalysts combined with the exothermic reaction can result in carbon deposition, the formation of local hotspots, and thermal runaways in the reactor that could lead to catalyst deactivation and operational safety issues. The CPOX requires further development of catalysts that can maintain long-term reactivity. For CPOX, the syngas production step remains a bottleneck in technical and economic aspects of the GTL process. Current CPOX studies mainly focus on developing catalysts that are active and stable over a long duration for high reaction temperature conditions. The hotspots in the reactor can generally be mitigated by co-feeding steam and/or CO_2. One of the few examples of the use of the CPOX process to produce syngas for industrial use is ConocoPhillips' GTL demonstration plant located in Ponca City, Oklahoma (USA). The demonstration plant started operation in 2003 and continued for two years.[52]

Autothermal Reforming and Two-Step Reforming

In the ATR process, both oxygen and steam are used to generate syngas.[53–56] The POX and SMR reactions occur in a single reactor such that the heat required by the endothermic SMR reaction is provided by the exothermic POX reaction. Although ATR can be designed to be a single-step process, two-step configurations of ATR are more widely used. In the two-step configuration, a pre-reformer is located upstream of the ATR reactor to partially oxidize the fuel with steam. In addition to its autothermal nature, ATR is advantageous over SMR for its ability to directly produce syngas with a H_2:CO ratio of 2:1, which is favorable for liquid fuel synthesis. The ATR process also consumes less oxygen compared to the POX process and less steam than to the SMR process. Moreover, coke deposition on the catalyst surface commonly occurs when the fuel is exposed to a metallic catalyst such as nickel or iron in a reducing environment with a low CO_2 and/or H_2O content, where it cannot gasify the carbon via the reverse Boudouard reaction, given in reaction (4.2.4), or the WGS reaction. Coke deposition on the catalyst reduces its carbon conversion efficiency and selectivity towards syngas and also causes its deactivation. Coke deposition can be controlled by maintaining a high CO_2 and/or H_2O content in the reforming product, but this results in a loss of syngas yield. Thus, there is a trade-off between preventing carbon deposition and obtaining a high syngas yield. Current syngas production processes generally use a CO_2 and H_2O combined content greater than 15% in order to suppress carbon deposition. This requires the use of downstream CO_2/H_2O separation units, which are capital intensive and result in a decreased fuel-to-syngas production efficiency:

$$CO_2 + C \leftrightarrow 2CO. \tag{4.2.4}$$

The single-step ATR process, shown in Figure 4.8, has not been used for large-scale operation while the two-step ATR process, shown in Figure 4.9, has been applied in a few commercial plants. Although the use of ATR processes in ammonia and methanol production industries dates back to the 1960s, recent interest in GTL technologies has driven its further development. An early application of ATR technology in the GTL process was the Oryx GTL plant.[53] The Oryx GTL project was a joint venture between

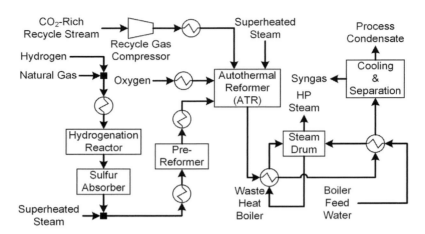

Figure 4.8 Single-step autothermal reforming technology for syngas production from natural gas.[54,55]

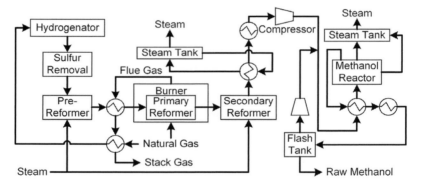

Figure 4.9 Two-step reforming for syngas production from natural gas.[56]

Qatar Petroleum and Sasol with a total investment of US$1 billion. It was designed to process 330 MMSCFD (9.3 million m³/d) of natural gas with a liquid fuel production capacity of 34,000 bbl/d. In this project, syngas was produced by the ATR technology developed by Haldor Topsøe. In a typical mode of this plant operation, natural gas is desulfurized by reacting it with hydrogen over a hydrogenation catalyst. The resulting H_2S is subsequently removed over a bed of zinc oxide. The desulfurized natural gas is heated and sent with steam under a steam:carbon ratio of typically 0.6:1 to an adiabatic pre-reformer containing nickel as the catalyst. The pre-reformer converts the carbon-aceous component of natural gas to syngas at relatively low temperatures to avoid high temperature coke formation. Operating the pre-reformer at lower temperatures is important because the natural gas feed contains hydrocarbon compounds other than methane which have a higher tendency, compared to methane, towards coke deposition. The pre-reformed gas is further heated, mixed with recycled gas, and introduced to the autothermal reformer. Gases are partially oxidized by oxygen in the top section of the autothermal reformer and then enter a catalyst bed where the methane reforming and WGS reaction occur.

4.2.3 Biomass Gasification

Other than fossil fuels as carbonaceous resources for producing syngas, biomass as renewable feedstock can also be used as carbonaceous resources for gasification.[57–59] If annual biomass production from forest and agricultural resources in the United States were increased to ~1.3 billion tons and used to produce liquid fuels, this could reduce petroleum consumption by nearly 30%.[60] Gasification enables the conversion of low-grade fuels such as wood, wood wastes, residues, and recyclables into combustible product gases, liquid fuel, and chemicals. These valuable bio-products can replace non-renewable fuels such as coal, natural gas, and liquid fuels.

Biomass gasification technologies have been commercialized, largely for power generation applications. The THERMIE demonstration project in Lahti, Finland, which started commercial operation in 1998, uses Foster Wheeler's atmospheric circulating fluidized bed gasification (ACFBG) process to convert biomass feedstock such as bark, wood chips, sawdust, and uncontaminated wood waste into syngas, as shown in Figure 4.10.[61]

The ACFBG gasifier is used in connection with a Benson-type once-through boiler, which consumes coal, natural gas, as well as biomass to produce 167 MW electricity and 240 MW heat for district heating. The ACFBG system consists of a gasification reactor, a cyclone separator to separate solid particles from gas, and a return pipe to bring the circulating material back into the bottom of the gasifier. Air for gasification is introduced by a high-pressure air fan through an air distribution grid at the bottom of the reactor below the bed of particles. The air velocity is high enough to fluidize the bed particles and convey some of them out of the reactor to the solid–gas separator, where most of the solids are separated from the gas and returned to the lower part of the gasifier. Hot product gas flows into an air preheater located downstream of the separator. The circulating solids contain char that is combusted with the fluidizing air, which generates the heat required for the pyrolysis process and subsequent gasification reactions. The circulating material also serves as a heat carrier and stabilizes the process temperatures. Ash is removed from the bottom with a water-cooled screw.

The biomass gasification plant constructed at the Electrabel Ruien power plant, which is the largest fossil fuel fired power plant in Belgium, was also based on this

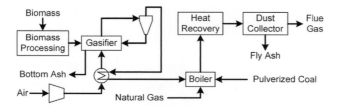

Figure 4.10 Foster Wheeler's atmospheric circulating fluidized bed biomass gasification system.[61]

technology.[62] The gasifier, in operation since 2003, was designed to use a variety of feedstock, like fresh and recycled wood chips, bark, and hard and soft board residues. Depending on the feedstock, the heat input to the boiler ranges from 45 to 70 MW_{th}. The 15 MW_{th} demonstration-scale bubbling fluidized bed (BFB) gasification plant at StoraEnso's Varkaus paper mill (Finland) was built by Foster Wheeler and is based on the same process. The other demonstration and commercial scale biomass to syngas gasification plants using ACFBG technology include the 28 MW_{th} Pietarsaari unit (Finland), the 20 MW_{th} Norrsundet unit (Sweden), the 15 MW_{th} Rodao unit (Portugal) and the 27 MW_{th} Karlsborg unit (Sweden).[63]

The Chiba waste recycling center in Japan recycles about 300 t/d of waste products using the Thermoselect gasification and reforming process with a capacity of 1.5 MW_{th}.[64] In the Thermoselect process, wastes are compacted without pretreatment, followed by drying and pyrolysis by indirect pyrolysis. The waste product is charged into a high-temperature reactor, where it melts by reaction with oxygen. This gas passes through the gas quenching and refining process and is recovered as clean synthesized fuel gas.

Biomass gasification is similar to coal gasification in that they both convert solid fuels to syngas. Therefore, biomass gasification faces similar challenges to those that occur in coal gasification, including high steam and oxygen consumption, low hydrogen content in syngas, and energy intensive purification requirements. Moreover, biomass has a lower heat of combustion compared to coal due to the relatively high oxygen content. Therefore, a significant amount of biomass must be completely burned to provide the endothermic heat for gasification reactions. As a result, the syngas yield from biomass gasification is lower than that from natural gas reforming and coal gasification.

4.3   Syngas Generation Metal Oxides in Chemical Looping

Metal oxides have been used for a variety of chemical looping applications.[65–68] The type of feedstock used, either in gaseous, liquid or solid form, affects both the design and operating conditions of the reducer. The chemical reactions in the chemical looping combustion (CLC) process with a feedstock $C_xH_yO_z$ can be described by reactions (4.3.1) through (4.3.3):

$$\text{Reducer}: \quad C_xH_yO_z + \frac{2x + \frac{y}{2} - z}{\delta} MeO_w \rightarrow x\,CO_2 + \frac{y}{2}H_2O + \frac{2x + \frac{y}{2} - z}{\delta} MeO_{w-\delta} \tag{4.3.1}$$

$$\text{Combustor}: \quad MeO_{w-\delta} + \frac{\delta}{2}O_2 \rightarrow MeO_w \tag{4.3.2}$$

$$\text{Overall reaction}: \quad C_xH_yO_z + \frac{2x + \frac{y}{2} - z}{2}O_2 \rightarrow xCO_2 + \frac{y}{2}H_2O. \tag{4.3.3}$$

Since the overall reaction in CLC, reaction (4.3.3), is exothermic, the heat required for the endothermic reaction, reaction (4.3.1), can be provided through the looping of oxygen carrier particles, which serve as heat transfer media. Apart from full oxidation (combustion) processes, chemical looping can be applied for partial/selective oxidation processes, including syngas generation, directly from a carbonaceous feedstock. The chemical looping configurations used for partial oxidation applications are generally referred to as chemical looping gasification (CLG) when using solid fuels and chemical looping reforming (CLR) when using gaseous fuels. The metal oxide oxygen carrier is key to successful CLG/CLR operation. Its basic properties have been elaborated in Chapter 2. The behavior for CLG/CLR applications is described below.

4.3.1   Selection of Oxygen Carrier

To operate the CLG/CLR process in a continuous cycle, the oxygen carriers need to be circulated between the reducer and the combustor. The thermodynamic properties of active metal oxide oxygen carriers are given in Chapter 2, as illustrated using the modified Ellingham diagram, shown in Figures 2.1 and 2.2. From these figures, metal oxides in the "syngas production" region are less oxidative compared to those in the "combustion" region and are more thermodynamically favored for the partial oxidation of carbonaceous fuels. Using metal oxides in the "syngas production" region as oxygen carriers, the methane partial oxidation reaction, reaction (4.3.4), is spontaneous, while the oxidation reactions of syngas, reactions (4.3.5) and (4.3.6), are thermodynamically constrained. Therefore, at thermodynamic equilibrium, the gaseous products are predominantly syngas with minimal CH_4, CO_2, and H_2O, even if the oxygen carrier is present in excess:

$$CH_4 + \frac{1}{\delta} MeO_w \rightarrow CO + 2H_2 + \frac{1}{\delta} MeO_{w-\delta} \tag{4.3.4}$$

$$CO + \frac{1}{\delta} MeO_w \rightarrow CO_2 + \frac{1}{\delta} MeO_{w-\delta} \tag{4.3.5}$$

$$H_2 + \frac{1}{\delta} MeO_w \rightarrow H_2O + \frac{1}{\delta} MeO_{w-\delta}. \tag{4.3.6}$$

Metal oxides in the "combustion" region, including NiO, Fe_2O_3, and CuO, are more oxidative. Thermodynamically, these metal oxides can spontaneously convert syngas into CO_2 and H_2O via reactions (4.3.5) and (4.3.6). As a result, at thermodynamic equilibrium, the main gaseous products are CO_2 and H_2O, and low syngas yield is achieved. However, CLG/CLR can still be conducted using metal oxides in the "combustion" region by reducing the ratio of the molar flow rate of the oxygen carrier to the carbonaceous feedstock below the minimum ratio required for effective CLC. Figure 4.11 shows the effect of the oxygen to fuel (CH_4) ratio on equilibrium gas product distribution using NiO as the oxygen carrier. As the oxygen to fuel ratio decreases from 4 to 1, selectivity towards CO increases while selectivity towards CO_2 decreases. Thus, a significant effect on the syngas yield can be achieved by adjusting the ratio of oxygen carrier to fuel in the reducer.

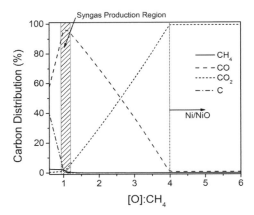

Figure 4.11 Effect of [O]:CH$_4$ ratio on product distribution using NiO/Ni.

The fact that metal oxides in the "combustion" region can be used for CLG/CLR applications indicates the widespread usefulness of CLC metal oxides also for gasification and reforming applications. Further, in order to be commercially viable, the raw materials used for producing oxygen carriers for CLC applications need to be low in cost. The considerations given below are also important:

(1)   From Figure 4.11, syngas selectivity is sensitive to the fuel to oxygen ratio. The [O]:CH$_4$ ratio, thus, needs to be well controlled to a value around 1, as given in the shaded area in the figure, in order to yield a steady production of syngas.

(2)   Changing the fuel to oxygen carrier ratio by increasing the fuel flow rate results in a decreased gas–solid contact time, leading to the incomplete conversion of fuel.

(3)   Since the lattice oxygen in the oxygen carrier is consumed by the fuel, metals can be formed as the reduced products of metal oxide in the reducer. Metals such as nickel and iron can catalyze the methane decomposition reaction, given in reaction (4.3.7):

$$CH_4 \rightarrow C + 2H_2. \tag{4.3.7}$$

Thus, carbon deposition on the metal oxide may occur, adversely affecting the syngas composition. Also, deposited carbon can be carried over to the combustor, where it is oxidized to CO$_2$ by air resulting in CO$_2$ emissions.

Using metal oxides in the "syngas production" region appears to be a more favorable approach for CLG/CLR applications. As shown in Figure 4.12, a metal oxide in this region, such as CeO$_2$, can convert CH$_4$ into a high purity syngas over a wide range of [O]:CH$_4$ ratios, given by the shaded region in the figure. However, compared to the metal oxides in the "combustion" region, these metal oxides are not as widely studied. Further, some of the metal oxides studied in the "syngas production" region contain expensive rare earth metals such as cerium, which are costly, negatively affecting the economic feasibility of the entire process. Other aspects of Figures 4.11 and 4.12 are discussed in Chapter 2.

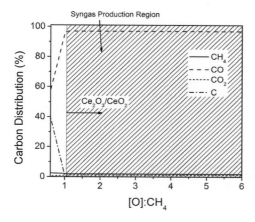

Figure 4.12 Effect of [O]:CH$_4$ ratio on product distribution using Ce$_2$O$_3$/CeO$_2$.

Since oxygen carrier selection plays a central role in determining the operating conditions and performance of CLG/CLR applications, experimental studies have been conducted to determine their behavior under reduction–oxidation cycles. Zafar et al. compared the behavior of NiO, CuO, Mn$_2$O$_3$, and Fe$_2$O$_3$ supported on SiO$_2$ synthesized by the impregnation method in a lab-scale fluidized bed reducer for CLR applications.[69] Their results show that Mn$_2$O$_3$/SiO$_2$, Fe$_2$O$_3$/SiO$_2$, and CuO/SiO$_2$ all cannot produce syngas with a concentration higher than 20%. Only NiO/SiO$_2$ shows the potential for syngas production with a high CH$_4$ conversion. They also observed that there are two distinct stages in methane conversion. In the first stage, CO$_2$ and H$_2$O are mainly produced, while in the second stage, CO and H$_2$ are produced with minimal CO$_2$ formation. Deactivation of the particles within the first ten cycles was found to be significant for NiO/SiO$_2$, Fe$_2$O$_3$/SiO$_2$, and Mn$_2$O$_3$/SiO$_2$, but less severe for CuO/SiO$_2$. For Fe$_2$O$_3$/SiO$_2$ and Mn$_2$O$_3$/SiO$_2$, the loss of activity was attributed to the formation of unreactive metal silicates. For NiO/SiO$_2$, the reduced activity was attributed to segregation and sintering of the Ni particles. A significant amount of carbon deposition was identified in the second stage of methane conversion and confirmed by a carbon balance based on the experimentally observed H$_2$:CO ratio being greater than 4, which is significantly higher than the ratio of 2 expected from the stoichiometry of the methane partial oxidation reaction.[69] Thus, this work suggested that NiO is a better candidate for an oxygen carrier that can be used for syngas production in fluidized bed reducers than other CLC oxygen carriers. Although the product gas composition was far from ideal and the reactions occurred in batch mode, this work indicated that syngas production was possible with both a high methane conversion and a high syngas selectivity.

4.4   Chemical Looping Using Fluidized Bed Reducers

A variety of CLG and CLR processes have been studied, where different oxygen carrier materials and different reactor configurations were used. CLG/CLR can be carried out

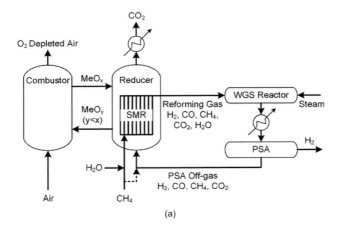

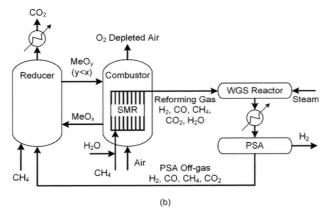

Figure 4.13 SMR–CLC process for H_2 production: (a) type a SMR embedded in reducer and (b) type b SMR embedded in combustor.[70,71]

in a circulating fluidized bed system, or more simply, in a fixed bed system. The SMR process discussed in Section 4.2.2 can be coupled with CLC for reforming process applications, whereby heat can be provided to the methane reforming reaction by embedding the tubular, packed bed SMR reactors in either the reducer or the combustor of the chemical looping system, as shown in Figure 4.13. Processes with this configuration are referred to as SMR–CLC processes. The SMR–CLC processes are discussed below along with other CLG/CLR for syngas production using fluidized bed reducers.

4.4.1   Steam Methane Reforming–Chemical Looping Combustion (SMR–CLC) Systems

As discussed in Section 4.2.2, the widely used SMR reaction for syngas generation from natural gas is highly endothermic. Therefore, SMR reactors are usually indirectly heated by being enclosed in a furnace where combustion with natural gas or other carbonaceous fuels occurs, resulting in CO_2 emissions. When hydrogen is the desired product, a WGS reactor is located after the SMR reactor to maximize the hydrogen yield. Subsequent hydrogen purification occurs through a pressure swing adsorption (PSA) system. The tail gas from the PSA system consists of CO_2 produced in the SMR and WGS reactors, unconverted CO and CH_4, and a small amount of H_2. This tail gas is used to fuel the furnace in a conventional SMR system. Flue gas from the furnace, if vented to the atmosphere without post-combustion CO_2 capture, results in CO_2 emissions. The use of post-combustion CO_2 technologies such as amine absorption units adds to both the capital and operating costs of hydrogen production.

The SMR–CLC process, which is a traditional SMR reactor embedded in one of the CLC reactors, was proposed for syngas generation from natural gas to simplify the carbon capture process. A general diagram of the SMR–CLC process is shown in Figure 4.13. Based on the CLC reactor in which the SMR reactor is embedded, there are two types of SMR–CLC systems: (1) Type a SMR reactor in the reducer, shown in Figure 4.13a; and (2) Type b SMR reactor in the combustor, shown in Figure 4.13b. Since the CLC process is typically operated at a temperature greater than the operating temperature of SMR, it can be used to provide heat for SMR reactions.

In the SMR–CLC process, the tail gas from the PSA unit is used as (part of) the fuel in the CLC system where it undergoes oxidation in the reducer reactor. The exhaust from the reducer is a mixture of CO_2 and H_2O, from which a pure CO_2 stream can be readily obtained for sequestration through condensation of H_2O. Hence, along with providing heat integration to the SMR process, the CLC–SMR system has the potential to provide an incremental cost benefit for CO_2 capture from an SMR.[70,71]

SMR–CLC Reactor Design

Experimental studies involving the SMR–CLC process are minimal, with most evaluations obtained from process and flow based simulations.[71] To date, there have been very few CLC processes demonstrated at the pilot scale. Because of the large surface area to volume ratio of the reactor, achieving autothermal operation in small-scale CLC reactor systems is difficult since the heat loss is significantly higher for small-scale systems as compared to large-scale systems. The small-scale test facilities thus require external heaters or burners to compensate for the heat loss. However, applying CLC merely for the purpose of external heating for the SMR process does not adequately manifest a novel reforming technology that is chemical looping based, as this reforming technology is fundamentally rooted on the already existing SMR.

Whether the SMR reactor is embedded in the reducer (Figure 4.13a) versus the combustor (Figure 4.13b) has a significant effect on the energy balance and performance of the SMR–CLC process. Because the reaction occurring in the reducer, reaction (4.3.1), can be either endothermic or slightly exothermic depending on the fuel and oxygen carrier used and the reaction in the combustor, reaction (4.3.2), is exothermic, the temperature in the combustor is typically higher than that in the reducer for CLC.[65,72]

In the type a SMR–CLC process, the CLC reducer provides the heat for both the endothermic SMR reaction and the reduction reaction of the metal oxides. This results in a reducer temperature in type a SMR–CLC processes that is lower than CLC alone under the same CLC operating conditions. The heat duty in the reducer can be balanced by either cycling the oxygen carrier at a higher temperature or increasing the oxygen carrier circulation rate. If the oxygen carrier is circulated at a higher temperature, the combustor must operate at a higher temperature than the reducer, which could be detrimental to the physical properties of the oxygen carriers, as higher temperatures in the combustor enhance sintering, and hence, deactivation of the oxygen carrier. If the oxygen carrier circulation rate is increased, the temperature differential between the combustor and the reducer is low. However, increasing the solid circulation rate increases the oxygen carrier attrition rate.

In the type b SMR–CLC process, the thermal energy released from the oxidation of the reduced oxygen carrier reaction provides the heat for the SMR reaction in the combustor. Thus, the temperature difference between the reducer and combustor is smaller for type b SMR–CLC as compared to type a SMR–CLC. A lower combustor temperature is advantageous since the rate of oxygen carrier sintering and deactivation is reduced.

The detailed energy balance of the SMR–CLC process was studied by Pans et al.[73] Two iron based oxygen carriers were used: (1) pure iron oxide, Fe_2O_3; and (2) an alumina supported iron oxide, Fe_2O_3–Al_2O_3. The effect of swinging between different oxidation states was analyzed for the H_2 yield from the SMR–CLC process, where pure Fe_2O_3 was reduced to Fe_3O_4 while the alumina supported Fe_2O_3 was reduced to FeO–Al_2O_3 ($FeAl_2O_4$). The cycling of Fe_2O_3 between these two different oxidation states affects the energy balance of the system since reaction (4.3.1) is endothermic for the Fe_2O_3/Fe_3O_4 but exothermic for the $Fe_2O_3/FeAl_2O_4$ system. Both type a and type b SMR–CLC processes were studied and compared under an autothermal operating condition. At 900 °C (1652 °F) in the reactor where the SMR reactor was embedded, the type b SMR–CLC system resulted in a higher hydrogen yield than type a for both Fe_2O_3/Fe_3O_4 and $Fe_2O_3/FeAl_2O_4$. However, the yield difference decreases as the solid circulation rate increases, as shown in Figure 4.14. For the Fe_2O_3/Fe_3O_4 redox pair, the temperature in the combustor is higher than the reducer for both type a and type b SMR–CLC. For the $Fe_2O_3/FeAl_2O_4$ redox pair, the combustor temperature is higher than the reducer temperature only for type a SMR–CLC, and the reducer temperature is higher than the combustor for type b SMR–CLC. This is beneficial for the reactions occurring in the reducer because it can enhance the kinetics of the reaction between the fuel and the oxygen carrier.

4.4.2   CLG/CLR With a Fluidized Bed Reducer

Unlike the SMR–CLC process which integrates the SMR process with the CLC process, the CLG and CLR processes produce syngas directly through partial oxidation of carbonaceous fuel through reduction of the oxygen carrier. The primary chemical looping reactions in the CLG/CLR process are

$$\text{Reducer}: \quad C_xH_yO_z + \frac{x-z}{\delta} MeO_w \rightarrow xCO + \frac{y}{2}H_2 + \frac{x-z}{\delta} MeO_{w-\delta} \qquad (4.4.1)$$

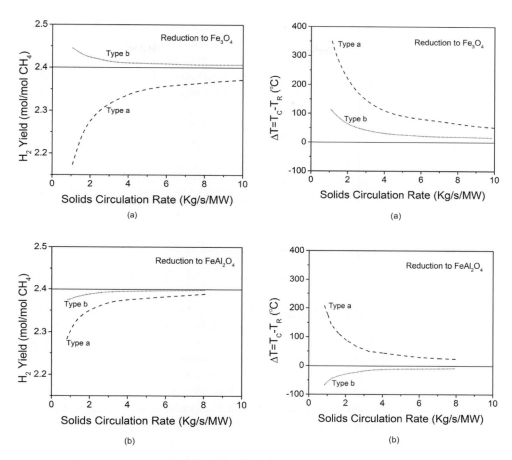

Figure 4.14 Performance of different SMR–CLC systems (type a and type b) with (a) pure iron oxide, Fe_2O_3 and (b) an alumina supported iron oxide, Fe_2O_3–Al_2O_3 oxygen carriers, where T_R and T_C are the temperatures for the reducer and the combustor respectively.

$$\text{Combustor :} \qquad MeO_{w-\delta} + \frac{\delta}{2}O_2 \rightarrow MeO_w \qquad (4.4.2)$$

$$\text{Overall reaction :} \qquad C_xH_yO_z + \frac{x-z}{2}O_2 \rightarrow xCO + \frac{y}{2}H_2 \,. \qquad (4.4.3)$$

For carbonaceous fuels with a high heating value, such as natural gas and coal, the overall reaction represented by reaction (4.4.3) is highly exothermic, which enables autothermal operation of the CLG/CLR processes. However, in the case of fuels with low heating values, such as biomass, maintaining autothermal operation may be challenging and is resolved through an efficient heat integration approach of the entire chemical looping process. Process simulations for both CLG and CLR that incorporate detailed heat integration schemes are discussed in Chapter 6.

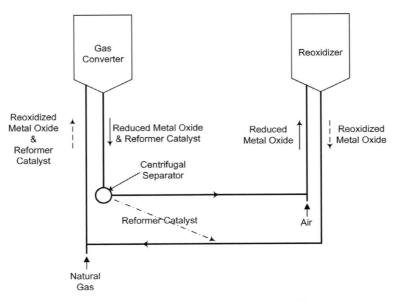

Figure 4.15 Welty process for methane to syngas conversion.[74]

The chemical looping approach to producing syngas in fluidized bed reactors from methane was first described in the Welty process by Welty et al. in a 1951 patent, even though the term "chemical looping" had not yet been adopted.[74] The Welty process consisted of two separate fluidized bed reactors, as shown in Figure 4.15.

In this process, oxygen carriers such as Fe_2O_3 or CuO partially oxidized methane in the gas converter reactor (reducer) to produce H_2 and CO. In the reoxidizer (combustor), the reduced oxygen carrier was re-oxidized with air. In addition to the oxygen carrier, a reforming catalyst such as nickel supported on either Al_2O_3 or MgO was also used to convert CH_4 into H_2 and CO through steam/dry methane reforming. The reforming catalyst was added to overcome the problem of methane over-oxidation to CO_2 and H_2O in the reducer. The initial design involved regenerating the entire oxygen carrier in the combustor. However, the regeneration step in the initial design oxidized the reduced oxygen carrier as well as the reforming catalyst from Ni to NiO, causing it to lose its catalytic activity. This drawback in the initial system design was overcome by the use of a solid–solid centrifugal separation method based on the density difference between the oxygen carrier and the nickel based catalyst. The solid–solid centrifugal separation ensured that the less dense nickel based catalyst remained at the top of the separator while the oxygen carrier was transported to the bottom of the separator. This solid–solid separator allowed for the oxygen carrier to be selectively transported to the combustor for regeneration, while keeping the nickel based catalyst in the reducer reactor. The system suffered from non-optimal heat integration as there were difficulties in maintaining an operating temperature difference between the reducer at 816–927 °C (1500–1700 °F) and the combustor at 871–982 °C (1600–1800 °F). In addition to heat integration difficulties, the Welty process was cumbersome to operate, and preserving the reactivity of the oxygen carrier and reforming catalyst presented technological complexities in order to sustain a high product yield. Between the work of Welty et al. and the early 2000s, little progress was made in further developing the CLG/CLR process until a similar concept was applied by Mattisson et al. to methane reforming, using fluidized beds.[75]

The typical designs for CLG/CLR processes have been based on a CLC process design using fluidized bed systems. Bubbling/turbulent fluidized beds or fast fluidized beds are used for the reducer or combustor. The overall looping system is configured with a circulating fluidized bed (CFB), where metal oxide oxygen carriers undergo reduction and oxidation reactions and continuously circulate between the reducer and combustor. These fluidized bed reactors for CLG/CLR processes have been developed with a fundamental understanding of the hydrodynamics of oxygen carrier transport in the reactor system. By properly controlling the fuel to oxygen feed ratio, the gaseous product composition can be shifted from a CO_2-rich gas stream in CLC to the CO-rich syngas desired for CLG/CLR. Experimental results from fluidized bed reducers conducted by two major research groups on reforming studies are given below.

300 W_{th} CLR Unit at Chalmers University of Technology

Research at Chalmers University of Technology (Chalmers) in Sweden has investigated CLR in a 300 W_{th} two-compartment fluidized bed reactor system for CLR using nickel based oxygen carriers.[76–79] The configuration of this reactor system is shown in Figure 4.16. The unit was initially designed for carrying out CLC experiments and was later adopted for CLR based systems.[79]

The design of the CLR system is derived from the reactor system for oil shale reforming proposed by Chong et al. in 1986, which featured solids exchange between two adjacent fluidized beds without gas mixing.[80] The 300 W_{th} unit consists of two adjacent chambers divided by a vertical wall with two slots, as shown in Figure 4.16.[81]

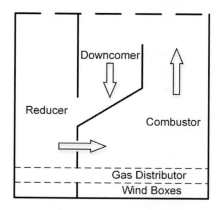

Figure 4.16 Chalmers 300 W_{th} fluidized bed chemical looping reforming set-up.[81] (Arrows in the figure indicate the direction of solids flow.)

The two chambers are operated under conditions to maintain a fluidized bed, but with different gas velocities to operate in different fluidization regimes. The combustor chamber is operated at a higher gas velocity than the reducer. In the combustor chamber, a fraction of the oxygen carrier particles is entrained by the gas flow and falls into a downcomer, which leads to the middle section of the reducer chamber. Operating the reducer at a lower gas velocity than the combustor results in a denser fluidized bed, corresponding to larger pressure drops. When the particles fall from the combustor into the reducer through the downcomer, the higher pressure at the bottom of the reducer drives the particles into the combustor to maintain continuous solids circulation. The particles are transferred from the reducer to the combustor through a slot at the bottom of the vertical wall between the two reactors, allowing for continuous solids circulation.

The 300 W_{th} unit was placed inside an electrically heated furnace to maintain the reactor temperature at a constant value. A water seal on the exit pipe of the reducer adjusted the pressure of the reducer by regulating the height of the water column. This was conducted to avoid dilution of the reducer gas by air leakage from the combustor. The reducer section of the two-compartment system had a square cross-sectional area of 25 mm × 25 mm (1 in × 1 in). The cross-section of the combustor was rectangular at the bottom with dimensions 25 mm × 40 mm (1 in × 1.6 in). The total height of the fluidized bed was 200 mm (7.9 in), with an enlarged section on top of each for gas–solid separation. The downcomer between the reducer and the combustor was 12 mm (0.5 in) wide. The slot connecting the bottom of the two reactors consisted of two walls, one in each reactor. The two walls minimized the gas leakage between the two reactors. The system used natural gas, with a composition equivalent to $C_{1.14}H_{4.25}O_{0.01}N_{0.005}$, as the fuel for the CLR experiments. Three kinds of Ni-based oxygen carrier particles, whose properties are given in Table 4.2, were tested for CLR operation.

For each type of oxygen carrier particle, experiments were performed with a reducer temperature of between 800 °C and 950 °C (1472–1742 °F), and varying air to fuel ratios. For each experiment, a steady state was maintained for 1–3 h. For CLC, the system required a low or modest fuel flow rate (0.2–0.75 L/min) and a high air flow rate (7–10 L/min), while a high fuel flow rate (0.8–1.5 L/min) and a moderate to high air flow rate (3.8–10 L/min) were used for CLR operation. Besides pure natural gas experiments, CLR experiments with feeds containing 30% CO_2 and/or 30% steam in the fuel gas were also tested.

Table 4.2 Ni-based oxygen carrier properties used in 300 W_{th} Chalmers's CLR unit.[79]

Oxygen carrier	Chemical composition	Production method	Size (μm)	Porosity (%)	Solids inventory (g)
N2AM1400	10% NiO on MgAl$_2$O$_4$	Freeze granulation	90–212	35	250
Ni18-αAl	18% NiO on α-Al$_2$O$_3$	Impregnation	90–212	53	180–250
Ni21-γAl	21% NiO on γ-Al$_2$O$_3$	Impregnation	90–250	66	170

Syngas generation by CLR was achieved by adjusting the ratio between the oxygen and fuel, which was indicated by the air factor (Ψ), as defined in equation (4.4.4):

$$\Psi = \frac{(O/C)_r - (O/C)_f}{(O/C)_c - (O/C)_f} \tag{4.4.4}$$

$$(O/C)_r = \frac{2y_{CO2,r} + y_{H2O,r} + y_{co,r}}{y_{CO2,r} + y_{CH_4,r} + y_{co,r}} \pm \tag{4.4.5}$$

$$(O/C)_f = \frac{2y_{CO2,f} + y_{H2O,f} + 0.01(1 - y_{H2O,f} - y_{CO2,f})}{y_{CO2,f} + 1.14(1 - y_{H2O,f} - y_{CO2,f})} \tag{4.4.6}$$

$$(O/C)_c = \frac{2y_{CO2,f} + y_{H2O,f} + 4.41(1 - y_{H2O,f} - y_{CO2,f})}{y_{CO2,f} + 1.14(1 - y_{H2O,f} - y_{CO2,f})}, \tag{4.4.7}$$

where y_A is the concentration of gas A and O/C is the oxygen to carbon ratio. The concentrations with subscripts "r", "f", and "c" refer to the gas composition of reducer gas product, fuel gas, and complete combustion, respectively. The constants "0.01", "1.14", and "4.41" correspond to the amount of oxygen in the natural gas, the amount of carbon in the natural gas, and the amount of oxygen needed for full combustion of the natural gas, respectively. When $\Psi = 1$, the ratio between fuel and air supply is equal to the stoichiometric amount required for full combustion, while when $\Psi < 1$, the oxygen supply is not sufficient for full combustion and syngas production would occur.

Typical results for the CLR experiments are given in Figures 4.17 through 4.20. The first consideration was to identify operating conditions that prevented carbon deposition. Carbon deposition not only reduces the syngas yield from methane and produces CO_2 emission in the combustor, but also weakens the mechanical strength of oxygen carriers and deactivates the particle reduction effects. In process applications, it will increase the variable operating cost by increasing the fresh oxygen carrier make-up rate. When using natural gas as the fuel feed without co-injection of CO_2 or H_2O, i.e. operating at a low Ψ value, a significant amount of CO_2 was detected in the combustor,

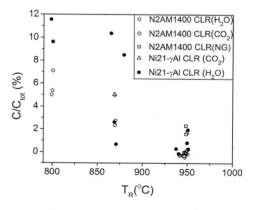

Figure 4.17 Percentage of fuel resulting in carbon deposition.[79]

indicating a considerable amount of carbon deposition occurred in the reducer. As shown in Figure 4.17, carbon deposition can be eliminated when the temperature is increased beyond 930 °C (1706 °F).

In the reducer, carbon deposition was expected since there was a significant amount of metallic nickel as a result of a low Ψ value, and metallic nickel is a catalyst for the methane decomposition reaction. Cho et al. reported that solid carbon formation depends strongly on the oxygen available from the nickel based oxygen carrier.[82] They observed that carbon deposition occurs rapidly when 80% of NiO is reduced to Ni. However in the CLR tests, even though there were instances where only 33–44% of the NiO was reduced to Ni, carbon deposition was still significant. Figures 4.18, 4.19, and 4.20 show the syngas production performance for a dry natural gas feed. The operating temperature for these points was greater than 930 °C (1706 °F). It was shown that the syngas quality is improved with reducing Ψ. However, the Ψ value can only be reduced

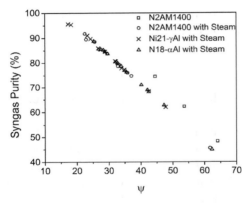

Figure 4.18 Syngas purity as a function of air factor, with and without steam injection for different oxygen carriers.[79]

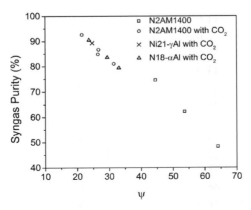

Figure 4.19 Syngas purity as a function of air factor, with and without CO_2 injection.[79]

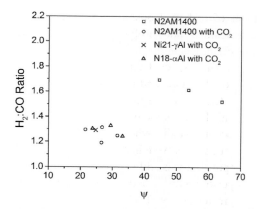

Figure 4.20 H_2:CO ratio as a function of air factor, with and without CO_2 injection.[79]

to a limited extent as carbon deposition increases beyond a certain value. The maximum syngas purity, defined by equation (4.4.8), was ~75%, with an H_2:CO ratio of 1.7:

$$\text{Syngas purity} = \frac{y_{CO} + y_{H_2}}{y_{CO} + y_{H_2} + y_{CO_2} + y_{H_2O} + y_{CH_4}} \times 100\%. \qquad (4.4.8)$$

This performance can be overcome by co-feeding steam or CO_2 to reduce or eliminate carbon deposition.

Also, when the system was operated at 950 °C (1742 °F) with $\Psi = 0.3$–0.35 and 30% steam or CO_2, a high syngas purity could be obtained from CLR. The addition of steam to the natural gas feed had two effects. First, a higher H_2 yield was expected due to the SMR reaction since Ni is a well-known SMR catalyst. Second, carbon deposition was reduced because of the carbon gasification reaction with steam. Figure 4.18 shows the comparison of system performance at above 930 °C, with and without steam injection for different oxygen carriers. The addition of steam has shown promising results for producing a high syngas purity as the reducer operates near the performance of a steam methane reformer.

Figures 4.19 and 4.20 show the performance difference of the reducer upon addition of 30% CO_2. As with steam injection, the addition of CO_2 also tends to decrease carbon deposition. Carbon deposition is reduced as the solid carbon is gasified by the CO_2 via reaction (4.2.4). The maximum syngas concentration achieved was ~90%, as shown in Figure 4.19, but the maximum H_2:CO ratio of 1.3, as shown in Figure 4.20, is significantly lower than the desired ratio of ~2 for cobalt based Fischer–Tropsch synthesis or methanol synthesis.

Overall, the oxygen carriers tested in Table 4.2 had good fluidization and flow characteristics with no significant agglomeration or defluidization reported. Of the three oxygen carriers tested, the N2AM1400 particle had the highest carbon formation resistance, as it showed minimal CO_2 in the combustor as compared to Ni18-αAl or Ni18-γAl under the same reaction conditions. For the N2AM1400 particle, the particle was almost identical in surface and bulk composition, density, porosity, and size distribution post-reaction as compared to fresh particles. The Ni18-αAl particle showed an increase in density (~9%) and a slight decrease in pore size.[79] The Ni21-γAl particle showed drastic changes after the redox reaction, where the apparent density increased by 40%.

140 kW$_{th}$ DCFB CLR Unit at Vienna University of Technology

Research at Vienna University of Technology (VUT) in Austria has studied CLR in a 140 kW$_{th}$ dual circulating fluidized bed (DCFB) reactor using nickel based oxygen carriers. The configuration of the DCFB is shown in Figure 4.21.[83,84]

The VUT DCFB reducer is a turbulent fluidized bed, and the combustor is a fast fluidized bed. The overall reactor system contains two particle circulation loops, three loop seals (upper loop seal, lower loop seal, and internal loop seal), and two cyclones, as shown in Figure 4.21. The internal loop seal, along with a cyclone, is situated above the reducer to enable local particle circulation within the reducer. This design enables the particles in the reducer to have a longer residence time, allowing for the reactions in the reducer to achieve equilibrium. The cyclone and the internal loop seal also enable the particles in the reducer to have a local particle circulation rate, which can be independent of the global solids circulation between the reducer and the combustor. The reducer and the combustor bottoms are interconnected by a lower loop seal that maintains a stable solids distribution between the two reactors. Steam was used as the fluidization gas in the loop seals as it was compatible for use in both the combustor and the reducer.

A pressure profile through the DCFB system was obtained for several operating conditions to probe the solids distribution in the system. A sample pressure profile of the VUT DCFB unit is shown in Figure 4.22, corresponding to the operation conditions given in Table 4.3.

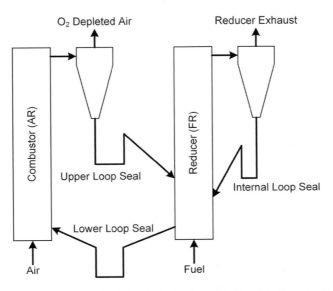

Figure 4.21 VUT 140 kW$_{th}$ dual circulating fluidized bed reactor for CLR.[84]

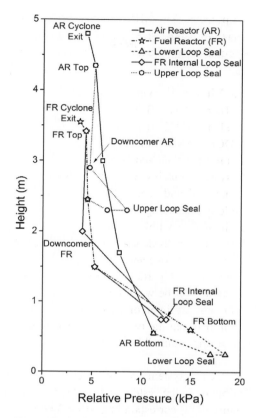

Figure 4.22 Pressure profile in the VUT DCFB system.[84]

Table 4.3 Experimental operating parameters for VUT's 140 kW$_{th}$ DCFB.[84]

Parameter	Value
Total solids inventory	65 kg
Fuel load	140 kW$_{th}$
Global air/fuel ratio	1.1
Reducer temperature	1173 K

The pressure profile in the combustor varied smoothly along the height of the reactor, which is typical of the fast fluidization regime. The pressure drop in the reducer was sharper in the lower section of the reactor, which reflects a dense-phase region at the bottom of the reducer. The overall pressure drop in the reducer was larger than the combustor because more solids were present in the reducer compared to the combustor. For each loop seal, a higher pressure at the gas inlet than the two outlets indicated that there was good gas sealing between the connected reactors.

The 140 kW$_{th}$ capacity DCFB reactor system for CLR applications used natural gas to produce syngas.[85,86] The oxygen carrier particles used in the experiments consisted of two different types of oxygen carrier materials that were mixed in a 50:50 weight ratio. The first oxygen carrier material was synthesized from NiO and Al$_2$O$_3$, which forms a mixture of NiO/NiAl$_2$O$_4$ (N-VITO) after sintering. The other oxygen carrier material was synthesized from NiO, Al$_2$O$_3$, and MgO, which forms a mixture of NiO/MgAl$_2$O$_4$–NiAl$_2$O$_4$ (N-VITOMg) after sintering. Forty percent of the weight of the oxygen carrier particle was active NiO. A description of the production method for oxygen carrier particles is provided in Linderholm et al. and Jerndal et al.[87,88] The mean particle size was 135 μm. The total solids inventory was 65 kg (143 lb) with 30 kg (66 lb) in the DCFB reactors and the balance being distributed in the cyclones and loop seals.[86]

Table 4.4 summarizes the experimental operating conditions. Three different operation temperatures were tested. Natural gas from the Viennese grid (98.7% CH$_4$) was introduced into the reducer and maintained at 140 kW$_{th}$ capacity during stable operation. The global stoichiometric air:fuel ratio, which is defined in equation (4.4.9), was decreased stepwise from 1.1, with an interval of 0.1, while the cooling duty of the reactor was adjusted accordingly to maintain the constant temperature:

$$\lambda = \frac{\dot{n}_{feed, AR} \times y_{O_2, feed, AR}}{2 \times \dot{n}_{feed, FR} \times y_{CH_4, feed, FR}}. \tag{4.4.9}$$

Figure 4.23 shows the syngas purity as a function of different air to fuel ratios for different design temperatures.[86] The syngas purity increases as the air to fuel ratio decreases for the same temperature. For the same air to fuel ratio, increasing the operating temperature of the reducer decreases the syngas purity. The reactions in the reducer are endothermic while those in the combustor are exothermic. The heat from the exothermic reaction in the combustor is transferred to the reducer using the oxygen carrier particles. The combustor operating temperature is designed to be higher than the reducer, such that the temperature difference is sufficient to obtain autothermal operation. The temperature difference between the two reactors is a function of multiple parameters, including the heat capacity of the oxygen carriers, gas product composition, and global air to fuel flow rate ratio.

Table 4.4 Operating conditions for 140 kW$_{th}$ VUT DCFB experiments.

Operating parameter	Parameter value
Air to fuel ratio (mol/mol)	0.5 to 1.1
Gas concentration (vol%) measured for the following species in the reducer reactor gas outlet	H$_2$O, H$_2$, CO, CO$_2$, CH$_4$
Gas concentration (vol%) measured for the following species in the combustor reactor	H$_2$O, O$_2$, Ar, N$_2$
Operating temperatures for chemical looping reactors	750 °C, 800 °C, 900 °C

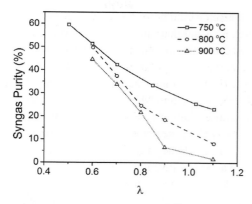

Figure 4.23 Syngas purity as a function of λ.[86]

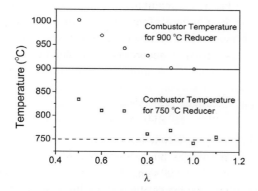

Figure 4.24 Temperature difference between combustor and reducer as a function of λ.[86]

Figure 4.24 shows the relation between the temperature of the combustor and the reducer as a function of different air to fuel ratios.[86] From Figure 4.24, a higher air to fuel ratio translates to a lower heat duty requirement in the reducer. When the air to fuel ratio is above 1.1, which operates in the CLC mode, the combustion heat generated is sufficient to provide the heat for the reducer. On the other hand, decreasing the air to fuel ratio results in a larger temperature difference requirement between the two reactors in order to achieve heat balance. The minimum air to fuel ratio was around 0.5.

Gas leakage between the reducer and the oxidizer was observed in the DCFB system during CLC experiments. This was a concern as the gas leakage could lead to a hazardous condition, with the reducing gas mixed with the oxidizing gases at high temperatures. It was determined that a leakage rate of no more than 0.5% volume of the gas from the combustor to the reducer was deemed acceptable in light of safety and process performance. However, leakage of loop seal gas, steam, was inevitable. As much as 5% volume of steam was detected in the combustor exhaust gas in CLR experiments. The leakage of loop seal steam could have a slight effect on decreasing the syngas concentration in the reducer product gas, but it was considered acceptable, given that the majority of the loop seal steam would have entered the combustor and exited in the combustor exhaust gas. It was reported that throughout the experiments, no CO_2 or CO was detected in the exhaust from the combustor, indicative of minimal carbon deposition in the reducer. These studies are among a few cases of CLR processes using fluidized bed reactors where no obvious carbon deposition was observed in the reducer without co-injection of steam or CO_2 with natural gas. However, this was achieved with low syngas purity ($<60\%$). Overall, an air to fuel ratio sufficient to prevent carbon deposition and a higher temperature of operation have the potential to yield optimal results for the DCFB system.

CLG Processes

CLG technologies can convert solid fuels, such as coal and biomass, to syngas using a fluidized bed reducer. However, CLG processes in the literature are focused mainly on biomass as a feedstock, rather than coal. Steam and/or CO_2 are usually employed as the gasification agent as well as serving as fluidization medium. A range of metal oxide based materials, including Fe, Ni, and Cu oxides are used as oxygen carriers. Studies at Guangzhou Institute of Energy Conversion (GIEC), Chinese Academy of Sciences (CAS) have examined CLG using biomass as the feedstock and the Fe-based materials as oxygen carrier in a 10 kW_{th} CFB unit.[89–91] A similar unit of 25 kW_{th} capacity using Ni and iron ore based oxygen carriers has been studied at Southeast University (SEU), China.[92,93] A typical syngas composition generated from these biomass CLG units is of 40% CO, 20% H_2, 20% CO_2, 10% CH_4, and 10% other gases. The carbon conversion efficiency, defined by the total amount of carbon detected at the gas outlets divided by the total amount of carbon fed into the system, is typically in the range 70–90%, indicating the potential accumulation of unconverted tar and char in the system. As a result of insufficient residence time for char gasification, carbon emission from the combustor increases with an increase in the feeding rate. Although elevated temperature can promote char gasification, the full accelerated oxidation of syngas can also take place, resulting in an increased CO_2 concentration and a decreased syngas yield in the gas product stream. The syngas yield also depends on the molar ratio of oxygen carrier to biomass fed into the reducer. An optimum syngas yield can be achieved at a moderate oxygen carrier to biomass molar ratio. The operation conditions, including temperature and oxygen carrier to biomass molar ratio, need to be carefully selected to reach a desired syngas property in a reducer operation.

Summary

To summarize Section 4.4.2, experimental studies on CLR with fluidized bed reducers indicate that fluidized bed reducers are capable of producing syngas from partial oxidation of methane. The syngas purity from these reducers can be adjusted from 0% (corresponding to full oxidation) to ~70% by adjusting the air factor Ψ from above 1 (for full oxidation) to ~0.4 (for syngas generation) without the addition of steam. Further, lowering the air to fuel ratio will result in significant carbon deposition catalyzed by reduced oxygen carrier, which lowers the syngas purity and adversely affects the chemical and physical properties of the oxygen carrier particles. To avoid carbon deposition, conversion of the oxygen carrier must be limited to a low level, leading to a high solid circulation rate for a given feedstock processing capacity. Carbon deposition can be alleviated by steam and/or CO_2 addition. With the steam and/or CO_2 addition, the air factor can be lowered and a higher syngas purity can be achieved. The CLG operation using fluidized bed reducers can produce a ~60% purity of syngas and a H_2:CO ratio of less than 0.5 from biomass. Feedstock conversion may not be complete due to the kinetic limitation concerning char gasification as well as the inherent channeling behavior of gaseous hydrocarbons produced in a fluidized bed reducer.

4.5 Syngas Generation with Moving Bed Reducer

As discussed in Sections 4.3 and 4.4, metal oxide oxygen carriers commonly used in CLC, such as iron oxide and copper oxide, pose challenges when used in a fluidized bed CLG/CLR reducer, because the oxygen carriers are thermodynamically capable of converting syngas into CO_2 and H_2O. Although syngas generation is possible with a high fuel to oxygen carrier ratio, CLG/CLR systems operating under that condition will suffer from low fuel conversion, low syngas purity, and carbon deposition. Among the metal oxides used for CLC processes, NiO has been more commonly used as the oxygen carrier for CLG/CLR in a fluidized bed reducer due to the high syngas yield resulting from its excellent reactivity. However, the toxicity and price of NiO may limit the feasibility of its use in large-scale CLG/CLR processes.

Alternatively, when a moving bed is used as the reducer reactor in CLG/CLR processes, use of an iron oxide based oxygen carrier can yield a high performance in both fuel conversion and syngas purity.[65,94,95] This makes use of a moving bed reducer reactor an attractive configuration option for CLG/CLR processes from both an economical as well as operational standpoint. This section will discuss the motivation for using a moving bed reactor as the reducer for syngas production using iron oxide based oxygen carriers, the criteria for designing a moving bed reducer reactor, and results from experimental studies at different scales of operation.

4.5.1 Iron Oxide Based Oxygen Carriers

Iron oxide based oxygen carriers have been widely studied for CLC applications. These oxygen carriers are particularly attractive for several reasons:

(1)   Iron oxide is abundantly available and low in cost.
(2)   Certain types of iron ores, like ilmenite, can be used directly as oxygen carriers in chemical looping processes.[72]
(3)   Iron oxide is non-toxic.

Among the three possible redox pairs for use as oxygen carriers, Fe_2O_3/Fe_3O_4, Fe_3O_4/FeO, and FeO/Fe, the Fe_2O_3/Fe_3O_4 redox pair is the most oxidative as, thermodynamically, Fe_2O_3/Fe_3O_4 can fully convert the hydrocarbon fuel fed to the reducer into CO_2 and H_2O. Also, the amount of unconverted CO/H_2 is significantly lower as compared to using NiO/Ni as the oxygen carrier redox pair. The Fe_3O_4/FeO and FeO/Fe redox pairs are comparatively less oxidative. As a result, CLC processes using iron oxide based oxygen carriers in a fluidized bed reducer use only the Fe_2O_3/Fe_3O_4 pair with no further reduction of the oxygen carrier. For CLG/CLR applications, however, Fe-based metal oxide needs to be reduced to the FeO/Fe redox pair, where it falls in the "syngas production" region in the modified Ellingham diagram in Figure 2.2. This indicates that the FeO/Fe redox pair inhibits thermodynamically the full oxidation of syngas into CO_2/H_2O. Furthermore, the addition of support materials such as TiO_2 and Al_2O_3 to an iron oxide based oxygen carrier may result in the formation of complex materials ($FeTiO_3$ and $FeAl_2O_4$) that are even less oxidative than FeO, resulting in an even higher syngas purity when compared to FeO alone. Therefore, if the iron oxide based oxygen carrier is sufficiently reduced to form a mixture of FeO and Fe in the CLG/CLR reducer reactor, a high syngas purity and low CO_2/H_2O concentration can be obtained.

A 1.5 mm pellet consisting of 60% Fe_2O_3/40% Al_2O_3 by weight was reduced in a TGA at 900 °C (1652 °F) using pure methane. Figure 4.25 shows the weight change of the pellet during methane reduction. Initially, the weight decreases rapidly due to the loss of oxygen from Fe_2O_3, but the weight eventually increases sharply, starting at approximately 89% of its original value. Based on the observed weight change, the reduction reaction can be divided into two stages: (1) reduction of Fe_2O_3 (weight loss); and (2) carbon deposition (weight gain). The weight loss observed in the first stage can be represented by reaction (4.5.1), and the weight increase in the second stage is caused by carbon deposition through reaction (4.5.2):

$$CH_4 + [O] \rightarrow CO + 2H_2, \quad (4.5.1)$$

where [O] represents the lattice oxygen in the iron oxide pellet. Carbon deposition is catalyzed by reduced metallic iron:

$$CH_4 \xrightarrow{Fe} C + 2H_2. \quad (4.5.2)$$

Avoiding carbon deposition is essential for optimal performance of the CLG/CLR process, which adds a constraint to the degree to which iron oxide based oxygen

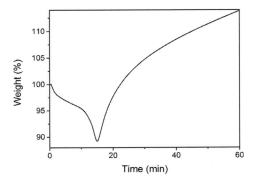

Figure 4.25 Weight change of 60% Fe_2O_3/40% Al_2O_3 pellet during methane reduction.

carriers can be reduced in the reducer reactor. Carbon deposition, as noted earlier, is undesirable.

The reduction pattern observed from TGA studies, as exemplified in Figure 4.25, can yield information for the selection of an appropriate reducer configuration. As the oxidation state of the oxygen carrier is related to its residence time in the reactor, it is closely connected to the extent of reactant reduction. From the discussion in Section 1.6.2, the residence time of the oxygen carrier particles in a fluidized bed reactor varies widely, but use of a cocurrent moving bed reactor as the reducer can readily control the solids residence time and hence reactant reduction to within a desired range. Based on TGA studies, it is noted that two types of oxygen carrier particles in the reducer are not desirable for the production of syngas: (1) under-reduced particles consisting of Fe_2O_3/Fe_3O_4 and Fe_3O_4/FeO; and (2) over-reduced particles consisting of Fe. The under-reduced particles will fully oxidize the reactant gas and syngas into CO_2 and H_2O, while the over-reduced particles will cause carbon deposition when they react with the reactant gas. Because the residence time of the oxygen carrier particles in a moving bed reducer reactor can be confined to within a narrow range, both over-reduction and under-reduction of oxygen carrier particles can be avoided. Therefore, the choice of oxygen carrier particle residence time is limited to a specific range that ensures only FeO/Fe particles exit the reducer. Additionally, the cocurrent configuration utilizing a moving bed reactor simulates a single stage contact and allows for syngas being in contact with reduced oxygen carrier particles consisting of FeO/Fe, which inhibits thermodynamically full oxidation of H_2:CO and therefore produces a high purity syngas.

4.5.2 CLR in Fixed Bed Reactors

The feasibility of syngas generation using a moving bed reducer reactor and iron oxide based oxygen carrier was first investigated in a fixed bed. In a cocurrent moving bed reducer reactor, oxygen carrier particles are fed from the top of the reactor and flow downwards with the reactant gas/product, exiting from the bottom of the reactor. This results in an oxidation state distribution of the particles, where particles at the top of the reactor have a higher oxidation state than those at the bottom. To mimic this oxidation state distribution of a cocurrent moving bed reducer reactor, a fixed bed reactor 15 in (38.1 cm) in length with an internal diameter of 0.5 in (1.27 cm) was packed with two layers of iron oxide based oxygen carriers with different oxidation states, as shown in Figure 4.26. The lower section of the reactor was filled with reduced oxygen carrier particles and the upper section was filled with oxidized oxygen carrier particles, at a ratio of 2.78:1. The specific weight distribution was chosen to impart a chosen residence time for the various oxidation states of iron, which was based on the kinetic weight change estimated from TGA experiments. CH_4 diluted by N_2 was injected into the top of the reactor using digital mass flow controllers. The fixed bed reactor was operated at 990 °C (1814 °F) using an electrical furnace. The outlet gas composition from the reactor was analyzed by a non-dispersive infrared (NDIR) gas analyzer as well as a gas chromatograph.

Gas Inlet (CH₄, N₂)

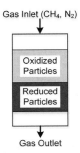

Figure 4.26 Schematic of a fixed bed set-up to mimic a cocurrent moving bed reactor.[96]

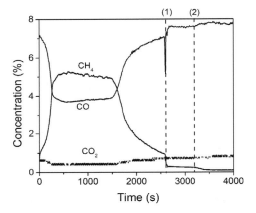

Figure 4.27 Product gas composition of a fixed bed reactor with CH_4 injection. (1) Reduced gas flow rate; (2) increased temperature.[96] (Reproduced from Luo S. et al., 2014, Energy Environ. Sci., 7, 4104–4117, with the permission of Royal Society of Chemistry.)

The methane conversion decreased from ~90% to less than 50% and remained low for ~20 min before the methane increased significantly again, reflecting fast, slow, and modest reaction kinetics stages in the Fe_2O_3 reduction process with CH_4, as shown in Figure 4.27. Once the conversion of methane had increased to about 90%, the flow rate of methane was decreased to increase its residence time. An immediate increase in methane conversion was observed and a steady production of syngas was maintained for 10 min. Then, the reactor temperature was increased to 1050 °C (1922 °F) to evaluate the effect of temperature on syngas production. A slight improvement in the methane conversion was achieved at the elevated temperature. During the period of high quality syngas generation, as quantified by an increase in the methane conversion, the $CO:CO_2$ mole ratio was ~10. The concentration of measured carbonaceous species in the gaseous product is also shown in Figure 4.27. Thermodynamic analysis using the modified Ellingham diagram indicates that methane should be fully converted at the experimental conditions used. However, incomplete methane conversion at 990 °C (1814 °F) and an improvement in methane conversion at a longer gas residence and/or at elevated temperature, imply the existence of a kinetic limitation to syngas production.

The kinetic limitation of syngas production was further explored in another fixed bed experiment in which the reduced iron oxide (FeO/Fe mixture) from the experiment described previously was subjected to oxidation by CO_2. 50% CO_2 (balance N_2) was introduced to the reactor for re-oxidation of the particles. About 1000 s after the start of the experiment, the input gas was switched to a pure CO_2 stream at the same flow rate, to examine the effect of kinetics on syngas yield. Reaction (4.5.3) and reaction (4.5.4) were the two primary reactions occurring during the experiment:

$$Fe + CO_2 \rightarrow FeO + CO \tag{4.5.3}$$

$$3FeO + CO_2 \rightarrow Fe_3O_4 + CO. \tag{4.5.4}$$

Figure 4.28 shows the concentration of carbonaceous species in the gaseous product, where a sharp change in product composition can be observed after a certain period of time. The shift from formation of CO_2 to CO illustrates the dependence of this reaction on oxygen carrier composition. As long as the oxygen carrier is a mixture of Fe and FeO, the CO concentration would be greater than CO_2, and a high purity syngas stream would be generated. However, when the oxygen carrier composition is a mixture of FeO and Fe_3O_4, the CO concentration immediately drops to a very low value. A change in CO_2 flow rate had little effect on the product stream concentrations, indicating that the product gas composition was controlled by thermodynamics rather than kinetics. As mentioned earlier, the gas composition was affected by kinetic factors like residence time and temperature in the fixed bed experiment with CH_4 injection. Therefore, the reaction between CH_4 and oxygen carrier can be considered to be the rate limiting step in syngas production. The experiments

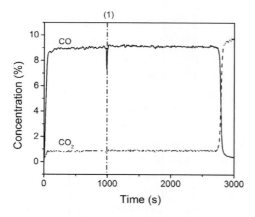

Figure 4.28 Product gas composition of a fixed bed reactor with CO_2 injection. (1) Increased CO_2 concentration.[96] (Reproduced from Luo S. et al., 2014, Energy Environ. Sci., 7, 4104–4117, with the permission of Royal Society of Chemistry.)

involving CH_4 reduction and CO_2 oxidation indicate that the reaction of methane with iron oxide based oxygen carriers is relatively slow and hence rate limiting. Also, it can be concluded that the product gas composition, the syngas purity, is determined by thermodynamics rather than kinetics.[96]

4.5.3   Reactor Design of CLG/CLR System With Moving Bed Reducer

As shown in Figure 4.29, the pressure distribution (left) and the corresponding CLG/CLR system (right) are given. The system comprises two reactors, i.e. the reducer and the combustor. The reactors are connected using non-mechanical gas sealing devices

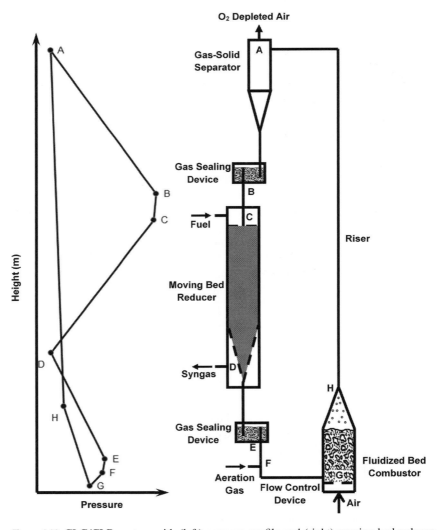

Figure 4.29 CLG/CLR system with (left) pressure profile and (right) moving bed reducer.

(see section on gas sealing device design) and flow control devices (see section on solids circulation control). The reducer is a moving bed reactor, in which the solid oxygen carrier particles and gaseous fuel both flow cocurrently downward. The combustor is a fluidized bed reactor and is connected to a riser that transports the oxygen carriers back to the reducer. During the operation, the fuel is introduced into the top of the reducer and converted to syngas at the bottom of the reducer.[95,96] Syngas is separated from the particles at the reducer outlet. The oxygen carrier particles that exit the reducer are then regenerated by air in the combustor and transported back to the top of the reducer via the riser. It is seen in the figure that, differently from the pressure distribution in the Vienna University of Technology unit given in Figure 4.22, the highest pressure in the system is located at point B, above the moving bed reducer, and the lowest pressures are located at points A and D, at the air exhaust and the syngas outlets.

Sizing the reactors and designing interconnecting gas sealing devices can be achieved through an understanding of the hydrodynamic properties of the gas and solid flow system. The flow system is affected by operating conditions, including temperature, fuel capacity, and residence times of gas and solid in each of the reactors, and these conditions need to be identified for the purpose of flow calculations based on system performance models. The product syngas composition is also part of the required information that determines the gas and solid flow rates in the reactors.

Reactor Sizing

The dimensions of the cocurrent moving bed reducer reactor are quantified based on the following criteria:

(1)   The volume of the reactor should be large enough to provide sufficient residence time for the reducer reactions for both solid and gaseous species, and constrained by equations (4.5.5) and (4.5.6):

$$V_r(1 - \epsilon) \geq F_s \tau_s \qquad (4.5.5)$$

$$V_r \epsilon \geq F_g \tau_g, \qquad (4.5.6)$$

where V_r is the volume of the reactor; ϵ is the voidage of the reactor; F_s is the volumetric flow rate of oxygen carrier particles; τ_s is the required residence time for the reducer reactions for oxygen carrier particles; F_g is the volumetric flow rate of gases; and τ_g is the required residence time for the reducer reactions for gases.

(2)   The cross-sectional area of the reactor should be large enough to avoid a high gas velocity in the reactor, with the criterion given in equation (4.5.7) to be satisfied:

$$\frac{F_g}{S_r} \leq 0.8 u_{mf}, \qquad (4.5.7)$$

where S_r is the cross-sectional area of the reducer; and u_{mf} is the minimum fluidization velocity of the oxygen carrier particles.

The sizing criteria for the combustor and riser are different from those of the reducer. As noted, the combustor reactor is a fluidized bed reactor where oxygen carrier particles are regenerated via the reaction with oxygen from air, and the purpose of the riser is to transport the re-oxidized oxygen carriers back to the reducer. Air introduced into the combustor thus serves the following purpose. (1) It provides oxygen for oxygen carrier regeneration; (2) it fluidizes the oxygen carrier particles in the combustor; (3) it entrains the oxygen carrier particles in the riser back to the reducer. The sizing of the combustor and riser is required to fulfill the following conditions:

(1)   At the designed air flow rate, the oxygen flow must be greater than the amount required by oxygen carrier regeneration. The required amount of oxygen can be determined based on the fuel capacity and expected syngas composition.

(2)   At the designed air flow rate, the gas velocity in the combustor should satisfy the criterion given in equation (4.5.8).

$$u_{mf} < \frac{F_{g,c}}{S_c} < u_t, \tag{4.5.8}$$

where u_{mf} is the minimum fluidization velocity of oxygen carrier particles; $F_{g,c}$ is the volumetric flow rate of gas in the combustor; S_c is the cross-sectional area of the combustor; and u_t is the terminal velocity of the oxygen carrier particles.

(3)   At the designed air flow rate, the gas velocity in the riser should be greater than the terminal velocity, with the criterion given in equation (4.5.9):

$$\frac{F_{g,c}}{S_r} > u_t. \tag{4.5.9}$$

(4)   The volume of the combustor should be large enough to provide sufficient residence time for combustor reactions for oxygen carrier particles in order for the particles to be completely re-oxidized, with the criterion in equation (4.5.10) to be satisfied:

$$V_c(1 - \epsilon_c) \geq F_s \tau_c, \tag{4.5.10}$$

where V_c is the volume of the combustor; ϵ_c is the voidage of the fluidized bed at the designed air flow rate; and τ_c is the required residence time for the combustor reactions for oxygen carrier particles.

Gas Sealing Device Design

Gas sealing devices are required between the reactors to avoid gases from one reactor interacting with gases from the other reactor. Specifically, their role is to allow solids to flow through the reactor system while reacting gases are not allowed to do so. It is important to have a perfect gas seal between the two reactors for two main reasons. First, gas leakage from the combustor to the reducer or vice versa may result in the formation of an explosive mixture in the system, which poses a safety hazard. Secondly, an inefficient seal would allow fuel to leak into the combustor and increase CO_2

emissions from the system thereby reducing carbon capture efficiency. Also, leakage of air into the reducer would result in formation of undesirable products that would contaminate the gaseous product stream from the reducer.

Automatic solid flow devices such as seal pots and loop seals, shown in Figure 4.30, are commonly used in chemical looping systems with fluidized bed reducers for gas sealing purposes. These devices cannot, however, provide a good control on solids flow (see section on solids circulation control), because they passively change their own operating conditions based on the specific device configuration and the operational condition of the system where the devices are located.[97]

A seal pot is essentially an external fluidized bed into which the straight dip leg discharges solid particles. The solid particles and the fluidizing gas for the seal pot are discharged to the desired downstream vessel through an overflow transport line, designed either as a downwardly angled pipe at the side, or an overflow dip leg in the middle of the fluidized bed. With a seal pot, the solids in the dip leg rise to a height necessary to handle the pressure difference between the solids inlet and the outlet.

A loop seal is a variation of the seal pot that places the solids inlet dip leg at the side of the fluidized bed in a separate solids supply chamber. This allows the solids return chamber to be operated independently of the solids supply chamber, which results in a smaller device size, a lower fluidization gas requirement, and a higher efficiency. The height and the diameter of the solids supply chamber as well as its distance from the solids return chamber can be adjusted based on the process requirements necessary for balancing the pressure and handling the solids flow. Independent lubricating gas can also be added to the different locations of the solids supply chamber to assist in the operation of the loop seal.

In chemical looping systems with a moving bed reducer reactor, gas sealing can be achieved by introducing a small amount of sealing gas into the middle of the standpipe

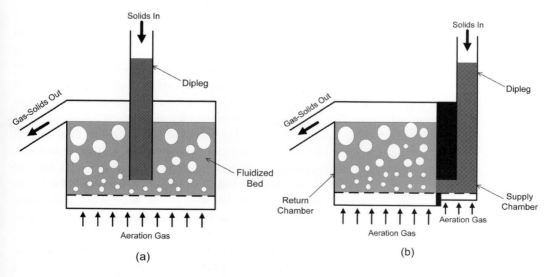

Figure 4.30 Automatic solid flow devices: (a) seal pot; (b) loop seal.

between the reducer and combustor reactor. The flow of the sealing gas will split into two towards both ends of the standpipe. The gas flow in the standpipe will induce a pressure drop, which prevents the gas in the reactor from flowing into the other reactor through the standpipe. The pressure drop can also be used to balance the pressure difference created by the gas flow in the two reactors.

Solids Circulation Control

The flow of oxygen carrier in chemical looping systems can be controlled either mechanically or non-mechanically. The use of mechanical valves to control the flow of solids was common during the early development of chemical looping processes, since they allowed a maximum flexibility over the control of solids. The mechanical valves could also provide effective gas sealing between the reactors despite the pressure difference at the two ends of the valves. Although mechanical valves have been part of a number of successful tests of continuous chemical looping processes, they have serious drawbacks. Since chemical looping processes circulate a large amount of oxygen carrier particles at high temperatures, the construction material of the valves and their internals would have to be able to withstand the high temperature. In addition, mechanical valves must also have a high abrasion resistance due to the large flow rate of solids.

In addition to the high solids flow rate and abrasive environment to be experienced by the mechanical valve, its repeated opening and closing during operation would accelerate the rate of wear and tear, leading to the possibility of mechanical failure occurring during operation. Also, since there would be solids flowing continuously through the valves, fines could accumulate in the valve seat, impeding its full closure, and hence leading to failure in ensuring an effective gas seal function. Mechanical valves would become cost prohibitive due to the expensive material of construction required to withstand the operating conditions, replacement frequency, and number of spares required due to the long lead time that would be associated with the purchase or maintenance of large mechanical valves. Further, a mechanical valve is difficult to scale up for commercial, long-term operation, as required in the chemical looping system. An attractive alternative to mechanical valves for solids flow control are the non-mechanical solids flow control devices. These devices are those that use aeration gases in conjunction with their geometric patterns to manipulate the solid particles flowing through them. They are widely used in industry due to their advantages over mechanical solids flow control devices. The non-mechanical solids flow control devices generally have no moving parts and thus have no issues with wear and tear, especially under extreme operating conditions, such as elevated temperatures and pressures. Also, these devices are normally inexpensive as they are constructed from ordinary system vessels. Because of their simplicity, non-mechanical solids flow control devices can be conveniently fabricated, avoiding the long delivery times associated with mechanical valves.

Figure 4.31 shows the common types of non-mechanical solids flow control devices, or non-mechanical valves, include the L-valve (Figure 4.31a) and J-valve (Figure 4.31b).[98] The principles of operation for these two types of valves are similar,

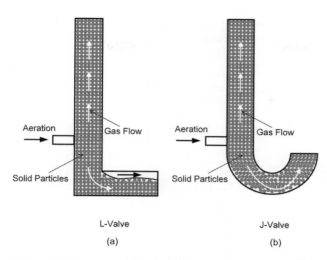

Figure 4.31 Non-mechanical valves, (a) L-valve; (b) J-valve.[98]

although their shape and the direction of solids discharge are different. Non-mechanical valves have limitations in their operating capability based on the physical properties of the solid. Non-mechanical valves function smoothly for particles belonging to Geldart groups B and D, but do not work well for particles belonging to Geldart groups A and C. Geldart group A particles generally retain gas in their interstices and remain fluidized for a substantial period of time even after fluidizing gas is removed. Therefore, they can pass through the non-mechanical valves even after the aeration gas flow is stopped. Thus, the solids flow rate cannot be controlled for Geldart group A particles. Geldart group C particles are very cohesive due to their relatively large inter-particle forces and thus are very difficult to flow using aeration gas in non-mechanical valves.

The aeration gas is generally added to the bottom portion of the standpipe section of a non-mechanical valve and it flows downwards through the bend. The slip velocity between the gas and the solid particles produces a frictional drag force on the particles in the direction of gas flow. When the drag force exceeds the force required to overcome the resistance to solids flow around the bend, the solids begin to flow through the valve. A certain minimum amount of gas flow is required before the start of solids flow through the non-mechanical valve. Above this threshold amount of gas that is required to initiate solids flow, the solids flow rate varies proportionately to the aeration gas flow rate. A quantitative relationship between the aeration gas velocity and the solids flux for particles of various sizes with a comparable density of ~2 g/cm^3 through an L-valve under ambient temperature is given in Figure 4.32.[99] These particles belong to Geldart groups B (180 μm, 260 μm, and 520 μm)[98] and group D (1500 μm).[99] It is seen that a higher aeration gas flow rate is needed for group D particles than for group B to reach the same solid flow rate. This is due to the greater drag force requirement for the solid particles of larger size to flow through the

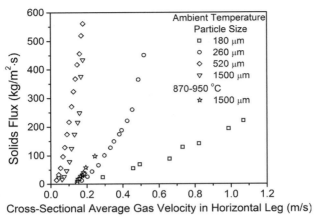

Figure 4.32 Solids flux as a function of average aeration gas velocity and flow temperature for particles of various sizes with a density of ~2 g/cm³ through an L-valve.[99]

non-mechanical valve. Figure 4.32 also indicates that, as the operating temperature increases to 870–950 °C, the group D curve becomes close to the group B curve under the ambient temperature.[99]

The actual aeration gas flow through the non-mechanical valve in a CFB system may be different from the amount of gas added externally through the aeration gas inlet port. It may be higher or lower than the amount of aeration gas externally introduced, depending on the operating conditions of the system.[97] In the case where the gas from the reactor leaks into the standpipe of a non-mechanical valve, the actual amount of aeration gas, Q_{ae}, would be the sum of the leaked gas flow into the standpipe, Q_{sp}, and the external aeration gas added from the aeration gas inlet port, Q_{ext}, as given in equation (4.5.11):

$$Q_{ae} = Q_{sp} + Q_{ext}.$$
(4.5.11)

If the gas from the aeration gas inlet port leaked upward through the standpipe into the reactor, then the actual amount of aeration gas would be obtained from equation (4.5.12):

$$Q_{ae} = -Q_{sp} + Q_{ext},$$
(4.5.12)

where the negative sign in front of Q_{sp} denotes a change in the direction of aeration gas leakage compared to the previous case.

The solids flow rate pattern after the bend in a non-mechanical valve is generally in the form of pulses of relatively high frequency and short wavelength. This pulsating flow creates pressure fluctuations of a fairly steady pattern in the non-mechanical valve. The pressure drop across the valve is high when the particle flow stops and low when the particle flow surges. Increasing the length after the bend increases the solids flow rate pulses, which increases the chaotic pattern of pressure fluctuations. The total pressure drop across the non-mechanical valve also increases with the length after the

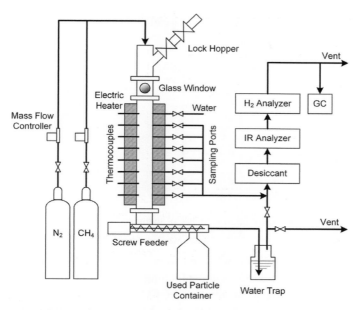

Figure 4.33 Bench-scale moving bed reactor system.[96]

bend. In some cases, additional gas is added to the section after the bend to prevent slug formation and to induce solids flow. However, this increases the total amount of external gas, and hence, the operating costs. Based on the above issues with a longer pipe length after the bend, it can be noted that the horizontal section of the non-mechanical valve should be as short as possible, to minimize the pressure fluctuations and the amount of aeration gas used.

4.5.4 Experimental Studies on Moving Bed Reducers for CLG/CLR Applications

Bench-Scale Reactor System

The reducer reactor performance for CLG/CLR applications can be illustrated based on experimental data obtained from a bench-scale moving bed reactor system, shown in Figure 4.33. The reactor system consists of a gas mixing panel, a moving bed reactor, and a gas analysis system.

The gas mixing panel enables inert and reactive gases such as N_2 and CH_4 to be mixed prior to their injection into the reactor. N_2 is used as a flushing gas and as an internal standard for determining the flow rate of the syngas generated from the gasification or reforming reactions from solid or gaseous feedstock, respectively. Steam is introduced to enhance H_2 generation in the reactor. All the gaseous and/or solid feedstocks are injected from the top of the reactor, flowing downward along with metal oxides in a cocurrent moving bed mode. Metal oxides used as oxygen carriers are spherical particles, 1.5 mm in diameter, consisting of an iron–titanium composite metal oxide (ITCMO). The moving bed reactor has a 2 in (5.08 cm)

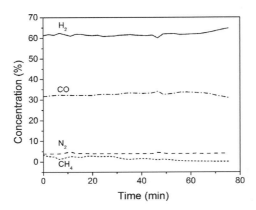

Figure 4.34 Syngas product distribution for CLR of methane.[96]

internal diameter with solids discharged from a screw feeder at the bottom of the reactor. Gaseous products are sampled along the reactor using a gas analysis system that quantifies the concentration of CO, CH_4, CO_2, O_2, and H_2. A tilted pipe with a lock hopper is located at the top of the reactor to allow pulverized solid feedstock to be mixed with metal oxides prior to being fed into the reactor. A glass window is located below the solid feeding pipe to monitor the solids inventory. The reactor operates under ambient pressure. Representative reducer performance results are given below and are illustrated in the order of chemical looping reforming using gaseous fuels and chemical looping gasification using coal and biomass.

CLR: Conversion of Gaseous Fuels

The CLR using gaseous feedstock with a gas composition of 90% CH_4 and 10% N_2 is conducted at 1000 °C (1832 °F). The Fe_2O_3 to CH_4 molar ratio used for this reaction is 0.8. In the reactor, the gas flows quickly through a slow moving bed of metal oxide oxygen carrier particles. The transient variation of outlet gaseous products can be marked by the initial formation of combustion products in transition to steady syngas products. The initial CO_2 formation at the reactor outlet is due to the presence of metal oxides that are in an oxidation state of Fe_2O_3 as they exit the reactor. The dominant equilibrium gaseous species when using Fe_2O_3 are CO_2 and H_2O. Although the stoichiometric ratio of Fe_2O_3:CH_4 at the inlet of the reactor system is configured for reactions to form CO and H_2, there is a time lapse for the metal oxide to become reduced to the oxidation state of Fe(II) and/or Fe(0) and exit the moving bed reactor. The dominant equilibrium gaseous species for these oxidation states become CO and H_2. These species thus represent the stable gaseous products from the reactor outlet at steady state. Figure 4.34 is the steady state syngas composition obtained from the reactor with an inlet Fe_2O_3:CH_4 molar ratio of 0.8, indicating stable syngas concentration generation in the moving bed reactor operation.

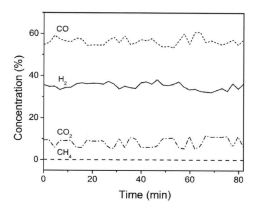

Figure 4.35 Syngas product distribution for CLG of coal only.

The syngas produced in the reactor consisted of 60% H_2 and 30% CO, and less than 10% of CO_2 and CH_4, as shown in Figure 4.34. The CH_4 conversion is at 95%, and syngas purity is higher than 85%. A higher CH_4 conversion and a higher syngas purity can be obtained by altering the reaction conditions, such as reaction time or reactant residence time, in this reactor system. A distinct difference in operation between the sub-pilot and bench-scale moving bed lies in their reactant residence time. When the sub-pilot system, shown in Figure 1.43, was operated at a temperature of 975 °C (1787 °F) with an Fe_2O_3:CH_4 molar ratio of 0.73 and a CH_4 flow rate of 10 standard l/m (SLPM), >99.9% CH_4 conversion with a H_2:CO ratio of 1.97 and syngas purity of 91.3% were obtained, as given in Figure 1.44. Comparing the operating conditions of the sub-pilot system to the bench-scale system, the sub-pilot system is operated at a slightly lower temperature and lower Fe_2O_3:CH_4 molar flow rate. The sub-pilot system, however, achieves a higher syngas conversion mainly due to its higher reactant residence time, i.e. 20% higher than the bench system. Evidently, the reactant residence time is a key factor to consider for optimization of moving bed reactor system performance.

CLG: Conversion of Coal

The CLG uses solid feedstock for reactions. With coal as feedstock, specifically sub-bituminous coal (Powder River Basin, PRB) and bituminous coal (Illinois #6), syngas can be produced in the presence of metal oxide oxygen carriers. Table 4.5 provides the proximate and ultimate analysis of these coals. In the CLG process, coal is pulverized and sieved to under 100 mesh (<150 micron), and mixed with the oxygen carrier particles before being fed into the reactor.

For PRB coal, experiments conducted at 1000 °C (1832 °F) using three different feed conditions are considered. In the first condition, only coal is introduced with the oxygen carrier particles. The mass ratio of oxygen carrier to coal was ~5. In the second condition, CH_4 is co-injected with coal and oxygen carrier particles. The purpose of CH_4 injection was to allow the H_2:CO ratio of the syngas to reach 1. The mass ratio of oxygen carrier to coal was ~7.7 and the mass ratio of CH_4 to

Table 4.5 Properties of coal used.

	PRB coal		Illinois #6 coal	
Proximate	%		%	
Moisture	13.5	as received	4.2	as received
Ash	7.9	dry	11.1	dry
Volatile matter	41.3	dry	38.5	dry
Fixed carbon	50.8	dry	50.4	dry
Ultimate	%		%	
Carbon	71.2	dry	70.2	dry
Hydrogen	4.5	dry	4.8	dry
Nitrogen	1.1	dry	0.9	dry
Oxygen (DIFF)	14.7	dry	9.9	dry

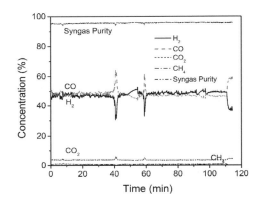

Figure 4.36 Syngas product distribution and syngas purity for CLG of PRB coal and methane.

coal was ~0.21. In the third condition, CH_4 and water are both co-injected with coal to allow the H_2:CO ratio to further increase to 1.8. The mass ratio of oxygen carrier to coal is ~22.22. Also, the mass ratio of CH_4 to coal is ~0.99 and that of water to coal is ~0.89.

Similar to CLR with CH_4, the product gas composition at the reactor outlet for CLG reactions also undergoes the transition from a CO_2/H_2O rich combustion gas to a CO/H_2 rich syngas. The syngas composition for the first condition (with coal only) is given in Figure 4.35. The average syngas composition is 57% CO, 37% H_2, and 8% CO_2, a H_2:CO ratio of ~0.65 and CO:CO_2 ratio of ~7. The syngas purity (on a dry basis) is well above 88%. Negligible CH_4 generation is observed in the product gas, indicating a complete conversion of coal volatiles. The carbon conversion in coal is estimated to be approximately 93%, corresponding to a 90% char conversion given a full conversion of coal volatiles.

CH_4 can be co-injected with coal to obtain a H_2 rich syngas from the CLG system, which corresponds to the second condition. Methane was added to coal to increase the H_2:CO ratio in the syngas to 1, as shown in Figure 4.36 for PRB coal and Figure 4.37

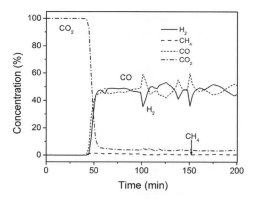

Figure 4.37 Syngas product distribution for CLG of Illinois #6 coal and methane.

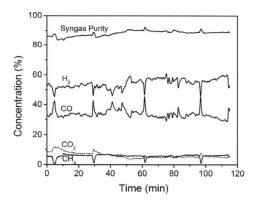

Figure 4.38 Syngas product distribution for CLG of PRB coal, methane, and steam.

for Illinois #6. For both coals, the syngas purity is above 95% on a dry basis with the $CO:CO_2$ ratio greater than 10 and minimal unconverted CH_4. The initial concentration in a CLG is also shown in Figure 4.37.

Co-injection of CH_4 and H_2O with coal, the third condition, can further increase the $H_2:CO$ ratio in the syngas. The amount of CH_4 and H_2O required for the desired syngas composition can be determined based on the thermodynamic relationship. As shown in Figure 4.38, the syngas from the reactor consists of 58% H_2, 32% CO, 5% CO_2, and 5% CH_4. With steam, methane, and coal injection, the $H_2:CO$ ratio of the syngas can be adjusted to ~1.8, a $CO:CO_2$ ratio of ~6, and syngas purity of greater than 85% on a dry basis.

Thermodynamics show that when steam is used to increase the H_2 content in the syngas, the oxygen carrier can only be reduced to the Fe(II) oxidation state instead of the Fe oxidation state, which can catalyze the decomposition of CH_4. As the reaction between CH_4 and the oxygen carrier is slower in the absence of Fe, CH_4 will react with the oxygen carrier with slower kinetics in the presence of steam. The kinetics can be improved under pressurized reaction conditions, but the reactant equilibrium conversion

Table 4.6 Two operating conditions and syngas product results for CLG of biomass.

Condition	1	2
Temperature	1000 °C	1000 °C
[O]:C	8.6	7.3
H_2O:C	1.16	1.13
Concentration (dry base)		
H_2 (%)	46.28	46.99
CO (%)	20.74	28.29
CO_2 (%)	32.98	24.72
H_2:CO	2.23	1.66

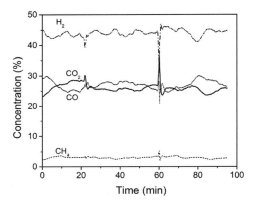

Figure 4.39 Syngas product distribution for CLG of biomass for reaction condition 2.

can be compromised by a pressure increase. Thus, for a given reaction residence time, a small amount of unconverted CH_4 can appear in the syngas when steam is added, which reduces the H_2:CO ratio to below the thermodynamic prediction. A comprehensive system analysis can determine the optimal operating condition.

CLG: Conversion of Biomass

The CLG of biomass can also generate syngas. The conversion of woody biomass pellets 3–10 mm (0.1–0.4 in) in length was conducted in a moving bed reducer reactor. The biomass pellets were premixed with the oxygen carrier. Addition of steam can increase the H_2 content in the syngas produced. Table 4.6 shows the two specific feed rate conditions for the oxygen carrier, steam and biomass. Here, C and [O] in the table corresponds to the carbon and oxygen contents in the biomass feedstock and metal oxide, respectively. The syngas yield for these two conditions varied with H_2:CO ratios of 2.23 and 1.66 and syngas purity (on a dry basis) of 67.02% and 75.28%, respectively. Details of the syngas concentration variation for condition 1 are given in Figure 1.51, while those for condition 2 are given in Figure 4.39.

To summarize Section 4.5.4, CLG/CLR processes with cocurrent moving bed reducers can efficiently convert gaseous and solid feedstock including natural gas, coal, and biomass into a high quality syngas using an iron based oxygen carrier. As a result of the uniform solid residence time and the absence of solid backmixing, the oxidation state of the oxygen carrier at the reducer outlet is controlled within the range that is thermodynamically favorable for syngas generation. The cocurrent flow pattern in the moving bed reducer ensures that the syngas is in contact with a specific oxidation state of the oxygen carrier, which regulates the thermodynamic composition of the syngas. The oxygen carrier conversion from a moving bed reducer is higher compared to that from fluidized bed reducers, yielding a lower solid circulation rate operation for a moving bed reducer for a given feedstock processing capacity. In addition, carbon deposition is inhibited in the moving bed reducer due to the limited formation of metallic iron. The H_2:CO ratio in the syngas can be adjusted by addition of H_2O and/or CO_2 to meet the downstream processing requirement, such as a H_2:CO ratio of 2:1 for Fischer–Tropsch synthesis or methanol production.

4.6 Chemical Looping CO₂ Neutral and Negative Processes

To eliminate carbon deposition on oxygen carriers, as in the 300 W_{th} CLR unit operation at Chalmers University of Technology, the co-feed of CO_2 was implemented, as discussed in Section 4.4. However, when there is no carbon deposition in the reactor operation, the co-feed of CO_2 can also be strategically deployed for the purpose of adjusting the purity of syngas and the H_2:CO ratio through donating an oxygen atom from CO_2 to the metal oxide lattice. In this section, the effect of CO_2 as feedstock co-fed with other carbonaceous feedstock for the gasification reaction is described in the context of the chemical looping CO_2 neutral or negative process operation. Further, the reducer modularization using CO_2 as feedstock for the benefit of syngas generation is elaborated.

4.6.1 CO₂ as Feedstock and Conversion in Process Systems

For a given chemical looping reducer that processes carbonaceous feedstock, its redox reactions and subsequent carbonaceous product formation will produce CO_2 that may or may not be required to be emitted to the exterior or to exit the process system under a steady state process system condition. The process system without CO_2 emission to the exterior while no fresh CO_2 feedstock is introduced to the reducer is regarded as a CO_2 neutral process. The process system without CO_2 emission to the exterior while fresh CO_2 feedstock is introduced to the reducer is regarded as a CO_2 negative process. It is noted that, in a broad sense, a CO_2 neutral process includes a general condition when the quantity of CO_2 input to a process system is equal to the quantity of CO_2 output from the process system under a steady state condition, while a CO_2 negative process includes a general condition when the quantity of CO_2 input to a process system is greater than the quantity of CO_2 output from the process system under a steady state condition. A CO_2 neutral or negative process may involve a CO_2 recycle in the process system. The chemical looping reducer that can perform either CO_2 neutral or CO_2 negative process conditions relies heavily on the ability of the chemical looping reducer to process CO_2 and other carbonaceous fuels as the feedstock. The reducer behavior in processing CO_2 is illustrated below, considering a cocurrent moving bed reactor that serves as a reducer.

When CO_2 is co-fed with hydrogen-rich fuel like natural gas, syngas can be produced with a H_2:CO of ~2, which is at a desirable level for liquid fuel production, or even higher with additional steam injection, as discussed in Section 4.5. When steam alone is used in conjunction with the chemical looping system, the H_2:CO ratio increases while the carbon efficiency decreases as CO is converted to CO_2. The decreased carbon efficiency can be represented by a decreasing S#, i.e. stoichiometric number representing the carbon efficiency, given in equation (4.6.1), which is negatively correlated to the level of CO_2 production:

$$S\# = \frac{y_{H_2} - y_{CO_2}}{y_{CO} + y_{CO_2}}. \tag{4.6.1}$$

A chemical looping reducer with CO_2 co-feed can minimize the conversion of CO to CO_2 thereby maximizing the value of S# and improving the carbon efficiency of the process. The increase in carbon efficiency with CO_2 co-feed for a reducer operation can be characterized using the CO_2 reaction parameter (CRP), defined by equation (4.6.2):

$$CRP = \frac{\dot{n}_{CO_2 \text{entering the reducer}}}{\dot{n}_{CO_2 \text{exiting the reducer}}}. \tag{4.6.2}$$

It is noted that $\dot{n}_{co_2}$ in equation (4.6.2) does not directly represent the feedstock CO_2 conversion in the reducer, rather it reflects the composite CO_2 outlet concentration effect due to overall reactions of carbonaceous feedstock and CO_2 introduced to the reducer. For a chemical looping process system with a CRP value of less than 1, the feedstock CO_2 introduced to the reducer may come in full from the recycle stream of CO_2 at the outlet of the reducer or other downstream processing units, while discharging the unutilized CO_2 to the exterior. When the value of CRP is greater than 1, the feedstock CO_2 to the reducer may come in part from the recycled stream along with fresh CO_2 make-up. A value of CRP greater than 1 corresponds to a CO_2 negative reducer operation condition that is necessary, under traditional downstream processing schemes, for the realization of a CO_2 negative process system.

Figure 4.40 shows a conceptual schematic of the proposed configuration using the chemical looping CO_2 recycle scheme with natural gas, steam, and recycled fuel gas from the downstream as reactants for liquid fuels production.[100] The process performance of the novel configuration shown in Figure 4.40 is demonstrated relative to a baseline case that uses conventional autothermal reforming (ATR) for syngas production, shown in Figure 4.41.[101] The baseline case given in Figure 4.41 utilizes 19,849 kmol/hr of natural gas as an input to the ATR syngas generation unit which produces a syngas containing 45,285 kmol/hr H_2 and 20,680 kmol/hr CO. This syngas has a H_2:CO ratio of 2.19, which

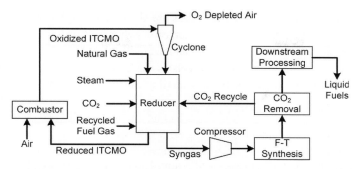

Figure 4.40 Chemical looping CO$_2$ recycle scheme for liquid fuels production.[100]

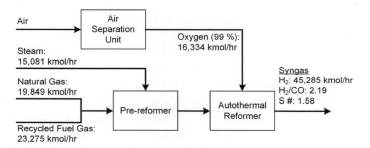

Figure 4.41 ATR baseline case for liquid fuels production using syngas from autothermal reforming.[101]

serves as an input to a cobalt based F–T process. In the ATR baseline case/process, 10% of the CO$_2$ is evolved from fuel usage for offsetting parasitic energy consumption and for generating H$_2$ for use in downstream liquid fuel upgrading, while the remaining 90% of CO$_2$ evolved is sequestrated. In the chemical looping baseline case/process, it is designed to match the syngas production and quality of the ATR baseline case, and utilizes 18,000 kmol/hr of natural gas for producing 45,285 kmol/hr H$_2$ and 20,680 kmol/hr of CO. The downstream processing in the chemical looping baseline is maintained to be consistent with the ATR baseline, resulting in 90% of CO$_2$ evolved being sequestered. In the chemical looping scheme given in Figure 4.40, all of the sequestered CO$_2$ can be recycled. The objective of the CLR process with CO$_2$ recycle for syngas production is to minimize consumption of natural gas and steam while maintaining the H$_2$:CO ratio, H$_2$ flow rate, and CO flow rate in the syngas, equivalent to the baseline. The recycled fuel gas stream is held constant, as it is a product of downstream processes, which are held constant. The S# must be greater than or equal to the chemical looping baseline process, and increasing the S# number increases the carbon efficiency of the process. These requirements are depicted in Table 4.7.

The results of the simulated CLR of methane using the ASPEN RGIBBS single stage module with CO$_2$ addition are presented below. The simulation is based on minimization of the free energy of the reactor products using an ASPEN RGIBBS single stage module. The RGIBBS module for the process simulations is set to a

Table 4.7 Syngas composition target for chemical looping for liquid fuels production.

H$_2$:CO	2.19
% of H$_2$ in baseline case	$\geq$ 100% (45,285 kmol/hr)
% CO$_2$ recycle	$<$ 90%
S# (H$_2$ – CO$_2$)/(CO + CO$_2$)	$\geq$ 1.58

temperature of 900 °C and a pressure of 1 atm. It is noted that a RGIBBS single stage module simulates the equilibrium condition at the outlet of the cocurrent moving bed reducer reactor.

In order to maximize the carbon efficiency of the system, a CO$_2$ recycle stream is introduced, as shown in Figure 4.40. The CO$_2$ in the recycle stream is obtained from the CO$_2$ separation from the outlet stream of the Fischer–Tropsch synthesis. The process targets for the syngas generation section are detailed in Table 4.7, with H$_2$:CO ratio 2.19, a H$_2$ flow rate of at least 45,285 kmol/hr, and S# greater than 1.58. The CO flow rate is determined by the H$_2$:CO ratio and H$_2$ flow rate. As discussed earlier, a CO$_2$ recycle of 90% serves as a standard operating condition for the chemical looping case with CO$_2$ recycle. Figure 4.42 summarizes these condition requirements.

Optimal results for the chemical looping case with CO$_2$ addition, shown in Figure 4.42, can be obtained at a natural gas flow rate of 15,500 kmol/hr and a steam flow rate of 11,000 kmol/hr, representing a 22% reduction in natural gas flow and 27% reduction in steam flow over the ATR baseline case, given in Figure 4.41. At this condition in, Figure 4.42, the H$_2$:CO ratio and H$_2$ flow rate match the ATR baseline case, as shown in the last row of Table 4.8. In Table 4.8, a set of different CRP values for the chemical looping reducer system with a constant natural gas flow rate (15,500 kmol/hr) but varied steam flow rates that match the syngas yield of the ATR baseline case are also given. In each case, the quantity of CO$_2$ input is coming entirely from the recycle CO$_2$ stream from the outlet of the reducer. At a natural gas price of $2/MMBTU, the reduction in the natural gas flow of 22% indicated above reflects a cost reduction for natural gas usage of $7,507.94 per hour, leading to an annual saving of $59.2 million for the CLR plant when it is operated at 90% capacity.[102–104]

When the CO$_2$ input to the reducer is coming from a 100% CO$_2$ recycled process stream and a fresh CO$_2$ external source, the CLR process is able to utilize more CO$_2$ than is being generated in the process stream. This process, as shown in Figure 4.43, is a CO$_2$ negative process.[100] Figure 4.44 shows the simulated effects of CO$_2$ input molar flow rates on H$_2$:CO ratios and the CO$_2$ concentration in syngas at various values of CRP and H$_2$O:CH$_4$ ratio. The figure includes CO$_2$ negative reducer operation conditions, represented by values of CRP greater than 1. Further, the figure can identify proper reducer operating conditions to yield a desired syngas property in H$_2$:CO and CO$_2$ product flow rate. As an example, shown by point A in the figure, a CO$_2$ negative reducer operation can take place at a H$_2$O:CH$_4$ ratio of 1 and a CRP of ~1.2 to yield a H$_2$:CO ratio of ~2.0.

Table 4.8 Flow rates that satisfy chemical looping requirements for liquid fuels production.

Input conditions			Syngas output conditions					
CO$_2$ in kmol/h	Natural gas kmol/h	Steam kmol/h	CO kmol/h	CO$_2$ kmol/h	H$_2$ kmol/h	H$_2$:CO	CRP	% H$_2$ of baseline
1050	15,500	6000	20,660	1432	45,256	2.19	0.73	100%
1460	15,500	7000	20,702	1835	45,350	2.19	0.80	100%
1880	15,500	8000	20,733	2246	45,410	2.19	0.84	100%
2300	15,500	9000	20,754	2661	45,454	2.19	0.86	100%
2720	15,500	10,000	20,768	3079	45,488	2.19	0.88	100%
2990	15,500	11,000	20,789	3320	45,510	2.19	0.90	100%

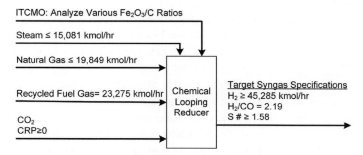

Figure 4.42 Example of a reducer for a chemical looping process with CO$_2$ introduced into the reducer reactor for syngas generation.

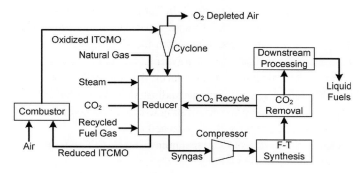

Figure 4.43 CO$_2$ negative chemical looping process system.[100]

Utilizing CO$_2$ at flow rates greater than the production rate eliminates direct sources of CO$_2$ emissions in the CLR-GTL process. The ability to consume CO$_2$ could be valuable in offsetting carbon emissions from other processes and would transform the CO$_2$ market. In addition, the carbon efficiency is drastically increased, and what was once a waste stream of the process is now utilized as a supplemental feedstock.

Table 4.9 Experimental verification of CO_2 negative chemical looping process conducted using moving bed reducer and using CO_2 as the feedstock.

	Experimental 1	Theoretical 1	Experimental 2	Theoretical 2
Input conditions				
CH_4 flow rate (SLPM)	1.2	1.2	1.2	1.2
H_2O flow rate (mL/min)	0	0	0.68	0.68
H_2O/CH_4 molar ratio	0	0	0.704	0.704
CO_2 flow rate (SLPM)	0.18	0.18	0.18	0.18
CO_2/CH_4 molar ratio	0.15	0.15	0.15	0.15
Fe_2O_3/CH_4 molar ratio	0.7	0.7	0.69	0.69
Reactor temperature (°C)	1010	1010	1039	1039
Output conditions				
% CH_4 conversion	96.1	99	95.6	99
H_2:CO molar ratio	1.61	1.65	2.06	2.17
CO/CO_2 molar ratio	14.1	13.02	13.5	9.55
CRP	1.57	1.82	1.51	1.37
% Syngas purity on dry basis	95.8	96.8	96.1	96.4

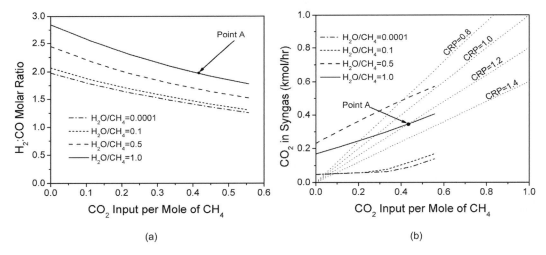

(a)   (b)

Figure 4.44 Simulation of the effect of CO_2 input rates on (a) H_2:CO ratios and (b) CO_2 in syngas at various values of CRP and H_2O:CH_4 ratio.[100]

Experimental verification of CO_2 negative reducer operation using a moving bed reducer, given in Section 4.5, with CO_2 as the feedstock, is presented in Table 4.9. The results obtained for two experiments performed at a pressure of 1 atm are exemplified for this verification. The reaction conditions for the first experiment are a Fe_2O_3/CH_4 molar ratio of 0.7, a CO_2/CH_4 molar ratio of 0.15, and a temperature of 1010 °C. Thermodynamically, under equilibrium, this experiment yields a CO:CO_2 ratio of 13.02, a H_2:CO ratio of 1.65, a CRP of 1.82, and a syngas purity on dry basis of 96.8%. The experimental results indicate a CO:CO_2 ratio of 14.1, a H_2:CO ratio of

1.61, a CRP of 1.57, and a syngas purity on dry basis of 95.8%. The conditions for the second experiment are a Fe$_2$O$_3$/CH$_4$ molar ratio of 0.69, a CO$_2$/CH$_4$ molar ratio of 0.15, a H$_2$O/CH$_4$ molar ratio of 0.704, and a temperature of 1039 °C. Thermodynamically, under equilibrium, this experimental condition yields a CO:CO$_2$ ratio of 9.55, a H$_2$:CO ratio of 2.17, a CRP of 1.37, and a syngas purity on dry basis of 96.4%. The experimental results obtained indicate a CO:CO$_2$ ratio of 13.5, a H$_2$:CO ratio of 2.06, a CRP of 1.51, and a syngas purity on dry basis of 96.1%. It is noted that compared to the first experiment, the second experiment involves steam addition to the reactor. The steam addition leads to a lower residence time for the methane flow in the reactor. The use of a higher temperature for the reaction in the second experiment is thus intended to enhance the kinetics in order to achieve a comparable methane conversion to that of the first experiment. Comparing the experimental values to the theoretical values for the output conditions for both cases indicates a reasonable, general matching of the values. Some specific deviations exist, however, which could be caused by the incomplete conversion of methane in both experiments. Incomplete methane conversion is due to the inadequacy of the residence time for the reaction resulting from reactor set-up. As indicated in Table 4.9, the molar flow rates of CO$_2$ at the reactor outlet for both experiments are considerably less than those at the reactor inlet. This result substantiates the feasibility of the CO$_2$ negative reducer operation and hence the CO$_2$ negative process concept as simulated for chemical looping technology applications.

4.6.2   Reducer Modularization and Product Yield Enhancement

When CO$_2$ is used as a feedstock in the reducer, there is a nonlinear relationship between the molar flow rate of CO$_2$ and H$_2$:CO ratio that is beneficial to product generation and can be explored through a modular approach to the reducer design. Given below is the thermodynamic calculation of this nonlinearity and the rationale for the reducer modularization approach.[105,106]

With CO$_2$ as feedstock for the reducer, the calculation based on the ASPEN RGIBBS single stage module reveals the nonlinearity relationship, in the variation between the CO$_2$:CH$_4$ ratios and the product H$_2$:CO ratios or H$_2$ molar flow rates at given H$_2$O:CH$_4$ ratios. The calculation is based on 1 mole of CH$_4$ at a temperature of 900 °C, a pressure of 1 atm, and an effective Fe$_2$O$_3$:CH$_4$ ratio of 0.40, with ITCMO as the oxygen carrier. Figures 4.45 and 4.46 show these relationships.[100] Such a nonlinearity relationship, induced by CO$_2$ addition to the reactor, can be explored to enhance the product yield from the reactor using a reactor modularization approach, as given in Figure 4.47.[100] In Figure 4.47, two reducers operated in parallel can be configured in a CO$_2$ recycle chemical looping system, where operation in one of the reducers is carried out under the condition represented by point A in Figure 4.45, while the other reducer is carried out under the condition represented by point B in Figure 4.45. Points A and B are arbitrarily selected to explore the nonlinearity effect on the product yield, while more precise point selection requires employment of an optimization scheme that maximizes the product yield from the process. The operation constraint for the process system is set for the net

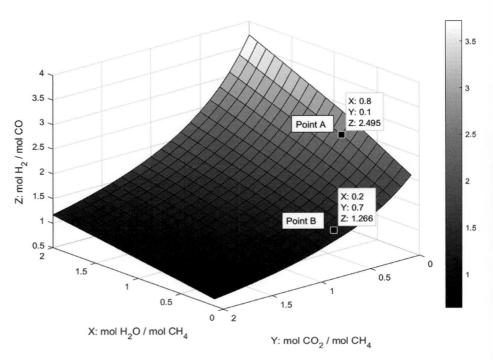

Figure 4.45 Simulated H_2:CO ratio as a function of H_2O:CH_4 and CO_2:CH_4 ratios at 900 °C and 1 atm for a cocurrent moving bed reactor and an effective Fe_2O_3:CH_4 ratio of 0.4.[100]

syngas yield at an effective H_2:CO ratio of 2. Specifically, the approach for reducer operation conducted under the conditions of point A is to maximize the H_2 production while operating the reducer at a H_2O:CH_4 ratio of 0.8, and CO_2:CH_4 ratio of 0.1. The point A operating condition results in a H_2:CO ratio of 2.495. The approach for reducer operation conducted under the conditions of point B is to maximize the CO production while operating the reducer at a H_2O:CH_4 ratio of 0.2, and CO_2:CH_4 ratio of 0.7. The point B operating condition results in a H_2:CO ratio of 1.266. The combination at a certain ratio of the syngas streams coming from these two reducers gives rise to a syngas product with a H_2:CO ratio of 2 from this chemical looping system. It is noted that as the reactant feed rates to each of the two reducers are different, the size of the reducers differs as well.

The nonlinearity effect obtained by operating two-reducer reactors as discussed above can be leveraged to reduce the natural gas consumption for an equivalent syngas production from a single reducer, illustrated using a specific example presented in Table 4.10. The two-reducer system, as indicated in the table, can be configured with one reducer processing 0.68 mol/s CH_4, 0.54 mol/s H_2O, and 0.07 mol/s CO_2; the other reducer processing 0.15 mol/s of CH_4, 0.03 mol/s of H_2O, and 0.10 mol/s CO_2. The products from each of these two reducers will contain a syngas with a H_2:CO ratio of 2.50 and 1.27, separately. The combined H_2:CO ratio from these two reducers will be 2.19, with a net H_2 flow of 2.04 mol/s and a net CO flow of 0.93 mol/s. The combined

Table 4.10 Comparison of the syngas yields for the two-reducer and the one-reducer system.

	Two-reducer system			One-reducer system
	Reducer 1 in modular system	Reducer 2 in modular system	Total of two-reducer system	
	Input conditions			
$CH_{4\ in}$ (mol/s)	0.68	0.15	0.83	1
$H_2O_{\ in}$ (mol/s)	0.54	0.03	0.57	0.23
$CO_{2\ in}$ (mol/s)	0.07	0.10	0.17	0
	Output conditions			
$H_{2\ out}$ (mol/s)	1.74	0.30	2.04	2.04
$CO_{\ out}$ (mol/s)	0.70	0.23	0.93	0.93
H_2:CO ratio	2.50	1.27	2.19	2.19

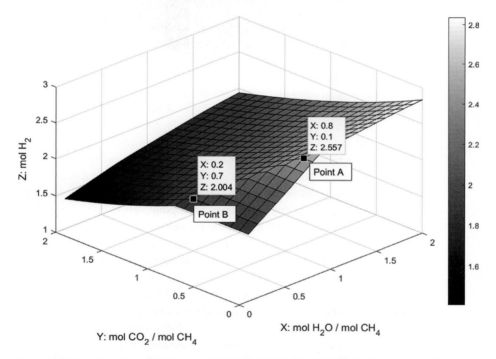

Figure 4.46 Simulated H_2 molar flow as a function of H_2O:CH_4 and CO_2:CH_4 ratios at 900 °C and 1 atm for a cocurrent moving bed reactor and an effective Fe_2O_3:CH_4 ratio of 0.4.[100]

natural gas consumption for the two-reducer modular system is 0.83 mol/s. In contrast, when this process is conducted with the same syngas output using one reducer instead, the natural gas consumption becomes 1 mol/s, which is 20.6% higher than the two-reducer system. It is noted that the steam and CO_2 used as inputs to the two-reducer system are higher than in the one-reducer system. However, the economic benefits from reduction in natural gas flow with a two-reducer system will outweigh the increased cost of higher steam and CO_2 input for a one-reducer system.

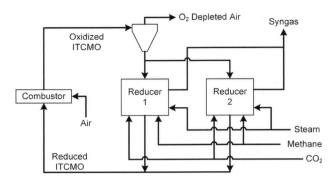

Figure 4.47 Chemical looping system operated with two reducers in parallel and a single combustor reactor with CO_2 input.[100]

The advantages of a modular CLR system can be further illustrated by an example of a scaled-up CLR system for syngas generation where reducers are integrated into a 50,000 bpd of GTL plant following the feed requirements or conditions given in Table 4.7 and Figure 4.42. The two-reducer process is configured with one reducer processing 8850 kmol/hr of natural gas with steam input of 7800 kmol/hr and the other reducer processing 6000 kmol/hr of natural gas with steam input of 3100 kmol/hr. The product syngas from one reducer and from the other reducer yield H_2:CO ratios of 2.98 and 1.52, separately. The combined H_2:CO ratio in the product syngas from these two reducers becomes 2.19. These modular process conditions can be compared to those in Table 4.8, where a single reducer is employed that requires to utilize 15,500 kmol/hr of natural gas. The natural gas flow for the two-reducer module in this particular comparison results in a reduction of natural gas flow by 650 kmol/hr while producing 50,000 bpd of liquid fuel product. It should be noted that the CO_2 addition to the one reducer, given in Table 4.8 (CRP = 0.9), yields a natural gas consumption of 15,500 kmol/hr, which is a reduction of 4349 kmol/hr or 22% over 19,849 kmol/hr in the ATR baseline case, given in Figure 4.41. The reducer configuration based on two-reducer modularization yields a further natural gas reduction of 650 kmol/hr, or 4.2% over 15,500 kmol/hr for a one-reducer system. The overall benefit of the addition of CO_2 to the reducer is thus significant.

The nonlinearity relationship induced by CO_2 addition to the reactor can be further explored using another reactor modularization configuration, as given in Figure 4.48.[100] In Figure 4.48, three reducers operating in parallel are configured in a CO_2 recycle chemical looping system, where operation in one of the reducers is carried out under the conditions represented by point A in Figure 4.49. The second reducer is operated under the conditions represented by point B, while the third reducer is operated under the conditions represented by point C in Figure 4.49. A three-reactor arrangement provides a higher degree of freedom in optimizing the product yield than does a two-reactor arrangement. Points A, B, and C are arbitrarily selected as an example to illustrate the three-reactor configuration effect on product yield. Again, the more precise point selection requires employment of an optimization scheme that maximizes the product yield from the process. The net syngas yield is set at an effective H_2:CO ratio of 2.

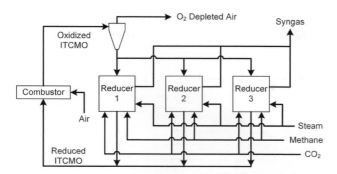

Figure 4.48 Conceptual schematic of chemical looping CO$_2$ recycle concept with three reducers operating in parallel with a single combustor reactor.[100]

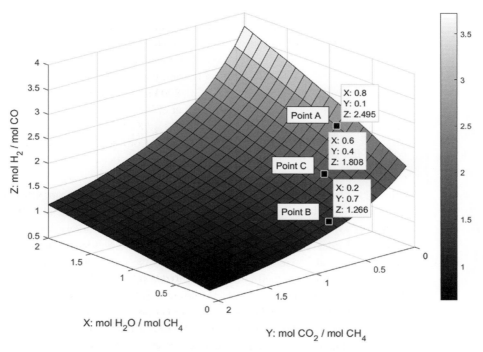

Figure 4.49 H$_2$:CO ratio as a function of H$_2$O:CH$_4$ and CO$_2$:CH$_4$ ratios at 900 °C and 1 atm for a cocurrent moving bed reactor at an effective Fe$_2$O$_3$:CH$_4$ ratio of 0.4.[100]

Specifically, the approach for reducer operation conducted under conditions corresponding to point A maximizes H$_2$ production while operating the reducer at a H$_2$O:CH$_4$ ratio of 0.8, and CO$_2$:CH$_4$ ratio of 0.1. The point A operating conditions result in a H$_2$:CO ratio of 2.495. The approach for reducer operation conducted under the conditions of point B is to maximize CO production while operating the reducer at a H$_2$O:CH$_4$ ratio of 0.2 and a CO$_2$:CH$_4$ ratio of 0.7. The point B operating conditions result in a H$_2$:CO ratio of 1.266. The approach for reducer operation conducted under the conditions of

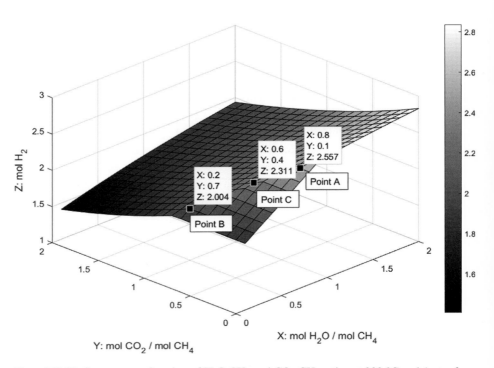

Figure 4.50 H_2 flowrate as a function of $H_2O:CH_4$ and $CO_2:CH_4$ ratios at 900 °C and 1 atm for a cocurrent moving bed reactor at an effective $Fe_2O_3:CH_4$ ratio of 0.4.[100]

point C is intended to modulate the effective $H_2:CO$ ratio from the three-reducer combined outlet to be equal to 2, while operating the reducer at a $H_2O:CH_4$ ratio of 0.6 and a $CO_2:CH_4$ ratio of 0.4. The point C operating conditions result in a $H_2:CO$ ratio of 1.808. Under certain operating conditions, a three-reducer modular system can enhance the product yield even more than does the two-reducer modular system. In a commercial syngas production operation, where the size of a reducer reactor vessel is restricted by such factors as fabrication costs and solids flow rates, a three-reactor modular system may be beneficial over a two-reactor modular system. This is because the three-reactor modular system can maintain the same fuel processing capacity and syngas production yield, while adhering to the size restrictions for a single reducer vessel in the module.

It should be noted that the two-reducer and three-reducer chemical looping processes with CO_2 recycle are examples of a multiple reducer system, and the concept can be expanded to "n" reducers.[100] While each of these "n" reducer reactors may perform a specific function, when in a process system, the performance of these reducers as a whole can be optimized for product yield considering temperature, pressure, reactor reaction time, $H_2:CO$ ratio, and $Fe_2O_3:CH_4$ ratio. Operation of multiple reducers with specific functions or operating conditions can be a way to reduce overall cost and maximize efficiency based on a given set of operating conditions.[100]

4.7   Concluding Remarks

Syngas is an important intermediate feedstock for the production of many downstream products such as electricity, hydrogen, chemicals, and liquid fuels. Traditional approaches for generating syngas have been through coal gasification or natural gas reforming, including such techniques as SMR, POX, and ATR. These reactions are endothermic and thus require heat that is obtained through combustion of the feedstock or inclusion of molecular oxygen obtained from an air separation unit in gasification or reforming reactions. Further, additional downstream processing of the syngas, such as use of the WGS reaction, is required to allow the H_2 and CO molar ratio in the syngas to be adjusted to a suitable value for downstream product generation. Generating syngas through traditional approaches is highly energy and capital intensive and can only be economical at large scales. Also, control of the emission of CO_2, generated during gasification, reforming, or the WGS reaction, may be needed in a carbon constraint situation. The emergence of chemical looping as a technology platform can significantly alter the syngas generation strategy, thereby alleviating the issues inherent in traditional approaches.

In chemical looping processes, gasification or reforming reactions occur in the reducer reactor, which is typically either a fluidized bed or a moving bed. The gas and solid contact modes in these two reactors are significantly different, resulting in vastly different compositions of the syngas produced. Experimental results indicate that, for a high methane to metal oxide oxygen carrier molar ratio, without steam addition, a fluidized bed reducer can produce a high purity of syngas while yielding incomplete methane conversion. On the other hand, for a low methane to metal oxide oxygen carrier molar ratio, without steam addition, a fluidized bed reducer can produce a low purity of syngas while yielding complete methane conversion. Carbon deposition can also occur in a fluidized bed. When nickel oxide is used as the oxygen carrier, a high methane conversion and high purity of syngas can be achieved. However, it will require the addition of steam in order to remove carbon deposition generated from the Ni catalytic reaction. To minimize carbon deposition and hence steam addition, nickel oxide conversion in the fluidized bed needs to be kept low. A high solids circulation rate is thus needed to meet the feedstock processing capacity requirement in this low nickel oxide conversion situation. Another characteristic of a fluidized bed is that, although it possesses excellent heat transport properties, gas reactants in bubble channeling flows can lead to incomplete gas conversion. Further, due to the presence of a wide distribution of residence times for metal oxide oxygen carriers in a fluidized bed, it is difficult to maintain a specific, desired metal oxide oxidation state for the reaction. These product yield and fuel conversion characteristics in the fluidized bed reducer can be appreciably improved by using a moving bed reducer, where the metal oxide reaction can be properly controlled to achieve the desired syngas generation. A gas–solid moving bed contact scheme in connection with a specific gas–solid flow direction, i.e. cocurrent downward gas–solid flow, can produce syngas at the condition near the thermodynamic limits.

The moving bed system can achieve high syngas purity and high methane conversion without carbon deposition and the need for steam addition, while maintaining a high metal oxide conversion and hence a low metal oxide circulation rate in the reactor system. Similar advantages exist for the moving bed reducer compared to the CFB reducer. A high feedstock conversion and product yield along with a high H_2:CO molar ratio can obviate the requirement for a WGS reactor in the moving bed based CLG and CLR processes with cocurrent gas–solid downward flows. A variety of chemical looping examples using the moving bed reducer reactor are presented in this chapter to illustrate the effect of the simplicity of the process through the intensification scheme. Such examples include the CLG process with coal which can achieve a carbon conversion of 93%, a H_2:CO ratio of 0.65, and a syngas purity of 89%. The H_2:CO ratio can be adjusted through co-injection of steam and methane to increase its value. The CLG process with biomass can obtain a H_2:CO ratio of 2 with a syngas purity of 70%. The CLR process in a moving bed reducer can achieve greater than 99.9% methane conversion, a H_2:CO ratio of 2, and a syngas purity of 91% at the sub-pilot scale.

CLC processes have inherently produced a pure stream of CO_2. A novel CO_2 negative chemical looping process with the ability to utilize the CO_2 it generates from the process and which also consumes excess CO_2 from an external source, reflects an important feature of a chemical looping gasification and reforming system in fossil energy conversion with full carbon emission control in neutral or negative conditions. Process simulations for a chemical looping system with 90% CO_2 recycle, and another case with 100% CO_2 recycle coupled with the addition of an external CO_2 source in a CO_2 negative operating condition, indicate that both of these cases can provide a significant efficiency improvement over a chemical looping system without CO_2 recycling. Process simulations for a moving bed reactor also indicate that 100% of the CO_2 generated from the system coupled with an additional 40% CO_2 from an external source can be consumed in a CLR process used to produce a syngas compatible with that from the Fischer–Tropsch synthesis for liquid fuels production. Further, significant economic advantages for using, cocurrently, multiple reducer reactors in a process that employs a CO_2 recycling scheme have been described.

References

1. Takanabe, K., "Catalytic Conversion of Methane: Carbon Dioxide Reforming and Oxidative Coupling," *Journal of the Japan Petroleum Institute*, 55, 1–12 (2012).
2. Shadle, L. J., D. A. Berry, and M. Syamlal, "Coal Conversion Processes," in Seidel, A., and Bickford, M., eds., *Kirk–Othmer Encyclopedia of Chemical Technology*, 3rd ed., John Wiley & Sons, Hoboken, NJ, pp. 224–377 (2002).
3. Higman, C. and M. van der Burgt, "Gasification Processes," Chapter 5 in *Gasification*, 2nd Edition, Gulf Professional Publishing, Houston, TX, pp. 91–191 (2008).
4. Wheeler, W., Apparatus for Lighting Dwellings or Other Structures, U.S. Patent 247,229 (1881).

5. Henry, W., "Experiments on the Quantity of Gases Absorbed by Water, at Different Temperatures, and under Different Pressures," *Philosophical Transactions of the Royal Society of London*, 93, 29–42 and 274–276 (1803).

6. Loomis, B., Process of Manufacturing Gas, U.S. Patent 404,209 (1889).

7. Rose, J. M., Process of Manufacturing Gas, U.S. Patent 415,546 (1889).

8. Lowe, T. S. C., Apparatus for Manufacture of Water-Gas, U.S. Patent 542,566 (1895).

9. Winkler, F., Manufacturing Fuel Gas, U.S. Patent 1,687,118 (1928).

10. Gumz, W., *Gas Producers and Blast Furnaces: Theory and Methods of Calculation*, John Wiley & Sons, Hoboken, NJ (1950).

11. Linde, C., Process of Producing Low-Temperatures, the Liquefaction of Gases, and the Separation of the Constituents of Gaseous Mixtures, U.S. Patent 727,650 (1903).

12. Linde, C., Apparatus for Producing Pure Nitrogen and Pure Oxygen, U.S. Patent 795,525 (1905).

13. Winkler, F., Plate Cylinder for Rotary Intaglio Printing, U.S. Patent 1,643,145 (1927).

14. Skinner, L. C., R. G. Dressler, C. C. Chaffee, S. G. Miller, and L. L. Hirst, "Thermal Efficiency of Coal Hydrogenation," *Industrial & Engineering Chemistry*, 41(1), 87–95 (1949).

15. U.S. Department of Energy, *Evaluation of the Performance of Materials and Components Used in the CO_2 Acceptor Process Gasification Pilot Plant*, Publication Number DOE-ET - 10253 T1, Pittsburgh, PA (1978).

16. U.S. Department of Energy, *Development of the Steam–Iron Process for Hydrogen Production*, Publication Number EF-77-C-01–2435, Pittsburgh, PA (1979).

17. Tri-State Synfuels Company, *Process Evaluation, Selection and Design,* Publication DOE/OR/20807—Tl-Vol.3, Houston, TX (1982).

18. Breault, R. W., "Gasification Processes Old and New: A Basic Review of the Major Technologies," *Energies*, 3(2), 216–240 (2010).

19. Renzenbrink, W., R. Wischnewski, J. Engelhard, and A. Mittelstädt, "High Temperature Winkler (HTW) Coal Gasification. A Fully Developed Process for Methanol and Electricity Production," in *Proc. 1998 Gasification Technologies Conference* (1998).

20. National Energy Technology Laboratory, Fluidized Bed Gasifiers, http://www.netl.doe.gov/research/coal/energy-systems/gasification/gasifipedia/winkler (Accessed on 17th August 2016).

21. Ariyapadi, S., KBR's Transport Gasifier – Technology Advancements & Recent Successes, https://www.netl.doe.gov/File%20Library/research/coal/energy%20systems/gasification/gasifipedia/05ARIYAPADI.pdf (2010).

22. National Energy Technology Laboratory, Fluidized Bed Gasifiers, http://www.netl.doe.gov/research/coal/energy-systems/gasification/gasifipedia/kbr (Accessed on 17th August 2016).

23. National Energy Technology Laboratory, Entrained Bed Gasifiers, http://www.netl.doe.gov/research/coal/energy-systems/gasification/gasifipedia/entrainedflow (Accessed on 17th August 2016).

24. Phillips, J., "Different Types of Gasifiers and their Integration with Gas Turbines," Section 1.2.1 in *The Gas Turbine Handbook*, United States Department of Energy/NETL, Morgantown, WV, pp. 67–77 (2006).

25. National Energy Technology Laboratory, Chemicals,http://www.netl.doe.gov/research/coal/energy-systems/gasification/gasifipedia/commercial-production (Accessed on 17th August 2016).

26. National Energy Technology Laboratory, IGCC Project Examples, http://www.netl.doe
.gov/research/coal/energy-systems/gasification/gasifipedia/tampa (Accessed on 17th August
2016).

27. CB&I, E-Gas Gasification Technology,http://www.cbi.com/getattachment/174013d4-ab9d-
4b53-8e6a-cc1b3a88d46c/E-Gas-Gasification-Technology.aspx (Accessed on 17th August
2016).

28. Amick, P., "E-Gas Technology 2013 Outlook," Presented at 2013 Gasification Technologies Conference, Colorado Springs, CO, October 13–16 (2013).

29. National Energy Technology Laboratory, Entrained Flow Gasifiers, http://www.netl.doe
.gov/research/coal/energy-systems/gasification/gasifipedia/egas (Accessed on 17th August
2016).

30. Mittasch, A. and C. Schneider, German DRP Patent 296,866 (1912).

31. Fischer, F. and H. Tropsch, "Conversion of Methane into Hydrogen and Carbon Monoxide,"
Brennstoff-Chemie, 9, 29–46 (1928).

32. Molburg, J. C. and R. D. Doctor, "Hydrogen from Steam-Methane Reforming with CO_2
Capture," Presented at the 20th annual International Pittsburgh Coal Conference, Pittsburgh,
PA, September 15–19 (2003).

33. Rostrup-Nielsen, J. R., "Syngas in Perspective," *Catalysis Today*, 71(3–4), 243–247 (2002).

34. Byrne, P. J., E. J. Gehr, N. J. Elizabeth, and R. T. Haslam, "Recent Progress in
Hydrogenation of Petroleum," *Industrial & Engineering Chemistry*, 24(10), 1129–1135
(1932).

35. Hawk, C. O., P. L. Golden, H. H. Storch, and A.C. Fieldner, "Conversion of Methane into
Hydrogen and Carbon Monoxide," *Industrial & Engineering Chemistry*, 24(1), 23–27 (1932).

36. Taylor, H. S., "The Purification and Testing of Hydrogen," Chapter X in *Industrial
Hydrogen*, The Chemical Catalog Company, New York, NY, pp. 171–200 (1921).

37. Prettre, M., *Catalysis and Catalysts*, Dover Publications, New York, NY (1963).

38. Rostrup-Nielsen, J. R., "Catalytic Steam Reforming," Chapter 1 in Anderson, J. R., and M.
Boudart, eds., *Catalysis, Science and Technology, Volume 5,* Springer-Verlag, Berlin,
Germany, pp. 1–117 (1984).

39. Barelli, L., G. Bidini, F. Gallorini, and S. Servili, "Hydrogen Production through Sorption-
Enhanced Steam Methane Reforming and Membrane Technology: A Review," *Energy*, 33
(4), 554–570 (2008).

40. Satterfield, C. N, *Heterogeneous Catalysis in Practice*, McGraw-Hill, New York, NY (1980).

41. Twigg, M. V., ed., *Catalyst Handbook*, Wolfe Publishing, London, United Kingdom (1989).

42. Roy, S., B. B. Pruden, A. M. Adris, J. R. Grace, and C. J. Lim. "Fluidized-Bed Steam
Methane Reforming with Oxygen Input," *Chemical Engineering Science*, 54(13–14),
2095–2102 (1999).

43. Methanex, Methanex Medicine Hat Proposed Expansion Project, http://www.ceaa.gc.ca/
050/documents/p80045/89784E.pdf (Accessed on 17th August 2016).

44. Hoek, A., "The Shell GTL Process: Towards a World Scale Project in Qatar," in *Proceedings of the 2006 DGMK Conference*, Dresden, Germany, October 4–6 (2006).

45. Fleisch, T. H., R. A. Sills, and M. D. Briscoe, "2002-Emergence of the Gas-to-Liquids
Industry: A Review of Global GTL Developments," *Journal of Natural Gas Chemistry*, 11
(1–2), 1–14 (2002).

46. Mansar, S., Pearl GTL: The Project, the Plant, & the Products, http://www.advancednrg
solutions.com/upload/SMansar_Pearl_GTL.pdf (Accessed on 17th August 2016).

47. Overtoom, R., N. Fabricius, and W. Leenhouts, "Shell GTL, from Bench Scale to World Scale," in *Proceedings of the 1st Annual Gas Processing Symposium*, Doha, Qatar, January 10–12 (2009).

48. Linde Group, Gas to Liquids (GTL): From Natural Gas to Clean Diesel, http://www.the-linde-group.com/en/clean_technology/clean_technology_portfolio/merchant_liquefied_natural_gas_lng/gas_to_liquids/index.html (Accessed on 17th August 2016).

49. JGC Corporation, GTL, http://www.jgc.com/en/03_projects/01_epc_energy_chemical/04_gtl_pj_01.html (Accessed on 17th August 2016).

50. Hickman, D. A. and L. D. Schmidt, "Production of Syngas by Direct Catalytic Oxidation of Methane," *Science*, 259(5093), 343–346 (1993).

51. Choudhary, T. V. and V. R. Choudhary, "Energy-Efficient Syngas Production through Catalytic Oxy-Methane Reforming Reactions," *Angewandte Chemie International Edition*, 47(10), 1828–1847 (2008).

52. Wright, H. A., J. D. Allison., D. S. Jack, G. H. Lewis., and S. R. Landis, "ConocoPhillips GTL Technology: The COPox™ Process," *ACS Preprints-Fuel Chemistry Division*, 48(2), 791–792 (2003).

53. Halstead, K., "Oryx GTL from Conception to Reality," *Nitrogen+Syngas*, 292, 43–50 (2008).

54. RHI-AG, Autothermal Reformer, http://www.rhi-ag.com/internet_en/products_solutions_en/eec_en/eec_agg_chp_en/eec_agg_chp_ar_en/ (Accessed on 17th August 2016).

55. ThyssenKrupp, Ammonia Plants, https://www.thyssenkrupp-industrial-solutions.com/en/products-and-services/fertilizer-plants/ammonia-plants-by-uhde/ (Accessed on 17th August 2016).

56. Haldor Topsøe, Large Scale Methanol Production from Natural Gas, http://www.topsoe.com/sites/default/files/topsoe_large_scale_methanol_prod_paper.ashx_.pdf (Accessed on 17th August 2016).

57. National Energy Technology Laboratory, Gasifiers & Gasification Tech for Special Apps & Alt Feedstocks, http://www.netl.doe.gov/research/Coal/energy-systems/gasification/gasifipedia/biomass-msw (Accessed on 17th August 2016).

58. National Energy Technology Laboratory, R&D for Gasifier Optimization/Plant Supporting Systems,http://www.netl.doe.gov/research/Coal/energy-systems/gasification/gasifipedia/specialapps (Accessed on 17th August 2016).

59. Jenkins, S., Biomass Gasification 101, http://www.netl.doe.gov/File%20Library/Research/Coal/energy%20systems/gasification/gasifipedia/Session-6-Jenkins-Biomass-Gasification-101.pdf (Accessed on 17th August 2016).

60. National Renewable Energy Laboratory, "From Biomass to Biofuel: NREL Leads the Way," NREL/BR-510-39436 (2006).

61. Nieminen, J. and M. Kivelä, "Biomass CFB Gasifier Connected to a 350 MW$_{th}$ Steam Boiler Fired with Coal and Natural Gas—THERMIE Demonstration Project in Lahti in Finland," *Biomass and Bioenergy*, 15(3), 251–257 (1998).

62. Ryckmans, Y. and F. Van den Spiegel, "Biomass Gasification and Use of Syngas as an Alternative Fuel in a Belgian Coal-Fired Boiler," in *Proceedings of the Conference on New and Renewable Energy Technologies for Sustainable Development*, Ponta Delgada, Azores, Portugal, June 24–26 (2002).

63. Maniatis, K., "Progress in Biomass Gasification: an Overview," Chapter 1 in Bridgwater, A. V., and Maniatis, K., eds., *Progress in Thermochemical Biomass Conversion,* Blackwell Science, Oxford, United Kingdom, pp. 1–31 (2001).

64. Yamada, S., M. Shimizu, and F. Miyoshi, "Thermoselect Waste Gasification and Reforming Process," *JFE Technical Report* 3 (2004).

65. Fan, L.-S., *Chemical Looping Systems for Fossil Energy Conversions*, John Wiley & Sons, Hoboken, NJ (2010).

66. Messerschmitt, A., Process of Producing Hydrogen, U.S. Patent 971,206 (1910).

67. Lane, H., Process of Producing Hydrogen, U.S. Patent 1,078,686 (1913).

68. Ishida, M., D. Zheng, and T. Akehata, "Evaluation of a Chemical-Looping-Combustion Power-Generation System by Graphic Exergy Analysis," *Energy*, 12(2), 147–154 (1987).

69. Zafar, Q., T. Mattisson, and B. Gevert, "Integrated Hydrogen and Power Production with CO_2 Capture Using Chemical-Looping Reforming – Redox Reactivity of Particles of CuO, Mn_2O_3, NiO, and Fe_2O_3 Using SiO_2 as a Support," *Industrial & Engineering Chemistry Research*, 44(10), 3485–3496 (2005).

70. Rydén, M. and A. Lyngfelt, "Using Steam Reforming to Produce Hydrogen with Carbon Dioxide Capture by Chemical-Looping Combustion," *International Journal of Hydrogen Energy*, 31(10), 1271–1283 (2006).

71. Adanez, J., A. Abad, F. García-Labiano, P. Gayan, and L. F. de Diego, "Progress in Chemical-Looping Combustion and Reforming Technologies," *Progress in Energy and Combustion Science*, 38, 215–282 (2012).

72. Lyngfelt, A. "Oxygen Carriers for Chemical-Looping Combustion," in Fennell, P. and B. Anthony, eds., *Calcium and Chemical Looping Technology for Power Generation and Carbon Dioxide (CO_2) Capture*, Woodhead Publishing, Oxford, United Kingdom, pp. 221–254 (2015).

73. Pans, M. A., A. Abad, L. F. de Diego, et al., "Optimization of H_2 Production with CO_2 Capture by Steam Reforming of Methane Integrated with a Chemical-Looping Combustion System," *International Journal of Hydrogen Energy*, 38(27), 11878–11892 (2013).

74. Welty, A. B. Jr., Apparatus for Conversion of Hydrocarbons, U.S. Patent 2,550,741 (1951).

75. Mattisson, T. and A. Lyngfelt, "Applications of Chemical-Looping Combustion with Capture of CO_2," in *Proceedings of the Second Nordic Minisymposium on Carbon Dioxide Capture and Storage*, Göteborg, Sweden, October 26 (2001).

76. Ryden, M., A. Lyngfelt, and T. Mattisson, "Synthesis Gas Generation by Chemical-Looping Reforming in a Continuously Operating Laboratory Reactor," *Fuel*, 85(12–13), 1631–1641 (2006).

77. Rydén, M., M. Johansson, A. Lyngfelt, and T. Mattisson, "NiO Supported on $Mg–ZrO_2$ as Oxygen Carrier for Chemical-Looping Combustion and Chemical-Looping Reforming," *Energy & Environmental Science*, 2, 970–981 (2009).

78. Johansson, E., T. Mattisson, A. Lyngfelt, and H. Thunman, "A 300 W Laboratory Reactor System for Chemical-Looping Combustion with Particle Circulation," *Fuel*, 85(10–11), 1428–1438 (2006).

79. Ryden, M., A. Lyngfelt, and T. Mattisson, "Chemical-Looping Combustion and Chemical-Looping Reforming in a Circulating Fluidized-Bed Reactor using Ni-Based Oxygen Carriers," *Energy & Fuels*, 22, 2585–2597 (2008).

80. Chong, Y. O., D. J. Nicklin, and P. J. Tait, "Solids Exchange between Adjacent Fluid Beds without Gas Mixing," *Powder Technology*, 47(2), 151–156 (1986).

81. Kronberger, B., E. Johansson, G. Löffler, T. Mattisson, A. Lyngfelt, and H. Hofbauer, "A Two-Compartment Fluidized Bed Reactor for CO_2 Capture by Chemical-Looping Combustion," *Chemical Engineering & Technology*, 27(12), 1318–1326 (2004).

82. Cho, P., T. Mattisson, and A. Lyngfelt, "Carbon Formation on Nickel and Iron Oxide-Containing Oxygen Carriers for Chemical-Looping Combustion," *Industrial & Engineering Chemistry Research*, 44(4), 668–676 (2005).

83. Pröll, T., K. Rupanovits, P. Kolbitsch, J. Bolhàr-Nordenkampf, and H. Hofbauer, "Cold Flow Model Study on a Dual Circulating Fluidized Bed (DCFB) System for Chemical Looping Processes," *Chemical Engineering & Technology*, 32(3), 418–424 (2009).

84. Pröll, T., P. Kolbitsch, J. Bolhàr-Nordenkampf, and H. Hofbauer, "A Novel Dual Circulating Fluidized Bed System for Chemical Looping Processes," *AIChE Journal*, 55(12), 3255–3266 (2009).

85. P. Kolbitsch, J. Bolhàr-Nordenkampf, T. Pröll, and H. Hofbauer, "Operating Experience with Chemical Looping Combustion in a 120 kW Dual Circulating Fluidized Bed (DCFB) Unit," *International Journal of Greenhouse Gas Control*, 4(2), 180–185 (2010).

86. T. Pröll, J. Bolhàr-Nordenkampf, P. Kolbitsch, and H. Hofbauer, "Syngas and a Separate Nitrogen/Argon Stream via Chemical Looping Reforming – A 140 kW Pilot Plant Study," *Fuel*, 89(6), 1249–1256 (2010).

87. Linderholm, C., T. Mattisson, and A. Lyngfelt, "Long-Term Integrity Testing of Spray-Dried Particles in a 10-kW Chemical-Looping Combustor using Natural Gas as Fuel," *Fuel*, 88(11), 2083–2096 (2009).

88. Jerndal, E., T. Mattisson, I. Thijs, F. Snijkers, and A. Lyngfelt, "NiO Particles with Ca and Mg based Additives Produced by Spray-Drying as Oxygen Carriers for Chemical Looping Combustion," *Energy Procedia*, 1(1), 479–486 (2009).

89. Huang, Z., F. He, Y. Feng, et al., "Synthesis Gas Production through Biomass Direct Chemical Looping Conversion with Natural Hematite as an Oxygen Carrier," *Bioresource Technology*, 140, 138–145 (2013).

90. Huang, Z., F. He, Y. Feng, et al., "Characteristics of Biomass Gasification using Chemical Looping with Iron Ore as an Oxygen Carrier," *International Journal of Hydrogen Energy*, 38(34), 14568–14575 (2013).

91. Wei, G., F. He, Z. Huang, et al., "Continuous Operation of a 10 kWth Chemical Looping Integrated Fluidized Bed Reactor for Gasifying Biomass Using an Iron-Based Oxygen Carrier," *Energy & Fuels*, 29(1), 233–241 (2014).

92. Huijun, G., S. Laihong, F. Fei, and J. Shouxi, "Experiments on Biomass Gasification using Chemical Looping with Nickel-Based Oxygen Carrier in a 25 kWth Reactor," *Applied Thermal Engineering*, 85, 52–60 (2015).

93. Ge, H., W. Guo, L. Shen, T. Song, and J. Xiao, "Biomass Gasification using Chemical Looping in a 25kWth Reactor with Natural Hematite as Oxygen Carrier," *Chemical Engineering Journal*, 286, 174–183 (2016).

94. Zeng, L., M. V. Kathe, E. Y. Chung, and L.-S. Fan, "Some Remarks on Direct Solid Fuel Combustion using Chemical Looping Processes," *Current Opinion in Chemical Engineering*, 1(3), 290–295 (2012).

95. Fan, L.-S., L. Zeng, and S. Luo, "Chemical-Looping Technology Platform," *AIChE Journal*, 61(1), 2–22 (2015).

96. Luo, S., L. Zeng, D. Xu, M. Kathe, E. Chung, N. Deshpande, L. Qin, A. Majumder, T.-L. Hsieh, A. Tong, Z. Sun, and L.-S. Fan, "Shale Gas-to-Syngas Chemical Looping Process for Stable Shale Gas Conversion to High Purity Syngas with H_2:CO Ratio of 2:1," *Energy and Environmental Science*, 7(12), 4104–4117 (2014).

97. Knowlton T. M., "Standpipes and Return Systems," in Grace, J. R., A. A. Avidan and T. M. Knowlton, eds., *Circulating Fluidized Beds*, Blackie Academic & Professional, London, United Kingdom (1997).

98. Knowlton, T. M., "Solids Transfer in Fluidized Systems," in *Gas Fluidization Technology*, Geldart, D., ed., John Wiley & Sons, New York, NY (1986).

99. Wang, D. and L.-S. Fan, "L-Valve Behavior in Circulating Fluidized Beds at High Temperatures for Group D Particles," *Industrial & Engineering Chemistry Research*, 54(16), 4468–4473 (2015).

100. Fan, L.-S., A. Empfield, M. Kathe, and E. Blair, Chemical Looping Syngas Production from Carbonaceous Fuels, U.S. Patent PCT/US2017/027241 (2017).

101. Goellner, J. F., V. Shah, M. J. Turner, N. J. Kuehn, J. Littlefield, G. Cooney, and J. Marriott, *Analysis of Natural Gas-to Liquid Transportation Fuels via Fischer-Tropsch*, United States Department of Energy/NETL, DOE/NETL-2013/1597, Pittsburgh, PA (2013).

102. U.S. Energy Information Administration, Natural Gas, http://www.eia.gov/naturalgas/weekly/ (Accessed on 17th August 2016).

103. de Klerk, A., "Gas-to-Liquids Conversion," Presented at Natural Gas Conversion Technologies Workshop of ARPA-E, U.S. Department of Energy, Houston, TX, January 13 (2012).

104. United States Department of Energy/National Energy Technology Laboratory, *Quality Guidelines for Energy System Studies: Specification for Selected Feedstocks*, United States Department of Energy /NETL, DOE/NETL-341/011812, Pittsburgh, PA (2012).

105. Kathe, M., A. Empfield, P. Sandvik, C. Fryer, Y. Zhang, E. Blair, and L.-S. Fan, "Utilization of CO_2 as a Partial Substitute for Methane Feedstock in Chemical Looping Methane-Steam Redox Processes for Syngas Production," *Energy & Environmental Science*, DOI: 10.1039/C6EE03701A (2017).

106. Kathe, M., C. Fryer, P. Sandvik, F. Kong, Y. Zhang, A. Empfield, and L.-S. Fan, "Modularization Strategy for Syngas Generation in Chemical Looping Methane Reforming Systems with CO_2 as Feedstock," *AIChE Journal*, DOI: 10.1002/aic.15692 (2017).

5 Catalytic Metal Oxides and Applications

A. Majumder, A. Wang, C. Chung, T.-L. Hsieh, and L.-S. Fan

5.1 Introduction

Selective oxidation reactions of carbon based feedstocks synthesize important chemicals and their precursors, such as alcohols, epoxides, aldehydes, and other organic compounds. They employ catalytic metal oxides, which are multifunctional, enabling materials, to fulfill three parallel roles: (1) provide active surface sites for the reactants without participating in the reaction; (2) reduce the activation energies of the reaction pathway; and (3) oxidize the reactants through the transfer of their lattice oxygen during reduction. The third role is akin to the well-known catalytic surface Mars–Van Krevelen mechanism, which is further discussed in this section and Section 5.6, that is characterized by the reaction products incorporating the compounds of the catalytic material lattice.[1] In these kinds of selective oxidation reactions, the catalytic metal oxides behave as "oxygen carriers," where oxygen from the crystal lattice bonds with the reactants. In these reactions, the surface of the metal oxide is reduced by the hydrocarbon reactant and can later be re-oxidized by gaseous oxygen.

The selectivity of a catalytic metal oxide can be controlled by multiple factors, such as structure, chemistry, electronic properties, composition, kinetics, and energies.[2] An ideal selective catalytic metal oxide needs to be tailored specifically for the application at hand. Hence, it becomes essential to understand the surface morphology, crystal structure, and the reaction mechanisms involved. The accessibility and mobility of lattice oxygen play a vital role in determining the reactivity and selectivity of the catalytic metal oxides. The structure of the catalytic metal oxide should allow for sufficient transportation of lattice oxygen toward the surface and needs to be able to sustain the vacancy created by the absence of oxygen. That is, the host structure needs to provide a sustainable pathway for the ions to diffuse in order to utilize the lattice oxygen. While participation of lattice oxygen facilitates the oxidation–reduction of the process, it does not necessarily result in high product selectivity. The selectivity of a given structure can originate from the bond strength between metal and oxygen, where a weak metal–oxygen bond will release oxygen that can fully oxidize the reactant and yield an undesired product.[3] An ideal catalytic metal oxide with readily accessible and mobile lattice oxygen, appropriate host structure, and suitable metal–oxygen bond strength has the potential to be both reactive and selective when reduced. That is, the catalytic metal oxide needs to have a low oxygen vacancy formation energy so that the lattice oxygen can transport to the surface, and the active sites on the surface need to render the partial oxidation reaction barrier lower than the full oxidation reaction barrier. In addition, the reduced catalytic metal oxide surface needs to be favorable for gaseous oxygen adsorption and dissociation for catalyst regeneration. These factors are essential for a catalytic metal oxide to be effective for selective oxidation.

As selective oxidation to a desired product requires the catalytic metal oxide to have active sites and mobile lattice oxygen, it is necessary to have a thorough understanding of the driving adsorption mechanisms. Various surface reaction models have been developed to describe the adsorption mechanisms associated with catalytic metal oxides. The most well known models to describe oxidation reactions are the Langmuir–Hinshelwood–Hougen–Watson (LHHW) mechanism, the Eley–Rideal mechanism, and the Mars–Van Krevelen mechanism, as illustrated in Figure 5.1a, 5.1b, and 5.1c, respectively. The LHHW mechanism occurs when the two reacting molecules adsorb onto the surface of the catalyst, and the adsorbed molecules react to form the product. The Eley–Rideal mechanism occurs when one of the reactant molecules adsorbs to the surface, and the adsorbed molecule reacts with the other gas phase reactant to form the product. The Mars–Van Krevelen mechanism is similar to the Eley–Rideal mechanism but the reactant molecule reacts with the lattice molecules, and this lattice-reacted molecule reacts with the other gas phase reactant to form the product.

This chapter presents additional examples on catalytic partial oxidation reactions using lattice oxygen transfer. The focus of discussion in each example is on the possible reaction mechanisms and the effect of metal oxide structure and morphology on product

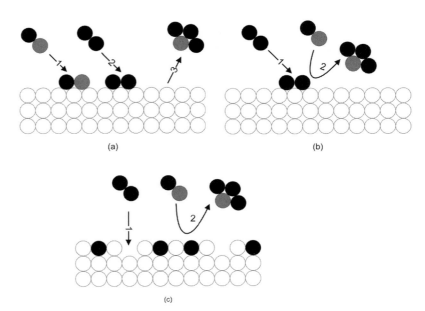

(a)   (b)

(c)

Figure 5.1 Major reaction mechanisms with selective oxidation such as (a) Langmuir–Hinshelwood–Hougen–Watson, (b) Eley–Rideal, and (c) Mars–Van Krevelen.[1]

selectivity and yield. The discussion begins with the example given in Section 5.2 which describes DuPont's novel two-step process for the production of maleic anhydride (MAN), a tetrahydrofuran intermediate, from butane. It illustrates different experimental scales of the process from the lab-scale to the commercial-scale set-up, along with the catalytic metal oxide performance, reactant conversions, and product selectivity at each scale. Syngas and/or hydrogen generation through solar-aided thermochemical processes using catalytic metal oxides is described in the second example, discussed in Section 5.3. In this case, the catalytic metal oxides serve the purpose, not only of carrying and releasing oxygen to the reactants for oxidation or releasing it directly as a product, but also as a heat transfer medium. Therefore, effective heat transfer becomes an extremely essential criterion during metal oxide selection and process development. The third example in Section 5.4 discusses the effect of the structural specificity of catalytic metal oxides on the selective oxidation of methane to formaldehyde, using molybdenum trioxide as an example. Specifically, the effect of its crystal planes and surface metal-oxygen sites on the redox mechanisms and product selectivity during methane oxidation is described. This is followed by illustration of the possible redox mechanisms involved in the partial oxidation of propylene over bismuth molybdate, as the fourth example in Section 5.5, and the involvement of lattice oxygen in several homogeneous gas phases and heterogeneous liquid phase partial oxidation reactions over other group IV catalytic metal oxides, as the fifth example, given in Section 5.6. As one proceeds through each example, it becomes evident how the previously mentioned factors of catalytic metal oxides are essential for each of these reactions. For the reactions in Section 5.4 to Section 5.6, the oxygen transfer and reaction mechanisms of the catalytic metal oxides that are presented can also be used for the oxygen carriers in chemical looping, but there is a distinction between the chemical looping concept and these selective oxidation reactions that proceed via lattice oxygen transfer. While chemical looping is based on the high temperature cyclic reduction and re-oxidation of oxygen carriers, these reactions do not require a separate regeneration step. The oxygen vacancies created by the lattice oxygen in the catalytic metal oxides as they oxidize the reactants are constantly replenished by gaseous oxidizing agents in the same reactor. It is noted that in Chapter 3, metal oxides used in the co-feed mode are referred to as metal oxide catalysts, based on their function as a conventional catalyst that does not donate oxygen during the reaction. In this chapter, the metal oxide used in the co-feed mode serves the function of a catalytic metal oxide that donates oxygen and also carries out the catalytic reactions.

5.2   DuPont's Process for Production of Tetrahydrofuran (THF)

Tetrahydrofuran (THF) is the starting compound for producing the polymer polytetramethylene ether glycol, which is a component of DuPont's Lycra® elastane fiber and Hytrel® copolyester elastomer products.[4] This section discusses methods for synthesizing THF and DuPont's process. Within DuPont's process, the main focus is on the first step of the THF synthesis process, the generation of maleic anhydride (MAN) from *n*-butane. The first step utilizes a vanadium phosphorus oxide (VPO) as the catalytic metal oxide and this section specifically focuses on the role of the catalytic metal oxide

VPO. The oxygen donating capacity of this catalytic metal oxide, possible reaction mechanism, synthesis, and the subsequent scaling up of the chemical looping application using this catalytic metal oxide from lab scale to commercial scale in carrying out MAN production are described. The challenges reported by DuPont during different stages of commercialization of MAN production are also discussed.

5.2.1　THF Synthesis Processes

One of the classical methods for THF production is the Reppe process, which was originally also used by DuPont.[5] In the Reppe process, acetylene and formaldehyde react to yield 2-butyne-1,4-diol, followed by hydrogenation to form 1,4-butanediol and acid-catalyzed dehydration of the saturated diol above 100 °C to form THF. Suitable catalysts include inorganic acids, acidic aluminum silicates, and rare earth oxides.[6]

THF can also be synthesized by other processes such as:

1.　Decarboxylation of furfural with a zinc–chromium–molybdenum catalyst, followed by hydroxylation to THF.
2.　Oxidation of butadiene to 1,4-diacetoxy-2-butene over a palladium–tellurium catalyst with acetic acid and a nitrogen–oxygen mixture.[6] This is followed by hydrogenation to 1,4-diacetoxybutane, and further hydrolysis to butanediol or THF.
3.　Catalytic hydrogenation of furan with a nickel catalyst.[6,7]

However, these processes are not viable for commercial THF production due to low selectivity and conversion, and therefore most commercial THF production is via the classical Reppe process. However, the carcinogenic nature of formaldehyde and the high cost of acetylene have made it imperative to look for improved alternatives.

5.2.2　DuPont's Method for THF Production

In the early 1980s, DuPont began investigating a two-step process for the production of THF from n-butane, shown in reaction (5.2.1) and reaction (5.2.2).[4] The process involved oxidation of n-butane to MAN in a fluidized bed reactor with an attrition-resistant VPO based catalytic metal oxide.[8,9] MAN is readily converted to an aqueous maleic acid solution, which is subsequently hydrogenated to THF over a rhenium-doped palladium catalyst:

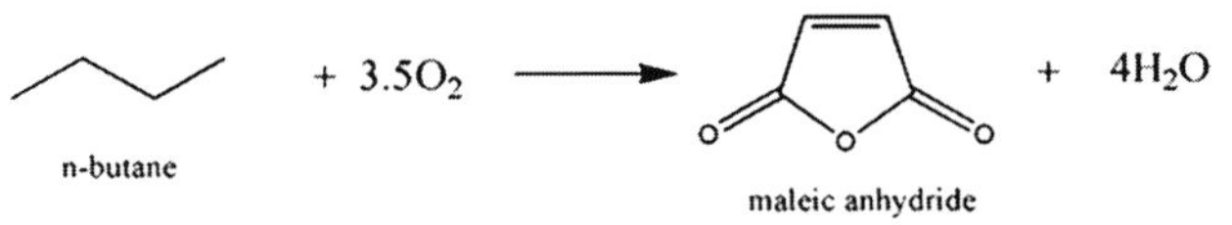

$$n\text{-butane} + 3.5O_2 \longrightarrow \text{maleic anhydride} + 4H_2O \qquad (5.2.1)$$

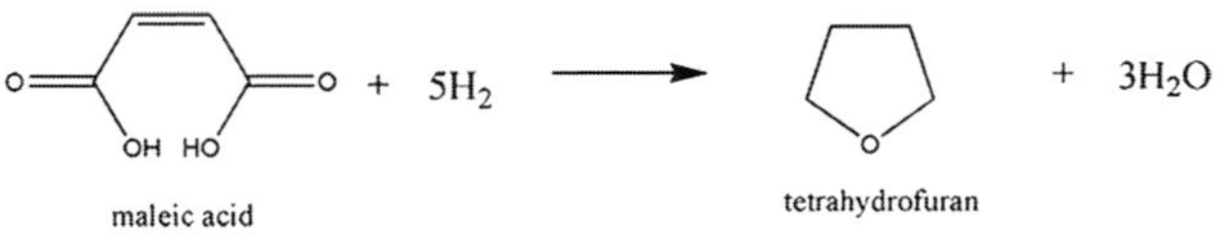

$$\text{maleic acid} + 5H_2 \longrightarrow \text{tetrahydrofuran} + 3H_2O \qquad (5.2.2)$$

The conventional method for MAN production was through the oxidation of benzene, but the carcinogenic nature of benzene and its limited resources compelled the necessity to seek an alternative feedstock for the commercial production of MAN. The capital cost for

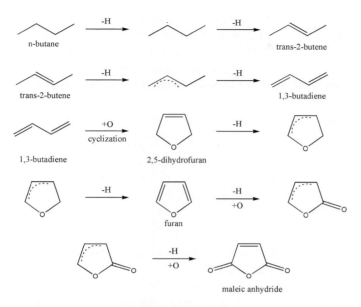

Figure 5.2 Mechanism for oxidation of *n*-butane to maleic anhydride.[13]

MAN production through the traditional route and DuPont's new route were comparable, and hence the difference lay mainly in the price of the feedstock.[10] There follows a list of the advantages of producing THF through the *n*-butane route (Figure 5.2).

1.   *n*-butane is easily available and inexpensive, which largely reduces the operating cost.
2.   *n*-butane is an environmentally safer option when compared to benzene.
3.   *n*-butane route uses all four carbon atoms of *n*-butane, whereas only four out of the six carbons of benzene are utilized, therefore, it is a better option because of the higher C conversion efficiency.[11]
4.   This route provides high selectivity for THF and high product purity.[12]
5.   The reduced number of steps results in a simplified process and hence fewer reactors are involved.

5.2.3   Catalytic Metal Oxide VPO

The highest activity for the oxidation of *n*-butane to MAN, reaction (5.2.1), has been found using the VPO catalyst. The activity of the catalytic metal oxide and its selectivity toward MAN greatly depends on the surface to volume ratio and the valence state of the vanadium ions in the metal oxide system. These parameters can be controlled by:

1.   The method of preparation of the catalytic metal oxide.[14]
2.   The composition of the catalytic metal oxide.[15]
3.   The operating conditions in the reactor, such as the residence time and the hydrocarbon concentration in the feed.[16]

The highest yields of MAN reported in the literature occurred when vanadium had an average valence between +4 and +5 (4.0–4.6), and the phosphorus to vanadium (P/V) ratio was in the range 0.9–1.6.[11,17] The P/V ratio dictates the selectivity toward MAN. It was found that the addition of P_2O_5 to the V_2O_5 catalytic system decreased the surface area and the acidity of the catalytic metal oxide. Thus, the addition of P_2O_5 decreased the activity of the catalytic metal oxide due to a decrease in the surface area, but a reduction in the acidic strength suppressed breakage of the C–C bond in butene and increased the selectivity for allylic oxidation. Hence the C_4H_8 to C_4H_6 conversion was promoted.[18,19]

For this reaction, the oxidation rate depended primarily on two factors: (1) oxygen chemisorption; and (2) surface oxygen reaction with butane gas. Thus, a reduction in surface area leads to a reduction in metal oxide activity. The apparent activation energy for the oxygen chemisorption was found to be larger than the surface reaction activation energy. At higher butane pressures (>2.7 kPa or 20 torr), oxygen chemisorption is the rate controlling step, while at lower butane pressures (<0.9 kPa or 7 torr), the surface reaction is the rate determining step.[20]

The VPO catalyst oxidizes n-butane to MAN by redox reactions on its surface layers. In its fully oxidized state, VPO selectively oxidizes n-butane to MAN through the formation of intermediate olefins, as shown in Figure 5.2, and the catalytic surface is reduced in the process, which is then re-oxidized by molecular oxygen. Figure 5.3 shows a simplified reaction pathway that the oxidation of n-butane to MAN is believed to follow.

This reaction pathway is dependent on the number and configuration of active oxygens in the vicinity of the adsorption sites on the catalytic surface. The active oxygen is chemisorbed onto the nearest adsorption sites, forming oxidizing sites for n-butane. The product yield and selectivity depends on the availability of the oxidizing sites. In the presence of sufficient oxidizing sites, the olefins quickly react to form MAN. When there is a lack of oxidizing sites, the olefins are desorbed from the catalytic surface without further reaction. When the number of oxidizing sites is insufficient, there are two situations:

1.   The existing literature reports the maximum yield of MAN occurs at temperatures at around 360 °C. When the temperature increases, the rate of parallel reaction increases, resulting in the formation of more carbon oxides as compared to MAN.

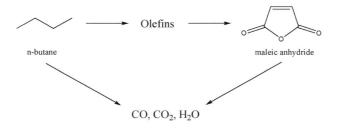

Figure 5.3 Simplified version of a possible reaction pathway for the oxidation of n-butane to maleic anhydride.[11]

The parallel reaction results in the total consumption of oxygen and thus, limits the formation of oxidizing sites for MAN synthesis.[21] As a result, the n-butane to MAN conversion is incomplete, and more olefins are found in the product.

2. Upon prolonged exposure in a reducing atmosphere, the surface layers of the metal oxide are over-reduced, resulting in loss of oxidizing sites. Thus, a formation of reduced layers on an oxidized core builds when the diffusion of lattice oxygen to the surface is very slow.[22]

At low concentrations of n-butane, the number of oxidizing sites is higher and thus, the olefins that are formed quickly react to form MAN. Since the reaction step from olefins to MAN is very fast, it becomes difficult to differentiate between the catalytic active sites for the dehydrogenation of n-butane and those necessary for oxidizing the olefins. It is believed that the reaction mechanism for the oxidation of n-butane to MAN is similar at low and high n-butane concentrations. The difference lies in the distribution of active oxidizing sites under these two conditions.

During synthesis of the catalytic metal oxide VPO, the P/V ratio plays a crucial role in determining the reactivity, selectivity, and attrition resistance of the metal oxide. VPO is synthesized beginning with a mixture of vanadium pentoxide, isobutyl alcohol, and benzyl alcohol, that is refluxed for 12 h, followed by the slow addition of 85% phosphoric acid and another 12 h of reflux. The mixture is then filtered and spray dried at 110 °C.[17] It was initially presumed that the P/V ratio in the catalytic metal oxide equals the P/V ratio used during synthesis. However, this was found to be true only over a limited range. As the P/V ratio increases beyond the range 0.9–1.1, the selectivity for MAN increases; however, the catalytic activity decreases.[18] At higher P/V ratios, the primary crystalline product formed is $(VO)_2H_2O(PO_3OH)_2$, while the majority of the excess phosphorus remained in solution and may form $VO(H_2PO_4)_2$. The changing P/V ratio during synthesis can, however, dramatically affect the crystallinity of the $(VO)_2H_2O(PO_3OH)_2$ phase. During calcination, the $VO(H_2PO_4)_2$ phase is converted to a $V^{5+}/V^{4+}/P^{5+}/O$ amorphous phase. The VPO particles synthesized by this method produced particles that were ideally shaped for a fluidized bed reactor but with very poor attrition resistance. To reduce the attrition rate, 30%–50% by weight of colloidal silica was added to the VPO particle before spray drying. After calcination, an attrition resistant particle was formed, with the silica uniformly distributed in the catalytic metal oxide particle. However, the presence of silica caused an unacceptable drop in selectivity toward MAN. As a result, DuPont developed a new synthesis technique to improve the attrition resistance without reducing product selectivity. A slurry of polysilicic acid was used with the metal oxide precursor, $(VO)_2H_2O(PO_3OH)_2$, to form a thin and durable silica shell on the final catalytic metal oxide particles. It was believed that the thin shell permitted the free passage of reactants and products and did not show any significant degradation of catalytic metal oxide properties. However, as the process was scaled, it became clear that the attrition resistant VPO particle failed to supply the required amounts of oxygen; therefore, molecular oxygen had to be either co-fed with n-butane or injected separately.

Figure 5.4 is a schematic of DuPont's chemical looping circulating fluidized bed (CFB) reactor for MAN production from *n*-butane.[22] The catalytic metal oxide VPO was oxidized with air in the fluidized bed regenerator, and *n*-butane was selectively oxidized to MAN in a separate riser reactor. The reduced metal oxide was separated from the product stream at the top of the riser, removed of any carbonaceous species in a stripper zone, and returned to the regenerator for re-oxidation. For one complete loop, the catalytic metal oxide VPO underwent one reduction/oxidation cycle.

5.2.4   Lab-Scale Tests

Lab-scale tests were conducted using a CFB reactor with a riser reactor 6 mm in diameter and 1.5 m in height. The results, shown in Figure 5.5, demonstrate that using

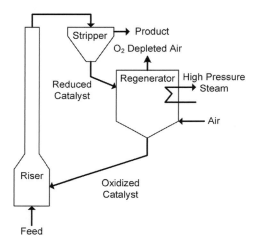

Figure 5.4 Schematic of chemical looping circulating fluid bed (CFB) reactor system for butane oxidation.[22]

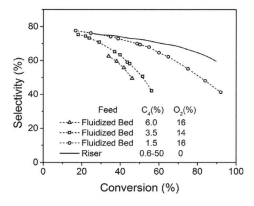

Figure 5.5 Selectivity versus conversion for butane oxidation in CFB versus a fluidized bed reactor.[22]

a CFB circumvented the need to introduce molecular oxygen in the riser.[22] However, this was found to be true only during the lab-scale tests. At larger scales, molecular oxygen was required in the riser to improve conversion and selectivity. The CFB showed higher selectivity to MAN as compared to the steady-state fluidized bed. In fixed and fluidized beds, due to the longer residence time, the catalytic metal oxide VPO was over-reduced, which decreased the selectivity to MAN. This could be avoided by streaming molecular oxygen into the fixed and fluidized bed reactors. The CFB riser allowed for shorter residence times of the VPO before recycling it to the regenerator zone. Thus, over-reduction and the addition of molecular oxygen in the riser could be minimized. In addition to attrition resistance, the catalytic metal oxide is required to have the capability for easy and fast oxygen donation. The oxygen requirement for this reaction is very high with a 3.5:1 oxygen to butane stoichiometric molar ratio. Therefore, the residence time in the riser had to be optimized depending on the extent of reduction and the ability of lattice oxygen contribution by the catalytic metal oxide.

5.2.5 Bench-Scale Tests

Catalytic metal oxide stability was further tested in a 6 mm riser facility composed of four interconnected quartz vessels placed inside an electric furnace, as shown in Figure 5.6, and operating conditions given in Table 5.1.[23] The unit consisted of a fluidized bed, a riser, a stripper, and a regenerator. The unit was charged with 1 kg of the

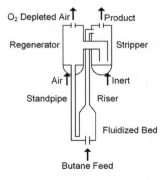

Figure 5.6 Bench-scale CFB system with a ¼ in riser reactor for butane oxidation.[23]

Table 5.1 Operating conditions for the bench-scale tests.[23]

Reaction temperature	360–400 °C
Butane feed concentrations	5–15 vol%
Air flow rate to the regenerator	2.1 L/min (STP)
Nitrogen flow rate to the stripper	0.3 L/min
Gas flow rate to the fluidized bed	2.6 L/min (STP)
Solid circulation rate	3.5 kg/hr

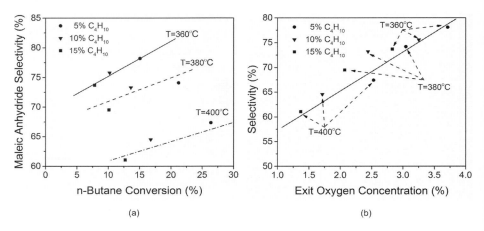

Figure 5.7 Performance evaluation of the catalytic metal oxide VPO in ¼ in riser with respect to maleic anhydride selectivity versus (a) butane conversion and (b) exit oxygen concentration.[23]

catalytic metal oxide – approximately 600 g in the regenerator, 300 g in the stripper, and 100 g in the fluidized bed/riser.

The bench unit was operated in both the redox and co-feed modes. For most co-feed conditions, the reactor was operated with 5.5 vol% oxygen. In the redox mode, all the oxygen during the reaction was provided by the metal oxide lattice. During most co-feed conditions, lattice oxygen provided approximately 50% of the total oxygen requirement during the reaction, with the remaining provided by molecular oxygen.

The deficiency in the oxygen transfer capacity of the metal oxide was observable during the bench-scale tests. Figure 5.7a and Figure 5.7b show the results from the tests conducted in the ¼ in riser.[23] The results suggest that MAN selectivity has a linear correlation to the oxygen exit concentration. Some studies report that increasing the oxygen concentration co-fed to the riser prevents the over-reduction of metal oxide, but may lower the selectivity.[24] However, there is some uncertainty associated with this because contradictory results have also been reported.[25] Another conclusion made from the bench-scale experiments was that high inlet butane concentrations resulted in a higher lattice oxygen contribution, often at the expense of selectivity.

5.2.6   Pilot Unit Tests

The pilot unit, shown in Figure 5.8, consisted of five vessels – fluidized bed, riser, riser stripper, regenerator, and regenerator stripper.[23] The fluidized bed was 0.3 m in diameter and 6 m in height, with a sparger injecting the recycle gas from the bottom. The riser had a diameter of 0.15 m and was 24 m in height. The regenerator had a diameter of 0.53 m and both the riser stripper, and the regenerator stripper had a diameter of 0.44 m. Oxidized VPO from the regenerator stripper flowed into the fluidized bed through a 3 m long vertical standpipe with a V-bend. The solid mass flux reached a maximum of about 220 kg/(m^{2}s) and the gas velocities reached 2 m/s. The

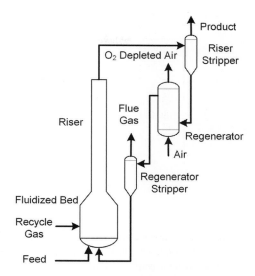

Figure 5.8 DuPont's pilot scale CFB reactor system for butane oxidation.[23]

stripping gas in the pilot unit was changed from nitrogen to steam in order to reduce the associated purge losses, but due to metal oxide agglomeration, DuPont eventually shifted to air as the stripping gas.

DuPont's pilot unit for MAN production was primarily focused on demonstrating operability and metal oxide performance in terms of selectivity, conversion, stability, attrition resistance, lattice oxygen contribution, and minimizing the metal oxide inventory. One of the initial problems during operation was solid circulation stability, due to the accumulation of fines and tuning issues with the standpipe. An issue addressed during pilot-scale testing was co-feeding of oxygen with n-butane. It was concluded from the bench-scale tests that co-feeding oxygen prevented metal oxide over-reduction. Safely injecting oxygen into a fluidized bed along with n-butane was a major problem due to flammability constraints. Oxygen was injected through spargers installed at multiple levels in the fluidized bed to bypass the flammability limitations. At high oxygen feed rates, multiple hotspots were observed with an increase in CO_2 concentration and a drop in O_2. The pilot unit was later modified by feeding in sufficient quantities of air in the regenerator stripper to re-oxidize the catalytic metal oxide in the stripper itself. This eliminated the regenerator completely, thereby reducing the catalytic metal oxide inventory by a factor of three and increased the MAN production to about 70 kg/hr per ton of catalytic metal oxide.

The lattice oxygen contribution was also one of the parameters studied during the pilot-scale tests. Since the partial pressure of butane in the pilot unit were much higher compared to the lab and bench scale, it was expected that the lattice oxygen contribution would be higher as well. Even though the contribution of lattice oxygen did increase with butane concentration, it did not increase by the same proportion. The residence time of the solids in the reducing zone of the 6 mm riser was about 1.5 min, whereas in the case of the pilot unit it was ~0.5 min. This was believed to be one of the reasons for

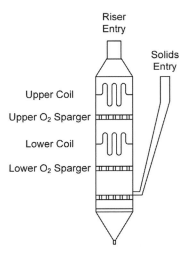

Figure 5.9 DuPont's commercial fluidized bed reactor configuration.[23]

the lower lattice oxygen contribution. Thus, the reduction time played an important role in the lattice oxygen contribution.

5.2.7   Commercial-Scale Tests

The commercial unit primarily consisted of the fluidized bed, riser, riser stripper, cyclone, and regenerator. The riser was 1.8 m in diameter and 28.5 m in height, the riser stripper was 4 m in diameter and 9 m in height, and the cyclone was 3 m in diameter. The schematic of the commercial fluidized bed reactor configuration is shown in Figure 5.9.[23] A grid plate distributor was used to inject fresh butane, along with a recycle stream to the 4.2 m diameter fluidized bed. Additional oxygen was fed to the fluidized bed at multiple levels through spargers. The fluidized bed and the regenerator were connected by a 1.2 m diameter standpipe. The standpipe cross-section changed from circular at the regenerator exit to elliptical at the fluidized bed entrance. The 11.5 m tall fluidized bed was equipped with cooling coils for heat removal. The solid circulation rate was controlled by two slide valves, one placed immediately below the cone of the riser stripper, and the other below the regenerator.

The commercial plant studies were aimed at optimizing the catalytic metal oxide inventory, minimizing attrition losses of the catalytic metal oxide, and maximizing heat transfer surface and uniform oxygen distribution in the fluidized bed and stable solid circulation.[23] In the commercial unit, DuPont managed to achieve a stable solid circulation stream. The overall pressure profile, shown in Figure 5.10a, was very different from that in the pilot unit, as the commercial unit had one fewer vessel, shorter standpipes, and only vertical flow of solids in the fluidized bed/riser. Figure 5.10b shows the concentration profile along the fluidized bed in the commercial unit.

From the pilot unit, the attrition rate was projected to be 5–15 kg/hr. The experimental attrition rate, however, showed a lower value, in the range 1–2.5 kg/hr over the

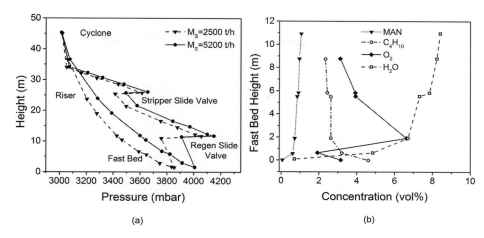

Figure 5.10 DuPont's commercial scale CFB system: (a) pressure profile; (b) concentration profile along the fluidized bed.[23]

first 100 days of operation. One of the concerns addressed during the commercial operation was maintaining the temperature profile in the fluidized bed and the regenerator, which depended on the solid circulation rate and reaction rates. As some uncertainty was associated with proper heat transfer in the system, both the fluidized bed and the regenerator at the commercial unit had cooling coils installed. The commercial unit was operated at a rate of 3500 kg/hr with an inlet butane concentration of 4.6 vol%, 3.1 vol% oxygen, and 0.7 vol% water vapor. At the exit of the fluidized bed, the gas concentration was 2.3 vol% butane, 1.9 vol% oxygen, 1 vol% MAN, and 8.4 vol% water. Based on the oxygen concentrations measured from various positions in the fluidized bed and the riser, it was concluded that the gas phase was not perfectly backmixed. The unit was operated at a standby circulation rate of 1600 kg/hr and a peak circulation rate of 6500 kg/hr. The maleic acid selectivity was similar to the pilot unit, ranging between 55 and 70 vol%. The commercial unit was operated at a higher solid circulation rate and hence had a lower residence time in the regenerator. Further, in order to reduce the butane purge losses, the butane feed concentration was lowered and the oxygen was doubled. As a result, the lattice oxygen contribution was reduced by a factor of 2.

The concept of DuPont's chemical looping process for *n*-butane oxidation to MAN for the production of THF is novel in light of the expected economic and environmental benefits of the process. The replacement of molecular oxygen with the catalytic metal oxide VPO, that served as the oxygen donor, helped circumvent the costly air separation system. The catalytic metal oxide VPO acted as an oxygen donor with feasible oxygen transport properties, conversions, and selectivity for MAN at small scales. Unfortunately, its commercial operation was hampered partly by the inadequacies in the chemical and mechanical viabilities of the catalytic metal oxide VPO, and their associated effects on the reaction kinetics of the particles. The operation was also affected by deficient transport properties of the VPO particles in performing the desired reaction and looping functions in a CFB system. It was known that in the regenerator, overheating led to agglomeration of the VPO particles that impeded or blocked proper particle flows. As effective oxygen donor behavior could not be maintained, the scale-up operation required increasing amounts of supplemental molecular oxygen in order to yield acceptable production of MAN, reflecting the inadequate oxygen donor property of the catalytic metal oxide. For a sustainable chemical looping operation, the catalytic metal oxide VPO must be able to provide the majority of the oxygen required for the reaction, with the remainder of it provided by the molecular oxygen. As the reaction occurs primarily on the surface of the catalytic metal oxide, the diffusion of the oxygen to the metal oxide surface for the reaction, as well as metal oxide regeneration, should readily be achievable. Since the reactant conversions and product selectivity are highly sensitive to the catalytic metal oxide conditions, a better understanding and control of the oxidation states of the catalytic metal oxide and its physical strength for operation in a fluidized system are deemed necessary.

5.3   Syngas and Hydrogen Generation with Solar Energy

Solar radiation is the most abundant source of energy in our planet's ecosystem. Considerable advances have been made in utilizing solar energy for industrial applications such as electricity generation via both photovoltaic (PV) and/or concentrating solar power (CSP) processes and hydrogen and syngas generation.[26–29] However, they are associated with technological and economic challenges. This section focuses on unique solar thermochemical applications involving lattice oxygen transfer of metal oxides for syngas or hydrogen production.

Solar thermochemical processes for syngas or hydrogen generation are an advanced hybrid process combining the reaction engineering of metal oxide lattice oxygen transfer with the process design of concentrating solar power systems. This metal oxide chemical looping and solar thermochemical combined cycle has the potential to reduce carbonaceous fuel consumption associated with conventional syngas/hydrogen production, and more importantly to allow for ultra-high temperature reactions for the thermolysis of water or carbon dioxide (CO_2). Under such a process design, the metal oxide material serves both as oxygen carrier and the heat transfer medium. In addition, the ability to produce hydrogen/syngas, which can be used as fuel or as an intermediate feedstock for a wide range of valuable chemical products, from renewable energy offers additional flexibility and marketability. The key technical challenge of this technology lies in the selection of a suitable metal oxide that must fulfill the following criteria: (1) lattice oxygen transfer properties with suitable redox thermodynamics; (2) chemical and physical stability of the material at the reaction temperature; (3) heat conduction properties; (4) reaction kinetics; and (5) economics. These criteria are similar to the properties required when selecting oxygen carriers, as discussed in Chapter 2.

As illustrated in Figure 5.11, syngas/hydrogen generation from solar energy can occur by two different approaches: (1) solar water splitting process, and (2) solar thermochemical process. The two processes are differentiated based on their feedstock, technological maturity, and required solar input. The solar thermochemical process for

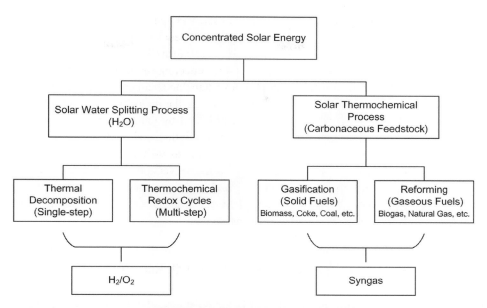

Figure 5.11 Solar thermal and solar thermochemical reaction schemes for syngas and hydrogen generation.

reforming/gasification to produce syngas replaces fossil fuels with solar energy to supply the energy for the endothermic heat of reaction. This design can potentially reduce the fossil fuel consumption in a reforming process by 20–35%. In addition, using a metal oxide based reforming catalyst as the heat transfer fluid has the potential to simplify the process flow and improve the economics. Because the reforming/gasification and solar thermal processes are both commercially ready, the key engineering challenge exists in the process integration and scale-up of this combined cycle design. In comparison, greater technical challenges are associated with solar water splitting process with respect to the temperature requirements of the dissociation of water into oxygen and hydrogen. The direct one-step reaction or thermal decomposition is only thermodynamically feasible at extremely high temperatures (>2200 °C). Because of the high temperature requirement, the material selection for reactor construction, the process configuration for better efficiency, and the optimal reaction engineering all require further research. Thus, this technology is not ready for commercialization in the near future. The indirect two-step reaction through thermochemical redox cycles using metal oxide material as an oxygen transfer intermediate represents a relatively attractive process because it renders the water splitting reaction thermodynamically feasible at more moderate temperatures. Existing research efforts have focused on the selection of a reactive and thermally/physically stable metal oxide material, particularly at moderate temperatures, that can facilitate the lattice oxygen transfer mechanism, as well as novel reactor designs that allow for improved heat transfer. In the following sections, ongoing research and progress made in these above mentioned solar thermal processes for the generation of hydrogen and/or syngas are elaborated.

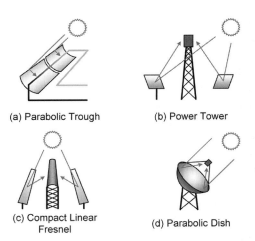

(a) Parabolic Trough   (b) Power Tower

(c) Compact Linear Fresnel   (d) Parabolic Dish

Figure 5.12 Concentrating solar power systems: (a) parabolic trough, (b) power tower, (c) compact linear Fresnel, (d) parabolic dish.[27]

5.3.1   Concentrating Solar Thermal Receiver Design

Concentrating solar power (CSP) systems typically consist of three main components: (1) solar collector field; (2) solar receiver; and (3) thermal conversion system. Hybrid solar thermochemical plants, such as the solar reforming combined cycle or solar water splitting processes, may incorporate additional reactor and thermal storage systems to enhance the system performance and capacity factor. The vast majority of existing CSP systems are aimed at producing electricity from solar energy. The four common types of CSP systems, as illustrated in Figure 5.12, can be described as: (a) parabolic trough; (b) power tower; (c) compact linear Fresnel; and (d) parabolic dish. Depending on the intensification factor of the solar radiation, the subsequent operating temperature, and the process scale, the four aforementioned CSP systems are targeted for different end applications. The parabolic trough and linear Fresnel systems have a lower operating temperature, in the range 400–500 °C, thus they are ideally suited for electricity generation via the conventional steam turbine cycle. The hybrid solar thermochemical processes operate at temperatures in excess of 900 °C that are only attainable in the power tower and dish systems.

The difference between the power tower and dish systems resides in its scalability. The power tower is a highly scalable system that has been demonstrated at 100 MW while dish systems are modular units that operate in the range 3–25 kW.[27] For the purpose of large-scale solar thermochemical processes, the power tower design is more feasible. The dish receiver on the other hand can provide distributed power/chemical production. Compared to power tower systems, the advantages of a dish receiver include its relatively low capital cost, a reduced complexity in overall design, and a relatively fast ramp rate that compensates for the intermittency of solar power. From a research perspective, various dish receiver designs have been demonstrated at the bench scale for different solar thermochemical reaction schemes.[30]

The power tower design exhibits the greatest potential for large-scale solar thermo-chemical applications owing to its high operating temperature and theoretical efficiency. This design utilizes a central receiver as a focal point to extract the concentrated solar radiation reflected from the heliostat field. Efficient utilization of solar energy requires a heat exchanger that can harness the high-quality thermal energy to drive an industrial process. Extensive research efforts have been exerted on solar receiver design and on heat transfer fluid (HTF) properties.[31–33] High temperature power tower system design necessitates careful consideration of the heat transfer medium with regard to: (1) the energy density of HTF; (2) the mechanical and chemical stability of the HTF at high temperatures; (3) the heat transfer coefficient between the HTF and solar radiation; (4) process reversibility during heating/cooling cycles; and (5) cost and environmental impact.[31] For the purpose of electricity generation, low-cost silica sand material is typically used as the HTF due to its thermal stability. For solar thermochemical processes, a metal oxide material can serve as both the reactive oxygen carrier and the HTF. The sintering effect of repeated thermal cycling on the stability and reactivity of the metal oxide material introduces additional complexity to the process design and material selection. The desirable properties of a suitable metal oxide material will be discussed in more detail in the ensuing sections.

Depending on the solar receiver design, the "active" heat exchange between the concentrated solar radiation and the heat transfer fluid can be achieved either directly or indirectly. Indirect heat transfer receivers contain a solar cavity that transfers heat via solar radiation to tubes. Heat transfer fluid is contained within the tubes, and heat transfer occurs across the walls of the tube through radiation/conduction. In this mode of operation, the indirect heat transfer receivers function as recuperative heat exchangers. The indirect heat transfer solar receiver was successfully demonstrated in the 10 MW_e Solar One and Solar Two projects under the development of Sandia National Laboratory.[34]

The direct heat exchanger design can potentially resolve the operational challenges mentioned earlier.[35] The direct solar receiver, also known as a "volumetric receiver," utilizes an array of three-dimensional porous interlocking shapes to absorb the solar flux in the depth of the structure. The heat absorbing porous material is made of metal or ceramic, to achieve the highest temperatures, such that a maximum temperature of 1000 °C, 1200 °C, and 1500 °C can be reached for metal, SiSiC ceramics, and SiC materials respectively. The direct heat exchange occurs as a working HTF passes through the volume to remove the heat from the solar absorption material via convection and radiation. Since the 1980s, this receiver design and its associated heat transfer mechanism for power/hydrogen generation applications has been investigated.[32,34,36–39] When a solid granular material is used as the HTF, the heat transfer rate depends strongly on the particle size and solid flow rate.[38,40] This type of receiver design offers higher efficiency and greater heat flux but suffers from varying heat distribution, which leads to the development of local hotspots. In addition, the free falling of solid particles may be subjected to operational instabilities under the influence of wind and other surrounding conditions.[37]

For simplicity, air is often used as the HTF. This design concept has been demonstrated in a number of pilot-scale projects ranging between 50 and 3000 kW.[35,41] For instance, the SolAir 3000 project, using a SiC ceramic radiation absorbing material, operated at the nominal temperature of 750 °C with mean solar fluxes of 520 kW/m^2 and peak power of 2950 kW$_{th}$. The SolAir 3000 project has been successfully scaled up to a 1.5 MW$_e$ solar thermal power tower plant in Germany that began delivering power to the grid in April 2009.[35] More recently, the possibility of using a solar granular material as the HTF instead of air has been investigated.[42,43] The granular heat transfer mechanism is found to be a function of the solid particle size, the shape of the heat absorbing material, and the contact pattern between the solids and the heat transfer surface. The hexagonal-shaped heat absorbing structure is suggested to be superior to a tubular-shaped structure as the accumulation of granular material on the transfer surface is mitigated. The direct heat transfer to granular HTF for solar thermal applications is considered to be on-going research, specifically with regard to the solid structural and morphological stability under thermal cycling.

The dish receiver designed for solar thermochemical applications consists of a parabolic dish that intensifies the solar flux, and a cavity section that absorbs the solar radiation. Inside the cavity chamber, a porous catalytic metal oxide material passively absorbs the solar flux by means of conduction and radiation. Once the metal oxide material reaches sufficiently high temperatures, the gaseous reactants are passed through the chamber such that the cooler gas stream and the endothermic reactions act as heat sinks for the solar flux. In the overall energy balance, the solar flux is balanced by the sensible heat of heating up the reactant species and the latent heat of reaction. This design of receiver has been utilized for a number of solar thermochemical systems involving the production of hydrogen and syngas.[30,35,44–46]

5.3.2   Solar Thermochemical Process

Syngas and hydrogen both serve as important energy carriers and as feedstock for chemicals production. The overwhelming majority of syngas and hydrogen are produced from fossil fuels, principally by catalytic reforming of methane, steam methane reforming (SMR), or autothermal reforming (ATR). Coal or biomass gasification also provides a viable avenue for syngas production. This section will focus on the methane reforming process.

One major drawback of the SMR and ATR processes is the parasitic energy consumption associated with the oxygen feedstock, from air separation units, which can account for up to 40% of the operating cost of the syngas plant.[47] Combining the solar thermal process with methane reforming eliminates the need for air separation units. The endothermicity of reforming reactions can be satiated by the solar energy. From a thermal energy storage (TES) perspective, the solar reforming combined cycle can be viewed as a heat conversion process that stores intermittent solar thermal energy as latent chemical energy. Under this design, the steam reforming catalyst can serve as the heat transfer fluid to absorb and transport the thermal energy throughout the process. Coupling chemical looping gasification (CLG) with the solar thermal process has garnered significant research interest in recent years. In the early stages of this type of solar thermochemical research, experiments involved only one endothermic reaction on a once-through basis. Ideally, the solar-heated metal oxide is reduced by a carbonaceous fuel such as natural gas, while partially oxidizing the hydrocarbon feedstock to syngas. The process is thus a combination of pure metal production and methane reforming. Steinfeld et al. conducted an appreciable amount of the pioneering work in this solar thermochemical scheme, as summarized in Table 5.2.[48,49] Iron and zinc oxides were the most studied metal oxides for solar thermochemical applications, as discussed in the following sub-section.

Iron

The iron/syngas solar co-production process based on a Fe_3O_4 and CH_4 system was proposed by Steinfeld et al. in 1993.[48] A thermodynamic analysis of the reaction system suggests that at 1 atm and temperatures above 1300 K, the equilibrium reaction components consist of solid metallic iron and a mixture of gaseous H_2 and CO in the ratio of 2:1, as shown in Figure 5.13.

The reduction of iron oxides (Fe_3O_4, FeO) by the reducing gases (H_2, CO) all depends critically on the reaction temperature and pressure:

$$Fe_3O_4 + 4CH_4 \leftrightarrow 3Fe + 8H_2 + 4CO. \tag{5.3.1}$$

In addition to the reforming reactions, a group of competing reactions can further complicate the reaction system:

$$CO_2 + H_2 \leftrightarrow CO + H_2O \tag{5.3.2}$$

$$C + CO_2 \leftrightarrow 2CO. \tag{5.3.3}$$

The direct contact of carbon with iron oxide can form intermediate products such as cementite, Fe_3C, which deactivates the metallic iron. Therefore, injection of CO_2 to induce the Boudouard reaction (reaction (5.3.3)) and minimize carbon deposition is critical to maintaining iron reactivity.

An iron oxide based two-step cyclic process for hydrogen and syngas production from water and methane can be illustrated in a chemical looping diagram, as given in Figure 5.14. In the first step of this process, the highly energy intensive reduction of Fe_3O_4 by methane to form syngas is driven by solar energy. In the second, regeneration step, metallic iron is oxidized by water to form hydrogen and Fe_3O_4. This reaction is exothermic and occurs at a lower temperature.

To examine this process, a fluidized bed reactor with a solar concentrator was constructed.[48] The reactor was a quartz tube of diameter 2 cm operated in a fluidized bed mode. The solar receiver, a 10 cm ID steel cylinder, was installed perpendicular to the reactor on the outside of it. A layer of specular reflective gold was electroplated onto the inner wall of the solar receiver in order to reflect infrared diffuse radiation, thus minimizing energy loss. A solar concentrator collected solar energy, and the reactor was located at the focus of the solar concentrator. A water-cooled steel plate with a circular aperture of 6 cm diameter was also attached to the solar receiver to prevent the radiation

Table 5.2 Studies relevant to solar thermochemical process.

Author	Year	Active material	Support	Reactor	Ideal reaction	Temperature	Solar source
Steinfeld et al.[48]	1993	Magnetite, Fe_3O_4	Silica	Fluidized bed reactor	$Fe_3O_4 + 4CH_4 \rightarrow 3Fe + 4CO + 8H_2$	800 °C	Two-stage solar concentrator, 1.1 kW
Steinfeld et al.[49]	1995	ZnO	Al_2O_3	Fluidized bed reactor	$ZnO + CH_4 \rightarrow Zn + CO + 2H_2$	930 °C	Parabolic receiver, 2.9 kW
Kodama et al.[50]	2000	WO_3 (SnO_2, In_2O_3, ZnO, V_2O_5, MoO_2, Fe_3O_4)	ZrO_2 (Al_2O_3, SiO_2)	Packed bed reactor	$CH_4 + \frac{1}{3}WO_3 \rightarrow \frac{1}{3}W + CO + 2H_2$ $W + 3H_2O \rightarrow WO_3 + 3H_2$	900 °C 800 °C	Electric furnace
Kodama et al.[51]	2002	Ni(II)-ferrite, $Ni_{0.39}Fe_{2.61}O_4$	ZrO_2	Packed bed reactor	Metal oxide + $CH_4 \rightarrow$ Reduced metal oxide + CO + $2H_2$ Reduced metal oxide + $H_2O \rightarrow$ Metal oxide + $2H_2$	900 °C 800 °C	Solar furnace simulator
He et al.[52]	2014	Fe_3O_4	Lanthanum strontium ferrite ($La_{0.8}Sr_{0.2}FeO_{3-\delta}$)	Fixed bed fluidized bed reactor	Metal oxide + $CH_4 \rightarrow$ Reduced metal oxide + CO + $2H_2$ Reduced metal oxide + $H_2O \rightarrow$ Metal oxide + $2H_2$	625–675 °C	Electric furnace

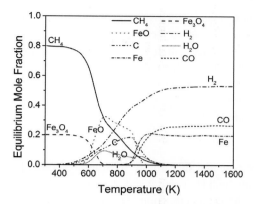

Figure 5.13 Thermodynamic equilibrium compositions for $Fe_3O_4 + 4CH_4$ reaction system at 1 atm.[48]

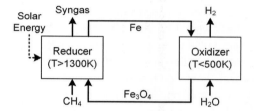

Figure 5.14 Schematic flow diagram for syngas and hydrogen production using Fe_3O_4–Fe redox pair at 1 atm.

from spilling. The capacity of the solar heating was estimated to be 1.1 kW for the reactor. This is a typical fluidized bed experiment set-up that utilizes solar thermal energy to provide heat for the endothermic reaction. The experiment demonstrated that fluidized particles were effective in absorbing solar radiation. The reaction between Fe_3O_4 and methane occurred in two stages. In the first stage, Fe_3O_4 was reduced to FeO, and more CO_2/H_2O than CO/H_2 was generated. In the second stage, FeO was reduced to Fe, and more CO than CO_2 was generated. The first stage had a higher reactivity than the second stage. The sintering and recrystallization of metallic iron contributed to the slower kinetics in the second stage. The conversion of the methane was ~20%.

Zinc

The solar based zinc and syngas co-production process was also proposed by Steinfeld et al. in 1995.[49] The reaction of interest is given in reaction (5.3.4):

$$ZnO + CH_4 \rightarrow Zn + CO + 2H_2. \tag{5.3.4}$$

The concept and advantages in the zinc process are the same as those in the iron process. The major difference is that zinc has a lower boiling point, 907 °C, at which metallic zinc in the product stream vaporizes. The zinc vapor is easily re-oxidized, lowering the overall process efficiency due to irreversible energy loss. To resolve this, a quenching unit is essential to rapidly cool down and collect the zinc vapor, impeding re-oxidation. Steinfeld et al. applied a water cooler above the fluidized bed, and sponge-structured solid deposition was observed.[49] The deposition was not of pure metallic zinc but also contained zinc oxide, which demonstrated that recombination had not been completely avoided. Even though the reactor system was not optimized for obtaining the best performance of the process, the feasibility of producing zinc and syngas simultaneously with solar power was demonstrated.

Iron–Nickel Composites

Kodama et al. investigated the potential of partially substituting iron oxide with other bivalent metal oxides including nickel, cobalt, and zinc oxide.[51] The unsupported particles were synthesized by the co-precipitation method with a constant dopant to iron molar ratio of 0.15, which led to a general chemical formula of $M_{0.39}Fe_{2.61}O_4$ (M = Ni, Co, Zn). Syngas production tests with methane reforming over the metal oxides were then carried out in a fixed bed reactor at 900 °C. Compared to pure Fe_3O_4, the nickel ferrite ($Ni_{0.39}Fe_{2.61}O_4$) particles provided a higher CO yield and selectivity of 22% and 72%, respectively, which were 13% and 9% higher than the former. The enhanced performance could be attributed to the high reactivity of Ni. However, severe sintering due to the high temperature reaction significantly hindered the water splitting reaction. In a later study by Kodama et al., ZrO_2 was then added to support the nickel ferrite particle to suppress the sintering effect.[53] The supported nickel ferrite was then used to carry out five cycles of the two-step process, which was methane reforming followed by water splitting. The recyclability of the composite was proven by X-ray diffraction (XRD) analysis. The XRD spectra showed peaks for reduced Ni–Fe alloy after the methane reforming step, and they completely disappeared after the water splitting step. Correspondingly, the peaks of oxidized ferrite on ZrO_2 appeared. The redox phase change indicated that the supported material had a superior recyclability than the unsupported, which showed XRD peaks of Ni–Fe alloy after the cyclic reactions. In the cyclic two-step processes, the first step gave methane conversions of around 46–58%, CO selectivity of around 47%, and a H_2:CO mole ratio of around 2.5, whereas the second step gave a steam conversion of around 20%.

5.3.3   Solar Thermal Water Splitting Process

Hydrogen production via the solar thermal water splitting process is a promising renewable technology that completely eliminates the need for a fossil fuel feedstock. In water splitting thermochemical cycles, water dissociates into hydrogen and oxygen either directly or via intermediate reactions involving lattice oxygen transfer to metal oxides. The mechanism of the latter reaction pathway is an example of the chemical looping process.

The traditional one-step water splitting thermochemical process follows the net reaction shown in reaction (5.3.5). The primary reaction products are hydrogen and oxygen, which can be used for the production of energy via fuel cell applications to form an inherently renewable cycle. As shown in Figure 5.15, the Gibbs free energy for

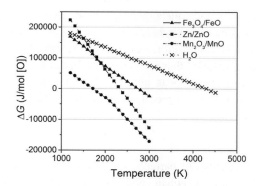

Figure 5.15 Gibbs free energy change with temperature for water splitting and metal oxide decomposition reactions.[29]

this reaction is zero at 4025 °C.[29] The dissociation conversion of water is relatively low, such that only 25% conversion can be observed at temperatures as high as 2227 °C:

$$H_2O \rightarrow H_2 + \tfrac{1}{2}O_2. \qquad (5.3.5)$$

The two-step water splitting thermochemical process involves an endothermic activation step (reaction (5.3.6)) to reduce the metal oxide, and an exothermic hydrolysis step (reaction (5.3.7)) to produce hydrogen and regenerate the metal oxide, as shown in reactions (5.3.6) and (5.3.7):

$$MO_x \leftrightarrow MO_{x-\delta} + \tfrac{\delta}{2}O_2 \qquad (5.3.6)$$

$$MO_{x-\delta} + \delta H_2O \leftrightarrow MO_x + \delta H_2. \qquad (5.3.7)$$

Existing research on the type of metal oxide materials suitable for this reaction scheme is summarized in Table 5.3. There exists great similarity between the proposed reaction scheme and the chemical looping reforming process. The differentiating factor resides in the fact that regeneration of metal oxide with water, although requiring a higher temperature, produces hydrogen as a by-product, and the heat input for this reaction step is supplied by solar energy instead of fossil fuel. The presence of metal oxide as an oxygen carrier material to facilitate the extraction of oxygen from water can effectively reduce the hydrolysis temperature.

Iron

The primary impediment of an iron based solar water splitting process involving an Fe_3O_4/FeO redox cycle resides in the thermal sintering at relatively high reduction temperatures. In an effort to lower the high reduction temperature, mixed metal oxides, substituting Fe_3O_4 with other metal oxides (Mn, Co, Mg, Ni, Zn, etc.), are examined. Fresno et al. reported the reactivity of commercially available mixed iron oxide materials, namely $NiFe_2O_4$, $Ni_{0.5}Zn_{0.5}Fe_2O_4$, $ZnFe_2O_4$, and $CuFe_2O_4$, for hydrogen production by the two-step thermochemical cycle.[61] In the samples tested, $NiFe_2O_4$ exhibited

Table 5.3 Studies relevant to solar thermal water splitting processes.

Author	Year	Active material	Support	Reactor	Ideal reaction	Temperature	Solar source
Kodama et al.[54]	2004	Magnetite, Fe_3O_4	Yttrium-stabilized zirconia, YSZ	Packed bed reactor	$(x/3)Fe_3O_4 + Y_yZr_{1-y}O_{2-y/2}(cubic) \rightarrow Fe_xY_yZr_{1-y}O_{2-y/2+x}(cubic) + (x/6)O_2$	1400 °C	Infrared furnace
					$Fe_xY_yZr_{1-y}O_{2-y/2+x}(cubic) + (x/3)H_2O = (x/3)Fe_3O_4 + Y_yZr_{1-y}O_{2-y/2}(cubic) + (x/3)H_2$	1000 °C	
Perkins et al.[55]	2008	ZnO	–	Aerosol reactor	$ZnO + heat \rightarrow Zn + 0.5O_2$	1600–1750 °C	Electric furnace
Gokon et al.[56]	2008	$NiFe_2O_4$	ZrO_2	Circulating fluidized bed reactor	$NiFe_2O_4 \rightarrow 3Ni_y\,Fe_{1-y}O + 0.5O_2$	1400 °C	Solar furnace simulator
				packed bed reactor	$3Ni_y\,Fe_{1-y}O + H_2O \rightarrow NiFe_2O_4 + H_2$	1000 °C	
Stamatiou et al.[57]	2010	ZnO, Fe_3O_4	–	TGA	Metal oxide + heat → reduced metal oxide + O_2	400~450 °C	TGA
					Reduced metal oxide + $xH_2O + (1-x)\,CO_2 \rightarrow$ Metal oxide + $xH_2 + (1-x)\,CO$	700~1000 °C	
Chueh et al.[44]	2010	Porous ceria monolithic brick, CeO_2	–	Radial flow reactor	$CeO_2 + heat \rightarrow CeO_{2-x} + 0.5xO_2$ $CeO_{2-x} + xCO_2 \rightarrow CeO_2 + xCO$ $CeO_{2-x} + xH_2O \rightarrow CeO_2 + xH_2$	1500 °C 800 °C	Cavity-receiver reactor
Furler et al.[58]	2011	Porous ceria felt, CeO_2	–	Radial flow reactor	$CeO_2 + heat \rightarrow CeO_{2-x} + 0.5xO_2$ $CeO_{2-x} + xCO_2 \rightarrow CeO_2 + xCO$ $CeO_{2-x} + xH_2O \rightarrow CeO_2 + xH_2$	1527 °C 827 °C	Cavity-receiver reactor
Muhich et al.[59]	2013	ALD Hercynite, $CoFe_2O_4$	ZrO_2	Stagnant flow reactor	$CoFe_2O_4 + 3Al_3O_3 + heat \rightarrow CoAl_2O_4 + 2FeAl_2O_4 + 0.5O_2$ $CoAl_2O_4 + 2FeAl_2O_4 + H_2O \rightarrow CoFe_2O_4 + 3Al_2O_3 + H_2$	1350 °C (isothermal)	Electric furnace
Scheffe et al.[60]	2014	Ceria powder	–	Aerosol reactor	$CeO_2 + heat \rightarrow CeO_{2-x} + 0.5xO_2$	1450–1550 °C	Electric furnace

the best reactivity for hydrogen yield and cyclic ability. Other approaches included doping iron with novel support materials. Kodama et al. first demonstrated the monoclinic zirconia supported ferrite (ferrite/m-ZrO_2) particles.[62] Later on, the cubic yttria stabilized zirconia (YSZ) was also proposed as a promising support material for Fe_3O_4.

Zinc

The Zn/ZnO redox cycle for water splitting reactions presents fewer challenges than that of the iron based redox cycle. Based on thermodynamic calculations, ZnO completely dissociates at around 2027 °C. The major technical challenge resides in separation of the decomposition products, as both Zn and O_2 are stable in a gaseous form at 2027 °C.[29] A sophisticated quenching and product separation scheme is required to achieve high hydrogen yield and favorable process economics. Reasonable reaction rates have been demonstrated by Steinfeld et al. at temperatures greater than 427 °C.[63] In addition, an energy analysis was performed on a theoretical ZnO/Zn process using heliostat solar concentrations of 5000 and 10,000 suns. Assuming equilibrium conditions, the maximum overall efficiency was found to be 25% and 36% for solar concentrations of 5000 and 10,000 suns, respectively.

Manganese

The Mn_2O_3/MnO based redox cycle has three steps instead of two, thus its maximum theoretical efficiency will be lower. However, it is offset by the greater feasibility in achieving maximum efficiency in each step of this three-step process. Specifically, the three reactions involved are shown in reactions (5.3.8), (5.3.9), and (5.3.10):

$$MnO + NaOH \leftrightarrow \tfrac{1}{2}H_2 + NaMnO_2 \qquad (5.3.8)$$

$$NaMnO_2 + \tfrac{1}{2}H_2O \leftrightarrow \tfrac{1}{2}Mn_2O_3 + NaOH \qquad (5.3.9)$$

$$\tfrac{1}{2}Mn_2O_3 \leftrightarrow MnO + \tfrac{1}{4}O_2. \qquad (5.3.10)$$

This proposed process cycle is favorable compared to the Mn_3O_4 cycle because a higher hydrogen yield per mass of oxide can be achieved. The theoretical efficiency of this three-step process is found to be 26–51% without considering the separation steps.[29] Compared to the zinc based redox process, the manganese based redox process has several advantages. First, the decomposition of Mn_2O_3 to form solid phase MnO and oxygen gas occurs at around 1526 °C, which is lower than that of ZnO decomposition. The subsequent separation of gaseous O_2 from solid MnO upon cooling is relatively straightforward.

Cerium

Recently, Furler et al. proposed ceria (CeO_x) as a metal oxide in the cyclic syngas production scheme.[58] Ceria is a metal oxide with a high oxygen ion conductivity, which releases molecular oxygen with partial oxidation capability at high temperature

($>$1400 °C). The reduced ceria can then react with steam or CO_2 to generate syngas ($<$1400 °C):

$$CeO_2 \rightarrow CeO_{2-x} + \tfrac{x}{2}O_2 \tag{5.3.11}$$

$$CeO_{2-x} + x\,H_2O \rightarrow CeO_2 + xH_2 \tag{5.3.12}$$

$$CeO_{2-x} + x\,CO_2 \rightarrow CeO_2 + xCO. \tag{5.3.13}$$

Other metals with similar characteristics include ZnO_2, SnO_2, etc. However, these volatile metals sublime and recombine with the released oxygen quickly during decomposition. The non-volatile ones such as the above mentioned iron oxide, on the other hand, suffer from slow kinetics. Ceria surpasses both disadvantages with desirable recyclability and high redox reactivity.

The solar reactor consisted of a cavity receiver with a 4 cm diameter circular aperture. The aperture was the entrance for the concentrated solar radiation and was closed by a quartz disk window. The cavity, or the main reacting chamber, contained a 62 mm ID, 85 mm OD, 100 mm height cylinder, which consisted of multiple porous ceria layers that directly absorbed the solar radiation. The addition of co-feeding of H_2O and CO_2 was conducted in both single and consecutive modes. In both modes, a purge of 2 L/min argon flow was always present to prevent the mixing of oxidizing and reducing gases and to dilute the reactants. To reach the thermal reduction temperature, a 3.6 kW radiative power was input with a mean solar concentration ratio of 2865 suns over the aperture.

Hercynite

Solar water splitting using iron based systems for hydrogen generation has been described. To illustrate the process, the syngas chemical looping (SCL) process, as discussed in Section 1.6.1 is again featured. In the SCL process, Fe_2O_3 entering the reducer reactor oxidizes a carbonaceous feedstock to CO_2 and H_2O while being reduced to a mixture of FeO/Fe, and the reaction is endothermic. In the oxidizer reactor, the FeO/Fe mixture reduces steam to hydrogen while being oxidized to Fe_3O_4, and the reaction is exothermic. This partially oxidized metal oxide is then oxidized to its full oxidation state in a combustor and then recycled back to the reducer. The optimal thermodynamic conditions for each reaction in the SCL process require that they each be conducted at different reaction temperatures. Thus, the overall process heat integration is essential. The heat integration is balanced by the circulation of solids through the three SCL reactors coupled with an efficient heat recovery system exterior to the SCL process.

In a typical solar chemical looping system for hydrogen generation, the reduction and oxidation reactions are usually also required to take place in different temperature windows in order to maximize the product yield. For example, the reduction and oxidation loop of ceria usually swings between 1500 °C and 900 °C.[44] A higher temperature is required to form vacancies in the perovskite-structured CeO_2 to release oxygen. When a fluidized bed reactor system is considered for a solar system, the design principle underlining the SCL process can be followed. However, when a fixed bed reactor system is considered, on the other hand, which involves switched feedstock, oxidation and reduction can be operated under the same or a similar temperature to circumvent the inefficiency associated with the cyclic heating and cooling requirements of a fixed bed system. However, since the metal oxide reduction using solar energy occurs at a very high temperature, an operation with a same or similar temperature necessitates a high temperature for steam reduction, one that is appreciably higher than is thermodynamically optimal. Further, the metal oxides for a solar based H_2 generation system can be formulated in such a manner that their reduced form can be re-oxidized in one step to their original oxidation state of metal oxides using steam, and hence avoiding the regeneration step as in the SCL process. In this re-oxidation reaction, however, excess steam well beyond the stoichiometric requirement is required.

A solar water splitting system that comprises the three key aspects indicated above, i.e. a fixed bed, a similar temperature swing for reduction and oxidation reactions, and a one-step re-oxidation scheme, is reported by Muhich et al.[59,64] Specifically, in this solar water splitting system, cobalt-doped hercynite is the metal oxide that participates in the reduction (reaction (5.3.14)) and the oxidation (reaction (5.3.15)) reactions:

$$CoFe_2O_4 + 3Al_2O_3 \rightarrow CoAl_2O_4 + 2\,FeAl_2O_4 + \frac{1}{2}O_2 \qquad (5.3.14)$$

$$CoAl_2O_4 + 2FeAl_2O_4 + H_2O \rightarrow CoFe_2O_4 + 3Al_2O_3 + H_2. \qquad (5.3.15)$$

Apart from the key metal oxide products indicated in the above equations, other metal oxide products may also be formed. In Figure 5.16, the Gibbs free energy of the iron oxide based reduction reaction is given as a function of temperature and shows that the melting point of Fe_2O_3 is lower than the temperature required for a spontaneous solar energy driven reduction reaction.

Cobalt-doped Fe_2O_3 structures show promise in decreasing the reduction reaction temperatures below the melting point of Fe_2O_3 and maintaining the non-volatile nature of ferrites.[68,69] After testing various materials for stabilizing the doped ferrite based materials, hercynite based materials were found to be able to maintain the solid phase stability and could be operated at lower reduction temperatures, as shown in Figure 5.17,[67,70,71] compared to those for iron based solar thermochemical cycles, as

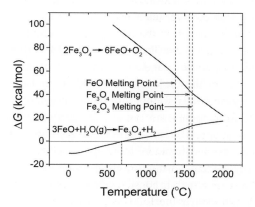

Figure 5.16 Gibbs free energy change of iron based solar thermochemical water splitting cycles.[65–67]

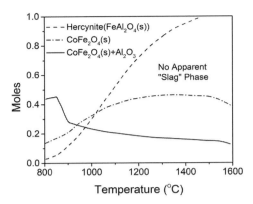

Figure 5.17 Molar variation of $CoFe_2O_4$, $FeAl_2O_4$, and $CoAl_2O_4$ with reduction temperature.[67,70,71]

shown in Figure 5.16. The thermodynamic rationale of this is discussed in the following paragraph. Figure 5.17 provides the equilibrium composition in moles of $CoFe_2O_4$, $FeAl_2O_4$, and $CoAl_2O_4$ materials with starting reactants of one mole of $CoFe_2O_4$ and five moles of Al_2O_3. The equilibrium solids composition of major components shown in Figure 5.17 can be used to calculate the H_2 production capacity for a given temperature.

Atomic layer deposition (ALD) was used to deposit CoO and Fe_2O_3 in a 1:1 ratio on the Al_2O_3 support, in order to form the desired $CoFe_2O_4$ stoichiometry on Al_2O_3. In this system, $CoFe_2O_4$ and Al_2O_3 are first reacted to form stable aluminates. During steam oxidation, the ferrite spinel and alumina phases are regenerated while H_2 is released. The advantage of this system is the fact that reduction via reaction (5.3.14) can begin at temperatures ~150 °C lower than ferrite or ceria, since $CoAl_2O_4$ and $FeAl_2O_4$ are both stable aluminates, which are more thermodynamically favorable over structural vacancies. The hercynite cycles were carried out under an isothermal condition of 1350 °C and can also be carried out over temperature swings from 1350 to 1000 °C and 1500 to 1200 °C. To illustrate the characteristics of hercynite, a temperature swing cycle of 1350 to 1000 °C was conducted with ceria for comparison. For each cycle, the thermal reduction occurred first to release oxygen, followed by introduction of steam for oxidation. Ceria generated less hydrogen than hercynite in the 1350/1000 °C cycles because of the non-optimized temperature for CeO_2. The 1350/1000 °C cycle of hercynite could hardly compare with the cycle at 1500/1200 °C and the isothermal 1350 °C cycle, as the former generated only one-third of the hydrogen of the latter two conditions. The rather poor performance of the former was attributed to the slower kinetics of the lower oxidation temperature of 1000 °C. According to the thermodynamic analysis, the oxidation reaction should be more spontaneous at low temperatures since the Gibbs free energy decreases while lowering the temperature. However, both the 1500/1200 °C and the isothermal 1350 °C cycle indicate the possibility of performing steam reduction at high temperatures.

Density functional theory (DFT) calculations predict an optimal structure for formulation of various hercynite based materials. DFT calculations for H_2 production capacities show that Fe and Co aluminates operating via a stoichiometric reaction mechanism have insufficient reducing power to split water. However, these materials can drive solar thermal water splitting (STWS) via a non-stoichiometric mechanism instead. Among the aluminates, due to the relative formation energies of oxygen vacancy sites, the relative H_2 production capacities are predicted to be 1, 0.7, and 2×10^{-4} for $FeAl_2O_4$, $Co_{0.5}Fe_{0.5}Al_2O_4$, and $CoAl_2O_4$, respectively. The experimental results of H_2 production during the steam oxidation step of water splitting cycles, conducted in a stagnant flow reactor, are shown in Figure 5.18.[59,64,70] There were 12 STWS cycles conducted for $FeAl_2O_4$ and $Co_{0.5}Fe_{0.5}Al_2O_4$; only the last six cycles were reported, as seen in Figure 5.18b, to represent long-term performance. $FeAl_2O_4$ demonstrated the highest H_2 production capacity in a STWS cycle, as was predicted theoretically. Although oxygen vacancy based $FeAl_2O_4$ material can produce ~500 micromole/g of H_2, the steam conversion associated with H_2 production requires significant improvement for this operation to be economically feasible on a commercial scale.

In conclusion, the continuous advancement of solar chemical looping technology has provided a bridge between existing research on metal oxide chemical looping and solar based renewable power generation. Integration of a concentrating solar thermal process allows for the process to operate at a higher temperature, thus achieving a high theoretical efficiency and driving certain energy intensive reactions that are thermo-dynamically favorable only at very high temperatures. The oxygen transfer mechanism of certain metal oxide materials enhances the overall thermodynamic requirements of the thermolysis of water or carbon dioxide. However, further research into the reactivity and stability of metal oxide material suitable for ultra-high temperature redox reactions, as well as process integration with respect to heat transfer between concentrating solar radiation and the heat transfer medium is necessary to drive the commercialization potential of this technology.

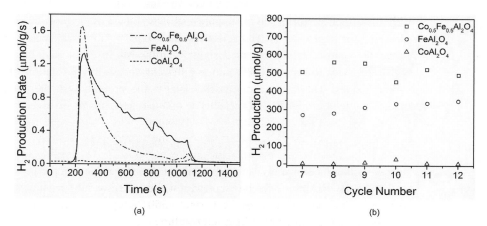

(a)   (b)

Figure 5.18 Experimental (a) rate of H_2 production during the steam oxidation step of a STWS cycle; and (b) total H_2 production over the steam oxidation step for six STWS cycles.[59,64,70]

5.4   Selective Partial Oxidation of CH$_4$ to HCHO

Methane is a highly abundant, low-cost carbon source, which is the feedstock for a number of high valued chemicals, such as formic acid (HCOOH), methanol (CH$_3$OH), and formaldehyde (HCHO). Formaldehyde is an important industrial chemical that is a precursor for the production of several resins used in the automobile and textile industries, in addition to its uses in disinfectants and biocides. Initially, CH$_4$ was converted to HCHO by a three-step process, where CH$_4$ was first converted to syngas via steam reforming over a nickel catalyst. The syngas was then converted to methanol over a copper–zinc catalyst, followed by air oxidation of methanol over a silver or iron molybdate catalyst to form HCHO. The importance of HCHO as an industrial precursor led to research on catalysts that could promote the direct oxidation of CH$_4$ to HCHO. Thermodynamically, carbon oxides are the favored products for a mixture of CH$_4$ and O$_2$ at equilibrium at above 427 °C. Hence, it was important to develop a catalyst that facilitated the formation of formaldehyde without catalyzing its further oxidation to carbon oxides. Increased interest in the direct oxidation of methane to formaldehyde led to investigation of various oxidants, catalyst formulations, and operating conditions. Catalytic metal oxides utilize lattice oxygen, similar to processes discussed in Chapters 2, 3, and 4, but have the added participation of the catalytic surface. Through better understanding of the lattice oxygen exchange and reaction mechanisms, as further discussed in this section, catalytic metal oxides can be designed for selective conversion to the desired products.

Catalytic selective oxidation of CH$_4$ to produce HCHO has been carried out with N$_2$O as well as O$_2$ as the oxidizing agent. Of the two, O$_2$ is the more popular oxidizing agent since N$_2$O has been less effective in re-oxidizing catalytic metal oxides during operation in redox cycles and has also resulted in lower selectivity for HCHO.[72,73] One-step partial oxidation of CH$_4$ to HCHO has been most commonly studied over supported catalytic metal oxides. Some of the active metal oxides include SiO$_2$, MgO, MoO$_3$, V$_2$O$_5$, WO$_3$, and ZnO, which have all been studied for their catalytic activity as well as for their performance as support materials.[74] Two of the most important criteria identified for improving HCHO selectivity and yield include understanding the role of active sites and effective reactor design, which assists in developing catalysts with a high selectivity towards HCHO and suppressing the further oxidation of HCHO to carbon oxides. This section discusses some of the properties in the catalytic metal oxides that lead to high HCHO selectivity and the mechanisms involved.

5.4.1   SiO$_2$ Supported V$_2$O$_5$ and MoO$_3$ as Catalytic Metal Oxides

Recently, the most widely studied catalysts for the selective partial oxidation of methane to HCHO have been based on MoO$_3$ and V$_2$O$_5$, with both catalysts considered to be most effective in the selective oxidation of CH$_4$ to HCHO.[74] They have been extensively studied for the effect of their different physicochemical properties, such as their structure, dispersion of the metal oxide species, density of active sites, and their reducibility on the selective oxidation of CH$_4$ to HCHO. Faraldos et al.[75] investigated

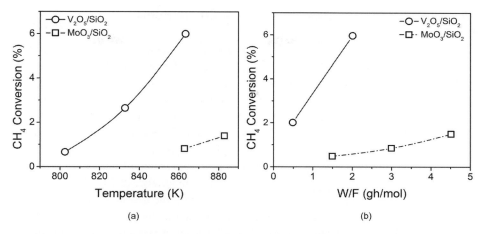

Figure 5.19 Influence of (a) reaction temperature and (b) residence time parameter (W/F) on the conversion of CH$_4$ for (a) W/F = 2.0 gh/mol; CH$_4$/O$_2$ = 11 M and (b) T = 590 °C; CH$_4$/O$_2$ = 11 M.[75]

SiO$_2$ supported V$_2$O$_5$ and MoO$_3$ in order to compare their activity towards CH$_4$ partial oxidation to HCHO. Figure 5.19 compares two catalytic metal oxide samples with a similar metal oxide surface concentration for their influence on reaction temperature (Figure 5.19a) and residence time parameter (Figure 5.19b) on CH$_4$ conversion. As seen from Figure 5.19a, it is evident that SiO$_2$ supported V$_2$O$_5$ exhibits a higher CH$_4$ conversion as compared to the methane conversion of supported MoO$_3$ under all conditions. The lower onset reaction temperature required by V$_2$O$_5$ as compared to MoO$_3$ to achieve the same conversion is also an indication of the higher reactivity of V$_2$O$_5$. Similar observations were reported from selective oxidation studies conducted on V$_2$O$_5$ and MoO$_3$.[76–78] It was also reported that with SiO$_2$ supported MoO$_3$, CO$_2$ is produced even at low CH$_4$ conversions, and the CO$_2$ selectivity remained constant over the entire range of CH$_4$ conversions studied. V$_2$O$_5$/SiO$_2$, on the other hand, formed no CO$_2$ at low CH$_4$ conversions and exhibited increasing CO$_2$ selectivity with increasing CH$_4$ conversion. This suggests that for MoO$_3$/SiO$_2$, CO$_2$ is the primary product, whereas it is formed by further oxidation of CO on V$_2$O$_5$/SiO$_2$.

Figure 5.20 shows the HCHO selectivity for both SiO$_2$ supported MoO$_3$ and V$_2$O$_5$ with similar metal oxide surface concentrations. As seen from Figure 5.20, for all CH$_4$ conversions, MoO$_3$/SiO$_2$ always exhibits a higher HCHO selectivity as compared to V$_2$O$_5$/SiO$_2$. This data suggests that there is no real correlation between HCHO selectivity and the reducibility of the two supported catalytic metal oxides, which is in agreement with other similar studies performed on the selective oxidation of CH$_4$.[79]

It has been reported that SiO$_2$ supported MoO$_3$ does not adsorb oxygen but some degree of adsorption occurs on SiO$_2$ supported V$_2$O$_5$.[79–81] Thus, in the case of vanadium oxide there is a higher interaction between the gaseous oxygen and the active sites. Isotopic studies have shown that selective oxidation of CH$_4$ to HCHO involves the incorporation of lattice oxygen from the catalytic metal oxides.[72,81–83] The vacancies

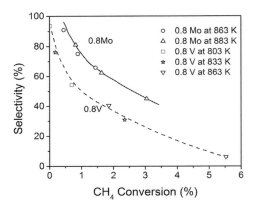

Figure 5.20 Selectivity of HCHO versus CH_4 conversion over MoO_3/SiO_2 at 863 K and 883 K, and over V_2O_5/SiO_2 at 803 K, 833 K, and 863 K.[75]

created on the metal oxide surface are filled by gaseous oxygen. In the case of SiO_2 supported V_2O_5, in addition to the lattice oxygen, surface-adsorbed oxygen species are also present, which makes the vanadium oxide more reactive and promotes further oxidation of HCHO to carbon oxides, resulting in a lower HCHO selectivity. Thus, while HCHO is the primary product for both SiO_2 supported MoO_3 and V_2O_5, the absence of adsorbed oxygen in molybdates suppresses the further oxidation of HCHO to CO and CO_2, leading to higher selectivity.

5.4.2   Crystal Planes and Active Sites of MoO₃

The catalytic conversion of a hydrocarbon molecule to a desired chemical is often highly sensitive to its structure. Specifically, selectivity of the product is dependent on the distribution of the crystal planes.[84] Catalysts that exhibit this kind of behavior fall under the category of catalytic anisotropy,[85] where the difference between crystal planes depends on the ratio of various active sites on each plane. The selectivity of a catalytic metal oxide can be improved by selecting suitable bond strengths and active sites. Establishing the relation between product selectivity and crystal planes/active sites is the first step in exploring the potential of catalytic metal oxides. MoO_3, which is one of the most effective catalytic metal oxides for selective oxidation of CH_4 to HCHO, exhibits such catalytic anisotropy. The conversion of CH_4 and the selectivity of its products, including HCHO, are presented here with respect to different crystal planes.

Smith and Ozkan[86] investigated the effect of structural specificity of unsupported MoO_3 on its reactivity and selectivity by preparing MoO_3 samples using temperature programmed techniques to preferentially expose different crystal planes. Figure 5.21 compares the selectivity and production rates of methane oxidation on MoO_3–C and MoO_3–R, which were synthesized using two different techniques to expose different crystal planes of MoO_3. These steady-state methane oxidation studies revealed that both the selectivity and production rate toward HCHO are significantly higher for MoO_3–C over MoO_3–R. MoO_3–C has a much higher concentration of (100) side planes, which

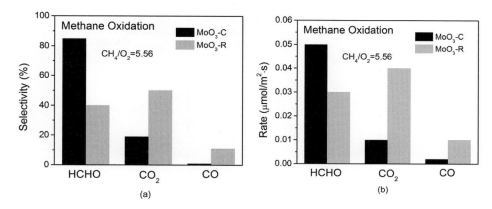

Figure 5.21 (a) Selectivity and (b) production rates of HCHO, CO$_2$, and CO over MoO$_3$-C and MoO$_3$-R.[86]

are concluded to be more selective towards HCHO, while the (010) basal planes, which are in larger concentrations in MoO$_3$-R, tend to form carbon oxide products. The HCHO production rate is also found to be higher for MoO$_3$-C, which further confirms the importance of the (100) side plane in HCHO formation. Laser Raman spectroscopy, used to identify the type of Mo and O sites in the two different MoO$_3$ structures, suggests that MoO$_3$-C has more exposed terminal Mo=O sites. Hence, the Mo=O sites are concluded to be the main surface species that promotes partial oxidation of CH$_4$. On the other hand, the bridged Mo–O–Mo sites, which have a higher density in MoO$_3$-R surface, promote the formation of complete oxidation products. Consequently, oxidation of HCHO over the two different MoO$_3$ catalytic metal oxides revealed that MoO$_3$-R is more active towards complete oxidation.

The reducibility of both types of site is characterized by subjecting both MoO$_3$-C and MoO$_3$-R to H$_2$ temperature programmed reduction (TPR) experiments. Table 5.4 compares the ratio of the H$_2$ consumption rate of both catalytic metal oxides in several temperature ranges. The values suggest that MoO$_3$-R is more readily reducible than MoO$_3$-C, meaning that the Mo–O–Mo sites are easier to reduce than the Mo=O sites. The differences in reducibility of these sites are reflected in the selectivity towards HCHO. Metal oxides that are more difficult to reduce have a higher energy barrier for the formation of complete and partial oxidation products (CO$_x$) and are more selective towards formation of HCHO. The difference, with respect to selectivity, between these two sites reveals the preferred structure of MoO$_3$ for selective oxidation of methane and provides insight when designing a MoO$_3$ catalytic metal oxide.

5.4.3 Reaction Mechanism with Isotopic Labeling Experiment

With the preferred planes and sites for partial oxidation of CH$_4$ identified, isotopic labeling experiments using $^{16}O_2$ and $^{18}O_2$ as reactive gases on both MoO$_3$-C and MoO$_3$-R samples revealed the reaction pathway.[87] The isotopic experiments

Table 5.4 Ratio of H_2 consumption in different temperature ranges.[86]

Temperature range (°C)	$\dfrac{\text{mol } H_2 \text{ consumed}/m^2(\text{MoO}_3\text{–C})}{\text{mol } H_2 \text{ consumed}/m^2(\text{MoO}_3\text{–R})}$
540–570	0.380
570–600	0.264
600–630	0.236
630–660	0.283
660–690	0.402
690–720	0.422

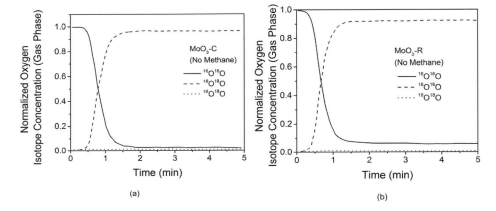

(a)   (b)

Figure 5.22 Normalized oxygen isotopic concentration in the absence of methane for (a) MoO$_3$–C and (b) MoO$_3$–R.[87]

investigated the interaction between oxygen and the metal oxide surface in the absence of CH$_4$, as well as the source of O atoms in the products during CH$_4$ oxidation. Figure 5.22 shows the concentration of oxygen isotopes in the first five minutes after $^{16}O_2$ is switched to $^{18}O_2$ in the absence of CH$_4$. After all of the $^{16}O^{16}O$ in the gas phase has exited the reactor, the remaining source of ^{16}O atoms is the chemisorbed surface oxygen atoms and lattice oxygen atoms. That is, the source of oxygen can only be from the Mo–O–Mo and Mo=O sites. Comparing the two catalytic metal oxides, the concentration of $^{16}O^{16}O$ was less for MoO$_3$–C than MoO$_3$–R. This suggests that MoO$_3$–R is more susceptible to exchange of O atoms than MoO$_3$–C. It was found that MoO$_3$–R contributed ~50% more ^{16}O in the gas phase than the contribution of MoO$_3$–C and at more than twice the rate. The high rate of desorption and exchange of lattice oxygen in the metal oxide surface, combined with the comparatively easier reducibility, allow for the surface Mo–O–Mo site to be more active to donate lattice oxygen than the surface Mo=O site. Such high activity promotes the formation of complete oxidation products with methane as the reactant.

Isotopic oxygen switching experiments with CH$_4$ were conducted to determine the source of oxygen in the products. Normalized CO$_2$ isotope concentrations, presented in

Figure 5.23, show that both $C^{16}O^{16}O$ and $C^{16}O^{18}O$ levels are higher than $C^{18}O^{18}O$ well after all the $^{16}O_2$ in the gas phase has exited the reactor. The levels of $C^{16}O^{16}O$ and $C^{16}O^{18}O$ are representative of the participation of desorbed oxygen atoms from MoO_3. Comparing the two catalytic metal oxides, the level of $C^{16}O^{16}O$ decreases faster for MoO_3–R than MoO_3–C. The contribution of ^{18}O atoms also increases faster for MoO_3–R. This suggests that ^{16}O from MoO_3–R depletes faster than the ^{16}O from MoO_3–C, and the newly introduced $^{18}O_2$ is utilized for formation of complete oxidation products at a higher rate. After ten minutes, the concentration of $C^{16}O^{18}O$ even exceeds the concentration of $C^{16}O^{16}O$ for MoO_3–R. Isotopes of H_2O also showed similar trends, where the rate of $H_2^{16}O$ depletion was also higher for MoO_3–R, confirming greater participation of lattice oxygen in the case of MoO_3–R. In the case of HCHO isotope concentrations, not only did $HCH^{16}O$ deplete faster for MoO_3–R than that for MoO_3–C, as observed in Figure 5.24, but the majority of HCHO formed was $HCH^{16}O$. This

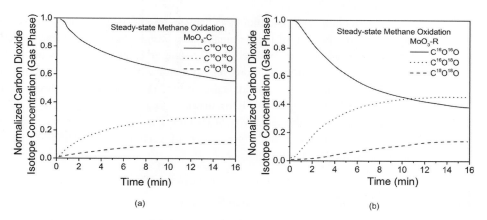

Figure 5.23 Normalized CO_2 isotope concentration for duration of 16 minutes after $^{18}O_2$ injection for (a) MoO_3–C and (b) MoO_3–R.[87]

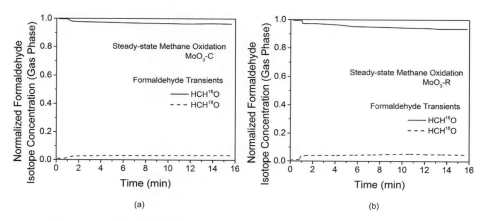

Figure 5.24 Normalized HCHO isotope concentration for duration of 16 minutes after $^{18}O_2$ injection for (a) MoO_3–C and (b) MoO_3–R.[87]

suggests that the reaction path is highly dependent on the oxygen atoms from the metal oxide and not the gaseous $^{18}O_2$. The total amount of ^{16}O in the gas phase for MoO_3–R is 240% more than the amount for MoO_3–C. The rate of ^{16}O participation is also higher for MoO_3–R, similar to the trend observed in the isotopic experiment without CH_4.

Based on the isotopic labeling studies, the oxygen that contributes to HCHO formation is mainly donated from the catalytic metal oxide itself with minimal contribution of gaseous oxygen to HCHO formation. However, the gaseous oxygen plays a role in replenishing the oxygen atoms on the surface. The total amount of ^{16}O present in the gas phase at the steady state is equivalent to 13–30 layers of MoO_3. The high percentage of ^{16}O in the products cannot simply be provided by oxygen atoms chemisorbed to the catalytic metal oxide surface. It is the bulk oxygen from the gaseous phase that diffuses through the catalytic metal oxide lattice to re-oxidize the reduced catalytic metal oxide, which explains the high percentage of ^{16}O in the products and its decrease over time after switching $^{16}O_2$ with $^{18}O_2$.

The reduction and oxidation mechanism on the catalytic metal oxide surface is dependent on the type of the terminal metal–oxygen bond. Laser Raman spectroscopy on the reduced and re-oxidized MoO_3 sample revealed that the bridged site with Mo–O–Mo termination is more readily oxidized by gaseous oxygen. Mo=O site replenishment, on the other hand, is preferred via oxygen diffusion through the catalytic metal oxide lattice. Since HCHO is mainly formed from ^{16}O, it is concluded that the Mo=O site is key to the selectivity of HCHO. The differences between the re-oxidation mechanisms and the relative population of each bond on MoO_3–C and MoO_3–R distinguish the HCHO selectivity.

5.4.4   Effect of MoO₃ Loading on Silica Support

In practice, the structural advantage of MoO_3 can be used to promote the activity of silica. Arena et al.[88] extensively studied the conversion of CH_4 to HCHO on SiO_2 supported MoO_3 by varying the amount of MoO_3 on silica and evaluating their interaction with CH_4. The notations adapted from this work are presented in Table 5.5 and will be used in the remainder of this section. The effect of MoO_3 loading has been studied using H_2 and CH_4 TPR and high temperature oxygen chemisorption (HTOC) experiments.

H_2-TPR profiles of MPS catalytic metal oxide are compared with PS and bulk MoO_3 in Figure 5.25. The shape and location of each peak indicate a noteworthy change on the

Table 5.5 Notations for the studies by Arena et al.[88]

MPO	Methane partial oxidation
PS	Precipitated silica
MPS 2	2 wt% MoO_3 on precipitated silica
MPS 4	4 wt% MoO_3 on precipitated silica
MPS 7	7 wt% MoO_3 on precipitated silica

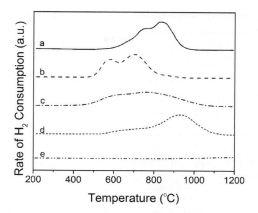

Figure 5.25 H_2-TPR profiles of (a) bulk MoO_3, (b) MPS 7, (c) MPS 4, (d) MPS 2, and (e) PS.[88]

catalytic metal oxide and provide information about its oxidation states. For PS, consumption of H_2 remained flat until 900 °C when it began gradually to increase. This indicates low activity between silica and H_2. With a small amount of MoO_3 present, such as MPS 2, the reduction temperature drastically decreased to 486 °C. However, significant H_2 consumption still did not occur until 800 °C. On increasing the MoO_3 loading further to 4% (MPS 4), the onset reduction temperature shifts toward a lower temperature, along with enhanced reduction at above 800 °C. The peaks broadened and occurred at below 800 °C. With 7% loading (MPS 7), two distinct peaks formed at 574 °C and 703 °C but the change in onset temperature ceased. For pure MoO_3, again, two peaks formed, but at different locations, 763 °C and 840 °C. These account for the oxidation state change of Mo^{6+} to Mo^{4+} and Mo^0.

For all MPS catalytic metal oxides, the onset temperature and peak locations are lower than PS and MoO_3. This suggests a positive interaction that results in different phases of MoO_3 on silica.[89] The first highly reducible polymolybdate-like phase is referred to as *Pm*. This phase forms only at ambient condition with adsorption of water on the surface. This phase reduces at below 400 °C. Since the MPS catalytic metal oxides were pre-treated at 600 °C prior to the TPR experiment, this phase does not contribute to the TPR profiles, which is evident from the high onset temperature. The second phase is recognized as MoO_3 crystallite dispersed on the silica surface. This phase experiences the reduction path of Mo^{6+} to Mo^{4+} to Mo^0, and is referred to as *Mc*. Therefore, the reduction of this phase can result in two distinct H_2 consumption peaks in the TPR profile. The last, isolated molybdates phase, referred to as *Im*, is formed when a strong interaction between MoO_3 and silica forms. The interaction is the hardest phase to reduce, with reduction occurring at above 800 °C and reducing in a single step from Mo^{6+} to Mo^0. This phase accounts for one H_2 consumption peak. Arena et al.[88] were able to determine the relative populations of each phase by deconstructing the TPR profiles into Gaussian peaks, using numerical methods. They concluded that *Im* is the dominant species at low loading, while *Mc* dominates at high Mo loading, which is reflected in Table 5.6.

Table 5.6 Concentration of *Mc* and *Im* species on MPS catalytic metal oxides.[88]

	%		$\mu mol\ g^{-1}$	
	Mc	*Im*	*Mc*	*Im*
MPS 2	22.7	77.3	31.5	107.4
MPS 4	67	33	186.1	91.7
MPS 7	92.9	7.1	451.6	34.5

Table 5.7 HTOC characterization data of MPS catalytic metal oxides.[88]

Catalytic metal oxide	O_2 uptake $\mu mol\ g_{cat}^{-1}$	O/Me %
PS	0.5	-
MPS 2	6.1	8.8
MPS 4	66.4	47.8
MPS 7	111.5	45.9

With the concentration of each phase identified, the reactivity of each phase with oxygen was explored with HTOC experiments. The results for PS and MPS catalytic metal oxides at 600 °C are presented in Table 5.7. For bare PS, oxygen intake is insignificant. *Im*-dominated MPS 2 yields almost no oxygen intake, while the *Mc*-rich MPS 4 and MPS 7 consume more oxygen and exhibit a higher dispersion value. Clearly, the *Mc* phase reacts more readily with gaseous oxygen than the *Im* phase.

Methane reactivity with each catalytic metal oxide was studied using CH_4-TPR experiments. The rate of oxygen consumption was calculated by performing an oxygen balance through the total summation of oxygen atoms in product molecules such as HCHO, CO, CO_2, and H_2O. The rate of lattice oxygen consumption, r*, is shown in Figure 5.26. Selectivity toward each product is shown in Table 5.8 for PS and MPS 4.

CH_4-TPR further confirmed the finding of the H_2-TPR studies. MPS 7, with its dominant *Mc* phase, exhibits the highest lattice oxygen consumption rate. MPS 2 has the lowest among all the MPS catalytic metal oxides. Bare PS support exhibits low reactivity toward CH_4 and only increases slightly with temperature. Formaldehyde and carbon are the only products formed in the absence of oxygen. For MPS catalytic metal oxides, reactivity with methane is slow, at 650 °C, but increases drastically at higher temperatures. The CH_4 conversion rate is inversely proportional to Mo loading. Oxygenated product formation also increased at beyond 650 °C. Three possible types of active sites on the catalytic metal oxide form: the exposed siloxane bridges that exist at low Mo loading, the Mo=O terminal bond of MoO_3 crystallite, and Mo–O–Mo bridges. Siloxane bridges have been shown to exhibit low reactivity with bare PS catalytic metal oxide. The production of oxygenated products was contributed mainly by MoO_3 crystallites, *Mc*, since they made up 67% of MoO_3 phases for MPS 4. *Im* phase does not participate in the reaction below 800 °C. At the same time, high HCHO selectivity is only observed at low temperatures. Hence, *Mc* phase is recognized as an important phase for

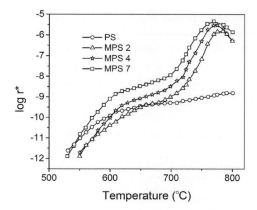

Figure 5.26 Rate of lattice oxygen consumption with CH_4-TPR for various catalytic metal oxides.[88]

Table 5.8 Methane partial oxidation on PS, and MPS 4 catalytic metal oxides, with r_0 representing rate of CH_4 conversion.[88]

Catalytic metal oxide	T_R	r_0	Selectivity (%)			
	(°C)	$\mu mol\ s^{-1}\ g^{-1}$	HCHO	CO	CO_2	C_2
PS	550	5.0×10^{-6}	100	0	0	0
	650	1.2×10^{-4}	100	0	0	0
	725	7.3×10^{-4}	82	0	0	18
	750	5.0×10^{-4}	5	8	0	85
	800	1.3×10^{-3}	2	0	0	98
MPS 4	550	3.0×10^{-6}	100	0	0	0
	650	5.0×10^{-4}	90	10	0	0
	725	0.3	2	78	10	10
	750	1.2	1	76	11	12
	800	1.8	2	66	4	28

HCHO production. Combined with the result obtained from pure MoO_3 crystals earlier in the section, Mo=O terminal sites must be the dominating sites on the *Mc* phase, resulting in the high selectivity for HCHO. This provides confirmation that both a high concentration of *Mc* phase and low operating temperature are favorable for HCHO selectivity.

For a structure-sensitive reaction such as CH_4 to HCHO, selectivity towards a target molecule is dependent on the environment of the participating lattice oxygen. For HCHO formation, terminal sites are found to be more effective than bridged sites on MoO_3. This remains true for both pure and SiO_2 supported MoO_3. Statistically, the oxidation states of reaction products are inversely proportional to the strength of the metal oxide bond on the surface sites. Lattice oxygen in Mo–O–Mo has a weaker bond than the terminal Mo=O bond in the crystallite, promoting the formation of complete oxidation products. As for Mo=O, the intermediate strength made it the perfect active site for formaldehyde, with oxidation number of 0. Similar to MoO_3, Irusta et al.[90]

showed that in the case of SiO_2 supported V_2O_5, the terminal V=O sites promote higher selectivity towards HCHO during the oxidation of CH_4, whereas, a higher concentration of the V–O–V sites on the surface leads to the formation of carbon oxides.

As mentioned earlier in the section, it takes both an understanding of the role of the active sites on the catalytic metal oxides and effective reactor design to achieve high HCHO selectivity during partial oxidation of CH_4. This section provides an understanding of metal oxide surface structure that promotes formation of the desired products by incorporation of lattice oxygen, while suppressing further oxidation. This is similar to the chemical looping gasification scheme, discussed in Chapter 4, where the oxygen carriers and reactor design are critical to the formation of syngas, of the desired quality, from methane. The reaction mechanism followed by the catalytic metal oxides is similar to that of the oxygen carriers. The main difference is that oxygen carriers in the chemical looping scheme are re-oxidized in a separate reactor by air, whereas here, the catalytic metal oxides are simultaneously regenerated in the same reactor using molecular oxygen. The selective oxidation reaction of CH_4 to HCHO could also be operated in the chemical looping mode, with cyclic reduction and regeneration of the catalytic metal oxides, and an effective reactor design.

5.5   Selective Partial Oxidation of Propylene

The catalytic oxidation of olefins using a bismuth molybdate based catalytic metal oxide has, since the 1960s, been extensively examined from the following aspects:

(1)   Reaction mechanism for oxidation
(2)   Participation of the lattice oxygen from the catalytic metal oxide
(3)   Performance of different phases of the catalytic metal oxide
(4)   Re-oxidation of the catalytic metal oxide
(5)   Post-catalytic reactions and products.

For propylene oxidation using bismuth molybdate catalytic metal oxides, the abstraction of an allylic hydrogen atom by gaseous oxygen adsorbed on the surface to form a symmetric π-allylic intermediate is the first and rate determining step.[91–93] Two reaction pathways have been proposed for the subsequent reactions. In the first, a second hydrogen atom is abstracted from either of the terminal carbon atoms of the allylic intermediate, followed by incorporation of a lattice oxygen atom from the catalytic metal oxide, thereby forming acrolein as the product. This is termed the redox pathway, since propylene reduces the catalytic metal oxide, which in turn acts as the oxidizing agent. In this reaction, the molecular oxygen serves only to replenish the oxygen in the catalytic metal oxide, or in other words, it re-oxidizes the reduced catalytic metal oxide. In the second pathway, there is direct interaction of the allylic intermediate with the molecular oxygen to form a peroxide or hydroperoxide species. This species can then either decompose to give acrolein and water or undergo homogeneous reactions to form other products, like propylene oxide. This is referred to as the peroxide pathway.[94] The α, β, and γ phases of bismuth molybdate exhibit different kinetics and mechanisms for the catalytic oxidation of propylene.[95] Also, it has been found that the catalytic metal oxide re-oxidation sites are different from the sites that provide oxygen for reaction with propylene.[96] The following sections describe the different aspects of catalytic propylene oxidation.

5.5.1 Redox Mechanism

The redox mechanism has been explored using isotopic oxygen tracers. To determine the source of oxygen in the products, the ^{18}O tracer has been used with ^{18}O either in the form of gaseous oxygen used for oxidation or in the form of ^{18}O-enriched catalytic metal oxides. Studies using isotopic gaseous oxygen for oxidation yielded products rich in ^{16}O with a lack of extensive incorporation of ^{18}O in the products. Figure 5.27 shows the results of the tracer experiment by Keulks, where isotopic gaseous oxygen was used.[97] It can be deduced from the experimental results that initially the products consisted mainly of $C_3H_4{}^{16}O$, and its concentration gradually began to decrease over time, while trace amounts of $C_3H_4{}^{18}O$ were only detected later. These results suggest that many subsurface layers of oxide ions from the catalytic metal oxide participate in the reaction. Oxide ions from the catalytic metal oxide react with propylene to form the oxidation products, which generates oxygen vacancies that are filled by the rapid diffusion of oxygen from the bulk catalytic metal oxide. This leads to the formation of oxygen vacancies in the bulk. Subsequently, gaseous oxygen is adsorbed onto the catalytic metal oxide surface and diffuses to the bulk to fill the lattice oxygen vacancies. This diffusion process is rapid and allows the ^{18}O to equilibrate throughout a number of layers before reaction. There were no indications of heterophase exchange and homophase equilibration reactions, since no evidence of $^{16}O_2$ and $^{16}O^{18}O$ in the gas phase was found.

The catalytic metal oxide composition or type has a significant influence on the oxygen diffusion mechanism. For example, bismuth molybdate has a layered structure, consisting of layers of $(Bi_2O_2)_n{}^{2+}$ and $(MoO_2)_n{}^{2+}$ connected by layers of $(O)_n{}^{2-}$.[98]

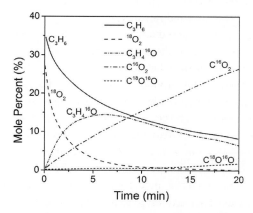

Figure 5.27 Gas phase composition variation with time for the reaction of C_3H_6 with $^{18}O_2$ at 425 °C.[97]

The active sites for propylene oxidation are believed to be anion vacancies in the oxygen boundary layers.[97] The oxygen that is incorporated into the products originated from these oxygen boundary layers. The transfer of oxygen to the products leads to reduction of the surface. The reduced surface, i.e. the anion vacancies, is re-oxidized by O^{2-} ions diffusing from the bulk of the catalytic metal oxide rather than the gas phase. The concentration of anion vacancies determines the rate of diffusion of lattice oxygen. The gas phase oxygen is absorbed in sites different from the propylene absorption sites, and subsequently diffuses through several layers to become equilibrated with oxygen ions in the bulk of the catalytic metal oxide. Thus, the concentration of $^{18}O^{2-}$ in the oxygen boundary layers is minimal, which explains the observed amount of ^{18}O in the products. ^{18}O concentration in the products gradually begins to increase as most of the ^{16}O in the catalytic metal oxide is consumed.

Oxygen diffusion within the catalytic metal oxide has been analyzed using two differently labeled γ-bismuth molybdates, one with ^{18}O-enriched bismuth oxide (γ-$Bi_2^{18}O_3 \cdot MoO_3$) and the second with ^{18}O-enriched molybdenum oxide (γ-$Bi_2O_3 \cdot Mo^{18}O_3$), and the results are shown in Figure 5.28. Figure 5.29 suggests that the oxygen atoms in acrolein are derived from the $(Bi_2O_2)_n^{2+}$ layers of the catalytic metal oxide only.[98] When propylene is catalytically oxidized over γ-$Bi_2^{18}O_3 \cdot MoO_3$, the concentration of ^{18}O in the acrolein product is initially high and then gradually decreases as the ^{18}O content in the $(Bi_2O_2)_n^{2+}$ layers depletes. For oxidation over γ-$Bi_2O_3 \cdot Mo^{18}O_3$, the ^{18}O concentration in acrolein is initially low, which indicates that oxygen in the product originating from the $(Bi_2O_2)_n^{2+}$ layers contains no ^{18}O. As the oxygen from the $(Bi_2O_2)_n^{2+}$ layers is consumed during reaction with the adsorbed propylene, it is rapidly replenished by the diffusion of ^{18}O from the $(MoO_2)_n^{2+}$ layers, which explains the subsequent increase in ^{18}O concentration.

Modified Redox Mechanism

There is a close relation between the participation of lattice oxygen and the concentration of catalytic metal oxide reduction and re-oxidation sites, which is explained by a

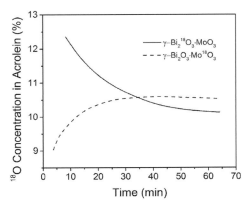

Figure 5.28 Time course of ^{18}O concentration in acrolein produced by the reduction of ^{18}O-labeled $Bi_2O_3 \cdot MoO_3$ catalysts with propylene.[98]

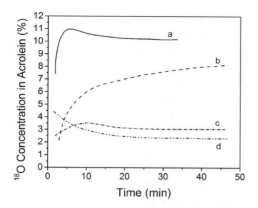

Figure 5.29 Time course of ^{18}O concentration in acrolein produced by the oxidation of propylene with $^{16}O_2$ over (a) -γ-Bi$_2$O$_3$·Mo^{18}O$_3$ (^{18}O 11%); (b) unlabeled catalyst with ^{18}O-enriched oxygen; (c) -γ-Bi$_2$O$_3$·Mo^{18}O$_3$ (^{18}O 4.5%); and (d) -γ-Bi$_2$^{18}O$_3$·MoO$_3$ (^{18}O 4.5%).[98]

modified redox pathway.[99] As mentioned earlier, the incorporation of lattice oxygen during oxidation of propylene and the re-oxidation of the catalytic metal oxide by gaseous oxygen occur at different sites on its surface.[98] Thus, if region A develops surface anion vacancies due to transfer of oxygen to the products, the vacancies will be filled by the diffusion of oxide ions from the bulk or region B. This leads to the formation of anion vacancies in region B, and these vacancies are filled with oxygen uptake from the gaseous phase. Under steady state conditions, the reduction and re-oxidation of catalytic metal oxides can be expressed in the form of a rate expression, as given in equation (5.5.1):

$$k_A N_A P_{C_3H_6} \theta_{fA} = k_B N_B P_{O_2}^{\frac{1}{2}} (1 - \theta_{fB}). \tag{5.5.1}$$

In equation (5.5.1), θ_{fA} and $(1-\theta_{fB})$ are the fractions of surface oxide ions and anion vacancies in regions A and B, respectively. N_A and N_B are the concentrations of the surface lattice oxygen in regions A and B, respectively, that participate in oxidation. k_A and k_B are the rate constants for the reduction step in region A and for the re-oxidation step in region B, respectively. If the surface reaction is the rate limiting step, i.e. the rate of diffusion of oxide ions from regions B to A is faster than the surface reaction, θ_{fA} and θ_{fB} can be approximated to be equal and are given as θ_f, with the rate expressions given by equation (5.5.2) and equation (5.5.3):

$$r = k_A N_A P_{C_3H_6} \theta_f \tag{5.5.2}$$

$$r = k_A N_A k_B N_B P_{C_3H_6} \frac{P_{O_2}^{\frac{1}{2}}}{k_A N_A P_{C_3H_6} + k_B N_B P_{O_2}^{\frac{1}{2}}}. \tag{5.5.3}$$

When $k_B \gg k_A$, equation (5.5.3) can be simplified to equation (5.5.4):

$$r = k_A N_A P_{C_3H_6}. \qquad (5.5.4)$$

Since propylene oxidation via redox mechanism occurs through transfer of lattice oxygen, the kinetics are expected to be first order in propylene and zero order in oxygen. The first-order kinetics over Bi–Mo catalytic metal oxides can be explained by equation (5.5.4). This modified redox mechanism provides a reasonable correlation between the extent of lattice oxygen participation and the number of active sites, by separating the reduction and re-oxidation sites on the surface and the bulk of the catalytic metal oxide.

5.5.2   Peroxide Mechanism

Under certain reaction conditions, propylene oxidation over bismuth molybdate leads to the formation of other products besides acrolein and CO_2, such as propylene oxide. The side reaction forming propylene oxide was found to occur only when the reactor had a large post-catalytic volume and/or the catalytic metal oxide packing had a large void fraction at higher propylene to oxygen ratios. This resulted in the possibility that surface initiated homogeneous reactions constituted the reaction pathway.[100] It is generally accepted that formation of the allylic intermediate is the first and the rate determining step. This allyl intermediate can then react with adsorbed surface oxygen to form peroxide or hydroperoxide species, which can then desorb into the gas phase and undergo further homogeneous reactions. Alternatively, the intermediate can desorb from the surface as an allyl radical; it first adds oxygen, followed by loss of a second hydrogen atom to form an allyl peroxide or an allyl hydroperoxide intermediate.[100] The peroxide intermediate can react further with propylene to form propylene oxide by a radical mechanism. It is independent of the more commonly accepted redox mechanism and can occur simultaneously.

5.5.3   Reaction Pathway for Oxidation Over Bismuth Molybdate Catalytic Metal Oxide

Both reaction pathways for propylene oxidation over bismuth molybdate catalytic metal oxides have been extensively studied, with experimental results that support the plausibility of both mechanisms.[97–100] The simultaneous occurrence of both mechanisms was studied using deuterated propylene and $^{18}O_2$ as tracers.[94] Using $^{18}O_2$ as a tracer helps in determining the source of oxygen participating in the formation of the products. The oxygen in the products could either come from a monolayer, or loss of chemisorbed oxygen from the catalytic metal oxide surface, or from multiple layers of oxygen in its lattice. Using deuterated propylene as a tracer assists in determining the rate controlling step and the sequence for the abstraction of a second hydrogen and addition of oxygen.

Extensive investigation of the reaction pathways has led to the development of a comprehensive mechanistic model of all reactions involved in the catalytic oxidation of propylene, as shown in Figure 5.30.[97–100]

Table 5.9 lists the theoretical and experimental distribution of acrolein from the oxidation of deuterated propylene over different phases of bismuth molybdate.[94] These results can be explained in relation to the reaction scheme shown in Figure 5.30. It is

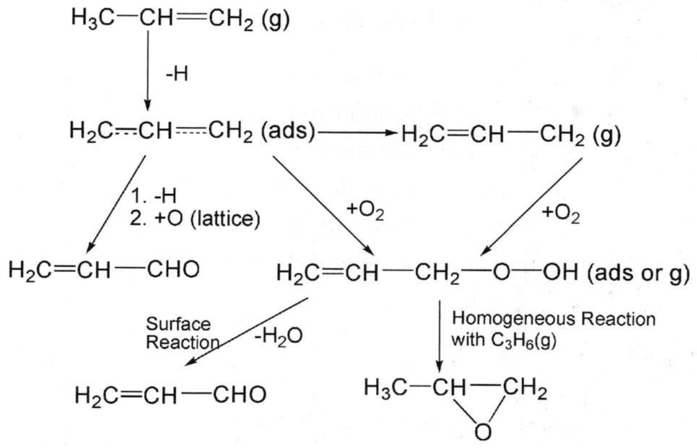

Figure 5.30 Reaction scheme for the oxidation of propylene over bismuth molybdate.[100]

generally accepted that the abstraction of the methyl hydrogen to form the symmetric allylic intermediate is the rate controlling step. Then the resulting deuterated species $CD_2–CD–CH_2$ can react further via either of the proposed mechanisms, i.e. the redox mechanism or the peroxide mechanism. If the reaction proceeds via the peroxide mechanism, each terminal carbon of the allylic species will have an equal probability of reacting with molecular oxygen. This leads to a 50–50 distribution of $CH_2=CD–CDO$ and $CD_2=CD–CHO$. However, if the reaction proceeds via the redox mechanism, there will be an additional hydrogen/deuterium abstraction from one of the terminal carbon atoms. Since it is more difficult to break a C–D bond as compared to a C–H bond, based on its isotopic effect, the deuterated carbon atom will react at a rate lower than the hydrogenated carbon atom. Based on theoretical calculations, the product distribution should be about 64% $CD_2=CD–CHO$ and 36% $CH_2=CD–CDO$. As seen in Table 5.9, the experimentally observed distribution in acrolein is in close agreement with the theoretical values. Based on these results, the redox mechanism can be considered to be the predominant reaction pathway in the catalytic oxidation of propylene over bismuth molybdate catalytic metal oxide.

5.5.4   Formation of Acrolein and Carbon Dioxide Over α, β, and γ-Bismuth Molybdate

The bismuth molybdate system has been detected in multiple phases, with superior catalytic activity and selectivity occurring in the following three phases: $\alpha\text{-}Bi_2Mo_3O_{12}$, $\beta\text{-}Bi_2Mo_2O_9$, and $\gamma\text{-}Bi_2MoO_6$. The results listed in Table 5.10 have been obtained by combining the data on catalytic propylene oxidation over the three different bismuth molybdate phases.[94,95,101] These results are used to differentiate between the redox and peroxide mechanisms based on the type of oxygen involved. In the redox pathway, lattice oxygen is extensively incorporated in the products, and hence a large number of oxide layers will participate in the reaction with propylene. The peroxide pathway uses chemisorbed molecular oxygen, and thus, only a monolayer or less of oxygen is involved.

Table 5.9 Oxidation of 2,3,3,3,-d_4-propylene: distribution of deuterated acroleins.[94]

	450 °C		350 °C	
	$CH_2=CD-CHO$ (%)	$CH_2=CD-CDO$ (%)	$CD_2=CD-CHO$ (%)	$CH_2=CD-CDO$ (%)
Theoretical distribution of acrolein via redox mechanism	64	36	69	31
Theoretical distribution of acrolein via peroxide mechanism	50	50	50	50
Observed distribution over:				
$Bi_2Mo_3O_{12}$	67	33	66	34
Bi_2MoO_6	65	35	64	36
$Bi_3FeMo_2O_{12}$	68	32	67	33

Table 5.10 Lattice oxide participation for propylene oxidation over α, β, γ phases of bismuth molybdate.[94,101]

		Percentage of total lattice O^{2-} participating in product formation	
Catalytic metal oxide	Temperature (°C)	C_2H_4O	CO_2
α-$Bi_2Mo_3O_{12}$	450	9	9
	400	4	4
β-$Bi_2Mo_2O_9$	450	98	98
	400	56	60
γ-Bi_2MoO_6	450	100	100
	400	45	45
USb_3O_{10}	450	0.5	0.5
	400	0.6	0.6

These studies were performed using $^{18}O_2$ as the tracer for oxidation of propylene over different phases of the catalytic metal oxide. The product gases were analyzed to determine the extent of lattice oxygen incorporation in the oxidation products. For a comparison, propylene oxidation was also studied over USb_3O_{10} catalyst, since the reaction pathway was known to proceed via the surface hydroperoxide mechanism, with no involvement of lattice oxygen.

From the results observed by Hoefs et al., multiple layers of lattice oxygen are involved in the formation of acrolein and carbon dioxide over all the three phases of bismuth molybdate.[101] Also, there was no distinction between the incorporation of lattice oxygen in acrolein and carbon dioxide, with both products showing almost the same amount of oxide layer participation. With the β and γ phases at 450 °C, since there was nearly complete participation of lattice oxygen, it can be concluded that only the redox mechanism is involved. In all other cases, there is less lattice oxygen participation due to the occurrence of a parallel mechanism where oxygen is adsorbed. Kinetic studies over the three different bismuth molybdate phases suggest temperature dependence of reaction kinetics. Above 400 °C, the α and γ phases demonstrated reaction kinetics that were first order in propylene and zero order in oxygen. However, below 400 °C, the reaction kinetics changed to zero or partial order in propylene and partial order in oxygen. Kinetics over the β phase closely resemble the kinetics over the α and γ phases. At temperatures greater than 425 °C, the β phase exhibited kinetics corresponding to the redox pathway. Apart from temperature dependence, oxidation kinetics are also heavily dependent on the pressure of the reactant. Thus, by selecting the proper range of propylene and oxygen partial pressures at lower temperatures, the reaction kinetics can be manipulated to resemble the kinetics typically associated with the redox mechanism.

5.5.5   Re-oxidation of Bismuth Molybdate

Studies using ^{18}O-enriched catalytic metal oxides, like those mentioned in Section 5.5.1, suggest that incorporation of oxygen into the products and re-oxidation of the reduced catalytic metal oxide occur at different sites. The oxide ions in the acrolein product come from the $(Bi_2O_2)_n^{2+}$ layers, which develop vacancies upon reaction with propylene. These vacancies are replenished by diffusion of oxygen from the $(MoO_2)_n^{2+}$ layers, which acts as the re-oxidation site assimilating oxygen from the gas phase into the lattice. However, there have also been studies indicating that the re-oxidation of bismuth molybdate catalytic metal oxide primarily occurs at a site associated with bismuth.[95] Extensive reduction of the catalytic metal oxide with propylene, in the absence of oxygen, converted Bi^{3+} to Bi^0 and Mo^{6+} to Mo^{4+}. On low temperature re-oxidation of the catalytic metal oxide (200 °C), the metallic bismuth is converted to Bi^{m+} $(0 < m < 3)$ and Mo^{4+} to Mo^{6+}. Re-oxidation of the reduced catalytic metal oxide at higher temperature (above 400 °C) completely oxidizes the catalytic metal oxide to Bi^{3+} and Mo^{6+}. The bismuth molybdate catalytic metal oxide thus has two different re-oxidation sites, depending on the re-oxidation temperature. Mo^{4+} is re-oxidized to Mo^{6+} during low temperature oxidation, but complete re-oxidation of Bi is only achieved when the high temperature sites are re-oxidized.

Formation of acrolein by lattice oxygen increases significantly after the high temperature sites are completely re-oxidized. The results, summarized in Table 5.11, indicate that acrolein formation is significantly lower when only the low temperature sites are re-oxidized.[96] Because re-oxidation of the low temperature sites has a low activation energy, it is easier to achieve re-oxidation even at temperatures below 225 °C. However, complete re-oxidation has a higher activation energy, hence it only occurs at higher temperatures and occurs on a surface plane different from where reduction occurs. According to these studies, there are two possible pathways for the re-oxidation of bismuth molybdate. The first involves direct re-oxidation of Bi^{m+} $(0 < m < +3)$ to Bi^{3+} by gaseous oxygen. The activation energy for this pathway includes the dissociation energy of oxygen and the activation energy for its diffusion into the bulk of the catalytic metal oxide. The second pathway involves interaction between Mo^{4+} and Bi^{3+}. The Mo^{4+} that remains even after the low temperature

Table 5.11 Reactivity of low and high temperature re-oxidized sites.[96]

Initial state[a]	Extent reduced[b] (%)	Products collected	
		Acrolein ($\times 10^{-3}$ mmol)	CO_2 ($\times 10^{-3}$ mmol)
Low temperature site re-oxidized	2.3	1.08	39.1
Both sites re-oxidized	2.3	7.9	47.1
Low temperature site re-oxidized	0.6	0.14	20.5
Both sites re-oxidized	0.6	4.8	8.3

Reaction conditions: T = 420 °C; total flow = 100 cm^3/min (20% C_3H_6, 80% N_2); catalyst weight = 615 mg.
[a] Indicates the level of re-oxidation of γ-Bi_2MoO_6 pre-reduced 6%.
[b] Weight loss from the initial; 2.3% reduction required 4 min, 0.6% reduction required 30 sec.

re-oxidation may be associated with shear structures that are possibly responsible for the fast diffusion of oxygen through the γ-phase bismuth molybdate.[96,102] These Mo^{4+} ions could be re-oxidized to Mo^{6+} by bismuth, according to reaction (5.5.5):

$$3Mo^{4+} + 2Bi^{3+} \rightarrow 3Mo^{6+} + 2Bi^0. \tag{5.5.5}$$

With gaseous oxygen present in the system, the Bi^0 will be re-oxidized by incorporating oxygen from the gaseous phase. Also, the activation energy will include activation energy for the shear structure rearrangement along with the dissociation energy for oxygen. With a number of different pathways proposed for the re-oxidation of bismuth molybdate catalytic metal oxide, it is impossible to conclude the exact mechanism responsible for the re-oxidation process.

In summary, various aspects of bismuth molybdate catalytic metal oxide, used for the partial oxidation of propylene, have been discussed. These include the reaction mechanisms involved during partial oxidation of propylene, involvement of lattice oxygen, mechanism for re-oxidation, and the effect of different phases of the catalytic metal oxide on the reaction. The partial oxidation of propylene using bismuth molybdate as the catalytic metal oxide has been considered because of the participation of lattice oxygen in the oxidation mechanism. While this section provides an in-depth analysis of the pathway for incorporating lattice oxygen into the products through different layers of the catalytic metal oxide, the overall redox mechanism is similar to the redox reactions involved in chemical looping processes.

5.6   Selective Oxidation of Alcohols

Selective oxidation of the hydroxyl group of alcohols is the foundation of many important industrial and fine chemical processes. Conventionally, stoichiometric amounts of Cr(VI) compounds or other inorganic oxidants were used for these selective oxidation processes, but resulted in the generation of heavy metal wastes. A second moderate catalytic process used hydrogen peroxide, H_2O_2, as the oxygen source, but peroxides are associated with various handling hazards. Catalytic selective oxidation with molecular oxygen gained attention due to its attractiveness from economic and environmental aspects. Also, its availability as an aqueous solution makes it a good choice for a strong oxidant for liquid phase selective oxidation reactions.

Based on the reaction conditions, oxidation reactions are believed to proceed through different mechanisms. At low temperatures, the reactions proceed via the peroxidic mechanism involving adsorbed radicals; at moderate temperatures the Mars–Van Krevelen mechanism gains prominence; and at high temperatures, reactions proceed via the free radical mechanism. These catalytic reactions with molecular oxygen are either electrophilic or nucleophilic in nature. During electrophilic oxidation, oxygen is activated to its electrophilic form (O^-, O^{2-}, or O_2^{2-}), which then attacks the reactant at the high electron density sites. The Mars–Van Krevelen mechanism for selective oxidation reactions is a nucleophilic oxidation process where the hydrocarbon is first activated, followed by reaction with nucleophilic oxygen, O^{2-}, or lattice oxygen from the catalytic metal oxides.

The kinetics for total oxidation reactions proceeding via the Mars–Van Krevelen mechanism depend heavily on the metal–oxygen bond strength of the catalytic metal oxide. Sections 5.6.1 and 5.6.2 present an insight into this mechanism for alcohols by discussing the selective oxidation of propylene on period IV metal oxides, which are typically used for selective oxidation of alcohols. Propylene oxidation over period IV metal oxides, however, mainly results in the formation of complete oxidation products like CO_2 and CO, instead of acrolein, and hence making it difficult to study selective oxidation reaction mechanisms. Thus, innovative techniques to study the selective oxidation mechanism using propylene over period IV metal oxides are developed and are discussed in Sections 5.6.1 and 5.6.2.

5.6.1   Selective Oxidation Mechanism of Propylene Over Period IV Metal Oxides

A different method was adopted to study the reactivity of lattice oxygen and its role in selective oxidations for the period IV metal oxides. The first step of hydrogen abstraction from propylene was bypassed by using allyl iodide to generate the allyl radicals in situ in the reactor. The C–I bond strength in allyl iodide is significantly lower than the C–H bond strength in propylene, which allows it to readily decompose into allyl radicals. This method was used to elucidate the mechanism of selective oxidation of propylene over period IV catalytic metal oxides. At low temperatures (between 100 °C and 200 °C), there was a high selectivity towards acrolein for oxidation over V_2O_5, Cr_2O_3, MnO_2, and CuO. At higher temperatures, the selectivity towards acrolein decreases while the selectivity towards complete oxidation products increases. For Fe_2O_3, NiO, and ZnO, the major product is 1,5-hexadiene, when the acrolein selectivity starts to decrease, but beyond a certain concentration of 1,5-hexadiene, benzene and propylene were seen in the products. Figure 5.31a shows the conversion and selectivity data for products during the oxidation of allyl iodide over V_2O_5 at various temperatures, while Figure 5.31b shows the product selectivity and conversion for oxidation of allyl iodide over Fe_2O_3.[103]

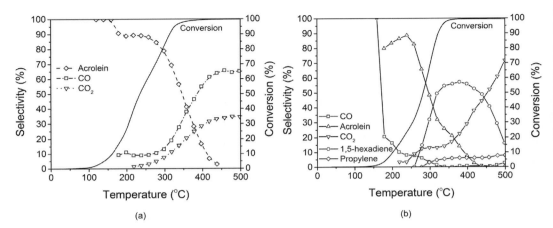

Figure 5.31 Allyl iodide oxidation conversion and selectivity over (a) V_2O_5 (b) Fe_2O_3.[103]

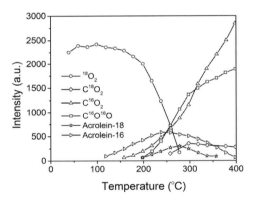

Figure 5.32 Product concentration for allyl iodide oxidation with $^{18}O_2$ over Cr_2O_3.[103]

5.6.2   Oxidation Mechanism of Propylene Over Period IV Metal Oxides with Isotopic ^{18}O

$^{18}O_2$ was used to study the mechanism of the oxidation reactions. Figure 5.32 displays the products for the selective oxidation of allyl iodide with $^{18}O_2$ over Cr_2O_3.[103] At low temperatures, in the presence of gaseous oxygen, acrolein-16 and $C^{16}O_2$ are formed first. This is a clear indication that the oxidation occurs with the oxygen from the catalytic metal oxide, because the molecular oxygen has the isotope ^{18}O rather than ^{16}O. Thus, oxidation over the period IV metal oxides proceeds via the Mars–Van Krevelen mechanism. As the temperature increases, ^{16}O from the metal oxide is depleted and is replenished by the isotopic molecular oxygen ^{18}O. This is also demonstrated from the increasing concentration of acrolein-18 and $C^{16}O^{18}O$ at higher temperatures. As the gas phase oxygen is depleted, almost all the lattice oxygen vacancies are replenished by

^{18}O, and C^{18}O$_2$ is observed in the products at even higher temperatures. Thus both selective and total oxidation of allyl iodide over the period IV metal oxides occur via the Mars–Van Krevelen mechanism.

When lattice oxygen is present, the hydrocarbons react preferentially with the lattice oxygen to form oxidation products. Based on the isotopic studies discussed in this section, both selective and complete oxidation products involve lattice oxygen from the catalytic metal oxides. The incorporation of lattice oxygen in the products is determined by a combination of two factors: (1) the inherent reactivity of the specific oxygen atoms; and (2) a lack of reactivity of neighboring atoms. But in the absence of lattice oxygen, when the metal oxide surface is reduced, the hydrocarbon intermediates react with the adsorbed gaseous oxygen. This usually occurs at elevated temperatures and is believed to occur via a pseudo Langmuir–Hinshelwood type mechanism, where no lattice oxygen is involved. This is discussed in Section 5.6.3, explaining the selective liquid phase oxidation of alcohols.

5.6.3   Heterogeneous Liquid Phase Selective Oxidation

The Mars–Van Krevelen mechanism is commonly observed in heterogeneous gas phase catalytic reactions, but this mechanism has also been observed during heterogeneous liquid phase oxidation of alcohols using molecular oxygen. During the selective oxidation of benzyl alcohol (PhCH$_2$OH) with molecular oxygen over a synthetic manganese oxide based catalytic metal oxide with an octahedral molecular sieve (OMS) structure, three possible mechanisms, as discussed in Section 5.1, for heterogeneous catalysis can be considered:

1.   The Eley–Rideal mechanism.
2.   The Langmuir–Hinshelwood–Hougen–Watson mechanism (LHHW).
3.   The Mars–Van Krevelen mechanism.

For the Eley–Rideal mechanism to be considered, one of the reactants must be in the gas phase. However, since oxygen is in dissolved form, this mechanism is not applicable. In the LHHW mechanism, both benzyl alcohol and oxygen will be adsorbed onto active sites on the surface of the catalytic metal oxide, followed by an irreversible surface reaction, yielding the oxidation products. This surface reaction is the rate determining step. The rate law for the liquid phase oxidation of benzyl alcohol is given in equation (5.6.1):

$$r_{ER} = \frac{kK_1K_2^{0.5}[PhCH_2OH][O_2]^{0.5}}{\left[1 + K_1[PhCH_2OH] + K_2^{0.5}[O_2]^{0.5}\right]^2},$$

(5.6.1)

where k is the rate constant for the surface reaction and K_1 and K_2 are the adsorption desorption equilibrium constants for PhCH$_2$OH and O$_2$, respectively. The dissolved O$_2$ concentration is considered constant, and equation (5.6.1) can be simplified to equation (5.6.2):

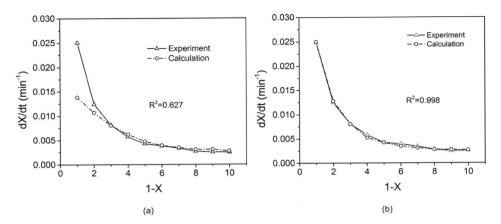

Figure 5.33 Fitting experimental data for the selective oxidation of benzyl alcohol with (a) the Langmuir–Hinshelwood–Hougen–Watson kinetic model; and (b) the Mars–Van Krevelen kinetic model.[104]

$$r_{ER} = \frac{a[PhCH_2OH]}{[b + c[PhCH_2OH]]^2}. \tag{5.6.2}$$

The model parameters can be obtained using a nonlinear regression algorithm. Figure 5.33 shows a comparison of the experimental and theoretical rates calculated using the simplified rate equation.[104] From Figure 5.33a, the fit is rather poor, which means that the probability for the LHHW mechanism being predominant is low.

In the Mars–Van Krevelen mechanism, Figure 5.33b, the lattice oxygen first oxidizes the hydrocarbon reactant molecule, and then the partially reduced catalytic metal oxide is re-oxidized by molecular oxygen. For the selective oxidation of benzyl alcohol, the rate of oxidation of the hydrocarbon reactant is given in equation (5.6.3):

$$r_1 = k_1[PhCH_2OH]\theta, \tag{5.6.3}$$

where θ is the fraction of surface oxide ions. The rate of re-oxidation of the surface is directly proportional to the surface remaining uncovered by oxygen and can be expressed as

$$r_2 = k_2[O_2]^n(1 - \theta). \tag{5.6.4}$$

At the steady state, both of these rates are equal and equation (5.6.5) is the combination of equation (5.6.3) and equation (5.6.4):

$$\beta k_1[PhCH_2OH]\theta = k_2[O_2]^n(1 - \theta), \tag{5.6.5}$$

where β, the stoichiometric coefficient for O_2, is 0.5.

The final rate expression for the Mars–Van Krevelen mechanism is given in equation (5.6.6):

$$r_{MVK} = \frac{1}{\dfrac{1}{k_1[PhCH_2OH]} + \dfrac{0.5}{k_2[O_2]^n}}, \tag{5.6.6}$$

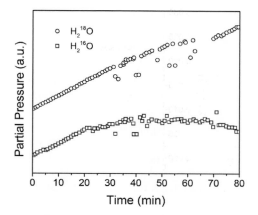

Figure 5.34 $H_2{}^{16}O$ and $H_2{}^{18}O$ concentrations during oxidation of benzyl alcohol using isotopic labeling.[104]

where k_1 and k_2 are rate constants for the two reaction steps, respectively. Equation (5.6.6) can be simplified to equation (5.6.7) with the equation parameters obtained by nonlinear regression:

$$r_{MVK} = \frac{a[PhCH_2OH]}{b + c[PhCH_2OH]}.$$ (5.6.7)

Figure 5.33b is a fit of the experimental and calculated data using the Mars–Van Krevelen rate law, and it suggests that the liquid phase oxidation of benzyl alcohol over the OMS catalytic metal oxide can be explained via the Mars–Van Krevelen mechanism.[104]

Oxidation with isotopic oxygen also confirmed the Mars–Van Krevelen mechanism. Figure 5.34 shows the concentration of $H_2{}^{16}O$ and $H_2{}^{18}O$ formed during the oxidation of benzyl alcohol over the OMS catalytic metal oxide using $^{18}O_2$.[104] The initial rates of formation of both $H_2{}^{16}O$ and $H_2{}^{18}O$ are identical. Since ^{16}O is present only in the catalytic metal oxide and the molecular oxygen contains only ^{18}O, lattice oxygen donation is involved. This supports the kinetic data backing the Mars–Van Krevelen mechanism. When all the ^{16}O in the lattice is consumed, the partial pressure of $H_2{}^{16}O$ gradually decreases, while that of $H_2{}^{18}O$ increases since the lattice oxygen vacancies are continuously replenished by gaseous oxygen. The OMS catalytic metal oxide used for these studies consists of intimately coupled Mn^{4+}/Mn^{2+} sites, which allow for electron exchange, leading to facile diffusion of gaseous oxygen from the surface to the bulk of the metal oxide and through the metal oxide lattice back to the surface to oxidize the reactant species.

Isotopic studies using $PhCD_2OH$ showed that there is a large kinetic isotopic effect because the C–D and C–H bonds differ in their bond energies. A large isotopic effect during these experiments indicates the formation of an intermediate with an electron deficient carbon center, which is the rate determining step. Reactions (5.6.8) and (5.6.9) show the first step of the Mars–Van Krevelen mechanism, where the hydrocarbon molecule is oxidized and the metal substrate undergoes reduction:[97]

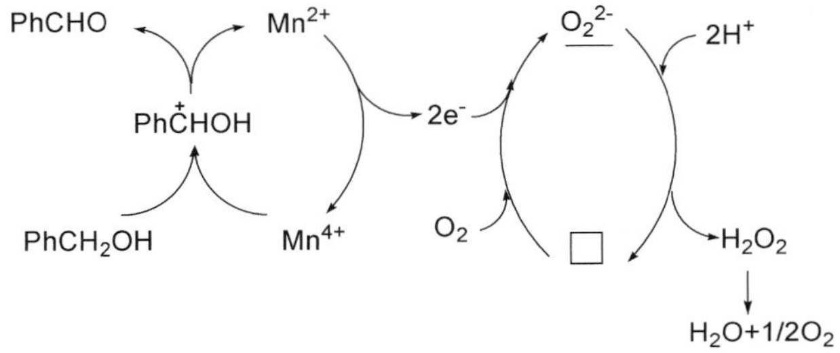

Figure 5.35 Overall alcohol oxidation mechanism.[104] Reprinted from Journal of Catalysis, 210, Makwana, V. D., Son, Y., Howell, A. R., and S. L. Suib, The Role of Lattice Oxygen in Selective Benzyl Alcohol Oxidation Using OMS-2 Catalyst: a Kinetic and Isotope-Labeling Study, 46–52, Copyright 2002, with permission from Elsevier.

$$PhCH_2OH + Mn^{4+} \xrightarrow{slow} PhC^+HOH + H^+ + Mn^{2+} \qquad (5.6.8)$$

$$PhC^+HOH \xrightarrow{fast} PhCHO + H^+. \qquad (5.6.9)$$

Studies have shown that Mn^{4+} is highly susceptible to reduction, which indicates rapid diffusion of lattice oxygen and higher oxidation activity. In the second half of the mechanism, Mn^{2+} undergoes re-oxidation by oxygen. Oxygen can be exchanged on the metal oxide surface by three mechanisms: R^0, R^1, and R^2. The R^0 mechanism is a pseudo Langmuir–Hinshelwood type mechanism in which the oxidation takes place via adsorption of molecules on the surface of the catalytic metal oxide and does not involve lattice oxygen. However, for the R^0 mechanism to prevail, much higher temperatures are required. The R^1 and R^2 mechanisms are Mars–Van Krevelen type mechanisms, in which oxygen from the gas phase replenishes the lattice oxygen vacancies. The relatively low temperatures used for these oxidation studies point towards the R^1 and R^2 mechanisms. Figure 5.35 shows the overall oxidation mechanism, as explained by the Mars–Van Krevelen mechanism.[104]

Selective oxidation of α-ketols, benzylic alcohols, and allylic alcohols were also studied using Ni–Al hydrotalcite as the solid oxidant catalyst. Hydrotalcites are homogeneous mixtures consisting of periodic heterobimetallic compositions of M^{+II} and M^{+III} ions. The Ni–Al hydrotalcite used has one aluminum atom for every two nickel atoms substituted alternately. Reaction (5.6.10) shows the reaction scheme for the liquid phase selective heterogeneous catalytic oxidation reaction.[105]

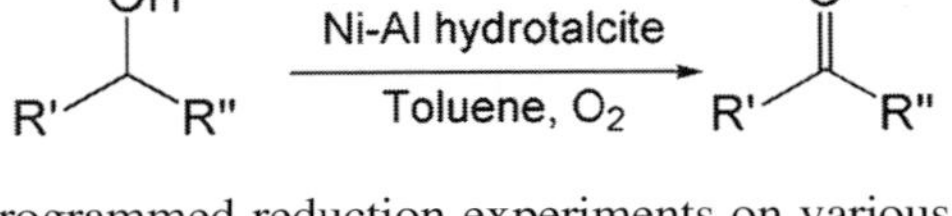

$$(5.6.10)$$

Temperature programmed reduction experiments on various benzyl alcohol oxidation and characterization studies on the catalyst point towards a reaction mechanism involving the formation of an alkoxide on interaction of the alcohol with a basic site in the brucite layer present in Ni–Al hydrotalcite. Further reaction of the hydrotalcite with molecular

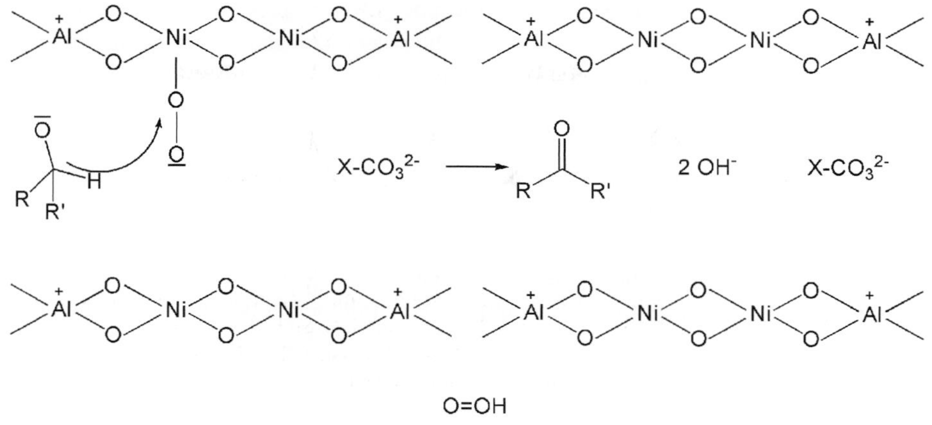

Figure 5.36 Proposed mechanism for oxidation over Ni–Al hydrotalcite.[105] Reproduced with permission from Choudary et al. Angewandte Chemie International Edition, 2001, 40(4), 763–766. Copyright ©2001 WILEY-VCH Verlag GmbH, Weinheim, Fed. Rep. of Germany.

oxygen leads to the formation of a peroxide. Hydride transfer between the alkoxide and the peroxide generates a carbonyl species, and hence regeneration of the basic site in the brucite layer. Another fresh alcohol molecule then reacts with the basic site in the brucite layer and the same cycle continues. Figure 5.36 gives a schematic explanation of the mechanism of selective alcohol oxidation over the Ni–Al hydrotalcite catalyst.[105]

In summary, Section 5.6 discusses a few other examples of partial oxidation reactions with catalytic metal oxides as functional materials. These selective oxidation reactions occur preferentially through incorporation of lattice oxygen from the metal oxides into the products. Understanding the mechanisms involved allows for the manipulation of metal oxide structures to promote diffusion of gaseous oxygen into the metal oxide lattice and then to the reactant molecules. It is only in the absence of lattice oxygen that oxidation via the exchange of adsorbed molecules becomes the predominant mechanism. Similar to the other examples discussed in this chapter, these catalytic metal oxides function like oxygen carriers in chemical looping processes, oxidizing reactants with lattice oxygen and replenishing the vacancies using gaseous molecular oxygen. The main difference lies in the fact that the diffusion of the lattice oxygen to the reactants and the replenishing of the vacancies with gaseous oxygen occur simultaneously in the same reactor. However, these partial oxidation reactions can be operated in the chemical looping mode as well, where reduction of the metal oxides by transfer of lattice oxygen and their regeneration occur sequentially in cycles in separate reactors.

5.7 Concluding Remarks

Selective partial oxidation reactions are used for the synthesis of a significant number of essential industrial chemicals. This chapter presents several representative examples of selective oxidation reactions forming different products and proposed reaction intermediates. Each section discusses the reaction mechanism in detail with an overview on its chemical looping application. In the reactions discussed in this chapter, the catalytic metal oxides function as oxygen carriers, transferring oxygen from the solid phase to the reactants via the catalytic metal oxide crystal lattice, in the form of lattice oxygen. For high conversions and product selectivity, the catalytic metal oxides require oxygen mobility through the lattice, a stable and sustainable host structure, and a moderate metal–oxygen bond strength.

DuPont's concept for the production of MAN from n-butane oxidation is novel owing to its expected economic and environmental benefits over the traditional process. The catalytic metal oxide VPO exhibited feasible oxygen transfer properties through its lattice, conversions, and selectivity for MAN formation. However, the desired oxygen carrying capacity of the catalytic metal oxide could not be sustained over time at progressively larger scales, leading to the requirement for an increased addition of supplemental molecular oxygen. The discontinuity of the commercial operation was due in part to inadequacy in sustaining the physical integrity and reactivity of the catalytic metal oxide VPO and hence its steady transport and reaction in the chemical looping system.

Syngas and/or H_2 generation via solar aided thermochemical processes use metal oxides as enabling materials. Some of the successfully studied materials include oxides of iron, zinc, and manganese. MoO_3 is a versatile catalytic metal oxide that is used for the conversion of methane/methanol to HCHO, propene/propylene to acrolein, and for the oxidation of other hydrocarbons. It exhibits catalytic anisotropy as the product selectivity is highly dependent on the distribution of the crystal planes. In the case of methane oxidation with MoO_3–C, which has a high concentration of (100) side planes, the exposed terminal Mo=O sites increase the selectivity for HCHO. On the other hand, the exposed terminal Mo–O–Mo sites in MoO_3–R which has a large concentration of (010) planes tend to favor CO_x production. The reduction and oxidation mechanisms of the catalytic metal oxide are heavily dependent on the type of metal–oxygen bond. The terminal Mo–O–Mo sites are more readily reduced and re-oxidized by gaseous oxygen, whereas the terminal Mo=O sites are less readily reduced and are re-oxidized preferably via lattice oxygen. Propylene oxidation to acrolein over bismuth molybdate catalytic metal oxide can occur either via a redox mechanism or a peroxide mechanism, which occur independently of each other. During the redox mechanism, oxygen in the product comes from the $(Bi_2O_2)_n^{2+}$ layers, which are rapidly replenished by oxygen diffusion from the $(MoO_2)_n^{2+}$ layers. Propylene oxidation over V_2O_5, Cr_2O_3, MnO_2, and CuO yields acrolein as the major product, while over Fe_2O_3, NiO, and ZnO, the selectivity for 1,5-hexadiene is higher. The product selectivity for the partial oxidation of alcohols over different metal oxides also depends on the catalytic metal oxide structure. Both homogeneous gas phase and heterogeneous liquid phase partial oxidation reactions over period IV metal oxides proceed via the Mars–Van Krevelen mechanism, where oxygen in the oxidation products comes from the catalytic metal oxide lattice.

Every example presented in this chapter reiterates the importance of catalytic metal oxide reactivity and performance for product yield and selectivity. In a sense, the catalytic metal oxides in these partial oxidation reactions act in a manner very similar to oxygen carriers in the chemical looping process. Once the catalytic metal oxide surface is reduced, it needs to be constantly replenished through gaseous oxygen. Consequently, these processes typically require co-injection of molecular oxygen in order to re-oxidize the catalytic metal oxide. However, co-injection of pure molecular oxygen to replenish the catalytic metal oxide lattice oxygen can increase the risk of explosion during pre-mixing and diminish the economic viability at commercial scale. It is these challenges that make the concept of chemical looping a desired solution, in which the reduced catalytic metal oxides can be regenerated in a separate reactor, i.e. the combustor, using simply air as in the chemical looping scheme. Chemical looping has been studied for the generation of electricity and such products as hydrogen, syngas, and other high valued chemicals and liquid fuels from carbonaceous fuels with inherent CO_2 capture. The concept can equally apply to the other partial oxidation examples discussed in this chapter for the production of high valued chemicals. With sufficient understanding of the interrelation of the physical and chemical properties of catalytic metal oxides, chemical looping partial oxidation is poised as a strong contender in the future for production of valuable chemicals.

References

1. Mars, P. and D. W. Van Krevelen, "Oxidations Carried Out By Means of Vanadium Oxide Catalysts," *Chemical Engineering Science*, 3, 41–59 (1954).
2. Suib, S. L., "Selectivity in Catalysis: an Overview," *ChemInform*, 24(48) (1993).
3. Ertl, G., H. Knözinger, and J. Weitkamp, eds., *Handbook of Heterogeneous Catalysis*, Wiley-VCH, Weinheim, Germany (1997).
4. Lerou, J. and P. Mills, *Precision Process Technology*, Springer, Berlin, Germany (1993).
5. Jenkins Jr., C. L., "Preparation of Tetrahydrofuran," U.S. Patent 4,124,600 (1978).
6. Muller, H., *Ullmann's Encyclopedia of Industrial Chemistry*, 6th edn., Wiley-VCH, Berlin, Germany (2003).
7. Prichard, W. W., "Conversion of Furan to 1, 4-Butanediol and Tetrahydrofuran," U.S. Patent 4,146,741 (1979).
8. Contractor, R., H. Bergna, H. Horowitz, et al., Butane Oxidation to Maleic Anhydride in a Recirculating Solids Reactor, *Studies in Surface Science and Catalysis*, 38, 645–654 (1988).
9. Contractor, R., J. Ebner, and M. J. Mummey, "Butane Oxidation in a Transport Bed Reactor–Redox Characteristics of the Vanadium Phosphorus Oxide Catalyst," *Studies in Surface Science and Catalysis*, 55, 553–562 (1990).
10. Wurzbach, G., "Benzene C_4 Cut as Alternative Raw-Materials for Production of Maleic Anhydride (Chemical Economics)," *Chemie Ingenieur Technik*, 45(22), 1297–1302 (1973).
11. Varma, R. and D. Saraf, "Selective Oxidation of C_4 Hydrocarbons to Maleic Anhydride," *Industrial & Engineering Chemistry Product Research and Development*, 18(1), 7–13 (1979).
12. Kanetaka, J., T. Asano, and S. Masamune, "New Process for Production of Tetrahydrofuran," *Industrial & Engineering Chemistry*, 62(4), 24–32 (1970).
13. Zhanglin, Y., M. Forissier, J. Vedrine, and J. Volta, "On the Mechanism of n-Butane Oxidation to Maleic Anhydride on VPO Catalysts. II. Study of the Evolution of the VPO

Catalysts Under n-Butane, Butadiene, and Furan Oxidation Conditions," *Journal of Catalysis*, 145(2), 267–275 (1994).

14. Kerr, R. O., "Oxidation of Aliphatic Hydrocarbons," U.S. Patent 3,156,707 (1964).

15. Nakamura, M., K. Kawai, and Y. Fujiwara, "The Structure and the Activity of Vanadyl Phosphate Catalysts," *Journal of Catalysis*, 34(3), 345–355 (1974).

16. Contractor, R. M., "Butane Oxidation to Maleic Anhydride," in *Proceedings of the 2nd International Conference on Circulating Fluidized Beds*, Oxford, UK (1988).

17. Contractor, R., H. Bergna, H. Horowitz, et al., "Butane Oxidation to Maleic Anhydride over Vanadium Phosphate Catalysts," *Catalysis Today*, 1(1), 49–58 (1987).

18. Cavani, F., G. Centi, and F. Trifiro, "Study of n-Butane Oxidation to Maleic Anhydride in a Tubular Flow Stacked-Pellet Reactor. Influence of Phosphorus on the Selectivity," *Applied Catalysis*, 15(1), 151–160 (1985).

19. Ai, M. and S. Suzuki, "Oxidation Activity and Acidity of V_2O_5–P_2O_5 Catalyst," *Bulletin of the Chemical Society of Japan*, 47(12), 3074–3077 (1974).

20. Escardin, A., C. Sola, and F. Ruiz, "Catalytic Oxidation of Butane to Maleic Anhydride. 1. Reaction Mechanism," *Anales De Quimica- International Edition*, 69(3), 385–396 (1973).

21. Centi, G., G. Fornasari, and F. Trifiro, "On the Mechanism of n-Butane Oxidation to Maleic Anhydride: Oxidation in Oxygen-Stoichiometry-Controlled Conditions," *Journal of Catalysis*, 89(1), 44–51 (1984).

22. Contractor, R. M., "Dupont's CFB Technology for Maleic Anhydride," *Chemical Engineering Science*, 54(22), 5627–5632 (1999).

23. Patience, G. S. and R. E. Bockrath, "Butane Oxidation Process Development in a Circulating Fluidized Bed," *Applied Catalysis A: General*, 376(1), 4–12 (2010).

24. Emig, G., K. Uihlein, and C. Häcker, "Separation of Catalyst Oxidation and Reduction – an Alternative to the Conventional Oxidation of n-Butane to Maleic Anhydride?" *Studies in Surface Science and Catalysis*, 82, 243–251 (1994).

25. Mallada, R., M. Menéndez, and J. Santamaría, "Use of Membrane Reactors for the Oxidation of Butane to Maleic Anhydride under High Butane Concentrations," *Catalysis Today*, 56(1), 191–197 (2000).

26. Xiao, L., S. Wu, and Y. Li, "Advances in Solar Hydrogen Production via Two-Step Water-Splitting Thermochemical Cycles based on Metal Redox Reactions," *Renewable Energy*, 41, 1–12 (2012).

27. Agrafiotis, C., H. von Storch, M. Roeb, and C. Sattler, "Solar Thermal Reforming of Methane Feedstocks for Hydrogen and Syngas Production – a Review," *Renewable and Sustainable Energy Reviews*, 29, 656–682 (2014).

28. Steinfeld, A., "Solar Thermochemical Production of Hydrogen – a Review," *Solar Energy*, 78(5), 603–615 (2005).

29. Perkins, C. and A. W. Weimer, "Likely Near-Term Solar-Thermal Water Splitting Technologies," *International Journal of Hydrogen Energy*, 29(15), 1587–1599 (2004).

30. Hirsch, D. and A. Steinfeld, "Solar Hydrogen Production by Thermal Decomposition of Natural Gas Using a Vortex-Flow Reactor," *International Journal of Hydrogen Energy*, 29(1), 47–55 (2004).

31. Kuravi, S., J. Trahan, D. Y. Goswami, M. M. Rahman, and E. K. Stefanakos, "Thermal Energy Storage Technologies and Systems for Concentrating Solar Power Plants," *Progress in Energy and Combustion Science*, 39(4), 285–319 (2013).

32. Hildebrandt, A. F. and K. A. Rose, "Receiver Design Considerations for Solar Central Receiver Hydrogen Production," *Solar Energy*, 35(2), 199–206 (1985).

33. Tan, T. and Y. Chen, "Review of Study on Solid Particle Solar Receivers," *Renewable and Sustainable Energy Reviews*, 14(1), 265–276 (2010).

34. Kolb, G. J., R. B. Diver, and N. Siegel, "Central-Station Solar Hydrogen Power Plant," *Journal of Solar Energy Engineering*, 129(2), 179–183 (2007).

35. Avila-Marin, A. L., "Volumetric Receivers in Solar Thermal Power Plants with Central Receiver System Technology: A Review," *Solar Energy*, 85(5), 891–910 (2011).

36. Jeter, S. M. and H. A. M. Al-Ansary, "High Temperature Solar Thermal Systems and Methods," *U.S. Patent* 9,377,246 (2016).

37. Meier, A., "A Predictive CFD Model for a Falling Particle Receiver/Reactor Exposed to Concentrated Sunlight," *Chemical Engineering Science*, 54(13), 2899–2905 (1999).

38. Rightley, M., L. Matthews, and G. Mulholland, "Experimental Characterization of the Heat Transfer in a Free-Falling-Particle Receiver," *Solar Energy*, 48(6), 363–374 (1992).

39. Siegel, N. P., C. K. Ho, S. S. Khalsa, and G. J. Kolb, "Development and Evaluation of a Prototype Solid Particle Receiver: on-Sun Testing and Model Validation," *Journal of Solar Energy Engineering*, 132(2), 021008 (2010).

40. Kim, K., N. Siegel, G. Kolb, V. Rangaswamy, and S. F. Moujaes, "A Study of Solid Particle Flow Characterization in Solar Particle Receiver," *Solar Energy*, 83(10), 1784–1793 (2009).

41. Romero, M. and J. González-Aguilar, "Solar Thermal CSP Technology," *Wiley Interdisciplinary Reviews: Energy and Environment*, 3(1), 42–59 (2014).

42. Morris, A., S. Pannala, Z. Ma, and C. Hrenya, "A Conductive Heat Transfer Model for Particle Flows over Immersed Surfaces," *International Journal of Heat and Mass Transfer*, 89, 1277–1289 (2015).

43. Al-Ansary, H., A. El-Leathy, Z. Al-Suhaibani, et al., "Experimental Study of a Sand–Air Heat Exchanger for Use With a High-Temperature Solar Gas Turbine System," *Journal of Solar Energy Engineering*, 134(4), 041017 (2012).

44. Chueh, W. C., C. Falter, M. Abbott, et al., "High-Flux Solar-Driven Thermochemical Dissociation of CO_2 and H_2O Using Nonstoichiometric Ceria," *Science*, 330(6012), 1797–1801 (2010).

45. Miller, J. E., M. D. Allendorf, R. B. Diver, et al., "Metal Oxide Composites and Structures for Ultra-High Temperature Solar Thermochemical Cycles," *Journal of Materials Science*, 43(14), 4714–4728 (2008).

46. Melchior, T., C. Perkins, A. W. Weimer, and A. Steinfeld, "A Cavity-Receiver Containing a Tubular Absorber for High-Temperature Thermochemical Processing Using Concentrated Solar Energy," *International Journal of Thermal Sciences*, 47(11), 1496–1503 (2008).

47. Rostrup-Nielsen, J. R., J. Sehested, and J. K. Nørskov, "Hydrogen and Synthesis Gas by Steam-and CO_2 Reforming," *Advances in Catalysis*, 47, 65–139 (2002).

48. Steinfeld, A., P. Kuhn, and J. Karni, "High-Temperature Solar Thermochemistry: Production of Iron and Synthesis Gas by Fe_3O_4-Reduction with Methane," *Energy*, 18(3), 239–249 (1993).

49. Steinfeld, A., A. Frei, P. Kuhn, and D. Wuillemin, "Solar Thermal Production of Zinc and Syngas via Combined ZnO-Reduction and CH_4-Reforming Processes," *International Journal of Hydrogen Energy*, 20(10), 793–804 (1995).

50. Kodama, T., H. Ohtake, S. Matsumoto, et al., "Thermochemical Methane Reforming Using a Reactive WO_3/W Redox System," *Energy*, 25(5), 411–425 (2000).

51. Kodama, T., T. Shimizu, T. Satoh, M. Nakata, and K. Shimizu, "Stepwise Production of CO-Rich Syngas and Hydrogen via Solar Methane Reforming by Using a Ni(II)–Ferrite Redox System," *Solar Energy*, 73(5), 363–374 (2002).

52. He, F., J. Trainham, G. Parsons, J. S. Newman, and F. Li, "A Hybrid Solar-Redox Scheme for Liquid Fuel and Hydrogen Coproduction," *Energy & Environmental Science*, 7(6), 2033–2042 (2014).

53. Kodama, T., N. Gokon, and R. Yamamoto, "Thermochemical Two-Step Water Splitting by ZrO_2-Supported $Ni_xFe_{3-x}O_4$ for Solar Hydrogen Production," *Solar Energy*, 82(1), 73–79 (2008).

54. Kodama, T., Y. Nakamuro, and T. Mizuno, "A Two-Step Thermochemical Water Splitting by Iron-Oxide on Stabilized Zirconia," *Journal of Solar Energy Engineering*, 128(1), 3–7 (2006).

55. Perkins, C., P. R. Lichty, and A. W. Weimer, "Thermal ZnO Dissociation in a Rapid Aerosol Reactor as Part of a Solar Hydrogen Production Cycle," *International Journal of Hydrogen Energy*, 33(2), 499–510 (2008).

56. Gokon, N., T. Mataga, N. Kondo, and T. Kodama, "Thermochemical Two-Step Water Splitting by Internally Circulating Fluidized Bed of $NiFe_2O_4$ Particles: Successive Reaction of Thermal-Reduction and Water-Decomposition Steps," *International Journal of Hydrogen Energy*, 36(8), 4757–4767 (2011).

57. Stamatiou, A., P. Loutzenhiser, and A. Steinfeld, "Solar Syngas Production from H_2O and CO_2 via Two-Step Thermochemical Cycles Based on Zn/ZnO and FeO/Fe_3O_4 Redox Reactions: Kinetic Analysis," *Energy & Fuels*, 24(4), 2716–2722 (2010).

58. Furler, P., J. R. Scheffe, and A. Steinfeld, "Syngas Production by Simultaneous Splitting of H_2O and CO_2 via Ceria Redox Reactions in a High-Temperature Solar Reactor," *Energy & Environmental Science*, 5(3), 6098–6103 (2012).

59. Muhich, C. L., B. W. Evanko, K. C. Weston, et al., "Efficient Generation of H_2 by Splitting Water with an Isothermal Redox Cycle," *Science*, 341(6145), 540–542 (2013).

60. Scheffe, J. R., M. Welte, and A. Steinfeld, "Thermal Reduction of Ceria within an Aerosol Reactor for H_2O and CO_2 Splitting," *Industrial & Engineering Chemistry Research*, 53(6), 2175–2182 (2014).

61. Fresno, F., R. Fernández-Saavedra, M. B. Gómez-Mancebo, et al., "Solar Hydrogen Production by Two-Step Thermochemical Cycles: Evaluation of the Activity of Commercial Ferrites," *International Journal of Hydrogen Energy*, 34(7), 2918–2924 (2009).

62. Kodama, T., Y. Kondoh, R. Yamamoto, H. Andou, and N. Satou, "Thermochemical Hydrogen Production by a Redox System of ZrO_2-Supported Co(II)-Ferrite," *Solar Energy*, 78(5), 623–631 (2005).

63. Steinfeld, A., "Solar Hydrogen Production via a Two-Step Water-Splitting Thermochemical Cycle Based on Zn/ZnO Redox Reactions," *International Journal of Hydrogen Energy*, 27(6), 611–619 (2002).

64. Muhich, C. L., B. D. Ehrhart, I. Al-Shankiti, B. J. Ward, C. B. Musgrave, and A. W. Weimer, "A Review and Perspective of Efficient Hydrogen Generation via Solar Thermal Water Splitting," *Wiley Interdisciplinary Reviews: Energy and Environment*, 5(3), 261–287 (2015).

65. Nakamura, T., "Hydrogen Production from Water Utilizing Solar Heat at High Temperatures," *Solar Energy*, 19(5), 467–475 (1977).

66. Kodama, T., "High-Temperature Solar Chemistry for Converting Solar Heat to Chemical Fuels," *Progress in Energy and Combustion Science*, 29(6), 567–597 (2003).

67. Muhich, C. L., B. D. Ehrhart, S. L. Miller, V. A. Witte, B. J. Ward, C. B. Musgrave, and A. W. Weimer, "Active and Flowable Doped-Hercynite Materials for Solar Thermal Redox Processing to Split Water," Presented at 2015 AIChE Fall Meeting, Salt Lake City, UT, November 8–13 (2015).

68. Scheffe, J. R., M. D. Allendorf, E. N. Coker, et al., "Hydrogen Production via Chemical Looping Redox Cycles Using Atomic Layer Deposition-Synthesized Iron Oxide and Cobalt Ferrites," *Chemistry of Materials*, 23(8), 2030–2038 (2011).

69. Scheffe, J. R., A. H. McDaniel, M. D. Allendorf, and A. W. Weimer, "Kinetics and Mechanism of Solar-Thermochemical H_2 Production by Oxidation of a Cobalt Ferrite–Zirconia Composite," *Energy & Environmental Science*, 6(3), 963–973 (2013).

70. Muhich, C. L., B. D. Ehrhart, V. A. Witte, et al., "Predicting the Solar Thermochemical Water Splitting Ability and Reaction Mechanism of Metal Oxides: a Case Study of the Hercynite Family of Water Splitting Cycles," *Energy & Environmental Science*, 8(12), 3687–3699 (2015).

71. Scheffe, J. R., J. Li, and A. W. Weimer, "A Spinel Ferrite/Hercynite Water-Splitting Redox Cycle," *International Journal of Hydrogen Energy*, 35(8), 3333–3340 (2010).

72. Banares, M., J. Fierro, and J. Moffat, "The Partial Oxidation of Methane on MoO_3/SiO_2 Catalysts: Influence of the Molybdenum Content and Type of Oxidant," *Journal of Catalysis*, 142(2), 406–417 (1993).

73. Barbaux, Y., A. Elamrani, and J. Bonnelle, "Catalytic Oxidation of Methane on MoO_3-SiO_2: Mechanism of Oxidation with O_2 and N_2O Studied by Surface Potential Measurements," *Catalysis Today*, 1(1), 147–156 (1987).

74. Hall, T. J., J. S. Hargreaves, G. J. Hutchings, R. W. Joyner, and S. H. Taylor, "Catalytic Synthesis of Methanol and Formaldehyde by Partial Oxidation of Methane," *Fuel Processing Technology*, 42(2), 151–178 (1995).

75. Faraldos, M., M. A. Bañares, J. A. Anderson, et al., "Comparison of Silica-Supported MoO_3 And V_2O_5 Catalysts in the Selective Partial Oxidation of Methane," *Journal of Catalysis*, 160(2), 214–221 (1996).

76. Spencer, N. D. and C. J. Pereira, "V_2O_5-SiO_2-Catalyzed Methane Partial Oxidation with Molecular Oxygen," *Journal of Catalysis*, 116(2), 399–406 (1989).

77. Parmaliana, A., F. Frusteri, D. Miceli, A. Mezzapica, M. Scurrell, and N. Giordano, "Factors Controlling the Reactivity of the Silica Surface in Methane Partial Oxidation," *Applied Catalysis*, 78(2), L7-L12 (1991).

78. Miceli, D., F. Arena, A. Parmaliana, M. Scurrell, and V. Sokolovskii, "Effect of the Metal Oxide Loading on the Activity of Silica Supported MoO_3 And V_2O_5 Catalysts in the Selective Partial Oxidation of Methane," *Catalysis Letters*, 18(3), 283–288 (1993).

79. Bañares, M. A., N. D. Spencer, M. D. Jones, and I. E. Wachs, "Effect of Alkali Metal Cations on the Structure of $Mo(VI)SiO_2$ Catalysts and its Relevance to the Selective Oxidation of Methane and Methanol," *Journal of Catalysis*, 146(1), 204–210 (1994).

80. Bielański, A. and J. Haber, "Oxygen in Catalysis on Transition Metal Oxides," *Catalysis Reviews Science and Engineering*, 19(1), 1–41 (1979).

81. Kartheuser, B., B. Hodnett, H. Zanthoff, and M. Baerns, "Transient Experiments on the Selective Oxidation of Methane to Formaldehyde over V_2O_5/SiO_2 Studied in the Temporal-Analysis-of-Products Reactor," *Catalysis Letters*, 21(3–4), 209–214 (1993).

82. Koranne, M. M., J. Goodwin, and G. Marcelin, "Oxygen Involvement in the Partial Oxidation of Methane on Supported and Unsupported V_2O_5," *Journal of Catalysis*, 148(1), 378–387 (1994).

83. Mauti, R. and C. A. Mims, "Oxygen Pathways in Methane Selective Oxidation over Silica-Supported Molybdena," *Catalysis Letters*, 21(3–4), 201–207 (1993).

84. Andersson, A. and S. Hansen, "Catalytic Anisotropy of MoO_3 in the Oxidative Ammonolysis of Toluene," *Journal of Catalysis*, 114(2), 332–346 (1988).

85. Volta, J., W. Desquesnes, B. Moraweck, and G. Coudurier, "A New Method to Obtain Supported Oriented Oxides: MoO_3 Graphite Catalyst in Propylene Oxidation to Acrolein," *Reaction Kinetics and Catalysis Letters*, 12(3), 241–246 (1979).

86. Smith, M. and U. Ozkan, "The Partial Oxidation of Methane to Formaldehyde: Role of Different Crystal Planes of MoO_3," *Journal of Catalysis*, 141(1), 124–139 (1993).

87. Smith, M. and U. Ozkan, "Transient Isotopic Labeling Studies Under Steady-State Conditions in Partial Oxidation of Methane to Formaldehyde over MoO_3 Catalysts," *Journal of Catalysis*, 142(1), 226–236 (1993).

88. Arena, F., N. Giordano, and A. Parmaliana, "Working Mechanism of Oxide Catalysts in the Partial Oxidation of Methane to Formaldehyde. II. Redox Properties and Reactivity of SiO_2, MoO_3/SiO_2, V_2O_5/SiO_2, TiO_2, and V_2O_5/TiO_2 Systems," *Journal of Catalysis*, 167(1), 66–76 (1997).

89. Arena, F. and A. Parmaliana, "Silica-Supported Molybdena Catalysts. Surface Structures, Reduction Pattern, and Oxygen Chemisorption," *The Journal of Physical Chemistry*, 100 (51), 1994–2005 (1996).

90. Irusta, S., A. Marchi, E. Lombardo, and E. Miré, "Characterization of Surface Species on V/ SiO_2 and V, Na/SiO_2 and their Role in the Partial Oxidation of Methane to Formaldehyde," *Catalysis Letters*, 40(1–2), 9–16 (1996).

91. Voge, H. H. and C. R. Adams, "Catalytic Oxidation of Olefins," *Advances in Catalysis*, 17, 151–221 (1967).

92. Sachtler, W., "The Mechanism of the Catalytic Oxidation of Some Organic Molecules," *Catalysis Reviews*, 4(1), 27–52 (1971).

93. Sampson, R. J. and D. Shooter, in Tipper, C. F. H. ed., *Oxidation and Combustion Reviews* Vol. 1, Elsevier, Amsterdam, Netherlands (1965).

94. Krenzke, L. D. and G. W. Keulks, "The Catalytic Oxidation of Propylene: VI. Mechanistic Studies Utilizing Isotopic Tracers," *Journal of Catalysis*, 61(2), 316–325 (1980).

95. Monnier, J. R. and G. W. Keulks, "The Catalytic Oxidation of Propylene: IX. The Kinetics and Mechanism over β-$Bi_2Mo_2O_9$," *Journal of Catalysis*, 68(1), 51–66 (1981).

96. Uda, T., T. T. Lin, and G. W. Keulks, "The Catalytic Oxidation of Propylene: VII. The Use of Temperature Programmed Reoxidation to Characterize γ-Bismuth Molybdate," *Journal of Catalysis*, 62(1), 26–34 (1980).

97. Keulks, G. W., "The Mechanism of Oxygen Atom Incorporation into the Products of Propylene Oxidation over Bismuth Molybdate," *Journal of Catalysis*, 19(2), 232–235 (1970).

98. Miura, H., T. Otsubo, T. Shirasaki, and Y. Morikawa, "Tracer Studies of Catalytic Oxidation by Bismuth Molybdate: II. Propylene Reduction of Labeled Catalysts and Catalytic Oxidation of Propylene," *Journal of Catalysis*, 56(1), 84–87 (1979).

99. Ono, T., T. Nakajo, and T. Hironaka, "Kinetic Features and Lattice-Oxygen Participation in Propene Oxidation over Bi–Mo Oxide and Some Mo Oxide Catalysts," *Journal of the Chemical Society, Faraday Transactions*, 86(24), 4077–4081 (1990).

100. Daniel, C. and G. W. Keulks, "The Catalytic Oxidation of Propylene: I. Evidence for Surface Initiated Homogeneous Reactions," *Journal of Catalysis*, 24(3), 529–535 (1972).

101. Hoefs, E. V., J. R. Monnier, and G. W. Keulks, "The Investigation of the Type of Active Oxygen for the Oxidation of Propylene over Bismuth Molybdate Catalysts Using Infrared and Raman Spectroscopy," *Journal of Catalysis*, 57(2), 331–337 (1979).

102. Grzybowska, B., J. Haber, W. Marczewski, and L. Ungier, "X-ray and Ultraviolet-Photoelectron Spectra of Bismuth Molybdate Catalysts," *Journal of Catalysis*, 42(3), 327–333 (1976).

103. Doornkamp, C., M. Clement, and V. Ponec, "Activity and Selectivity Patterns in the Oxidation of Allyl Iodide on the Period IV Metal Oxides: The Participation of Lattice Oxygen in Selective and Total Oxidation Reactions," *Applied Catalysis A: General*, 188(1), 325–336 (1999).

104. Makwana, V. D., Y. Son, A. R. Howell, and S. L. Suib, "The Role of Lattice Oxygen in Selective Benzyl Alcohol Oxidation Using OMS-2 Catalyst: a Kinetic and Isotope-Labeling Study," *Journal of Catalysis*, 210(1), 46–52 (2002).

105. Choudary, B., M. L. Kantam, A. Rahman, C. Reddy, and K. K. Rao, "The First Example of Activation of Molecular Oxygen by Nickel in Ni–Al Hydrotalcite: a Novel Protocol for the Selective Oxidation of Alcohols," *Angewandte Chemie International Edition*, 40(4), 763–766 (2001).

6 Process Simulations and Techno-Economic Analyses

M. Kathe, W. Wang, and L.-S. Fan

6.1 Introduction

A process simulation is the representation of a thermal or chemical transformation by a mathematical model that is solved to provide information about mass and energy requirements, equipment performance, and overall process feasibility.[1] They are system level models used for economic calculations, energy efficiency analyses, and process comparisons. Component level information such as reactor geometry, mixing and hydrodynamics, and transport phenomena, as well as molecular level information such as surface area, reaction mechanism, and electronic interactions, are not included.[2] Beginning with FlowTran, developed by Monsanto in the 1950s and 1960s, several process simulation software packages have since been developed, including ProSim, PRO/II, gPROMS, ChemCAD, and Aspen Plus®, and they play an important role in process development by allowing for the evaluation of a commercial scale plant based on available experimental results.

The process simulation software that is used extensively throughout academic research and industrial operations is Aspen Plus®. Aspen was developed in the late 1970s at Massachusetts Institute of Technology for the United States Energy Research and Development Administration (now United States Department of Energy or USDOE).[3] The project objective was to develop a computer program to aid in performing process evaluations; mainly process economics. Today, Aspen has evolved into Aspen Plus® and, as part of the AspenOne package software suite, has become the essential plant and process design software program. Aspen Plus® provides complete flexibility in designing a process simulation. Every new process begins by specifying the component species and the property method used for calculating the component species thermodynamic properties. Together, these two required specifications display the strength of the Aspen Plus® software program, due to its extensive chemical databank, available methods for calculating thermodynamic properties, and ability to adjust property values when necessary. Once the component species and property methods have been selected, Aspen Plus® assumes control and calculates the thermodynamic properties based on user input, and only allows the inputs required to satisfy the degrees of freedom. The chemical looping partial oxidation examples and the results presented in this chapter were obtained using Aspen Plus® (AspenTech), and they are available on the website as supplemental material.

This chapter focuses on developing and comparing a chemical looping based coal to methanol, gas to liquids, and biomass to olefins process to a conventional one, the baseline. Section 6.2 provides a background on process simulations and describes the conventional coal to methanol, gas to liquids, and biomass to olefins processes. Section 6.3 focuses on the chemical looping approach to partial oxidation and its integration through syngas production into a coal to methanol plant, a gas to liquids plant, and a biomass to olefins plant. Section 6.4 provides the modeling parameters and modified parameter calculations for iron based chemical looping partial oxidation processes using Aspen Plus®, with a specific focus on the reducer. The general definitions, equations, and techno-economic analysis assumptions that are common to all process examples are also provided. In Section 6.5, the coal to syngas (CTS) process and its integration through syngas generation into a coal to methanol plant is developed and analyzed. Section 6.6 describes the shale gas to syngas (STS) process and its integration through syngas generation into a liquid fuels plant. Section 6.7 describes a chemical looping based biomass to syngas (BTS) process for eventual olefins production through syngas and methanol intermediates. Section 6.8 provides performance and techno-economic analysis results for other processes, including a modified CTS process using a coal and natural gas mixture for syngas generation to produce methanol, a modified STS process where the reducer reactor is a fluidized bed instead of a moving bed, and the CTS process for electricity generation.

6.2 Conventional Partial Oxidation Process

The conventional method of upgrading carbonaceous feedstock is an indirect route where syngas is generated and then converted to a final product. In both the coal to methanol (CTM) process and gas to liquids (GTL), the final product, methanol and liquid fuel, respectively, is formed indirectly from syngas, which allows the overall process to be broadly divided into two sections: syngas generation and product generation. The USDOE has published reports for the CTM and GTL process detailing the performance, efficiency, and economics of these partial oxidation technologies, and they represent the baseline case used for comparison to the chemical looping syngas generation technology in a CTM and a GTL process. For the biomass to olefins (BTO) process, the National Renewable Energy Laboratory (NREL) has published reports detailing the overall process and economics of biomass to chemicals and serves as the baseline case for chemical looping syngas generation process comparison when integrated in a BTO process. The design principles for these conventional technologies are also illustrated in detail for all the major operations and steps in processing carbonaceous feedstock to final products using syngas as an intermediate. A perspective is provided that describes the process changes necessary for implementing carbon capture.

6.2.1 Coal to Methanol (CTM) Process

The main method for methanol production is through syngas generation followed by catalytic conversion to methanol. The syngas is typically obtained through natural gas

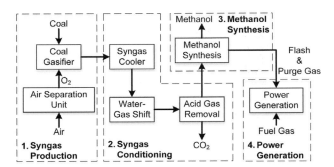

Figure 6.1 Flow diagram of a coal gasification plant for methanol production process.[4]

reforming but coal gasification represents about 10% of the methanol produced from syngas.

The coal to methanol process can be divided into four sections, based on function, with Figure 6.1 showing the block flow diagram of the CTM process. They are: (1) syngas production; (2) syngas conditioning; (3) methanol synthesis; and (4) power generation. The function, design basis, mass and energy balance, and performance of each section are provided in the baseline report and summarized in this section.[4] The resulting environmental performance, energy efficiency, and techno-economic analysis form the baseline results and are used for comparison with the chemical looping approach.

Syngas Production

Coal gasifiers are classified as entrained flow, fluidized bed, or moving bed, based on the gas–solid contact pattern. In the CTM process, a Shell gasifier, which is an entrained flow gasifier using dry coal and oxygen as the feed, is used. In an entrained flow gasifier, coal with a particle diameter of less than 0.24 in (6 mm) and an oxidant enter cocurrently from the top of the gasifier. The high temperature, high pressure, and extremely turbulent flow conditions inside the Shell gasifier result in a high carbon conversion (>97%), obtained with a residence time in the order of seconds. The product composition is dictated by the coal to oxygen ratio, gas–solid contact mode, and residence time. Figure 6.2 shows a typical configuration for the Shell coal gasification process (SCGP) for which dried coal is used.

In the CTM process, sub-bituminous Powder River Basin (PRB) coal is the feed-stock. Since PRB coal has a high moisture content, it is first dried to 6 wt% moisture before entering the coal gasifier. The dried coal is then pressurized using lock hoppers and nitrogen from the air separation unit (ASU) and fed into the coal gasifier with oxygen from the ASU to produce synthesis gas (syngas), a gaseous mixture mainly comprised of H_2, CO, CO_2, and H_2O.[5–8]

The SCGP operates at a pressure of 350 psi (2.4 MPa), temperature of 2800 °F (1540 °C), and uses steam as a moderator for the exothermic reaction between coal and oxygen. A unique aspect of the SCGP lies within the gasifier vessel wall, where it is lined with an inner membrane that acts as a heat exchanger to produce low pressure

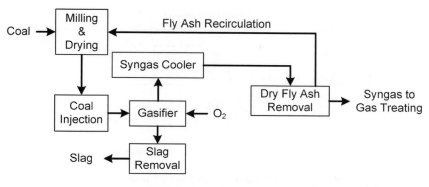

Figure 6.2 Typical process configuration for Shell coal gasification process.[4-6]

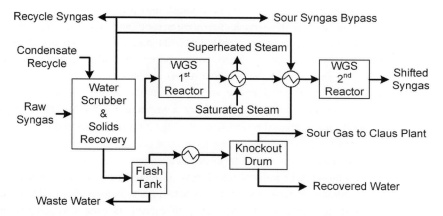

Figure 6.3 Simplified block flow diagram of syngas conditioning section.[4-6]

steam. Additionally, the temperature gradient that is maintained by flowing water outside of the membrane wall leads to a molten ash layer on the inside of the refractory lined vessel, which results in a longer lifetime than other gasifiers.

Syngas Conditioning Block

Coal has an intrinsic hydrogen to carbon molar ratio of between 0.3 and 0.7, which is well below the approximate value of 2 required for most product synthesis, and it contains heavy metals along with additional impurities poisonous to product synthesis catalysts. The syngas conditioning block, shown in Figure 6.3, tailors the syngas produced from coal gasification to be suitable for product generation by adjusting the H_2:CO molar ratio and removing impurities. Depending on the final product, the syngas quality and specifications differ, with Table 6.1 providing the syngas properties required for selected products. The stoichiometric number (S#), defined in equation (6.2.1), is a

Table 6.1 Syngas characteristics necessary for product synthesis.[9]

Product	Liquid fuels	Liquid fuels	Methanol	Methanol	Fuel gas	Fuel gas
Technology	F–T: iron catalyst	F–T: cobalt catalyst	Vapor phase	Liquid phase	Boiler	Turbine
H_2:CO	~ 0.6	~ 2.0	> 1.7	> 1.7	n/a	n/a
S#	n/a	n/a	> 1.7	> 1.7	n/a	n/a
Sulfur (ppm)	< 1	< 1	< 1	< 1	n/a	< 25
Particulates	Low	Low	Low	Low	Low	Low
Temperature °F	572–752	392–572	212–392	212–392	>932	>932
(°C)	(300– 400)	(200–300)	(100–200)	(100– 200)	(>500)	(>500)
Pressure psi	290–430	290–435	700–1176	725 (5)	Low	5800 (40)
(MPa)	(2–3)	(2–3)	(5–8)			

parameter representing the carbon efficiency in methanol synthesis. In Equation (6.2.1), $\dot{n}_{H_2}$, $\dot{n}_{CO}$, and $\dot{n}_{CO_2}$ represent the molar flow rates of each component in the syngas:

$$S\# = \frac{\dot{n}_{H_2} - \dot{n}_{CO_2}}{\dot{n}_{CO} + \dot{n}_{CO_2}}. \tag{6.2.1}$$

Specifically for methanol production, the H_2:CO molar ratio and the S# should both be greater than 1.7.[9] To increase the hydrogen content of the raw syngas exiting the SCGP, the raw syngas is split into three streams, with the split fraction determined by constraining the H_2:CO ratio of the syngas entering the methanol synthesis reactor to 2:1.[4] 60% volume of the syngas enters a two-stage water–gas shift (WGS) reactor and the remaining 40% bypasses the WGS reactors and enters the acid gas removal (AGR) unit. The H_2:CO ratio is increased from 0.37 to 2 through the WGS reaction, shown in reaction (6.2.2):[8–11]

$$CO + H_2O \rightarrow CO_2 + H_2 \; \Delta H = -41.1 \text{ kJ/mol (298 K).} \tag{6.2.2}$$

The WGS reaction is a catalytic, equilibrium-limited chemical reaction. Depending on the catalyst's tolerance towards sulfur, the WGS reaction is classified as either sweet shift or sour shift. The sour shift, operating between 450 °F (232 °C) and 550 °F (288 °C), uses a cobalt–molybdenum based catalyst and has the ability to handle sulfur compounds like H_2S and COS. The sweet shift is further classified as high temperature or low temperature shift. The high temperature shift uses a chromium or copper promoted iron based catalyst, operating between 550 °F (288 °C) and 900 °F (482 °C), whereas low temperature shift uses a copper–zinc–aluminum catalyst operating between 400 °F (204 °C) and 500 °F (260 °C).

For coal gasification applications, the sour shift is used due to the presence of H_2S and COS. Specifically, the raw syngas obtained from the syngas cooling section is heated to 530 °F (277 °C) before entering the first WGS reactor. The exothermic heat of reaction leads to a temperature rise resulting in an outlet gas temperature of 900 °F

(482 °C). This shifted syngas is cooled to 450 °F (232 °C) in inter-stage coolers and then enters the second stage, low temperature WGS reactor. The extent of reaction in the second stage is minimal as only the unconverted CO from the first stage is converted, resulting in a slight temperature increase to 600 °F (316 °C) at the outlet of the second stage WGS reactor. The syngas exiting the WGS system is cooled in several steps using a low temperature gas cooling unit for maximum heat recovery. The water is condensed in knock-out drums while also removing almost all of the ammonia present in the syngas. This water is stripped in a vacuum drum to recover ammonia as sour gas and sent to the Claus unit for further treatment.

The catalyst used for methanol synthesis has a low sulfur tolerance (<1 ppmv H_2S), so the shifted syngas is mixed with the raw syngas bypass prior to entering the AGR unit. AGR systems used for CO_2 removal from syngas streams use physical absorption based solvents, like polyethylene glycol or methanol, and are licensed under the trade names Selexol and Rectisol, respectively. Selexol is a trade name for a polyethylene glycol based system licensed by UOP that achieves H_2S removal to 25 ppmv in a single pass. Rectisol is a trade name for a chilled methanol based system licensed by Linde and Lurgi that can achieve deep sulfur removal to less than 1 ppmv H_2S in a single pass. Since methanol has higher vapor pressure compared to polyethylene glycol, a lower operating temperature, between −40 °F (−40 °C) and −80 °F (−62 °C), is required for Rectisol as compared to Selexol, which operates near ambient temperatures. Because of its deep sulfur cleaning properties, Rectisol is the industry standard for an AGR system for syngas prior to chemical generation even though Rectisol is more expensive than Selexol.[12]

A typical Rectisol process flow diagram is shown in Figure 6.4. If the process does not involve carbon capture, the pure CO_2 stream generated from a Rectisol unit is released into the atmosphere. If CO_2 removal is integrated into the process, the pure CO_2 stream is compressed to 2220 psi (15.3 MPa) and transported for sequestration. In either case, the H_2S-rich stream is sent to a Claus unit, shown in Figure 6.5, for sulfur production. The major reactions in the Claus process are shown in reactions (6.2.3) and (6.2.4):

$$H_2S + \tfrac{3}{2}O_2 \rightarrow H_2O + SO_2 \qquad (6.2.3)$$

$$2H_2S + SO_2 \rightarrow 2H_2O + 3S \text{ (elemental sulfur as } S_2, S_6, S_8). \qquad (6.2.4)$$

The acid gas stream exiting the Rectisol process is removed of water in knock-out drums prior to entering a reaction furnace. The furnace uses air or oxygen from an ASU to convert one-third of the H_2S to SO_2. The acid gas exiting the reaction furnace contains H_2S and SO_2 in a molar ratio of 2:1. Reaction (6.2.4) is equilibrium limited with a typical one-pass efficiency of around 75% and three to four stages required for an H_2S conversion greater than 95%. Water and elemental sulfur produced during each stage are removed in a condenser, and the unconverted H_2S and SO_2 enter the next converter for further reaction. An oxygen blown Claus process is used for both the non-CO_2 capture case and 90% CO_2 capture case, as oxygen is readily available.

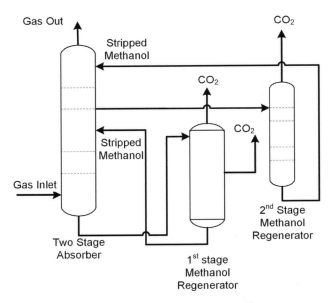

Figure 6.4 Rectisol process for acid gas removal.[12]

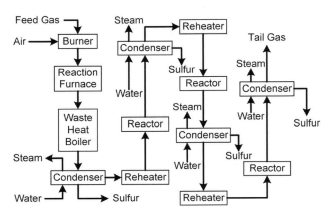

Figure 6.5 Process diagram of a three-stage air-blown Claus process.[12]

Methanol Production Block

The typical methanol synthesis reactor operates at elevated pressure, at around 800 psi (5.5 MPa). The clean syngas from the Rectisol unit is converted to methanol through reactions (6.2.5–6.2.7) using a copper–zinc catalyst:

$$2H_2 + CO \rightarrow CH_3OH \qquad \Delta H = -90.7 \text{ kJ/mol } (298 \text{ K}, 6 \text{ MPa}) \qquad (6.2.5)$$

$$CO_2 + 3H_2 \rightarrow CH_3OH + H_2O \quad \Delta H = -40.9 \text{ kJ/mol } (298 \text{ K}, 6 \text{ MPa}) \qquad (6.2.6)$$

$$CO + H_2O \rightarrow CO_2 + H_2 \qquad \Delta H = -49.8 \text{ kJ/mol } (298 \text{ K}, 6 \text{ MPa}). \qquad (6.2.7)$$

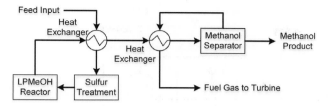

Figure 6.6 Liquid phase methanol (LPMeOH) process.[14]

Reactions (6.2.5–6.2.7) are exothermic and favor low operating temperatures for high conversions, but higher temperatures favor kinetics. Pressure is another important consideration, where higher pressures favor reactions (6.2.5–6.2.6) towards methanol production. The methanol yield is limited by the availability of H_2, which is overcome through the copper–zinc catalyst possessing water–gas shift activity allowing for in situ H_2 production via steam injection.

The industrial processes for methanol synthesis can be classified as either liquid phase or vapor phase. The liquid phase methanol (LPMeOH) process, shown in Figure 6.6, uses a slurry bubble column reactor. The slurry is composed of an inert mineral oil/powdered catalyst and transfers the exothermic heat of reaction to an internal tubular heat exchanger to produce low pressure steam. When compared to the vapor phase methanol synthesis, the presence of a liquid phase heat transfer fluid allows for improved control over operating conditions and improved process reliability, as large changes in process flows are quickly dampened.

The baseline case uses the vapor phase methanol synthesis process due to a greater presence of commercial operations.[13–15] The vapor phase methanol process consists of multiple trains of catalytic fixed bed reactors. First, compressed syngas is mixed with a fraction of the unconverted syngas recovered from the methanol synthesis section, and the combined stream is pre-heated to a temperature of 400 °F (204 °C) and enters the fixed bed reactors operating at a temperature of 450 °F (232 °C) and 735 psia (5 MPa). The fixed bed reactors are water cooled to maintain a near isothermal operating temperature.

The syngas quality from the gasifier is conditioned such that it is tailored for methanol production, which is then converted to olefins in a methanol to olefins (MTO) process. Sulfur is then removed from the tar-reformed syngas using a zinc oxide bed. The remaining purge and flash gas is combusted in a gas turbine to produce power to offset the parasitic energy requirements of the process.

Power Production Block

A natural gas combined cycle (NGCC) is used to generate the additional electricity required to fulfill the methanol plant auxiliary load.[16] The actual configuration of the power production block varies depending on the level of carbon capture. In the case of no carbon capture, the natural gas is combusted in a combined cycle configuration, and the flue gas is vented to the atmosphere after heat recovery. For 90% CO_2 capture, the CO_2 emissions from the flue gas are captured using a monodiethanolamine chemical

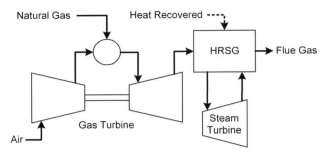

Figure 6.7 Power production block with no carbon capture.[16-18]

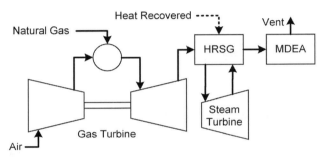

Figure 6.8 Power production block with 90% carbon capture.[16-18]

absorption solvent, described in further detail in Section 6.2.2. The power production block configuration for non-carbon capture and 90% carbon capture are shown in Figure 6.7 and Figure 6.8, respectively.

The individual blocks described are necessary components of a coal to methanol plant. Depending on various factors, like heat integration network, process licensor, site and plant characteristics, and market supply/demand conditions, variations within the individual blocks can be applied. The various blocks are integrated for two configurations: (1) without carbon capture, referred to as methanol baseline–B (MB–B) and shown in Figure 6.9; and (2) with 90% carbon capture, referred to as methanol baseline–A (MB–A) and shown in Figure 6.10.[4]

6.2.2 Natural Gas to Liquid Fuel (GTL) Process

The gas to liquids process is in commercial operation with facilities such as Sasol's Oryx plant and Shell's Pearl plant, producing several thousand barrels of liquid fuels per day. First, natural gas is reformed to syngas and then converted into liquid fuels using the Fischer–Tropsch process. The overall process is shown in Figure 6.11 and discussed in sections based on function.

The GTL process can be divided into three sections based on function, with Figure 6.12 showing the block flow diagram of the GTL process. They are: (1) syngas production; (2) liquid fuels production; and (3) product recovery and

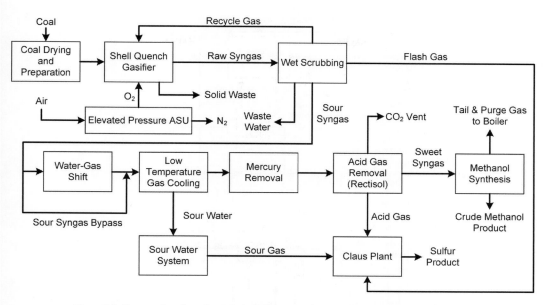

Figure 6.9 Conventional coal to methanol process block diagram with no carbon capture (MB-B).[4,17,18]

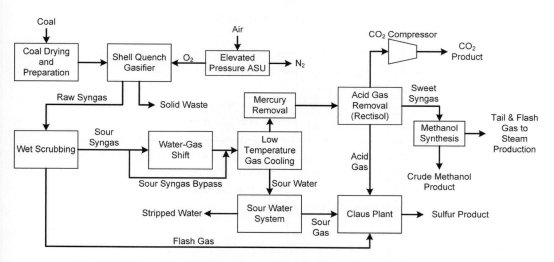

Figure 6.10 Conventional coal to methanol process block diagram with 90% carbon capture (MB-A).[4,17,18]

upgrading. The function, design basis, mass and energy balance, and performance of each section are provided in the baseline report and summarized in this section.[19] The resulting environmental performance, energy efficiency, and techno-economic analysis form the baseline results and are used for comparison with the chemical looping approach.

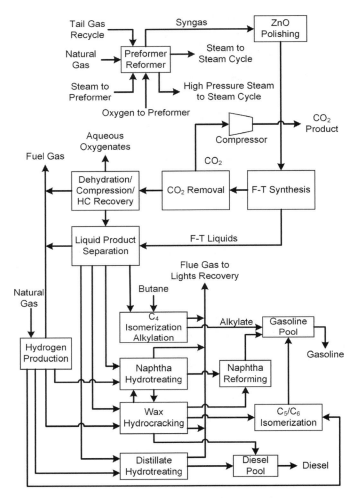

Figure 6.11 Process flow diagram of the GTL process.[19]

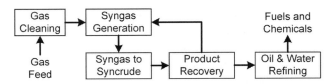

Figure 6.12 Baseline gas to liquid block flow diagram.[19]

Syngas Production

Figure 6.13 is an expanded view of the syngas generation section, which is derived from Haldor-Topsoe's autothermal reforming process. Natural gas is split for hydrogen production and syngas production. The hydrogen is produced in a steam methane

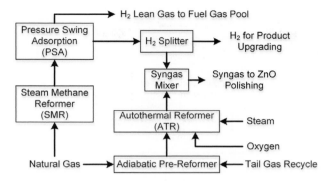

Figure 6.13 Syngas generation in the conventional, baseline GTL process.[19]

reformer system and primarily used for downstream upgrading purposes, but also to offset any variations in the syngas quality from the autothermal reformer.[19,20] An adiabatic pre-reformer located before the autothermal reformer (ATR) converts natural gas mixed with tail gas recycle from the liquid fuels Fischer–Tropsch product generation section to syngas and methane from the higher hydrocarbons. The pre-reformer operates at 842 °F (450 °C) using a nickel on supported alumina reforming catalyst, which prevents coke formation that would occur in the autothermal reformer from the higher hydrocarbons.

The methane and syngas from the pre-reformer, 95% purity oxygen from a cryogenic ASU, and steam enter the autothermal reformer operating at 1935 °F (1057 °C) and 24 atm (2.4 MPa) to produce syngas with zero net heat generation by combining the endothermic steam methane reaction and exothermic partial oxidation reaction. Operating the autothermal reformer at high pressure eliminates the compression energy required for the Fischer–Tropsch reactors and avoids carbon deposition that is unavoidable below 12 atm (1.2 MPa) regardless of steam concentration or burner design. The ASU exploits the Joule–Thomson effect to achieve the low temperatures required for cryogenic distillation; however, this occurs at the expense of the need for air compression to a minimum of 6 atm (0.6 MPa) with energy recuperated through the use of a turbo-expander in conjunction with the Joule–Thomson effect. Because of the operating temperatures and pressures of the ASU, it is both energy and capital intensive, requiring at least 200 kWh/tonne oxygen at 95% purity and atmospheric pressure, with increasing energy requirements as purity and pressure increase. In the baseline GTL process, the ASU is a standalone piece of equipment not integrated with the GTL plant for additional potential energy and economic benefits.

Impurities in the raw syngas, mainly sulfur compounds but also chlorides, from the autothermal reformer are removed in a zinc oxide (ZnO) polishing unit. Sulfur and chloride compounds will react with the zinc oxide to form zinc sulfide and zinc chloride, respectively. The ZnO polishing unit is necessary to protect the Fischer–Tropsch catalysts from poisoning. Entering the Fischer–Tropsch reactors, the syngas has a H_2:CO ratio of 2.19 and an S# of 1.59.

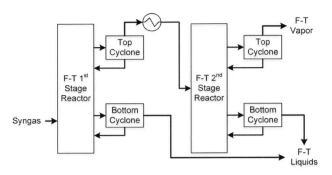

Figure 6.14 Fischer–Tropsch reactor section.[19,21]

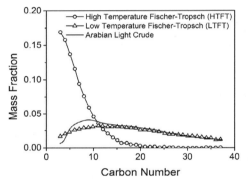

Figure 6.15 Typical product distribution as a function of carbon number for F–T technologies.[19]

Liquid Fuels Production

The remainder of the block flow diagram in Figure 6.12 is the product generation section, representing liquid fuels production via Fischer–Tropsch (F–T) reaction and product upgrading. Historically, F–T reactors have used iron based catalysts for both high and low temperature synthesis. More recently, low temperature, cobalt based catalysts have been developed and are currently employed in Sasol's Mossel Bay, Sasol/Qatar Petroleum's Oryx, Sasol/ChevronTexaco/Nigerian National Petroleum Company's Escravos, and Shell's Bintulu and Pearl facilities. The F–T reactors in Figure 6.14 rely solely on the cobalt based catalyst in a two-stage slurry bed reactor to produce 50,000 barrels/day of liquid fuel that consists of gasoline (31%) and diesel (69%). The complete product distribution for various commercial F–T technologies can be estimated using an Anderson–Schulz–Flory distribution, as shown in Figure 6.15, and varies depending on the catalyst, operating conditions, and technology used.

In the F–T reactor section for liquid fuels production, syngas is first split into two equal streams and enters the first stage F–T slurry reactor with syngas as the gas phase, F–T wax as the liquid phase, and cobalt based catalyst as the slurry phase. The first stage F–T reactor operates at 482 °F (250 °C) and 22 atm (2.2 MPa), which is very similar to the 428 °F (220 °C) and 24 atm (2.4 MPa) specified in an independent

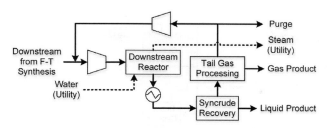

Figure 6.16 Product upgrading section downstream of F–T in a GTL plant.[19,21]

analysis performed by Bechtel and Amoco.[21] The F–T reaction is highly exothermic, releasing 165 kJ per reactant unit of syngas, where one reactant unit of syngas is 2 moles of H_2 and 1 mole of CO. The heat is removed by generating 11 atm (1.1 MPa) steam, consistent with Bechtel and Amoco's analysis.[21] The parallel first stage F–T reactors are identical in conversion, 61% CO, and product separation.

The gas phase (F–T vapors) and liquid phase (F–T wax) product exiting the first stage F–T reactor are first separated using a cyclone and then further separated in a three-stage separator by condensing out liquid water and F–T liquids from the solids (catalyst). The F–T liquids are separated from liquid water, where the liquid water proceeds to water treatment and the F–T liquids are further processed, as discussed in the Product Recovery and Upgrading section. The majority of the solid phase, catalyst, is recycled with a small fraction disposed to remove spent catalyst and allow for an input of fresh catalyst. The remaining unconverted syngas from the two parallel first stage reactors are combined and become the reactant feed to the second stage F–T reactor, which operates at 482 °F (250 °C) and 21 atm (2.1 MPa). Again, the second stage F–T reactor operating conditions are nearly identical to the design specified by Bechtel and Amoco, 428 °F (220 °C) and 21 atm (2.1 MPa).[21]

Product Recovery and Upgrading

The product recovery and upgrading section is shown in Figure 6.16. All the liquids from the F–T reactors are separated in a single fractionation column into one vapor stream and three liquid streams: naphtha <350.6 °F (<177 °C), middle distillates 350.6–649.4 °F (177–343 °C), and wax >649.4 °F (>343 °C) based on boiling point temperature range, up to 649.4 °F (343 °C). The vapor stream is cooled using air and water, which produces a three-phase liquid–liquid–gas mixture that is separated using a three-phase separator. The remaining vapor stream is mixed with the tail gas recycle stream and fuel gas stream, the hydrocarbon liquids are recycled back into the fractionation column, and the water is sent to wastewater treatment.

The vapor from the second F–T reactor is subjected to further separations into a CO_2 stream using chemical absorption via monoethanolamine (MEA), an oxygenates stream, a condensed water stream, and the product, a pure hydrocarbon stream.

The removal of CO_2 using an amine based solvent has been in commercial application since the 1930s for natural gas sweetening. Of the amine based solvents, MEA has been the most prevalently studied, licensed by Fluor Daniel under the trade name

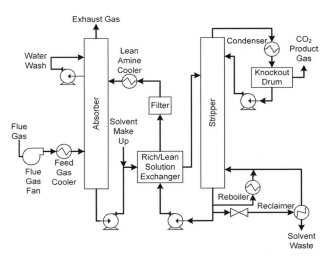

Figure 6.17 Schematic of a typical MEA process.[12,19,21]

Econamine Process. A typical MEA design is shown in Figure 6.17. First, a CO_2-lean, MEA stream is contacted countercurrently with a CO_2-rich gas stream in an absorption column operating between 104 °F (40 °C) and 140 °F (60 °C), with lower temperatures favored to minimize evaporative solvent losses. The CO_2-lean MEA stream contains between 10% and 30% weight MEA, with the average being 20%, and 0.1 to 0.2 mol CO_2/mol MEA. Above 20% MEA loading, the MEA solvent must be added with corrosion inhibitors to protect the absorption column material of construction, typically carbon steel, from pitting and galvanized corrosion.

At the exit of the absorber, the CO_2 loading of the MEA solvent increases to ~0.5 mol CO_2/mol MEA. The CO_2 is recovered from the MEA solvent in another absorption column, the stripper (or regenerator), using steam with pressures and temperatures ranging between 1.5 atm (0.15 MPa) and 2 atm (0.2 MPa) and 212 °F (100 °C) and 284 °F (140 °C), with more typical values being between 212 °F (100 °C) and 248 °F (120 °C). In the stripper, the CO_2 that is chemically bound to MEA is desorbed to regenerate the CO_2-lean MEA solvent that is pumped into the absorber to complete the cycle. The CO_2 gas from the stripper exits with the steam. After steam is condensed to liquid water, a pure CO_2 gas stream is produced. The general design principles of the MEA process can be applied to any other chemical absorption solvent processes, like the monodiethanolamine (MDEA) based process used in the power production block of Section 6.2.1.

The hydrocarbon stream is separated similar to the vapor stream for liquid recovery. The remaining vapor stream after hydrocarbon separation, the vapor stream from the liquid recovery section, and the gases produced during hydrocarbon upgrading – Naphtha hydrotreating, wax hydrocracking, and distillate hydrotreating – are all mixed and cooled to 113 °F (45 °C) and moderate pressure to recover C_{3+} hydrocarbons that undergo C_4 isomerization to produce gasoline.

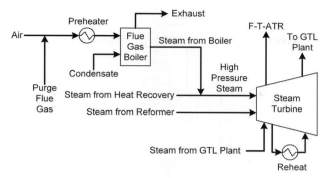

Figure 6.18 Schematic of the power and steam generation unit.[19,21]

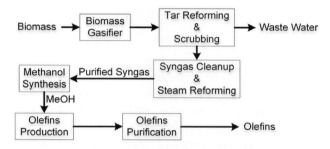

Figure 6.19 Outline of biomass to olefins process.[22]

Power Production

A detailed schematic of the power production block is shown in Figure 6.18. The lighter hydrocarbons compose the tail gas recycle, with a 5% purge to prevent accumulation of inert gases. The C_{3+} hydrocarbons are recovered from various downstream processing units and recycled as tail gas. The 5% tail gas purge stream is combusted in a power generation cycle and is termed fuel gas.

6.2.3 Biomass to Olefins (BTO) Process

The biomass to olefins process represents a future possibility for converting a renewable feedstock into a high value product. The BTO process is constructed using already existing and demonstrated processes that are integrated to convert biomass to syngas into olefins using methanol as an intermediate.

The BTO process is shown in Figure 6.19. An NREL baseline report is used for material flows, energy efficiency calculations, and techno-economic analysis for biomass gasification.[22–24] The syngas clean-up and methanol production section are similar to the CTM process described in Section 6.2.1. The power production section is similar to those shown in Figures 6.7 and 6.18, and the description of these common sections can be found in Sections 6.2.1 and 6.2.2. The syngas production from biomass, tar

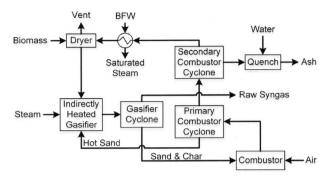

Figure 6.20 Block flow diagram of an indirectly heated biomass gasifier system.[22–24]

reforming, and methanol to olefins production sections are described in further detail in this section.

Syngas Production

Figure 6.20 shows the biomass feed preparation and indirectly heated gasifier for syngas generation.[22–24] Biomass is dried to ~5% w/w moisture before being injected to the indirectly heated gasifier operating at 1652 °F (900 °C) and under slight pressure. The gasifier is termed indirectly heated as the heat is supplied through the circulation of an inert heat transfer material, olivine particles. A series of cyclones after the gasifier separate the solid particulates and char from the syngas. The majority of the olivine, char, and ash is separated in the cyclone and then enters the char combustor where the char is combusted to reheat the olivine. The raw syngas from the primary gasifier cyclone is sent to the tar reformer. The off-gas from the char combustor is separated using cyclones and a direct contact water scrubber. The hot gas is used to generate steam in a heat recovery steam generator (HRSG) and used for biomass drying.

An alternative to the indirectly heated gasifier is a directly heated gasifier where an ASU provides the oxygen for biomass gasification and steam is used to moderate the temperature. The design of these gasifiers follows those used for coal gasification, as described in Section 6.2.1.

Tar Reforming

During gasification, up to 20% of the biomass is converted into tars, which are long-chained poly-aromatic hydrocarbons. The production of tars is unique to biomass based systems and can be removed through condensation using a water scrubber. However, biomass gasification has low carbon efficiencies, therefore cracking the tars into lower carbon hydrocarbons is preferred over condensing them in a water scrubber and subsequent wastewater treatment. Figure 6.21 shows the tar reformer section, which consists of a tar reformer, purification, and compression. The tar reformer is a bubbling fluidized bed that uses hot catalytic particles and steam to crack the hydrocarbons into syngas, and ammonia into nitrogen and

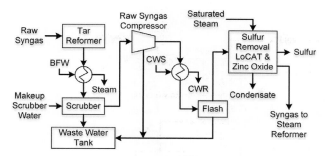

Figure 6.21 Schematic of typical tar reforming section in a BTO process plant.[21–24]

hydrogen. A scrubber then removes the particulates, ammonia, and residual tars from the syngas stream. The syngas stream, still containing methane and light hydrocarbons, is compressed to 450 psi (3.1 MPa) and purified over a zinc oxide bed, as described in Section 6.2.2. The remaining hydrocarbons in the syngas stream are converted using a steam methane reformer (SMR) with a nickel based catalyst at 1652 °F (900 °C).

Syngas to Methanol

The syngas to methanol section for the BTO process follows the CTM process described in Section 6.2.1 with one slight variation. After sulfur removal, the sulfur-free gas stream is reformed in a tubular steam reformer to adjust the S# of the syngas such that after CO_2 removal it is close to 1.92. The tubular steam reformer and methanol reactor are dependent upon each other, as the methanol reactor produces both methanol and a stream of purge and flash gases. A fraction of the purge gas stream from the methanol reactor is mixed with the syngas from the tubular steam reformer to adjust the S# to 1.7 prior to entering the methanol reactor.

Methanol to Olefins

Olefins can be produced from methanol via the methanol to olefins (MTO) process. Figure 6.22 shows a simplified MTO process diagram. The crude methanol produced is reacted with a proprietary catalyst in either a fluidized bed or a series of fixed bed reactors. The MTO fluidized bed reactor operates at 662 °F (350 °C) and a pressure of 45 psi (0.3 MPa). UOP/HYDRO's proprietary catalyst system indicates a near complete methanol conversion and ~80% selectivity for ethylene and propylene. The spent catalyst is then regenerated in a combustor where carbon deposits are burnt off with air. The product stream of the MTO reactor is separated using a distillation column, with the final product consisting of polymer grade ethylene, polymer grade propylene, and a mixture of lighter hydrocarbons. The various blocks are integrated for a baseline configuration with carbon capture, as shown in Figure 6.23.[25–28]

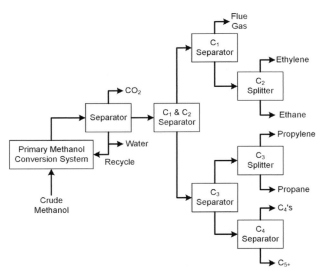

Figure 6.22 Schematic of a generic MTO process processing crude methanol.[25]

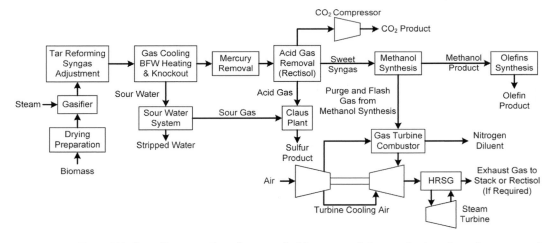

Figure 6.23 Overall process flow diagram of a biomass to olefins production plant via syngas and methanol as intermediates.[22–24]

6.3   Chemical Looping Approach to Partial Oxidation

The thermodynamic motivation behind a chemical looping approach for partial oxidation is provided in Chapters 2 and 4. This section introduces three chemical looping processes using three carbonaceous feedstocks, coal, natural gas, and biomass, for syngas production that are based on this thermodynamic motivation. Section 6.3.1 describes coal gasification using the coal to syngas process, Section 6.3.2 describes the process

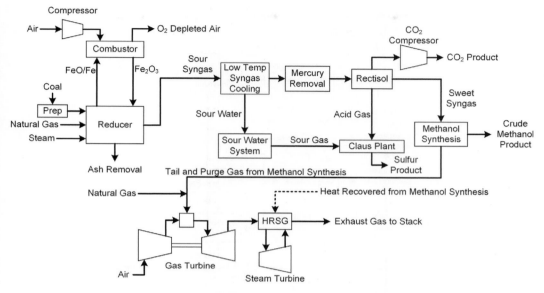

Figure 6.24 Coal to methanol block diagram using the CTS process to produce syngas.[29]

converting natural gas to syngas via the shale gas to syngas process, and Section 6.3.3 presents the process producing syngas from biomass via the biomass to syngas process.

6.3.1 Coal to Syngas (CTS) Process

The CTS process employs an iron oxide based composite particle to react with coal to produce syngas. To obtain the desired H_2:CO molar ratio in the syngas from the CTS process, steam is injected along with coal. Figure 6.24 shows the CTS process integrated into the CTM process for methanol production. Syngas generation from the CTS process replaces the coal gasifier and ancillary equipment in the conventional, baseline CTM process, discussed in Section 6.2.1. The CTS process utilizes a unique combination of a specific ratio of coal, steam, and oxygen carrier to improve carbon utilization over the conventional process, shown in Figure 6.9 and Figure 6.10.[29]

6.3.2 Shale Gas to Syngas (STS) Process

In a broad sense, the feedstock used for the STS process includes natural gas, methane, and other gaseous carbonaceous fuels, in addition to shale gas. The STS process produces syngas from natural gas by selectively donating lattice oxygen from iron oxide based composite particles.[30] Figure 6.25 shows the STS process integrated with a liquid fuels production plant. The syngas generation unit differs from the conventional, baseline GTL process, introduced in Section 6.2.2, but the syngas quality is nearly identical, which allows the liquid fuels production process to be equivalent to the one shown in Figure 6.11. Steam and natural gas react with iron oxide based composite

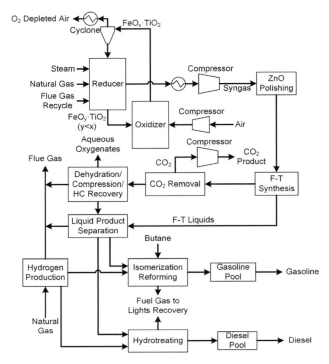

Figure 6.25 STS process and H_2 production coupled with F–T complex to produce 50,000 bpd of liquid fuels.[19,30]

particles to produce syngas in the reducer. The lattice donation of oxygen from the iron oxide leads to a greater control of the selectivity in natural gas oxidation. The H_2:CO ratio is flexible and can be adjusted by altering the iron oxide to natural gas, steam, and flue gas recycle ratio, as well as the reaction residence times. For an equivalent syngas specification (H_2:CO ratio), the STS process requires less steam per mole of carbon input and has a higher selectivity based on the S# compared to the conventional process.[30]

6.3.3   Biomass to Syngas (BTS) Process

In the BTS process, shown in Figure 6.26, an iron oxide based composite particle converts biomass to syngas in a cocurrent moving bed reducer. Biomass is pyrolyzed to release tars and volatiles, which are partially oxidized with lattice oxygen from the metal oxide to form syngas. The devolatilized char or fixed carbon in the biomass is also partially oxidized to syngas with steam and CO_2, where the lattice oxygen from the metal oxide catalyzes the gasification reactions. Steam is introduced into the reducer to adjust the syngas composition. The cocurrent moving bed reducer coupled with the highly reactive oxygen carrier produces high purity syngas, thus eliminating the tar reformer and steam reformer in the conventional biomass gasification process.

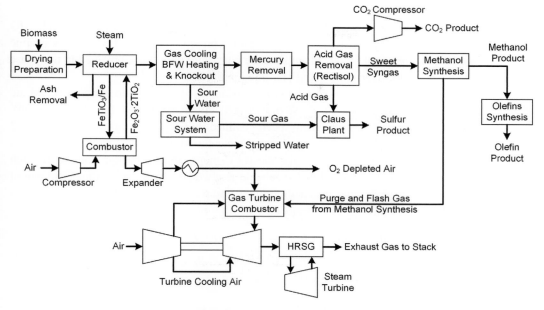

Figure 6.26 BTS process coupled with intermediate methanol produced for olefins synthesis.[4,26,29,30]

6.4 Process Modeling and Analysis Parameters

The CTS, STS, and BTS processes for syngas generation and their extension into the CTM, GTL, and BTO processes for methanol, liquid fuels, and olefins production represent independent applications. However, the process simulation parameters, process development procedures, and process analyses all share common properties and methods. This section provides the process simulation and techno-economic analysis definitions and background information that are shared for all the processes given in this chapter. Section 6.4.1 defines the process simulation modeling parameters used in Aspen Plus® for the process simulation set-up, initial design, and final process integration. Section 6.4.2 defines the economic terms and baseline parameters necessary to complete the techno-economic analysis.

6.4.1 Aspen Plus® Process Simulation Parameters

Chemical looping processes developed for partial oxidation utilize either an iron–titanium composite metal oxide (ITCMO) or iron–aluminum composite metal oxide (IACMO) for oxygen donation. The ITCMO and IACMO particle results in a better syngas selectivity and yield than single metal oxide particles. $FeTiO_3$ and $FeAl_2O_4$ are the stable, reduced forms in a titanium-rich ITCMO and an aluminum-rich IACMO, respectively. Thermodynamic analysis coupled with experimental results has confirmed the ITCMO system to be accurately represented by a Fe_2TiO_5–$FeTiO_3$ redox pair and the IACMO system by a $Fe_2Al_2O_6$–$FeAl_2O_4$ redox pair.[29,30] For thermodynamic analysis and process modeling, Fe_2TiO_5 (Fe_2O_3 + TiO_2) is used to represent ITCMO and $Fe_2Al_2O_6$ (Fe_2O_3 + Al_2O_3) is used to represent IACMO. The thermodynamic conditions for metal oxides given here are consistent with those obtained in the phase diagram by FactSage simulation, presented in Section 1.6.2.

The process simulations involving iron oxide based chemical looping processes are completed using Aspen Plus® v8.8 with a common base used for all simulations. Table 6.2 provides the component species defined for the process simulations. Table 6.3 and equations (6.4.1–6.4.7) define the parameters used to calculate the thermodynamic values for component species, and clearly demonstrate the background information contained within the Aspen Plus® databanks that would require an inordinate amount of time to locate and solve for multiple temperature ranges without the aid of process simulation software:

$$G_i^\propto = a_{n,i}^\propto + b_{n,i}^\propto T + c_{n,i}^\propto (T \ln T) + d_{n,i}^\propto T^2 + e_{n,i}^\propto T^3 + f_{n,i}^\propto T^4 + g_{n,i}^\propto T^{-1} + h_{n,i}^\propto T^{-2}$$

(6.4.1)

$$H_i^\propto = a_{n,i}^\propto - c_{n,i}^\propto T - d_{n,i}^\propto T^2 - 2\, e_{n,i}^\propto T^3 - 3 f_{n,i}^\propto T^4 + 2g_{n,i}^\propto T^{-1} + 3h_{n,i}^\propto T^{-2}$$ (6.4.2)

$$S_i^\propto = -b_{n,i}^\propto T - c_{n,i}^\propto (1 + \ln T) - 2\, d_{n,i}^\propto T^2 - 3e_{n,i}^\propto T^3 - 4 f_{n,i}^\propto T^4 + g_{n,i}^\propto T^{-1} + 2h_{n,i}^\propto T^{-2}$$

(6.4.3)

$$C_{p,j}^\propto = -c_{n,i}^\propto - 2\, d_{n,i}^\propto T - 6e_{n,i}^\propto T^3 - 12f_{n,i}^\propto T^4 - 2g_{n,i}^\propto T^{-1} - 6\, h_{n,i}^\propto T^{-2}$$ (6.4.4)

$$H_i^\propto (T) - H_i^{ref,\propto}(T^{ref}) = \int_{T^{ref}}^{T} C_{p,j}^\propto dT$$ (6.4.5)

$$S_i^\propto (T) - S_i^{ref,\propto}(T^{ref}) = \int_{T^{ref}}^{T} C_{p,j}^\propto dT$$ (6.4.6)

$$G_i^\propto (T) = H_i^\propto (T) - TS_i^\propto (T).$$ (6.4.7)

The parameters "a" to "h" in Table 6.3 are the values for the Barin equations in equations (6.4.1–6.4.4), and T_1 and T_2 are the temperature ranges for which these values have been calculated. The use of temperature beyond these values is calculated by the use of linear extrapolation, shown in equations (6.4.5–6.4.7).[31–33]

The simulation model parameters defined in this section form the basis of thermodynamic databanks for simulating chemical looping processes. Table 6.4 lists the property methods that define thermodynamic parameter calculation. In Sections 6.5, 6.6, and 6.7, the process modeling simulation and techno-economic analysis for three integrated processes, i.e. coal to methanol, natural gas to liquid fuels, and biomass to olefins are discussed. In Section 6.8, other processes are presented and include a coal/natural gas mixture to methanol, GTL using a fluidized bed reducer, and electricity generation using the CTS process.

Table 6.2 List of chemical species.

Type: Solid	
Aluminum oxide: alpha-corundum (Al_2O_3)	Iron-dialuminum tetraoxide ($FeAl_2O_4$)
Carbon-graphite (C)	
Silicon carbide (SiC)	Iron titanium oxide ($FeTiO_3$)
Titanium dioxide (TiO_2)	Di-iron titanium pentoxide (Fe_2TiO_5)
Iron (Fe)	Tri-iron carbide (Fe_3C)
Ferrous oxide (FeO)	Iron monosulfide (FeS)
Hematite (Fe_2O_3)	Iron disulfide pyrite (FeS_2)
	Magnetite (Fe_3O_4)

Type: Conventional	
Argon (Ar)	Formaldehyde (CH_2O)
Hydrogen (H_2)	Acetaldehyde (C_2H_4O)
Oxygen (O_2)	Acetone (C_3H_6O)
Water (H_2O)	Ammonia (NH_3)
Carbon monoxide (CO)	Chlorine (Cl_2)
Carbon dioxide (CO_2)	Hydrogen chloride (HCl)
Nitrogen (N_2)	Cl^- (Chlorine ion)
Nitric oxide (NO)	CO_3^{2-} (Carbonate ion)
Nitrogen dioxide (NO_2)	COS (Carbonylsulfide)
Sulfur (S)	C_2H_6O (Dimethylether)
Di-atomic sulfur (S_2)	H_3O^+ (Hydronium ion)
Hexa-atomic sulfur (S_6)	HCN (Hydrogen cyanide)
Octa-atomic sulfur (S_8)	HCO_3^- (Bicarbonate ion)
Sulfur dioxide (SO_2)	HS^- (Hydrogen sulfide ion)
Methane (CH_4)	NH_2COO^- (Carbamate)
Ethane (C_2H_6)	NH_4^+ (Ammonium ion)
Propane (C_3H_8)	OH^- (Hydroxyl ion)
n-butane (C_4H_{10})	S^{2-} (Sulfide ion)
Methanol (CH_4O)	Propanol (C_3H_8O)

Type: Nonconventional
Coal
Ash

6.4.2 Techno-Economic Analysis Procedure

The techno-economic analysis for the processes given in this chapter follows the general principles of traditional engineering economics, where cost estimates are divided into two main subcategories, capital costs and operating and maintenance costs, and the sum forms the total plant cost. The definitions for the terms used and the cost estimation approach for each term are provided below. From the definitions, the procedure for completing the techno-economic analysis is obtained with process-specific assumptions provided in each respective section.

Table 6.3 Barin equation parameters.

Components	Fe_3O_4	Fe_2O_3	Fe	H_2S	S	FeS	$Fe_{0.877}S$	C
Units	°C	°C	°C	°C	°C	°C	°C	°C
T_1	576.8	686.8	626.8	344.85	95.15	137.8	324.85	526.8
T_2	1596.8	826.8	756.8	824.85	100.85	324.8	526.8	1,726.8
Units	**J/kmol**	**J/kmol**	**J/kmol**	**J/kmol**	**J/kmol**	**J/kmol**	**J/kmol**	**J/kmol**
a	-9.6×10^8	-5.6×10^{10}	-9.8×10^9	-6.3×10^7	-1.5×10^7	-1.2×10^8	-1.3×10^8	$-6,526,612$
b	535,583	3.5×10^8	1.08×10^8	318,487	422,552	481,567	580,701	84,137
c	$-50,897$	-4.7×10^7	-1.6×10^7	$-44,015$	$-68,354$	$-72,358$	$-90,018$	$-11,137$
d	-36.2	20,106	11,275	-7.5	59.2	0	38.5	-9.2
e	0	-1.7	-1.5	0	0	0	-0.0066	0.00147
f	0	0	0	0	0	0	0	-1.25×10^{-7}
g	-4.3×10^{10}	1.6×10^{13}	1.15×10^{12}	0	0	0	0	2.01×10^8
h	0	0	0	0	0	0	0	4.83×10^{10}

Table 6.4 Reactor model set-up.

Parameter	Setting
Reactor module type	RGIBBS
Stream class	MIXCINC
Thermodynamic and physical databank (in order)	Combust, Inorganic, Solids, Aqueous, Pure 22
Method filter	Common
Base method	PR-BM (Peng-Robinson Base method)
Free water method	Steam tables (Steam-TA)
Additional property methods	AVAILMAX, EXERGYMS, GMX, DGMX

Process, Plant, and Project

The plant is the physical representation of the process and covers the entire process from inlet feedstock preparation to purified product. The techno-economic analyses presented in this chapter are for a plant constructed on a green-field site, where all cost estimates are based on a new installation for land and equipment. The term "project" refers to the overall economics for a given process, from initial design cost estimate for individual components to the final cost required for construction and operation over the expected lifetime of the plant inclusive of all cost components. When the term project is used in the context of process economics and techno-economic analysis, it refers to the cost requirements for complete design, construction, and operation of the plant and the required price of final product for profitability analysis.

Capital Cost Estimates

Total capital cost estimates include equipment cost, freight, bulk materials, equipment installation labor and erection (direct and indirect), materials and labor for construction of buildings, supporting structures, site improvements, engineering, construction management, start-up services, and process and project contingency.

The chemical looping equipment costs are obtained from sub-pilot and pilot-scale demonstration processes. Engineering, construction management, home office and fees expenses are calculated as a fixed percentage of the total estimated project cost. Process contingency contains the costs that are associated with technical unknowns based on stage of the development cycle. The chemical looping technology is demonstrated at a pilot scale at the National Carbon Capture Center, which corresponds to a process contingency of 15%. Project contingency covers costs that will be incurred in installing the project but which are not accounted for in the equipment cost estimate. A project contingency of 15% is used in cost estimates for the techno-economic analysis of chemical looping technology.

Operating and Maintenance (O&M) Costs

Operating and maintenance costs include material and labor, administrative and labor support, consumables, waste disposal, fuel, electricity, and co-product or by-products credit. O&M costs consist of fixed O&M, which is independent of production quantity, and variable O&M, which is proportional to production quantity. A breakdown of the cost components considered is discussed below.

Operating labor cost is a fixed cost that is required in order to operate the plant on a daily basis. The number of personnel required to operate the chemical looping process is identical to the number of personnel required for the baseline case. Labor administration and overhead charges are assessed at a rate of 25% of the O&M labor. For general consumables, costs are provided by the baseline report and de-escalated to June 2011 dollars. The annual consumables cost accounts for the annual capacity factor (90%) and is adjusted through equation (6.4.8).

$$\text{Annual Cost} = (\text{Hourly Consumption Rate})(8760 \text{ hours/yr})$$
$$(\text{Capacity Factor})(\text{Unit Cost}). \tag{6.4.8}$$

Net power imports or exports are priced at \$60/MWh. By-product quantities and costs are determined similarly to the other consumables. Transport, storage, and monitoring (TS&M) costs for CO_2 are based on a Midwest location at \$11/tonne CO_2, which is identical to the baseline case. The economic analysis accounts for additional costs associated with the project. Additional costs are project-specific related costs that have not been accounted for in either the capital or O&M costs. They include all the necessary investments prior to project start-up, improvements to infrastructure necessary for the project, and fees necessary to comply with existing regulations. The additional costs are 18% of the total capital cost.

Economic Analysis Metrics

To compare chemical looping process economics to the baseline economic results, capital and O&M cost estimates along with global economic assumptions are used to determine the following economic metrics: first year required selling price (RSP), capital and O&M component contribution, TS&M, and cost of CO_2 captured. The first year RSP is the first year selling price required to equal the return on equity based on the internal rate of return for a 30 year plant life and the assumed financial structure and escalation of the methanol selling price. The RSP is calculated using equations (6.4.9–6.4.10).

Required Selling Price (RSP)

$$RSP = \frac{\begin{array}{c}\text{first year}\\\text{capital charge}\end{array} + \begin{array}{c}\text{first year}\\\text{fixed operating}\\\text{costs}\end{array} + \begin{array}{c}\text{first year}\\\text{variable operating}\\\text{costs}\end{array}}{\text{Annual Net Production}} \qquad (6.4.9)$$

$$RSP = \frac{(CCF)(TOC) + OC_{FIX} + (CF)(OC_{VAR})}{(CF)(PR)}, \qquad (6.4.10)$$

where

RSP = required selling price, which is the revenue received by the producer during the plant's first year of operation (expressed in base-year dollars).

CCF = capital charge factor, which is based on finance structure and set to 0.237.

TOC = total overnight capital costs, expressed in base-year dollars.

OC_{FIX} = the sum of all fixed annual operating costs, expressed in base-year dollars.

OC_{VAR} = the sum of all variable operating costs, expressed in base-year dollars.

CF = capacity factor.

PR = production rate, which is the total production when operating for 8760 hours.

CO_2 Capture Cost = cost of CO_2 captured ($/tonne CO_2), calculated using equation (6.4.11):

$$CO_2 \text{ Capture Cost} = \frac{COE_{Capture} - COE_{No\ Capture}}{CO_2\ Captured_{Per\ Net\ Output}}, \qquad (6.4.11)$$

where

$CO_2\ Captured_{Per\ Net\ Output}$ = amount of CO_2 captured per unit of production.

6.5   Coal to Methanol (CTM) Process Modeling

Industrial methanol production methods from syngas vary significantly.[34] To minimize performance and cost variations from the production method and isolate cost differences to the syngas production method, the syngas production rate and quality using CTS are matched to the syngas composition in the baseline methanol case, described in Section

6.2.1, and the equipment is sized to the production rate. This section describes the chemical looping based CTS process for the production of high hydrogen syngas suitable for crude methanol production.

The IACMO chemistry is simulated for syngas production in the CTS process, and optimization of the CTS process is conducted with a process model that focuses on the CTS reactors, auxiliary equipment, and design input parameters.

6.5.1   CTS Process Design for Methanol Production

Design Concept

Performance modeling described in this section focuses on controlling the syngas composition from the CTS process and integrating it into the CTM process by identifying the specific requirements of the downstream technology given in Table 6.1. For methanol production, the S#, defined in equation (6.2.1) is the specific requirement for syngas composition to be equal to 1.7. The CTS process is designed to produce syngas to match the S# from the conventional CTM process.

The metal oxide particles provide oxygen for coal gasification to syngas in the reducer reactor. The reduced metal oxide particles are re-oxidized in the combustor reactor using oxygen from the compressed air stream. The re-oxidation reaction is exothermic, which increases the temperature of the depleted air and the metal oxide. The high temperature depleted air from the combustor preheats the inlet air to maintain the temperature in the combustor. Further, the depleted air is transported through a particulate control device to remove any fines, and then an expander to recover a portion of the electricity required to compress the air.

A sensitivity analysis of the CTS process is performed by adjusting reactor inlet parameters and observing the effect on syngas quality and heat balance. Figure 6.27 provides a block diagram showing the components considered in this optimization process. The optimal condition for the CTS process is obtained through a sensitivity

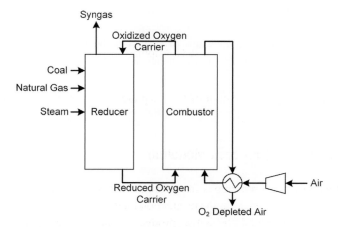

Figure 6.27 Block diagram for optimization of the CTS process.

analysis of the reactor operating conditions and reactor inputs, and observing their effects on syngas composition and reducer heat duty. The methanol catalyst possesses a very low tolerance towards sulfur, less than 0.1 ppmv H_2S, while the synthesis requires a high H_2:CO ratio syngas. In cases where the CTS process is able to match the S# of 1.7 directly from the reducer, a high temperature sulfur clean-up such as a zinc oxide bed can be used. This combination bypasses the use of an air separation unit, water–gas shift reactors, and acid gas removal. In other cases, the CTS reducer conditions are adjusted using a combination of lattice oxygen to carbon and steam to carbon ratio such that a H_2:CO molar ratio with an S# for methanol production is obtained. This optimization strategy requires CO_2 removal to obtain the appropriate S# and H_2S removal to obtain the sulfur tolerance specifications of the methanol catalyst. In this integration strategy, the ASU and the WGS unit are bypassed, but a Rectisol unit is used for adjusting the syngas composition by removing CO_2 and H_2S. Based on the CTM process requirements, there are four optimization objectives for the CTS syngas composition: (1) producing syngas suitable for methanol synthesis (S# of 2); (2) maximizing process efficiency – minimizing feedstock consumption; (3) providing sufficient heat to maintain heat balance; and (4) obtaining at least 90% carbon capture.

Reactor Operating Conditions: Syngas Composition

The initial configurations developed employ a ZnO bed and adjust the syngas composition from the reducer to avoid using WGS and Rectisol. Process simulation results on these configurations indicate that achieving the necessary S# while obtaining sufficient heat from the reactions is feasible with a support material weight percentage greater than 98%.

The operating conditions are identified for the reducer by specifying a net heat duty of zero. Figure 6.28 shows the reducer set-up for satisfying this condition. Coal injection rate is adjusted such that the carbon molar flow rate into the reducer equals 1 kmol/hr. The solids inlet temperature is specified as 2156 °F (1180 °C) and steam is supplied to the reducer at 932 °F (500 °C), and a temperature swing in the reducer exists as the net reaction is endothermic. The reducer performance is analyzed at a fixed Al_2O_3 support weight percentage of 75% and varying the Fe_2O_3:C and H_2O:C ratio. Figure 6.29 shows the outlet temperature of solids in the reducer as a function of Fe_2O_3:C and H_2O:C ratios. The first performance constraint restricts the temperature

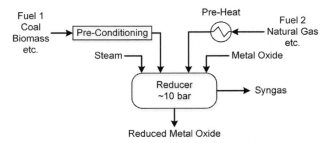

Figure 6.28 Reducer reactor block diagram for CTS process investigation.

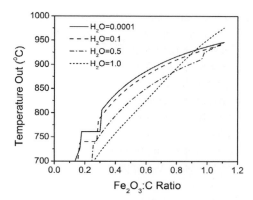

Figure 6.29 Temperature of solids exiting reducer as a function of Fe_2O_3:C ratio.

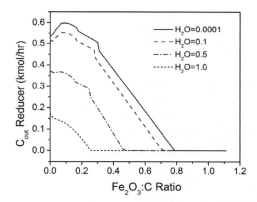

Figure 6.30 Unconverted carbon exiting reducer as a function of Fe_2O_3:C ratio.

swing such that the solids temperature exiting the reducer, which is equal to the solids temperature inside the reducer, is high enough to provide reasonable kinetics. The minimum temperature of the outlet is fixed at 1562 °F (850 °C).

Another constraint for reducer operating condition selection is no solid carbon formation at the outlet, which is a function of the Fe_2O_3:C and H_2O:C ratio. Figure 6.30 shows the solid carbon exiting the reducer per mole of carbon input from the coal. The amount of unconverted carbon decreases as the Fe_2O_3:C and H_2O:C ratio increases. The minimum Fe_2O_3:C ratio for which no unconverted carbon exists is around 0.8 for a H_2O:C ratio of 0.001, and decreases to 0.5 for a H_2O:C ratio of 0.5.

For a constant Fe_2O_3:C ratio above 0.8, the H_2 and CO_2 concentrations increase while the CO concentration decreases with increasing H_2O:C ratio, as shown in Figure 6.31, Figure 6.32, and Figure 6.33. Figures 6.31–6.33 show the flow rates of H_2, CO, and CO_2 in syngas, respectively, per mole carbon in coal as a function of Fe_2O_3:C and H_2O:C ratios. Initially, a Fe_2O_3:C ratio is chosen such that there is no unconverted carbon in the reducer solids outlet. Further, the temperature swing on the

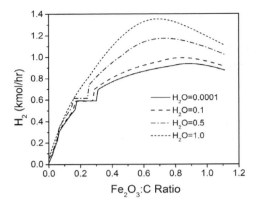

Figure 6.31 H_2 in syngas as a function of Fe_2O_3:C ratio.

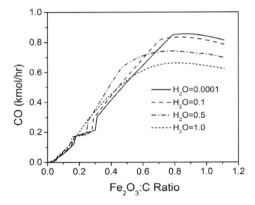

Figure 6.32 CO in syngas as a function of Fe_2O_3:C ratio.

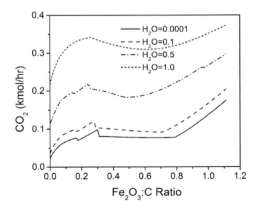

Figure 6.33 CO_2 in syngas as a function of Fe_2O_3:C ratio.

solids is chosen such that there is a reasonable outlet solids temperature. An operating condition is selected such that the H_2:CO ratio from the reducer is near the S# required for methanol synthesis while maximizing the CO content per mole of carbon from coal. This ensures the S# of syngas downstream of the Rectisol unit is suitable for methanol synthesis allowing for bypass of the WGS unit.

Reactor Operating Conditions: Pressure

The CTS operating pressure depends on two factors: (1) net compression power required for maintaining the pressure of the looping reactors; and (2) kinetics and thermodynamics of a pressurized operation. As discussed in Section 2.2.5, increasing the pressure increases the reaction kinetics between ITCMO and methane. It is expected that a similar increasing relationship can be extended to the reaction between IACMO and coal. However, since the gasification of carbon to syngas is a volume expansion reaction, the thermodynamic equilibrium shifts towards the reactant side with increasing pressure. The operating pressure for the chemical looping reactor system is also driven by the operating energy consumption for a particular pressure operation, as indicated in Section 6.2.2. Typical syngas to methanol reactors operate between 700 and 1176 psi (5 and 8 MPa). There are two extremes in terms of operating pressure for the CTS process, near atmospheric and near methanol reactor pressure. Operating at a pressure near the methanol reactor pressure increases air compression energy but decreases syngas compression energy, while operating at near atmospheric pressure reduces the air compression energy requirements but increases the syngas compression energy. The minimum compression energy required is between the minimum and maximum operating pressures.

Figure 6.34 shows the simulation set-up used to analyze and minimize the operating energy requirements of the CTS process by coupling the air stream and the natural gas inlet stream with the depleted air and syngas stream, respectively. The unique methodology of coupling the air stream and the syngas stream with the incoming natural gas

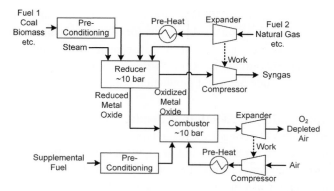

Figure 6.34 Block flow diagram for investigation of compression duty as a function of operating pressures.

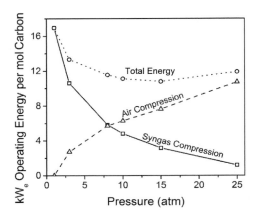

Figure 6.35 Net compression energy requirements for the CTS compressors and expanders.

stream has advantages for compression energy requirements and heat integration design.

The gas turbine has a certain let-down that can provide a portion of the inlet air compressor duty. The effective net compressor duty is then the remaining difference between the original compressor duty and that supplied by the gas turbine. A similar analysis for compensating a fraction of the load on the syngas compressor includes coupling the high pressure fuel stream and the syngas stream exiting the reducer. This compensates for a fraction of the syngas compression energy requirements, with the effective net compression energy being the balance required for syngas compression.

The reducer side streams reduce the pressure on the inlet lines, and the energy gained is used to compensate for a portion of the syngas compression requirements on the outlet side. The let-down process itself follows the Joule–Thomson cooling effect and requires additional pre-heating of the inlet fuel. Depending on reactor operating pressure and oxygen carrier composition, an additional heater may be used to pre-heat the fuel entering the reducer. The combustor operates in reverse as the let-down is on a high temperature air stream. This naturally heats the inlet air stream and could be heated further if necessary. Figure 6.35 shows the results for the net operating energy requirement as a function of operating pressure for a downstream syngas pressure of 30 atm (3 MPa). An operating pressure of 10 atm (1 MPa) is chosen for integrating the CTS process into a full-scale crude methanol production unit.

CTS Process Integration for Methanol Production

Based on the conditions described above, a typical range of operating conditions obtained by setting the reducer temperature to a certain value is shown in Figure 6.36. Figure 6.36 provides the syngas composition as a function of the Fe_2O_3:C ratio and is used to determine the optimal condition for methanol synthesis. A Fe_2O_3:C ratio of 0.8, a reducer temperature of 1832 °F (1000 °C), and a H_2O:C ratio of 1.98 are the conditions used for scaling the CTS process into a methanol plant.

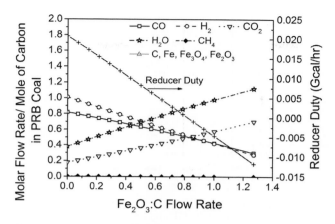

Figure 6.36 Optimization scan for CTS process.

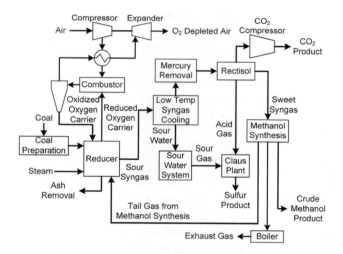

Figure 6.37 Coal to methanol process block diagram using the CTS process for syngas production.

A block flow diagram showing an overview of the CTS-CTM process based on operating conditions obtained from Figure 6.36 is given in Figure 6.37. The CTS-CTM process is considered in two steps: (1) conversion of the feedstock to syngas with a composition suitable for methanol production; and (2) conversion of syngas to methanol. Aside from replacing the ASU and gasifier with the CTS process, other significant deviations from the conventional methanol process include the elimination of WGS reactors and a power generation section.

For the CTS-CTM process in Figure 6.37, syngas is generated from PRB coal through the partial oxidation of coal by lattice oxygen donation from the IACMO in the CTS reducer. In the reducer, the IACMO is reduced to a lower oxidation state. Additionally, steam is added to enhance H_2 production. The reduced IACMO is then regenerated in the combustor through air oxidation. The produced syngas is cooled and

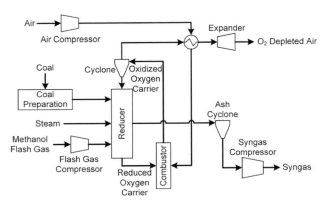

Figure 6.38 Aspen Plus® process model for the CTS process.

then sent to acid gas scrubbers to remove H_2S and 90% of the CO_2, resulting in a sweet syngas that is used to produce methanol. Flash gas from the methanol process is recycled back into the reducer, and the purge gas is combusted in a boiler to produce steam. The following subsections provide details for integrating the CTS process into a full-scale methanol production system.

CTS Reactors

For syngas production, chemical looping replaces the gasifier, ASU, and WGS reactors in the baseline coal to methanol process with 90% CO_2 capture, shown in Figure 6.9. A block flow diagram of the CTS process, as modeled in Aspen Plus®, is shown in Figure 6.38. The syngas from the reducer is cooled and conditioned while the reduced IACMO particles are re-oxidized.

Low temperature Syngas Cooling and Conditioning

As part of syngas conditioning for methanol synthesis, the syngas is heat exchanged to produce superheated steam. The cooled syngas is conditioned with a water scrubber and compressed to decrease downstream equipment size, enhance the CO_2 capture process, and obtain the necessary pressure for methanol synthesis. The main difference in the low temperature syngas cooling and compression section for the CTS process compared to the baseline process is an additional compression step. This is necessary since the CTS process operates at a lower pressure than the gasifier, and operating the AGR at low pressures requires a large equipment size capable of handling the gas volume. An additional consideration is that a minimum syngas pressure of 700 psi (5 MPa) is required for the methanol synthesis reactor. Therefore, a syngas compressor is added after the syngas cooler and prior to the mercury removal system. This compressor increases the syngas pressure from 140 psia (1 MPa) to 500 psia (3.5 MPa). In addition to reducing the equipment size, the pressure increase also improves the efficiency of the Rectisol physical solvent.

Acid Gas Removal, Sulfur Recovery Unit, and CO_2 Compression

The methanol solvent from the absorber is stripped in two stages of flashing via pressure reduction. The acid gas leaving the first stage solvent regenerator is suitable for processing in a Claus plant, described in Section 6.2.1. The regenerated solvent from the first stage is virtually free of sulfur compounds but contains CO_2. The second stage of absorption removes the remaining CO_2. The rich solvent from the bottom of the second stage of the absorber is stripped in a steam heated regenerator and returns to the top of the absorption column after cooling and refrigeration. To provide CO_2 capture, the CO_2 stream is compressed to 2200 psig (15 MPa) in a multiple-stage, intercooled compressor for pipeline transportation to the storage site.

Methanol Synthesis

The methanol production process is identical to the baseline in order to confine all process changes to the syngas generation section. This allows the differences in the techno-economic analysis results to be a result of converting the traditional coal gasifier to the CTS process. The CO_2-lean syngas with a H_2:CO ratio of 1.7:1 from the AGR process is compressed from 500 psia (3.4 MPa) to 755 psia (5.2 MPa). The compressed syngas is mixed with recycled gas, heated to 400 °F (204 °C), and routed to the methanol reactor. The reactor is steam cooled to facilitate a near isothermal operation at 475 °F (246 °C) and 735 psia (5 MPa). In-line blowers, coolers, and knock-out drums are used within the synthesis loop to maintain reactor pressure and remove crude methanol. To limit CO_2 emissions, the flash gas is injected into the reducer reactor. The purge gas, which contains inerts such as N_2, is sent to a boiler where it is combusted to generated steam and then vented to the atmosphere.

Overall Process Performance

The performance results for CTS cases are summarized and compared to the baseline results in Table 6.5.

Table 6.5 Overall performance of the CTS process in a 10,000 tonne/day methanol plant.

	MB-B	MB-A	CTS
Mass flows (lb/hr)			
As received coal	1,618,190	1,618,190	1,395,457
Oxygen from ASU containing 95% O_2	1,010,968	1,010,968	NA
Steam to gasifier, quench, shift reactors, CTS	1,533,584	1,533,584	1,624,318
Clean syngas for methanol production	1,183,080	1,183,080	1,025,106
Tail gas from Claus unit	61,476	61,476	50,089
Captured CO_2 (no capture for MB-B)	0	1,569,410	1,302,138
Electrical loads (kW$_e$)			
Total gross power	320,680	390,170	20,830
Total net power **	**12,280**	**21,480**	**−323,504**

Notes ** Negative value indicates power purchase required.

NO_x Emissions

The CTS process yields low thermal NO_x formation by avoiding excessive temperatures in the combustor and strong reducing conditions in the reducer. The exhaust from the boiler is a potential source of NO_x emissions, but the boiler design incorporates low NO_x burners to limit NO_x formation and achieve emission standards.

Mercury

To achieve 90% mercury capture, a sulfur impregnated activated carbon bed is used. This technology has been shown to have a removal efficiency of 95% based on Eastman Chemical's coal to methanol plant operation in Kingsport, Tennessee. Similar to the baseline case, a 95% removal efficiency is used for mercury removal from the CTS process.

Particulate Matter (PM)

The primary source of PM is the result of attrition of the IACMO particles and entrained fly ash. To mitigate these emissions, gas–solid separators at the outlet of the reducer and combustor are used as particulate control devices for PM removal to follow EPA emission standards.

Solid Waste – Ash/Spent IACMO

Fly ash from coal and attrition products from the IACMO are the primary solid wastes discharged from the CTS process. These two streams represent the fines circulating with the bulk metal oxide particles and are separated from coarser particles using a cyclone. The fines can be further separated into fly ash and attritted metal oxide since the particle size of fly ash is smaller than the attritted metal oxide. Fly ash is disposed of in landfill while the attritted metal oxide is recycled and reformed into fresh oxygen carrier. Based on the coal flow into the CTS reducer, 450,373 ton/yr of fly ash is disposed from the chemical looping system and 20,696 ton/yr of attritted IACMO is reprocessed.

Carbon, Sulfur, Waste Water, and Make-up Water Balances

The CTS carbon balance and sulfur balance are given in Table 6.6. The CTS water balance is provided in Figure 6.39.

6.5.2   CTS-CTM Techno-Economic Analysis

Cost Estimation Results

The cost estimating methodology is provided in Section 6.4.2 with the methanol required selling price calculated from equations (6.4.9) and (6.4.10). The results of the techno-economic analysis are used for comparison purposes between the baseline reference and the CTS process for methanol production[29,35,36].

Total plant costs for the CTS cases are presented in Table 6.7. There is a significant capital cost reduction using the CTS process compared to coal gasification, where a 27% reduction is obtained when comparing cases with CO_2 capture, MB-B and CTS. The methanol required selling price using the CTS process with carbon capture is still lower than the reference case without carbon capture. This is related to the higher

Table 6.6 Carbon and sulfur balance for the CTS-CTM process.

Carbon balance			
Carbon in, kg/hr		**Carbon out, kg/hr**	
Coal	316,919	Ash	0
		Combustor flue gas	0
		CO_2 to pipeline	161,636
		Claus vent	0
		Fired boiler	8,334
		Crude methanol	146,648
		Miscellaneous losses	301
Total	**316,919**	**Total**	**316,919**
Sulfur balance			
Sulfur in, kg/hr		**Sulfur out, kg/hr**	
Coal	4,605	Sulfur product	4,582
Natural gas	0	Miscellaneous losses	23
Total	**4,605**	**Total**	**4,605**

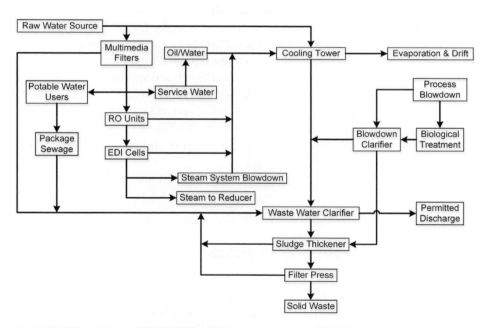

Figure 6.39 Water balance for the CTS-CTM process producing 10,000 tpd crude methanol.

efficiency and lower capital costs associated with the CTS process. When carbon capture costs are considered, the CTS process results in a negative carbon capture cost, which demonstrates its attractiveness. The RSP and component breakdown for the two reference cases, MB-A and MB-B, and the CTS process are given in Figure 6.40.

Table 6.7 Comparative summary of capital and operating costs and cost of methanol production for MB-A and MB-B cases and CTS case.

Case	MB-B	MB-A	CTS
Total plant costs (2011 MM$)	4,586	4,775	3,497
Total as spent capital (2011 MM$)	6,580	6,852	5,003
Capital costs ($/gal, 2011$)	1.18	1.23	0.89
Fuel cost ($/gal, 2011$)	0.24	0.26	0.18
O&M costs ($/gal, 2011$)	0.23	0.23	0.16
CO_2 TS&M costs ($/gal, 2011$)	0.00	0.06	0.05
Electricity cost ($/gal, 2011$)	0.00	0.00	0.14
Required Selling Price ($/gal, 2011$)	**1.64**	**1.78**	**1.41**

Table 6.8 Summary of parameters, costs, and corresponding ranges used in sensitivity study.

Parameter/Cost	Nominal	Min	Max	Range basis
Electricity price ($/MWh)	100%	−17%	225%	Maximum price equivalent to price with CO_2 capture
CTS equipment cost	100%	50%	200%	Variations

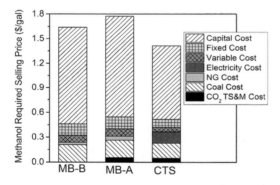

Figure 6.40 Cost breakdown of RSP for methanol production.

Overall, the CTS-CTM process provides an economically competitive route for the production of high H_2 syngas suitable for methanol production.

Sensitivity Studies

A sensitivity analysis of the CTS-CTM process under varying economic conditions is examined to determine the important parameters affecting the RSP of methanol. By identifying the parameters most sensitive to the RSP, the future focus of the CTS-CTM process can capitalize on reducing those costs. In addition, the sensitivity analysis identifies the parameters that can be adjusted that result in the lowest possible RSP. In the sensitivity studies, the parameters and range investigated must be limited to a reasonable value, summarized in Table 6.8.

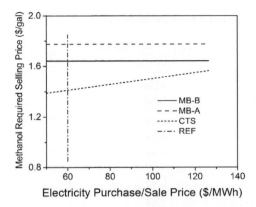

Figure 6.41 Sensitivity of methanol RSP to electricity price.

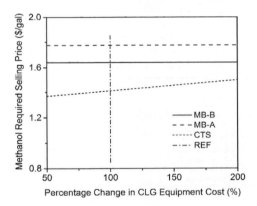

Figure 6.42 Sensitivity of methanol RSP to CTS equipment cost.

The RSP sensitivity to electricity price is given in Figure 6.41. An increase in electrical costs from the reference value to 225% is the expected electricity price increase when switching from power generation without carbon capture to power generation with carbon capture, and the RSP for methanol using the CTS-CTM process increases by $0.20/gal.

Figure 6.42 shows the RSP sensitivity to CTS equipment costs from −50% to +100%. Increasing the cost of the CTS equipment by 100% only results in a $0.10/gal increase in the methanol RSP, which is not sufficient to change the economic favorability of the options. Overall, chemical looping technology can reduce the methanol RSP by 21% with 90% carbon capture. Specifically, the capital cost investment for the syngas generation section is reduced by over 50% and the plant capital cost investment by 28%. Chemical looping technology with 90% carbon capture is able to improve the coal consumption efficiency by 14%, resulting in a methanol RSP lower than the baseline non-capture case.

6.6   Natural Gas to Liquid Fuel (GTL) Process Modeling

The GTL process modeling uses the gas to liquids process described in Section 6.2.2 as a baseline case. This section evaluates the STS process to produce syngas for liquid fuels production via F–T technology. The STS process utilizes the ITCMO particle, discussed in Sections 1.6.2 and 2.2.5, as the metal oxide to convert natural gas to syngas. The unique combination of a cocurrent downward moving bed reducer reactor and ITCMO particles enables the STS process to eliminate the ASU and high operating temperature of the autothermal reformer, while also significantly reducing steam consumption as compared to natural gas reforming. In this section, a techno-economic analysis of the STS process integrated into a 50,000 bpd liquid fuel production facility is performed to identify the critical parameters that affect the overall economics.

6.6.1   STS Process Design for Liquid Fuels Production

Design Concept

The basis for the STS process is one mole of methane entering the reducer. The chemical reactions between methane and the ITCMO particle are simulated using a single stage RGIBBS reactor in Aspen Plus®. This section provides insight into the various facets of integrating the STS process into a 50,000 bpd liquid fuel plant.

The overall process for natural gas to liquids follows the same baseline as described in Section 6.2.2. In the baseline case, the reactants to the autothermal reformer consist of natural gas, oxygen, steam, and recycled tail gas, consisting mainly of light hydrocarbons, with the material flows shown in Figure 6.43. 5% of the recycle gas stream is fuel gas used for power production. The STS process is developed by maintaining the recycle fraction at 5%.

Based on the syngas production from the ATR, an optimization approach that involves the parameters of H_2O:C molar ratio, Fe_2O_3:C molar ratio, temperature swing, natural gas pre-heat temperature, and fuel gas pre-heat temperature is developed, as shown in Figure 6.44. The optimization problem is defined such that the syngas

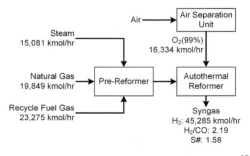

Figure 6.43 Baseline autothermal reforming case.[19,21]

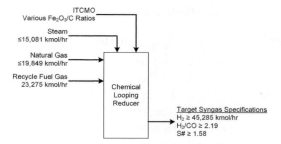

Figure 6.44 STS process optimization set-up.

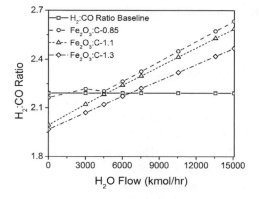

Figure 6.45 H_2:CO ratio for various H_2O:C_{input} ratios for a natural gas flow rate consistent with the baseline case.

production flow rate is equivalent to the baseline case, with manipulation of the variables previously specified such that the flow rate of natural gas and steam is minimized.

The initial value for the natural gas flow rate is identical to the baseline case, 19,849 kmol/hr. A sensitivity analysis on the H_2:CO ratio, S#, and H_2 flow rate is then performed to observe their effect on syngas generation. The range of natural gas flow rates acceptable for the STS process is identified, based on the criteria that the syngas flow rate should be equivalent to the baseline case, while the natural gas and steam flow rate are lower as compared to the baseline. The results are summarized in Figure 6.45 and Figure 6.46. The solid line in Figure 6.45 and Figure 6.46 represents the H_2:CO molar ratio and H_2 flow rate from the baseline case, respectively. Any of the conditions above the black line can be used for further simulations. Figure 6.45 and Figure 6.46 also provide the performance of Fe_2O_3:C ratio as a function of H_2O:C ratio when the natural gas flow rate is equal to the baseline case, and the ITCMO to carbon molar ratio is varied. Further, the conditions governing temperature swing and pressure are adjusted to an autothermal operation.

Following the methodology used to obtain Figures 6.45 and 6.46, the set of conditions from the STS reducer that satisfies or betters the baseline ATR case are compiled

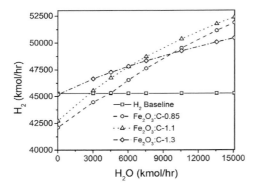

Figure 6.46 Hydrogen flow rates (kmol/hr) for varying steam flow rates from the STS reducer.

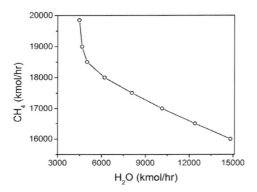

Figure 6.47 Methane input as a function of steam input for the STS reducer.

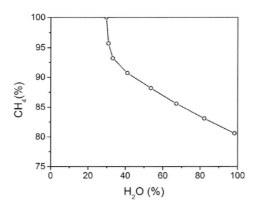

Figure 6.48 Methane of the baseline as a function of % steam of the baseline.

in Figure 6.47, which shows the steam necessary for operating the STS reducer at a fixed natural gas flow rate. Any of the points plotted in Figure 6.47 can be used for further analysis. Figure 6.48 shows the natural gas and steam flow rates as a percentage of the baseline ATR natural gas flow and steam flow rates. The complete mass and

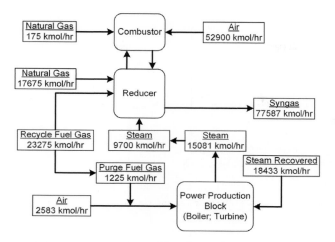

Figure 6.49 Process flow diagram for STS process integrated into 50,000 bpd GTL plant.

energy balance of the STS process for a specific case that satisfies the necessary constraints is shown in Figure 6.49 and used for further analysis.

Process Performance

The STS process is integrated into the 50,000 bpd liquid fuels plant given in Section 6.2.2. In order to compare the STS process for liquid fuels production with the baseline, the defining parameters for 50,000 bpd liquid fuels are constant between the two processes, mainly syngas production in terms of H_2:CO ratio and H_2 flow rate, but also additional parameters. For example, in the baseline case, natural gas is mixed with a tail gas recycle stream consisting mainly of light hydrocarbons (C_1–C_4). The tail gas flow rate is also added to the reducer to maintain the carbon emissions per unit of liquid fuel produced identical to the baseline. The STS-GTL process and the baseline are then compared in terms of natural gas utilization, steam utilization, and carbon efficiency.

The overall STS-GTL process is shown in Figure 6.50. The air compressor and expander are coupled to produce electricity through an input–output shaft with an isentropic efficiency of 0.80. Based on experimental results, cyclones are modeled with a gas–solid separation efficiency of 99% and an ITCMO particle purge rate of 0.02%.

Overall Process Performance

Table 6.9 provides the performance results for the STS-GTL process and comparison to the baseline results. Compared to the conventional GTL process with ATR, the STS process requires 11% v/v less natural gas feed to produce an equivalent amount of liquid fuel. Further, the STS process eliminates the need for the ASU. These combined benefits decrease the parasitic energy requirements of the syngas generation unit by 60% (kW_e basis), resulting in more than twice the net power output generated from a conventional GTL plant. The ITCMO particle in a cocurrent downward gas–solid

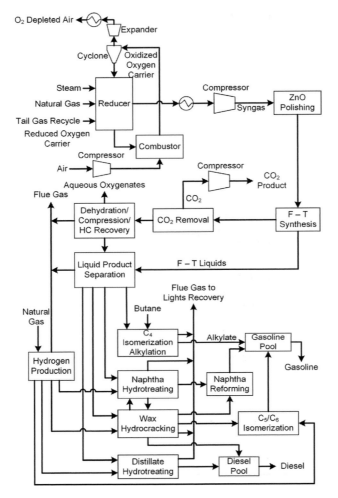

Figure 6.50 Overall chemical looping system gas to liquids process for 50,000 bpd plant.

contact pattern allows the STS process to achieve a high selectivity for syngas generation, with less than 3% v/v CO_2 produced from natural gas with no carbon deposition, while improving the natural gas consumption efficiency by 10%.

NO_x Emissions

NO_x emissions in the baseline case are limited by using a low NO_x burner and dilution by humidification, with emissions around 0.001 kg/bbl, which is in accordance with New Source Performance Standards (NSPS) of 15 ppmvd (15% O_2). The NO_x emissions from the STS-GTL process are minimal, as both the reducer and the combustor operate at lower temperatures than conventional boilers. NO_x emissions from the power production unit are identical to the baseline case since the capacity and configuration remain unchanged.

Table 6.9 Overall performance of the STS process in a 50,000 bpd GTL plant.

Component	Base case	STS (10 atm)	STS (7 atm)
Nat gas flow (kg/hr)	354,365	317,094	317,267
H_2O/C in natural gas	0.68	0.25	0.25
H_2:CO	2.19	2.178	2.18
Stoichiometric number (S#)	1.59	1.96	1.99
Butane feed flow (kg/hr)	18,843	18,843	18,843
Diesel fuel (bbl/day)	34,543	34,543	34,543
Gasoline (bbl/day)	15,460	15,460	15,460
Total liquids (bbl/day)	50,003	50,003	50,003
Electrical loads (kW_e)			
Total gross power	303,700	303,700	303,700
Net plant power	40,800	179,050	196,950

Table 6.10 Carbon balance for STS-GTL process producing 50,000 bpd liquid fuel.

Carbon in (stream)	(kg/hr)	Carbon out (stream)	(kg/hr)
Natural gas	236,544	Diesel	147,524
Air	316	Gasoline	61,908
Butane	15,576	Air	306
		CO_2	38,498
Total	252,436	Stack gas	4,200
		Total	252,436

Table 6.11 Water balance for the STS process and comparison to the baseline case.

Process values (bbl/hr H_2O)/(bbl/hr F–T liquids)	Baseline	STS
Water demand	4.72	2.265
Internal recycle	−0.38	−0.38
Raw water withdrawal	4.23	1.885
Process water discharge	0.99	0.99
Raw water consumption	3.24	0.895

Particulate Matter and Solid Waste

The particulate emissions and solids handling strategy are equivalent to the CTS process described in Section 6.5.2.

Carbon and Waste Water and Make-up Water Balances

The carbon balance and water balance for the STS-GTL process are provided in Table 6.10 and Table 6.11, respectively. The STS-GTL process reduces the water demand by over 50% and is a result of the reduction in steam consumption for syngas production.

Table 6.12 Cost summary for STS-GTL plant producing 50,000 bpd liquid fuel.

Component (2011 $)	ATR	STS
Total plant cost ($×1000)	2,750,000	1,880,000
Total as-spent cost ($×1000)	4,310,000	3,250,000
Capital costs ($/(bbl/day))	86,000	65,000
Capital costs ($/bbl)	51.5	41.8
O&M costs ($/bbl)	27.29	27.23
Fuel costs ($/bbl)	16.9	12.6
Electricity cost ($/bbl)	−1.15	−5.1
Required selling price ($/bbl)	94.54	76.53
West Texas intermediate crude oil competitive price ($/bbl)	**60.5**	**48.9**

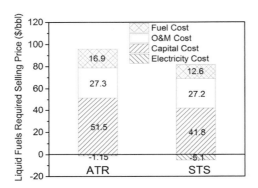

Figure 6.51 Cost component breakdown for liquid fuels RSP for baseline and STS-GTL.

6.6.2   STS-GTL Techno-Economic Analysis

Cost Estimation Results

The cost estimating methodology has been described in Section 6.4.2. Total plant capital costs for the STS-GTL process are given in Table 6.12, and the liquid fuels RSP with cost component breakdown is given in Figure 6.51. Conventional GTL plants require a high initial capital investment, with the ATR and ASU representing over 38% of the total plant cost. The baseline total plant cost of a 50,000 bbl/day GTL plant using ATR is estimated to be $86,000/bbl, which results in a required liquid fuel selling price greater than crude oil. The STS process for syngas production results in a capital cost reduction of 72.5%. The dramatic decrease in syngas generation capital cost requirements leads to the STS-GTL process producing liquid fuels for $65,000/(bbl/day), a 25% decrease compared with liquid fuels produced from ATR. At a natural gas price of $2/MMBTU, a capital charge factor of 0.237, the current state-of-the-art GTL plant can only be economically competitive when West Texas intermediate crude oil prices are greater than $60/bbl. A substantial reduction in the capital cost allows the STS-GTL process to remain economically competitive, even when West Texas intermediate crude oil prices are as low as $48/bbl, a 20% reduction over the conventional GTL process.

Table 6.13 Summary of parameters and corresponding costs range used in sensitivity study.

Parameter/Cost	Nominal	Min	Max	Range basis
Electricity price ($/MWh)	100%	100%	221%	Maximum price equivalent to price with CO_2 capture
Natural gas price ($/MMBTU)	100%	100%	600%	Low price – observed during periods and in regions with excess supply. High price – typical world market price or periods with limited supply

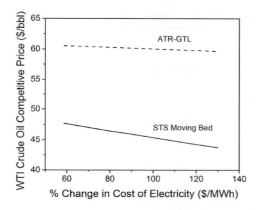

Figure 6.52 Sensitivity of crude oil price as a function of electricity cost for GTL plant.

Sensitivity Studies

The objective of the sensitivity study is equivalent to that of the CTS-CTM sensitivity study. Table 6.13 provides the parameters investigated and the range of the sensitivity study.

Figure 6.52 shows the RSP sensitivity to electricity price. An increase in electricity costs leads to an increase in competitiveness of both the ATR and STS technology. The conventional GTL plant is rather insensitive to the cost of electricity, as the required cost of crude oil decreases from $60/bbl to $59/bbl when the electricity price increases by 221%. The increase in electricity costs is more beneficial to the STS-GTL process as crude oil decreases from $49/bbl to $45/bbl. The benefits from an increase in the electricity selling price are higher for the STS-GTL process as the operating energy savings are significantly higher than the conventional GTL process.

Figure 6.53 shows the RSP sensitivity to variation in the natural gas cost. The RSP for the conventional GTL technology increases by 90%, $54/bbl absolute, with an increase of 600% in the natural gas cost. Correspondingly, the STS-GTL process increases by 82%, $40/bbl absolute. The STS-GTL process is competitive with a lower crude oil selling price due to its higher carbon efficiency.

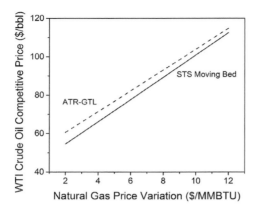

Figure 6.53 Sensitivity of crude oil price as a function of natural gas cost for GTL plant.

6.7   Biomass to Olefins (BTO) Process Modeling

The BTO process modeling follows the process described in Section 6.2.3 as a baseline case. The BTS process uses ITCMO particles in a cocurrent moving bed reducer to produce a high quality syngas from biomass, tars, and/or volatiles. The resulting syngas from the BTS process has an S# in the order of 2, after acid gas removal using Rectisol. This section describes the BTS process for methanol production to be used as an intermediate in the MTO process. A techno-economic analysis for the BTS-BTO process is also provided.

6.7.1   BTS Process Design for Olefins Production

Design Concept

The BTS process is used to produce syngas and methanol for eventual olefins production using the MTO process. The syngas quality and requirements for methanol production were given in Section 6.2.1 and Section 6.5.1. The BTS and CTS processes are similar for syngas generation, but the inherent differences in physical properties between biomass and coal require a sensitivity analysis to be performed for the BTS process to determine the optimal reducer conditions. The BTS process sensitivity analysis varies reducer inlet parameters over a range of values to determine their effect on syngas quality and heat balance.

The block diagram in Figure 6.54 shows the components considered in the BTS process optimization. Figure 6.55 is the process simulation optimization block. Since the methanol catalyst possesses a very low tolerance towards sulfur, while the synthesis requires syngas with a high H_2:CO ratio, the BTS reducer conditions are analyzed with CO_2 removal after syngas generation. The optimization objectives for the BTS-BTM process are identical to those of the CTS-CTM process, described in Section 6.5.1, since the syngas requirements for methanol synthesis do not change and the design philosophy for the BTS process is identical to the CTS process for methanol production.

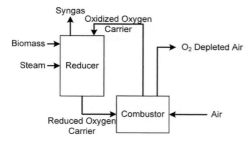

Figure 6.54 Block diagram for BTS process optimization.

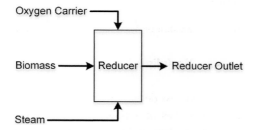

Figure 6.55 BTS reducer block diagram for sensitivity analysis and optimization.

Reactor Operating Conditions: Syngas Composition

The method for determining the operating conditions is identical to the CTS-CTM process described in Section 6.5.1. The operating conditions for the reducer are identified by specifying a net heat duty of zero. The solids temperature into the reducer is fixed at 1832 °F (1000 °C) and the steam supplied to the reducer at 932 °F (500 °C) results in a temperature swing in the reducer. Figure 6.28 shows the set-up of the reducer for this condition.

To analyze the ITCMO temperature swing that occurs from the combustor to the reducer, the TiO_2 support weight percentage is fixed at 80%, while the Fe_2O_3:C ratio is varied from 0 to 1.5. Figure 6.56 shows the ITCMO reducer outlet temperature as a function of Fe_2O_3:C and H_2O:C ratios. The first performance constraint restricts the temperature swing such that the temperature of the solids exiting the reducer is high enough to provide reasonable kinetics. The minimum temperature at the reducer outlet is set to 1562 °F (850 °C), based on experimental results.

Another constraint for reducer operating condition selection is the absence of solid carbon formation at the outlet. Figure 6.57 plots the solid carbon exiting the reducer per mole of carbon input from the coal. The unconverted carbon decreases as the Fe_2O_3:C and H_2O:C ratios increase. The operating condition is selected such that for a specified Fe_2O_3:C ratio, there is no solid carbon deposition at the reducer outlet.

Finally, an operating condition is selected such that the H_2:CO ratio after CO_2 removal is near the S# required for methanol synthesis, while the CO:C molar ratio is

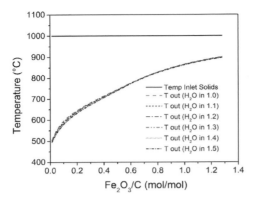

Figure 6.56 Temperature of ITCMO at reducer outlet as a function of Fe_2O_3:C ratio. A black and white version of this figure will appear in some formats. For the color version, please refer to the plate section.

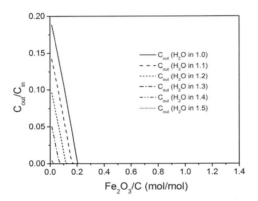

Figure 6.57 Unconverted carbon exiting the reducer as a function of Fe_2O_3:C ratio.

maximized. Satisfying these two conditions ensures that the syngas is suitable for methanol synthesis without the need for a WGS reactor. For a constant Fe_2O_3:C ratio, the H_2:CO ratio content increases with increasing H_2O:C ratio, as shown in Figure 6.58. Figure 6.59 shows the CO in syngas per mole of carbon input from biomass as a function of Fe_2O_3:C and H_2O:C ratios.

Reactor Operating Conditions: Pressure

The BTS process operating pressure is chosen by minimizing the operating energy consumption. Figure 6.60 shows the process simulation set-up used to analyze and minimize the operating energy requirements of the BTS process by coupling the air inlet stream with the depleted air stream. Following the optimization concept used for the CTS process, described in Section 6.5.1, the BTS operating pressure is 10 atm (1 MPa).

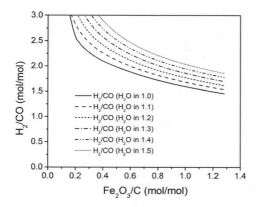

Figure 6.58 H_2:CO ratio of syngas as a function of Fe_2O_3:C and H_2O:C ratios.

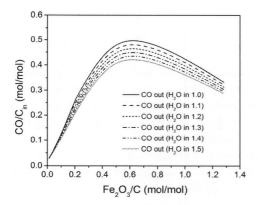

Figure 6.59 CO in syngas as a function of Fe_2O_3:C ratio.

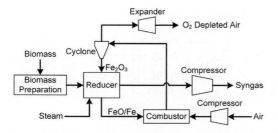

Figure 6.60 Process simulation for compression duty as a function of operating pressure.

BTS Process Integration for Methanol Production

The BTS process is designed to match the 95 MMgal/yr methanol production rate of the baseline indirect gasification case, described in Section 6.2.3. The economic analysis follows the CTS-CTM and STS-GTL process design approach, where changes to the baseline case configuration are minimized and restricted to syngas

Table 6.14 Process comparison for the indirectly heated gasifier case with BTS chemical looping case.

Case	Indirectly heated gasifier	BTS
As-received biomass (mtpd)	4000	3550.8
Dry wood chips (mtpd)	2000	1775.4
% Reduction	0%	11.2%
Methanol (MMgal/yr)	95	95
H_2O input (lbmol/hr)	11,569.6	8924.3
Carbon efficiency (%)	32.4	36
Syngas specifications		
H_2:CO	2.12	2.12
S#	1.92	1.93
Electrical loads (kW_e)		
Total gross power	23.3	23.3
Net power	−7.4	+0.9
Thermal efficiency (% LHV basis)		
Biomass to methanol product	48.1	54.2

generation. The BTS process uses ITCMO as the metal oxide particle for biomass gasification in the reducer and is regenerated using oxygen from ambient air in the combustor. The BTS process eliminates the ASU, biomass gasifier system, tar reformer unit, and tubular steam reforming reactor, leading to a net plant carbon efficiency improvement of 3–4 percentage points. The syngas produced from the BTS reducer is cooled and subjected to CO_2 removal using an AGR unit. The overall syngas generation from the BTS process is adjusted to match the syngas stream specifications and quantity of the baseline case, prior to mixing with the recycle purge stream from methanol synthesis.

Table 6.14 summarizes the process simulation results. For the same quantity and quality of syngas from a BTS process, the biomass requirement decreases by 11% as compared to the baseline case. Further, as compared to an indirect gasification system, the BTS process requires 22% less steam. From an energy standpoint, the BTS process eliminates the auxiliary power consumption associated with incomplete char combustion and the reforming unit. The power generation schematic of the BTS process is kept similar to the baseline for this analysis. Overall, the BTS process can provide significant economic advantages from an operating cost standpoint, as is seen from the 6% point increase in the LHV thermal efficiency. This analysis as presented is conservative in nature and neglects the additional energy gain from the BTS combustor unit. Based on the BTS process performance results given in Table 6.14, the BTS process is integrated into a methanol production plant, and the methanol is then converted to olefins using the MTO process described in Section 6.2.3.

6.7.2   BTS-BTO Techno-Economic Analysis

The cost estimating methodology is provided in Section 6.4.2 with the methanol required selling price calculated from equations (6.4.9) and (6.4.10), and the

Table 6.15 Economic analysis of RSP of methanol produced from BTS process.

Parameter	Indirect gasification	BTS
Total plant cost (2011 million $)	79.3	60.1
Total as-spent cost (2011 million $)	296	224
O&M costs ($/gal, 2011 $)	0.74	0.69
Methanol required selling price ($/gal, 2011 $)	1.77	1.62

Table 6.16 Economic analysis of RSP of olefins (ethylene and propylene) produced from BTS process through methanol as an intermediate.

Methanol to olefins economics			
MTO yields with UOP MTO-100 catalyst[37,38] (output/kg of methanol)			
0.21 kg ethylene, 0.145 kg propylene, 0.04 kg C_4s, 0.01 kg C_5s, 0.03 kg coke, 0.56 kg/hr H_2O			
@ natural gas cost of $2/MMBTU methanol: $100/metric tonne (density: 6.6 lb/gal @ 25 °C), ethylene: $ 500/metric tonne, propylene: $400/metric tonne			
Parameter	Indirect gasification	BTS	BTS comments
Ethylene required selling price	$6.39/lb	$5.84/lb	Scaled from reported yields of methanol catalysts and market prices for ethylene
Propylene required selling price	$7.40/lb	$6.77/lb	Scaled from reported yields of methanol catalysts and market prices for propylene

BTS-BTM economic results are summarized in Table 6.15. The methanol produced using the BTS-BTM process has a required selling price of $1.62/gal, as opposed to the $1.77/gal used for a conventional indirect biomass gasification process.

In the BTS-BTO process, methanol is an intermediate. Once methanol is produced, it is converted to olefins using the MTO process. Table 6.16 provides the yields of a MTO-100 catalyst. The RSP for methanol is used as the input to calculate the RSP for olefins produced from the MTO process. The $1.62/gallon methanol RSP from the BTS-BTM process translates to an ethylene required selling price of $5.84/lb, as opposed to $6.39/lb from the conventional biomass gasification process. The BTS-BTM process can also produce propylene at $6.77/lb, as opposed to $7.40/lb, which shows the BTS process has the potential to provide significant economic advantages for biomass gasification to value-added chemicals. The BTS process can be used to provide significant cost savings in the production of value-added chemicals like methanol, ethylene, and propylene.

6.8   Other Processes

This section analyzes unique alternative process configurations for the CTS and STS processes, discussed in Sections 6.3.1 and 6.6.2, respectively. Specifically, the CTS

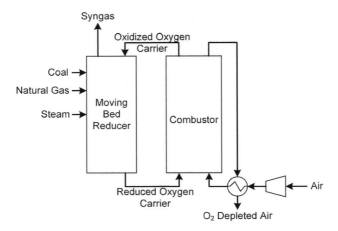

Figure 6.61 Block diagram for optimization of CTS co-feed system.

process integrated into a CTM plant is provided in Section 6.5.1. Section 6.8.1 integrates the CTS and STS processes for methanol production using a co-feed of natural gas and coal (50% HHV basis). The overall design principles for the CTS process are similar to those used in Section 6.5. The STS process is modeled in Section 6.8.2 for a fluidized bed reducer utilizing ITCMO particles. The fluidized bed reducer is modeled using experimental results presented in Section 4.4.2. As a final example, Section 6.8.3 details producing electricity from syngas using the CTS process and demonstrates the versatility of the chemical looping platform. Other processes presented in the following sections also serve to illustrate the versatility of process simulations as a useful tool for holistic process development.

6.8.1   Coal and Natural Gas Co-feed in a CTM Configuration

The analysis of a coal and natural gas co-feed in a CTM configuration is motivated by the increased carbon efficiency associated with natural gas for methanol production. Since natural gas has a higher hydrogen to carbon ratio (~4) as compared to coal (~1), the carbon efficiency of the process increases with increasing natural gas fraction. A natural gas and coal mixture with a 50% HHV contribution from each fuel is used as an example. The 50% HHV ratio allows coal to be consumed at a high rate while improving carbon efficiency. The potential carbon and thermal efficiency improvement for a CTM process with natural gas co-feed is compared with the economic impact of utilizing natural gas for methanol production.

Process Modeling of Co-feed System

The process modeling set-up for the co-feed CTS process with natural gas is shown in Figure 6.61, and process integration into a methanol production plant follows a similar design to that of a coal-only case, the CTS process given in Section 6.5.1. The co-feed system is analyzed to identify suitable Fe_2O_3:C and H_2O:C ratios for which autothermal operation of the CTS process can operate. The co-feed CTS process is integrated into a

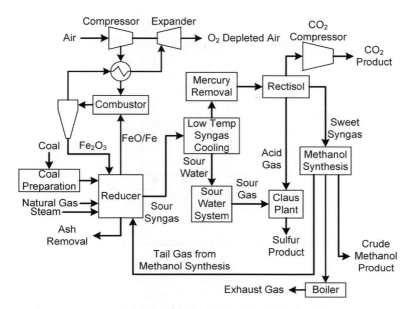

Figure 6.62 Overall block diagram for the CTS-A co-feed system.

methanol production plant with a second sub-model developed to balance steam generation against the steam loads in the system, and include the process steam necessary for adjusting the syngas composition in the co-feed CTS process and power generation. The steam system is designed such that the purge gas boiler functions as an evaporator to generate saturated steam that is subsequently superheated to be used as process steam. Additional superheating from the boiler generates steam for the steam turbine and power generation.

The overall block diagram for the co-feed CTS-CTM process is shown in Figure 6.62. The overall mass and energy balances are adjusted such that the net plant output is 10,000 mtpd methanol. The integrated process performance for the co-feed case is presented in Table 6.17 as case CTS-A, and shows that a natural gas co-feed improves the overall process by reducing water, steam, and air consumption, and sulfur and carbon emissions. The co-feed case requires less parasitic energy for coal and ash handling, which is expected since less coal is being consumed. An improvement in carbon efficiency with natural gas co-feed leads to a higher syngas quality from the CTS reducer, resulting in a lower CO_2 concentration in the syngas and correspondingly a smaller AGR unit, CO_2 compressor, syngas compressors, and syngas cooling loads. The results from the process efficiency standpoint alone show significant benefits associated with co-feeding natural gas to the reducer.

Techno-Economic Analysis

The cost estimating methodology is described in Section 6.4.2. Total plant capital costs for the CTS-A case are presented in Table 6.18. The CTS-A case has a lower capital cost investment compared to the CTS case due to a higher carbon efficiency. The capital and operating costs are used to obtain the overall cost of methanol

Table 6.17 Overall performance of the coal and natural gas co-feed CTS-A process in a 10,000 tonne/day methanol plant.

	MB-B	MB-A	CTS	CTS-A
Mass flows (lb/hr)				
As-received coal	1,618,190	1,618,190	1,395,457	718,631
Oxygen from ASU containing 95% O_2	1,010,968	1,010,968	NA	NA
Steam to gasifier, quench, shift reactors, CTS	1,533,584	1,533,584	1,624,318	693,587
Air to direct-fired boiler	121,518	121,518	181,009	606,106
Clean syngas for methanol production	1,183,080	1,183,080	1,025,106	1,039,864
Tail gas from Claus unit	61,476	61,476	50,089	25,589
Captured CO_2 (no capture for MB-B)	0	1,569,410	1,302,138	663,393
Electrical load (kW$_e$)				
Total gross power	320,680	390,170	20,830	31,491
Total net power **	**12,280**	**21,480**	**−323,504**	**−245,692**

Table 6.18 Comparative summary of capital and operating costs.

Case (2011 MM$)	MB-B	MB-A	CTS	CTS-A
Total plant costs (2011 MM$)	4,586	4,775	3,497	2,996
Total as-spent capital (2011 MM$)	6,580	6,852	5,003	4,291
Capital costs ($/gal, 2011$)	1.18	1.23	0.89	0.81
Fuel cost ($/gal, 2011$)	0.24	0.26	0.18	0.39
O&M costs ($/gal, 2011$)	0.23	0.23	0.16	0.14
CO_2 TS&M costs ($/gal, 2011$)	0	0.06	0.05	0.03
Electricity cost ($/gal, 2011$)	0	0	0.14	0.11
Required selling price ($/gal, 2011$)	1.64	1.78	1.41	1.48

production and required selling price, as shown in Table 6.18 and Figure 6.63. The CTS-A case results in a 5% increase in the methanol RSP as compared to the CTS case despite having a higher carbon efficiency. While the co-feed of natural gas improves the syngas quality and results in reduction of reactor size, capital cost investment requirements, water demands, and auxiliary loads, the benefits are offset by the cost of purchasing natural gas from the market. The parameters for which the two cases show a different trend include the cost of natural gas, cost of coal, and capacity factor.

Sensitivity Studies

A sensitivity analysis on the capacity factor, natural gas price, and coal price was conducted to determine their effect on the methanol RSP. Table 6.19 provides the parameters and range for each. From Figure 6.64, the methanol RSP increases drastically as capacity factor decreases, and is less sensitive at higher capacity factors and highly sensitive at lower capacity factors. This is expected since the capital contribution

Table 6.19 Summary of parameters and costs and the corresponding ranges used in sensitivity study.

Parameter/Cost	Nominal	Min	Max	Range basis
Capacity factor	90%	30%	90%	
Natural gas price ($/MMBTU)	100%	33%	196%	Low price – observed during periods and in regions with excess supply High price – typical world market price or periods with limited supply
Coal price, $/ton	100%	55%	164%	Regional variation in coal price

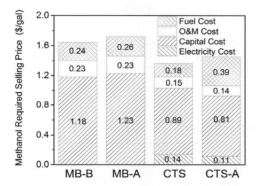

Figure 6.63 Cost breakdown of methanol production for baseline and CTS cases.

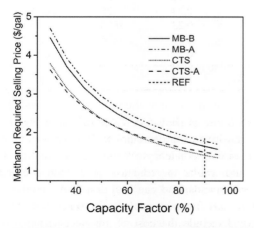

Figure 6.64 Sensitivity of methanol RSP to capacity factor.

to the RSP is distributed over a lower annual production when the capacity factor is decreased. When natural gas is the feedstock for methanol production, the RSP is less sensitive to the capacity factor than coal, because the capital costs provide a smaller contribution to the RSP. This smaller increase for the natural gas cases results in a reversal in the economic ordering of the CTS and CTS-A options at 50% capacity factor.

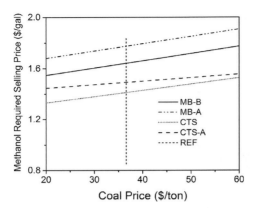

Figure 6.65 Sensitivity of methanol RSP to coal price.

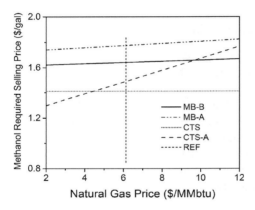

Figure 6.66 Sensitivity of methanol RSP to natural gas price.

The MB-A, MB-B, and CTS cases use coal as the only feedstock and show similar sensitivities to the methanol RSP with respect to coal costs, as shown in Figure 6.65. The methanol RSP is rather insensitive to the coal price, increasing by approximately 12% when the coal price increases from −60% to +160% of the reference cost. The CTS-A case uses coal and natural gas as the feedstock with each contributing 50% HHV, which is why it shows a lower sensitivity to coal cost than the other cases.

The methanol RSP sensitivity to the natural gas price is shown in Figure 6.66. For cases MB-A, MB-B, and CTS, natural gas is used only for electricity generation in an NGCC cycle. For these cases, the RSP is almost completely insensitive to the natural gas price since its consumption is minimal. The RSP for cases utilizing natural gas as a feedstock has a greater sensitivity to natural gas cost, increasing by $0.30/gal for the CTS-A case and $0.60/gal for the baseline case. The increase in sensitivity is due to the feedstock cost contributing a greater portion to the RSP than for electricity generation. This sensitivity results in case CTS-A becoming more favorable than CTS when the natural gas price is less than 70% of the reference value.

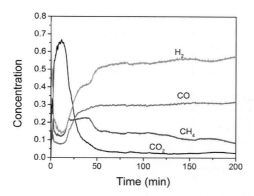

Figure 6.67 Syngas generation from natural gas in a fluidized bed reducer using ITCMO.[40] A black and white version of this figure will appear in some formats. For the color version, please refer to the plate section.

6.8.2   Fluidized Bed Syngas Generation in a GTL Configuration

In a moving bed reactor, the flow can be considered to be plug flow and can be described using the properties of plug flow behavior, specifically a constant residence time.[39] This fluid flow behavior in a cocurrent downward flow moving bed reactor allows it to be accurately modeled using a single stage RGIBBS reactor in Aspen Plus®. However, in a fluidized bed reactor, solid phase backmixing leads to a wide residence time distribution. This section outlines the process model development of a fluidized bed reducer reactor, provides the performance results when integrated into a GTL process, and compares the economics of a fluidized bed reducer for syngas generation STS-GTL process to that of the STS-GTL process economics discussed in Section 6.6.2.

Process Modeling of a Fluidized Bed Reactor

Experimental results obtained for methane conversion to syngas in a fluidized bed reactor, shown in Figure 6.67, show that syngas with a H_2:CO ratio of 1.8 and a CO: CO_2 ratio of 12 can be produced from methane in a fluidized bed reactor. However, the methane conversion is limited to 70% due to gas channeling and backmixing. Using design concepts for modeling a chemical looping fluidized bed reducer to generate syngas from natural gas, a fluidized bed reducer model is developed that closely matches the experimental results. The experimental results obtained are the best performance achievable by a fluidized bed and are used for modeling the fluidized bed STS-GTL process and comparing the economics of a fluidized bed reducer to a moving bed reducer.

In the STS-GTL process, the fluidized bed reducer replaces the ATR for syngas generation. The design concept for integrating the fluidized bed STS process into the GTL plant is similar to that described in Section 6.6.1, where the syngas quality and quantity is adjusted to match the 50,000 bpd for the GTL plant. The fluidized bed performance is analyzed at the same Fe_2O_3:C ratio as that in

Table **6.20** Overall performance of the STS process in a 50,000 bpd GTL plant.

Component	Base case	STS (10 atm) moving bed	STS (10 atm) fluidized bed
Natural gas flow, kg/hr	354,365	317,094	452,992
Natural gas flow, kmol/hr	20,451	18,300	26,143
H_2O/C_1	0.68	0.25	0.25
H_2/CO	2.19	2.178	2.18
S#	1.59	1.96	~ 1.90
O_2/C_1	0.73	0.48	0.48
Butane feed flow, kg/hr	18,843	18,843	18,843
Diesel fuel, bbl/day	34,543	34,543	34,543
Gasoline, bbl/day	15,460	15,460	15,460
Total liquids, bbl/day	50,003	50,003	50,003
Electrical load (kW$_e$)			
Total gross power	303,700	303,700	303,700
Net plant power	40,800	179,050	150,480

Section 6.6, with the exception that the fluidized bed reducer is modeled using the results shown in Figure 6.67. The H_2:CO ratio is adjusted to the same as that used for the moving bed simulation. The recycle fuel gas stream is assumed to be completely converted, equivalent to the moving bed STS process in Section 6.6. These assumptions represent the optimal performance possible for a fluidized bed reactor.

Process Performance

In Table 6.20, the performance results for the fluidized bed STS-GTL process are summarized and compared to the moving bed STS-GTL process. For the same 50,000 bpd liquid fuel output, the fluidized bed reducer consumes 43% more natural gas than a moving bed reducer, which results in the carbon efficiency for the fluidized bed STS-GTL process being lower than the baseline case. The fluidized bed reducer requires more energy to operate the amine scrubber and air compressor than the moving bed reducer, thus the net plant power output is reduced by 16%.

Techno-Economic Analysis Results

Table 6.21 summarizes the total plant capital costs for the STS-GTL process. The detailed costs for the baseline case and moving bed STS-GTL process were discussed in Section 6.6.2. As a comparison, the total plant cost for the baseline GTL plant is approximately $86,000/bbl, which results in a required liquid fuel selling price greater than crude oil. The fluidized bed STS-GTL process total plant cost is $68,505/(bbl/day) as compared to the $65,000/(bbl/day) for a moving bed reducer. At a 20% IRR and a natural gas price of $3/MMBTU, the fluidized bed STS process is competitive when the crude oil is priced at $69/bbl, whereas the moving bed STS process is competitive at $45/bbl.

Table 6.21 Cost summary for fluidized bed STS-GTL plant.

Component (2011 $)	ATR	STS moving bed	STS fluidized bed
Total plant cost ($×1,000)	2,750,000	1,880,000	1,941,332
Total as-spent cost ($×1,000)	4,310,000	3,250,000	3,330,422
Capital costs ($/(bbl/day))	86,000	65,000	68,505
Capital costs ($/bbl)	51.5	41.8	
O&M costs ($/bbl)	27.29	7.23	
Fuel costs ($/bbl)	16.9	12.6	
Electricity cost ($/bbl)	−1.15	−5.1	
Required selling price ($/bbl)	94.54	76.53	
Crude oil competitive price ($/bbl)	**60.5**	**48.9**	**69**

6.8.3   CTS Process in an IGCC Configuration

The current commercial technology for an integrated gasification combined cycle (IGCC) plant uses a coal gasifier and an ASU to produce syngas that can be used for power production and/or chemicals production. Electricity generation using an IGCC has been analyzed with and without carbon capture.[41] In a specific configuration with non-carbon capture, the syngas generated from coal gasification is removed of sulfur using a modified Selexol process prior to combustion in a gas turbine. The net plant efficiency for this configuration is 39% (HHV), which results in a total plant cost (TPC) of $1987/kW$_e$ and a first year cost of electricity (FYCOE) of 76.3 mills/kWh. For 90% CO_2 capture, WGS reactors and a conventional 2-stage Selexol process remove the CO_2 and sulfur and are additional equipment added to the non-carbon capture IGCC process. For 90% CO_2 removal, the combustion fuel is syngas that is predominantly H_2 (>90% v/v) and the balance being CO (10% of carbon input, < 10% v/v). Compared to IGCC without carbon capture, the additional equipment costs required for CO_2 removal result in a 6.4% decrease in the net plant efficiency to 32.6%, an increase in the TPC by 36% to $2711/kW$_e$, and FYCOE by 38% to 105.6 mills/kWh.

An emerging technology that has the potential to provide significant improvements over current cryogenic distillation ASUs for coal gasification is the ion transfer membrane (ITM) technology being developed by Air Products.[42–45] ITM uses a proprietary, non-porous mixed metal oxide conducting membrane that produces an oxygen ion gradient to separate oxygen from air at temperatures in the range 1292 °F (700 °C) to 1832 °F (1000 °C). A 30–35% reduction in capital costs over cryogenic ASU technology can be achieved for an equivalent oxygen production when the ITM technology is integrated into an IGCC. When ITM replaces current cryogenic ASU in an IGCC with 90% CO_2 capture, the overall net plant efficiency increases by 0.8% and the FYCOE decreases by 6–7 mills/kWh. ITM for oxygen generation has economic advantages over cryogenic distillation, but these are limited as maintaining a high temperature is required for optimal membrane performance.

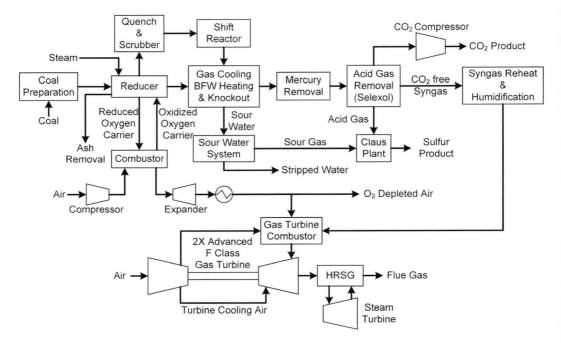

Figure 6.68 Block flow diagram of CTS-IGCC process for electricity generation.

The baseline IGCC case gasifies Illinois #6 coal, a bituminous coal, in a GEE gasifier since it achieves the highest process efficiency and lowest cost of electricity in a carbon-constrained scenario. The GEE gasifier uses a coal slurry feed and uses a 0.42 H_2O:C ratio for gasification, along with 95% pure oxygen from an ASU. This results in a syngas composition of 0.71 moles CO, 0.68 moles H_2, and 0.28 moles CO_2 per mole of carbon in as-received coal.

Design Concept

The CTS process has been described in Section 4.5 (as a CLR system) and Section 6.3.1. For electricity generation, an ITCMO particle is used. For the CTS-IGCC process, a cocurrent downward flow moving bed reducer gasifies coal in the presence of steam to produce syngas, while reducing the ITCMO particle. The gas–solid contact pattern between dry coal, ITCMO, and steam in the reducer is designed based on thermodynamics to control the syngas yield precisely such that it is suitable for electricity generation and has been verified from bench-scale experiments.

Figure 6.68 shows the CTS-IGCC process for electricity generation. Coal, ITCMO particles, and steam are the inlet streams to the reducer, with syngas and reduced ITCMO particles produced. The syngas is then cooled and mercury removed. This syngas has sulfur and CO_2 removed in a two-stage Selexol unit. The CO_2 capture process is designed for 90% CO_2 removal from the syngas stream, and the CO_2 is compressed for sequestration. The acid gas free syngas is then combusted in a gas turbine, contributing approximately 10% (heating value based) of the total syngas used for combustion in an IGCC electricity generation cycle. The reduced ITCMO particles (mainly Fe/FeO) from the reducer are partially re-oxidized with steam in a countercurrent moving bed oxidizer to produce pure H_2, which provides the remaining 90% (heating value based) of the total syngas used as fuel in the IGCC gas turbine. The partially oxidized ITCMO particles are fully re-oxidized with air in a third reactor, the combustor. The exit stream from the combustor contains a mixture of coal ash, oxidized ITCMO particles, and depleted air. A primary gas–solid separator separates the oxidized ITCMO particles from the depleted air, where the ITCMO particles enter the reducer to complete the cycle, and the depleted air is delivered to a second stage gas–solid separator where ash is removed from the depleted air. The high temperature, depleted air stream fulfills two roles. First, it is used to generate steam for the steam turbine cycle. Second, a fraction of the stream is used directly as the diluent for the gas turbine in the IGCC electricity generation cycle. For IGCC applications with no carbon capture, there is a 2% increase in efficiency when the CTS process replaces the gasifier. The CTS-IGCC process replaces the coal gasifier while removing the cryogenic ASU and WGS reactors for carbon capture applications.

Process Modeling of the CTS-IGCC Process

The process modeling set-up for the CTS process follows the process model provided in Section 6.5.1 and is described here in the context of electricity generation in IGCC applications. To produce a syngas suitable for IGCC electricity generation, with 90% carbon capture, the important CTS process parameters include the H_2O:C ratio, Fe_2O_3:C ratio, temperature swing for autothermal operation, and operating pressure. All parameters are examined and optimized to produce a syngas with carbon in gaseous form. The gaseous carbon distribution is approximately 90% CO, with the remainder being CO_2.

The critical advantage of using a CTS reducer instead of a GEE gasifier for coal gasification resides in the fundamental thermodynamics of ITCMO particles with coal and steam in the cocurrent moving bed reactor system. The CTS process controls the oxygen transfer in the coal gasification reaction and produces greater than 0.9 moles CO per mole of as-received Illinois #6 coal. If a steam quantity equivalent to the GEE gasifier is added to the CTS reducer, 0.7 moles H_2 per mole of carbon in as-received Illinois #6 coal is obtained, while maintaining the CO content of the syngas. This leads to an advantage in the overall process efficiency for both the 90% carbon capture and non-carbon capture cases. The syngas quality comparison is provided in Table 6.22.

Table 6.23 shows the overall improvement in the efficiency in an IGCC scenario when replacing the GEE gasifier with the CTS process. The CTS process produces syngas from an equivalent amount of coal at a thermal efficiency of 82%, as opposed to the baseline GEE efficiency of 77%. The gas turbines in the CTS-IGCC process are identical to the baseline case with a thermal to electric conversion efficiency of 37.5%. The CTS-IGCC process produces 249.5 kW_e per mole of carbon of total gross power, as compared to 239 kW_e for the baseline case. The total auxiliary power requirement for the CTS-IGCC process is also significantly lower than the baseline case due to a superior quality of syngas, resulting in a smaller AGR unit for CO_2 separation. The net power efficiency increases by three percentage points in the non-capture scenario and by four percentage points in a 90% carbon capture scenario over the baseline case.

Table 6.22 Overall integrated performance of the CTS-IGCC process.

Parameter	GEE gasifier[41]	CTS reducer
Gasifier/reducer input		
$\left(\dfrac{\text{moles of } H_2O}{\text{moles of carbon in as-received coal}}\right)$	0.426	0.01–0.4
Gasifier/reducer output		
$\left(\dfrac{\text{moles of } H_2}{\text{moles of carbon in as-received coal}}\right)$	0.678	0.48–0.70
$\left(\dfrac{\text{moles of CO}}{\text{moles of carbon in as-received coal}}\right)$	0.707	0.91–0.93
$\left(\dfrac{\text{moles of } CO_2}{\text{moles of carbon in as-received coal}}\right)$	0.270	0.09–0.06

Table 6.23 Comparison of process efficiency (per mole of carbon).

Parameter value	GEE gasifier	GEE gasifier	CTS reducer	CTS reducer
% carbon capture	0	90	0	92
Thermal efficiency for syngas production (%)	77.5	77.4	80.2	80.1
Total gross power (kW)	239.56	235.40	249.5	249.5
Net power (kW)	199.21	174.23	216.18	193.76
Net power efficiency (%)	39.00	34.11	42.32	37.93

Table 6.24 Comparative summary of capital and operating costs.

Case (2011 MM$)	GEE- baseline	CTS-IGCC
% carbon capture	90	92
Total plant costs (2011 $/kW)	3,387	2,768
Total as-spent capital (2011 $/kW)	4,782	3,908
Capital costs ($/MW, 2011$)	74.2	60.6
Fuel cost ($/MW, 2011$)	30.7	27.6
O&M costs ($/MW, 2011$)	30.4	27.1
Required selling price ($/MW, 2011$)	**135.4**	**115.3**

Techno-Economic Analysis

The capital and operating costs are used to obtain the overall cost of electricity production and the required electricity selling price for both the CTS-IGCC process and the baseline case, summarized in Table 6.24. The first year COE for the CTS-IGCC process is the lowest of any gasifier system utilizing IGCC for electricity production with 90% carbon capture. The CTS-IGCC process reduces the capital cost investment by 18% compared to the GEE-IGCC baseline case, due to a higher carbon efficiency. This change is the main driver behind the 15% reduction in electricity RSP for the CTS-IGCC process as compared to the baseline IGCC case.

6.9   Concluding Remarks

The continually increasing complexity in the design of chemical plants emphasizes the critical need for process simulations. Using manual calculations to design a process from feedstock to product that incorporates a mass and energy balance while calculating thermodynamic properties using higher-order equations of state, temperature profiles in heat exchangers, and product composition in chemical reactors, is a near impossibility given the complex interrelations among all unit operations within a process. Further, refinement of a process for the purpose of increasing process intensification, decreasing overall process energy requirements, and performing a sensitivity analysis demands the use of process simulation software.

This chapter highlights the steps necessary to develop and evaluate new processes, not only to demonstrate their viability, but also to prove that the new process is an improvement over the conventional, baseline process. The first step in process development is understanding the baseline process in order to establish what are the most important factors contributing to process inefficiencies and process economics. Moreover, the established baseline process provides the details and metrics necessary for developing a new process and its associated process simulation that can be compared to and evaluated against the baseline process. In this chapter, the baseline processes include coal to methanol via coal gasification, natural gas to liquid fuels via autothermal reforming, biomass to olefins via biomass gasification, and coal to electricity via coal gasification, whereas the new processes include coal to methanol via the CTS process, natural gas to liquid fuels via the STS process, biomass to olefins via the BTS process, and coal to electricity via the CTS process, which are all based on chemical looping partial oxidation.

The second step in developing a new process is to define a process goal. For chemical looping partial oxidation processes, one of the primary goals is to produce syngas from carbonaceous feedstock – coal, natural gas, woody biomass – as an intermediate for product generation at a higher cold gas efficiency and lower cost, compared to the baseline. Another goal for process systems integration is to determine the effect of syngas production from chemical looping. This is achieved by developing an Aspen Plus® process model for chemical looping partial oxidation, based on experimental results and optimized on a thermodynamic rationale. The Aspen Plus® model expands on the thermodynamic approach for autothermal operation, as given in Chapter 4. The syngas production from the chemical looping system is tailored to match the performance specification in the baseline cases.

Once the chemical looping partial oxidation model for syngas production matches that of the baseline with respect to flow and quality, the third step focuses on incorporating product generation – methanol, liquid fuels, or olefins – into the model. The overall process balance is optimized to ensure that steam conditions, water balance, and pollutant emissions are comparable with the baseline process, to confirm that the performance is commercially applicable. Based on the results of the process simulation, equipment sizing calculations are performed and optimized with hydrodynamic and fabrication considerations to obtain a conservative cost estimate from the fabrication vendor. The conservative cost estimate is calculated by increasing the fabrication vendor cost estimate by 50%. The capital and operating and maintenance cost estimates for the process are then completed based on existing standards. The plant performance and cost-estimating results are used to determine the overall plant economic performance. The economic assessment determines the first year cost of production, cost of electricity, product required selling price, and cost of CO_2 captured. Sensitivity studies are performed to identify parameters that have a significant impact on the production costs, reflecting on the opportunities for future cost reductions and on parameters that are sensitive and can be changed resulting in the most favorable operation option.

Detailed economic analyses indicate that capital costs for chemical looping partial oxidation are always lower than the existing technologies. For the case of methanol production from coal, chemical looping processes with carbon capture are able to reduce the methanol production cost to be competitive with current market prices from natural gas without carbon capture. A chemical looping gas to liquids plant is able to reduce the capital costs associated with syngas production by more than 50%, while decreasing the feedstock and steam consumption by more than 30%. Converting biomass to olefins through syngas and methanol intermediates using chemical looping is able to reduce the cost of ethylene/ propylene production by 8.5% over indirect biomass gasification. Using chemical looping for converting coal to syngas can produce electricity as well as chemicals. For electricity production, chemical looping coal gasification can reduce the cost of electricity by 15% as compared to the conventional IGCC electricity method. The economic values can be more accurate when large-scale demonstration data are available.

References

1. Motard, R. L., M. Shacham, and E. M. Rosen, "Steady State Chemical Process Simulation," *AIChE Journal*, 21(3), 417–436 (1975).
2. Puigjaner, L. and G. Heyen, eds., *Computer Aided Process and Product Engineering*, Wiley-VCH, Weinheim, Germany (2006).
3. Evans, L., J. Boston, H. Britt, P. Gallier, P. Gupta, B. Joseph, V. Mahalec, E. Ng, W. Seider and H. Yagi, "ASPEN: An Advanced System for Process Engineering," *Computers & Chemical Engineering*, 3, 319–327 (1979).
4. Goellner, J. F., N. J. Kuehn, V. Shah, C. W. White, and M. C. Woods, *Baseline Analysis of Crude Methanol Production from Coal and Natural Gas*, United States Department of Energy/NETL, DOE/NETL-341/013114, Pittsburgh, PA (2014) .
5. Higman C. and M. van der Burgt, "Gasification Processes," Chapter 5 in Higman C., and M. van der Burgt, eds., *Gasification*, 2nd edn., Gulf Professional Publishing, Burlington, MA, 91–191 (2008).
6. Tennant, J. B., "Gasification Systems Program," http://www.netl.doe.gov/File%20Library/ research/coal/energy%20systems/gasification/gasifipedia/DOE-Gasification-Program-Overview .pdf (2012).

7. Darde, A., R. Prabhakar, J. Tranier, and N. Perrin, "Air Separation and Flue Gas Compression and Purification Units for Oxy-Coal Combustion Systems," *Energy Procedia*, 1(1), 527–534 (2009).

8. Smith, A. R. and J. Klosek, "A Review of Air Separation Technologies and Their Integration with Energy Conversion Processes," *Fuel Processing Technology*, 70(2), 115–134 (2001).

9. Ciferno, J. P. and J. J. Marano, *Benchmarking Biomass Gasification Technologies for Fuels, Chemicals and Hydrogen Production*. United States Department of Energy/NETL, Pittsburgh, PA (2002).

10. Twigg, M. V., *Catalyst Handbook*, 2nd edn., Manson Publishing, London, UK (1996).

11. Lee, S., *Encyclopedia of Chemical Processing vol. 5*, Taylor & Francis, New York, NY (2006).

12. Korens, N., D. R. Simbeck, and D. J. Wilhelm, *Process Screening Analysis of Alternative Gas Treating and Sulfur Removal for Gasification*, United States Department of Energy/ NETL, Pittsburgh, PA (2002).

13. Cheng, W.-H. and H. H. Kung, *Methanol Production and Use*, Marcel Dekker, New York, NY (1994).

14. Heydorn, E. C., B. W. Diamond, and R. D. Lilly, *Commercial-Scale Demonstration of the Liquid Phase Methanol (LPMeOH) Process*, United States Department of Energy/NETL, DOE/NETL-341/061013, Pittsburgh, PA (2003).

15. Biedermann, P., T. Grube, and B. Höhlein, eds., *Methanol as an Energy Carrier*, Forschungszentrum Jülich, Jülich, Germany (2006).

16. Kuehn, N. J., K. Mukherjee, P. Phiambolis, et al., *Current and Future Technologies for NGCC Power Plants*, United States Department of Energy/NETL, DOE/NETL-341/061013, Pittsburgh, PA (2013).

17. Chou, V. H., J. L. Haslbeck, N. J. Kuehn, et al., *Cost and Performance Baseline for Fossil Energy Plants Volume 3a: Low Rank Coal to Electricity: IGCC Cases*, United States Department of Energy/NETL, DOE/NETL-2010/1399, Pittsburgh, PA (2011).

18. Pinkerton, L., E. Varghese, and M. Woods, *Updated Costs (June 2011 basis) for Selected Bituminous Baseline Cases*, United States Department of Energy/NETL, DOE/NETL-341/ 082312, Pittsburgh, PA (2012).

19. Goellner, J. F., V. Shah, M. J. Turner, et al., *Analysis of Natural Gas-to Liquid Transportation Fuels via Fischer-Tropsch*, United States Department of Energy/NETL, DOE/NETL-2013/1597, Pittsburgh, PA (2013).

20. Rath, L. K., *Assessment of Hydrogen Production with CO_2 Capture Volume 1: Baseline State-of-the-Art Plants*, United States Department of Energy/NETL, DOE/NETL-2010/1434, Pittsburgh, PA (2010).

21. Choi, G. N., S. S. Tam, J. M. Fox III, and J. J. Marano, "Baseline Design/Economics for Advanced Fischer–Tropsch Technology," Presented at the Coal Liquefaction and Gas Conversion Contractors Review Meeting, Pittsburgh, PA (1993).

22. Zhu, Y., S. A. Tjokro-Rahardjo, C. Valkenburg, L. J. Snowden-Swan, S. B. Jones, and M. A. Machinal, *Techno-Economic Analysis for the Thermochemical Conversion of Biomass to Liquid Fuels*, Pacific Northwest National Labs, PNNL-19009, Richland, WA (2011).

23. Basu, P. *Biomass Gasification and Pyrolysis: Practical Design and Theory*, Academic Press, Burlington, MA (2010).

24. Worley, M. and J. Yale, *Biomass Gasification Technology Assessment: Consolidated Report*, National Renewable Energy Laboratory, NREL/SR-5100–57085, Golden, CO (2012).

25. Eastman Chemical Company, and Air Products and Chemicals, Inc., Project Data on Eastman Chemical Company's Chemicals-From-Coal Complex in Kingsport, TN, http://www.netl .doe.gov/File%20Library/research/coal/energy%20systems/gasification/Eastman-Chemicals-from-Coal-Complex.pdf (2003).

26. Chen, J. Q., A. Bozzano, A. Glover, B. Fuglerud, and T. Kvisle, "Recent Advancements in Ethylene and Propylene Production using the UOP/hydro MTO Process," *Catalysis Today*, 106(1), 1–4 (2005).

27. Vora, B. V., T. L. Marker, P. T. Barger, H. R. Nilsen, S. Kvisle, and T. T. Fuglerud, "Economic Route for Natural Gas Conversion to Ethylene and Propylene," *Studies in Surface Science and Catalysis*, 107, 87–98 (1997).

28. Wilson, S. and P. Barger, "The Characteristics of SAPO-34 which Influence the Conversion of Methanol to Light Olefins," *Microporous and Mesoporous Materials*, 29(1–2), 117–126 (1999).

29. Kathe, M., D. Xu, T.-L. Hsieh, J. Simpson, R. Statnick, A. Tong, and L.-S. Fan, *Chemical Looping Gasification for Hydrogen Enhanced Syngas Production with in-situ CO_2 Capture*, United States Department of Energy, OSTI: 1185194 (2015).

30. Luo, S., L. Zeng, D. Xu, et al., "Shale Gas-to-Syngas Chemical Looping Process for Stable Shale Gas Conversion to High Purity Syngas with a H_2:CO Ratio of 2:1," *Energy & Environmental Science*, 7(12), 4104–4117 (2014).

31. Reid, R. C., J. M. Prausnitz, and B. E. Poling, *The Properties of Gases and Liquids*, McGraw-Hill, New York, NY (1987).

32. AspenTech. The Aspen Physical Property System Combustion Data Bank, JANAF Thermochemical Data.

33. Aly, F. A. and L. L. Lee, "Self-Consistent Equations for Calculating the Ideal Gas Heat Capacity, Enthalpy, and Entropy," *Fluid Phase Equilibria*, 6(3–4), 169–179 (1981).

34. Aasberg-Petersen, K., C. S. Nielsen, I. Dybkjaer, and J. Perregaard, Large Scale Methanol Production from Natural Gas, http://www.topsoe.com/sites/default/files/topsoe_large_scale_methanol_prod_paper.ashx_.pdf (Accessed on 18th September 2016).

35. United States Department of Energy/NETL, *Cost Estimation Methodology for NETL Assessments of Power Plant Performance*, United States Department of Energy/NETL, DOE/NETL-2011/1455, Pittsburgh, PA (2011).

36. Gaspar Inc. Welding & Fabrications, https://www.gasparinc.com/ (Accessed on 18th September 2016).

37. Marker, V. V., *"Natural Gas to High-Value Olefins,"* in *Proceedings of the 15th World Petroleum Congress*, Beijing, China, October 12–17 (1997).

38. Honeywell UOP, UOP MTO Methanol to Olefins Process, http://www.uop.com/mto-process-flow-scheme/ (Accessed on 18th September 2016).

39. Trambouze, P. and J.-P. Euzen, *Chemical Reactors, from Design to Operation*, Editions Technip, Paris, France (2004).

40. Xu, D., T.-L. Hsieh, and L.-S. Fan, Unpublished data (2014).

41. United States Department of Energy/National Energy Technology Laboratory, "Cost and Performance Baseline for Fossil Energy Plants Volume 1: Bituminous Coal and Natural Gas to Electricity. Revision 2a," United States Department of Energy /NETL, DOE/NETL-2010/1397, Pittsburgh, PA (2013).

42. Stiegel, G. J., S. J. Clayton, and J. G. Wimer, "DOE's Gasification Industry Interviews: Survey of Market Trends, Issues and R&D Needs," *Proceedings of Gasification Technologies Conference*, San Francisco, CA, October 7–10 (2001).

43. United States Department of Energy/National Energy Technology Laboratory, *Current and Future IGCC Technologies: A Pathway Study Focused on Non-Carbon Capture Advanced Power Systems R&D using Bituminous Coal: Volume 1*, United States Department of Energy/NETL, DOE/NETL-2008/1377, Pittsburgh, PA (2008).

44. United States Department of Energy/National Energy Technology Laboratory, *Current and Future IGCC Technologies: Volume 2: A Pathway Study Focused on Carbon Capture Advanced Power Systems R&D using Bituminous Coal*, United States Department of Energy/NETL, DOE/NETL-2009/1389, Pittsburgh, PA (2009).

45. Repasky, J. M., L. L. Anderson, E. E. Stein, P. A. Armstrong, and E. P. Foster, "ITM Oxygen Technology: Scale-Up Toward Clean Energy Applications," in *Proceedings of the 29th International Pittsburgh Coal Conference*, Pittsburgh, PA, October 15–18 (2012).

Index

design equations, 275–276
gas–solid contact in reducer, 31, 38–40, 269, 429
pressure profile, 274–275

nanostructure formation. *See* particle morphology
natural gas
  for ethylene production. *See* oxidative coupling of methane
  for liquid fuels production. *See* gas to liquids (GTL) process
  reforming. *See* conventional natural gas reforming
  reforming in fluidized bed reducer. *See* chemical looping reforming in a fluidized bed
  reforming in moving bed reducer. *See* shale gas to syngas (STS) process
  reforming with chemical looping combustion. *See* chemical looping combustion
nickel oxide. *See* active metal oxide
non-mechanical seals and valves, 276–281

OCM. *See* oxidative coupling of methane
oxidative coupling of methane
  ARCO gas to gasoline (GTG) process, 27
  carbon oxides formation, 180–182
  catalyst performance, 175–176, 184, 186–187, 192, 195
  catalyst properties, 184–186, 192, 195
  catalytic metal oxide performance, 188–191
  catalytic metal oxide properties, 188–189
  chemical looping kinetics, 207–208
  chemical looping process configuration, 47–48, 176–177, 187–188, 215
  chemical looping process simulation, 215–219
  chemical reactions, 176
  co-feed kinetic models, 204–207
  co-feed process configuration, 176
  co-feed reaction kinetics, 195–204
  ethylene formation, 178–180
  purification and processing, 214–215, 219–221
  reactor, 210–214
oxidizer reactor. *See* hydrogen production
oxygen carrier particle
  active metal oxide. *See* active metal oxide; bimetallic active metal oxide
  attrition. *See* attrition
  cost, 7
  morphology. *See* particle morphology
  recyclability, 68–70
  supports. *See* supports for metal oxide
oxygen vacancy. *See* ionic diffusion

particle morphology
  cobalt–cobalt oxide, 90
  copper oxide–nickel oxide, 101
  core–shell. *See* core–shell morphology
  iron oxide–nickel oxide, 99–101

iron–titanium complex metal oxide (ITCMO), 94–96, 101
iron–iron oxide, 15–16, 87–88
manganese–manganese oxide, 90
nanostructure formation, 15–16, 87–88, 91, 94, 97–101
nickel–nickel oxide, 88–89
perovskites
  catalyst, 192–194
  catalytic metal oxide, 207–208
  oxygen carrier, 67–68, 159
  support for metal oxide, 65–66, 96–97, 157–158
phase diagrams, 31, 38–40, 123–136
process integration and economics
  biomass to syngas (BTS) to methanol, 46–47
  biomass to syngas-biomass to olefins (BTS-BTO) process, 418–423
  coal to syngas-coal to methanol (CTS-CTM) process, 44–45, 396–409, 424–428
  coal to syngas-integrated gasification combined cycle (CTS-IGCC) process, 431–434
  economic assumptions, 393–396
  shale gas to syngas–gas to liquids (STS-GTL) process, 40–42, 410–417, 429–430
propylene to acrolein
  catalytic metal oxide performance, 355
  chemical reactions, 346–347
  effect of catalytic metal oxide phase, 351–353
  oxidation mechanism, 353–354
  reaction mechanism, 356–357
  reduction mechanism, 347–351

reaction mechanism
  Eley–Rideal, 179, 196, 308, 357
  Langmuir–Hinshelwood, 199
  Langmuir–Hinshelwood–Hougen–Watson (LHHW), 308, 357
  Mars–van Krevelen, 179, 308, 355, 357
reducer reactor, 22–24, 28–29, 31–32, 35, 38–40, 42, 45, 55, 60, 255–256, 258–261, 264–269, 271, 274–278, 287–288, 293–298, 390, 397–404, 418–420, 429–430, 432–433
reforming
  natural gas. *See* conventional natural gas reforming; chemical looping combustion
  natural gas in a fluidized bed reducer. *See* chemical looping reforming in a fluidized bed
  natural gas in a moving bed reducer. *See* shale gas to syngas (STS) process

SCL process. *See* chemical looping combustion
shale gas to syngas (STS) process
  experimental results, 40, 271–274, 281–283
  for liquid fuels production. *See* process integration and economics
  process configuration, 40, 389–390
  process simulation, 410–413